Mathematical Foundations for Deep Learning

Mathematical Foundations for Deep Learning bridges the gap between theoretical mathematics and practical applications in artificial intelligence (AI). This guide delves into the fundamental mathematical concepts that power modern deep learning, equipping readers with the tools and knowledge needed to excel in the rapidly evolving field of artificial intelligence.

Designed for learners at all levels, from beginners to experts, the book makes mathematical ideas accessible through clear explanations, real-world examples, and targeted exercises. Readers will master core concepts in linear algebra, calculus, and optimization techniques; understand the mechanics of deep learning models; and apply theory to practice using frameworks like TensorFlow and PyTorch.

By integrating theory with practical application, *Mathematical Foundations for Deep Learning* prepares you to navigate the complexities of AI confidently. Whether you're aiming to develop practical skills for AI projects, advance to emerging trends in deep learning, or lay a strong foundation for future studies, this book serves as an indispensable resource for achieving proficiency in the field.

Embark on an enlightening journey that fosters critical thinking and continuous learning. Invest in your future with a solid mathematical base, reinforced by case studies and applications that bring theory to life, and gain insights into the future of deep learning.

Mathematical Foundations for Deep Learning

Mehdi Ghayoumi
State University of New York

CRC Press
Taylor & Francis Group
Boca Raton London New York

CRC Press is an imprint of the
Taylor & Francis Group, an **informa** business
A CHAPMAN & HALL BOOK

Designed cover image: Mehdi Ghayoumi

First edition published 2026
by CRC Press
2385 NW Executive Center Drive, Suite 320, Boca Raton, FL 33431

and by CRC Press
4 Park Square, Milton Park, Abingdon, Oxon, OX14 4RN

CRC Press is an imprint of Taylor & Francis Group, LLC

ISBN: 978-1-032-69073-5 (hbk)
ISBN: 978-1-032-69072-8 (pbk)
ISBN: 978-1-032-69074-2 (ebk)

DOI: 10.1201/9781032690742

Typeset in Times
by Newgen Publishing UK

Contents

Preface

Mathematical Foundations for Deep Learning is a guide to the key mathematical principles behind modern deep learning techniques. I hope this book brings clarity to these essential concepts in artificial intelligence (AI), enhancing both your theoretical understanding and practical skills.

In this book, we explore important mathematical areas crucial for deep learning, such as linear algebra, calculus, probability theory, and more. Each chapter balances theory with practice, offering examples and exercises to strengthen your grasp of the material. We delve into the mathematics that power neural networks, optimization algorithms, and various deep learning architectures, aiming to connect complex theory with real-world applications.

The book is organized into 12 chapters, each focusing on a specific area of mathematics as it relates to deep learning. We start with foundational concepts to ensure that all readers, regardless of their background, have the tools needed to tackle more advanced topics.

Our journey begins with linear algebra and multivariate calculus, the building blocks of deep learning models. These chapters lay the groundwork for understanding how data is represented and manipulated in neural networks. We then move on to probability theory and optimization, exploring how models learn from data and how their performance can be improved.

In later chapters, we introduce subjects like information theory, graph theory, and differential geometry, which play important roles in designing and operating deep learning systems. We also cover advanced topics like topology, harmonic analysis for convolutional neural networks, and dynamical systems for recurrent neural networks, showing how these areas contribute to the latest research and applications in AI. Finally, we discuss quantum computing and its potential impact on the future of deep learning.

At the core of this book is a detailed look at how these mathematical foundations come together in the practical building of deep learning models. By understanding the underlying math, you'll be better equipped to solve complex problems and innovate in the field.

I wrote this book for a wide audience, from students new to deep learning to experienced professionals wanting to deepen their understanding of the math behind the models they use. My goal is to make the content accessible to everyone, with clear explanations, practical examples, and exercises to help you learn and apply what you've read.

Thank you for joining me on this journey into the mathematical heart of deep learning. I hope this book not only expands your knowledge but also inspires you to push the boundaries of what's possible in AI.

Happy reading!
Mehdi Ghayoumi
Beverly Hills, CA, USA
August 2024

Acknowledgments

The creation of this book has been deeply influenced by the extraordinary individuals in my life. While many have contributed, none have been more significant than my beloved parents, to whom I offer my deepest gratitude.

First and foremost, I dedicate this book to my dear mother, **Khadijeh Ghayoumi**. Your unwavering belief in me, endless love, and constant support have been the foundation of my strength throughout this journey. This book is a humble tribute to the profound impact you've had on my life. The valuable lessons you've taught me continue to guide my path, and your inspiring resilience reminds me daily to persevere, no matter the challenges that arise.

In loving memory of my father, **Aliasghar Ghayoumi**, I also dedicate this work. Though you are no longer with us, your spirit continues to light my way. Your legacy of integrity, perseverance, and wisdom remains a guiding force in my life. This book honors the values you embodied, and I hope it stands as a testament to your lasting influence.

To both of you, my parents, I owe the principles that have shaped my academic journey: the relentless pursuit of knowledge, the importance of perseverance, and the virtue of humility. This book reflects those values and the love and guidance you've given me. I hope it makes you proud.

To my extended family, I express my heartfelt thanks. Your unwavering belief in me and your constant encouragement have been crucial in achieving this goal.

I also wish to sincerely thank my colleagues, mentors, and collaborators. Your wisdom, expertise, and guidance have been invaluable as I navigated the complexities of academia. The intellectual camaraderie we've shared has played a pivotal role in bringing this work to completion.

From the bottom of my heart, thank you.

About the Author

Mehdi Ghayoumi is a distinguished Assistant Professor at the Center for Criminal Justice, Intelligence, and Cybersecurity at the State University of New York (SUNY) Canton. His academic career is marked by excellence and leadership, reflecting a deep commitment to both teaching and research. Previously, as a Research Assistant Professor at SUNY Binghamton, he spearheaded innovative projects at the Media Core Lab, driving forward research and development in emerging technologies. At Kent State University, his exceptional teaching earned him the prestigious Teaching Award for two consecutive years, 2016 and 2017.

Ghayoumi has been instrumental in developing several courses in fields such as machine learning, data science, robotics, and programming. His broad and impactful research interests include Machine Learning, Machine Vision, Robotics, Human–Robot Interaction (HRI), and privacy. Focusing on the creation of practical and viable systems for real-world environments, his current multidisciplinary research spans HRI, manufacturing, biometrics, and healthcare.

Deeply involved in the academic community, Ghayoumi serves on technical program committees for numerous high-profile conferences and workshops. He is also a member of the editorial boards for several respected journals in machine learning, mathematics, and robotics. His influence reaches some of the most prestigious conferences in these domains, including ICML, ICPR, NeurIPS HRI, FG, WACV, IROS, CIBCB, and JAI.

Ghayoumi's contributions are substantial and far-reaching. His research has been presented at leading conferences and published in top-tier journals, earning recognition for its depth and practical relevance. He has made significant strides in advancing HRI, Robotics Science and Systems (RSS), and machine learning applications, demonstrating his ongoing commitment to pushing the boundaries of knowledge and technology.

1 Introduction

1.1 INTRODUCTION

Welcome to *Mathematical Foundations for Deep Learning*, a journey into the core of mathematics and its impact on artificial intelligence (AI). Over the past 10 years, deep learning has changed industries and how we use technology. This book aims to make these complex ideas easier to understand by focusing on the main mathematical principles behind deep learning. Whether you're a computer scientist looking to deepen your knowledge, an artist exploring where creativity meets technology or just someone who's curious, this book will show how mathematics is the backbone of AI. You'll learn how abstract concepts turn into real-world applications through clear examples, such as the role of linear algebra in neural networks and the use of calculus in optimization algorithms. This chapter gives an overview of the main topics we'll cover. It highlights how these mathematical ideas help build and improve deep learning models, preparing you for the detailed exploration ahead.

1.2 IMPORTANCE OF MATHEMATICS IN DEEP LEARNING

Deep learning is a field that combines many different areas, and mathematics plays a big role in it. But why is math so important for deep learning? Simply put, mathematics provides the foundation for everything in deep learning. It helps organize complex information, creates structure from random data, and gives meaning to the numbers we work with. If you want to design, understand, or improve deep learning algorithms, having a strong grasp of these mathematical ideas is not just useful but essential. Mathematics makes it possible to build effective models that can learn from data and make accurate predictions.

1.2.1 Structuring Chaos

Deep learning is about finding meaningful patterns hidden in large amounts of data. However, as there's so much data and it can be quite random, finding these patterns can feel like searching for a needle in a haystack. This is where mathematics comes in to turn chaos into something we can manage and understand. Mathematical principles, especially linear algebra, provide systematic ways to process and represent data. Concepts like vectors, matrices, and tensors are fundamental in deep learning. They allow us to represent and manipulate data efficiently. This structured approach makes it easier to use computational techniques that speed up data analysis. For example, matrix operations let us transform entire datasets into a single, coherent action, greatly simplifying the data processing phase. When deep learning algorithms extract insights or patterns from data,

DOI: 10.1201/9781032690742-1

mathematics, particularly statistics and probability, provides the essential tools to measure these discoveries. Measures like mean, median, standard deviation, and correlation give us the language to describe and assess patterns, turning abstract data into clear insights. Moreover, mathematics plays a key role in optimizing deep learning models. Based on calculus, algorithms like gradient descent adjust model parameters step by step to minimize errors. This optimization process enables deep learning models to learn from data and improve their accuracy over time. Mathematics also offers strategies to handle outliers and prevent overfitting. Techniques like regularization help balance a model's complexity, improving its ability to generalize and leading to more reliable predictions in real-world applications.

1.2.2 Imposing Rules on Randomness

Deep learning often involves dealing with uncertainty, like the random starting weights in neural networks. This brings up an important question: How can we use this randomness to improve our model's predictions? Mathematics, with its clear yet flexible rules, offers a structured way to handle this uncertainty. A neural network begins by setting initial weights, which greatly affect how well it performs. These weights are usually given random values within a specific range, but this randomness isn't without control. Mathematical principles guide the choice of this range and how the initial weights are distributed, ensuring they're set up properly for effective learning. As the network trains, it adjusts these weights based on the errors it makes, a process where mathematics truly shines. Math provides systematic methods for updating these weights, which is crucial for learning effectively from randomness. At the heart of learning in neural networks is backpropagation, an algorithm based on the chain rule from calculus. Backpropagation calculates the gradient of the loss function with respect to the network's weights. This gradient shows how to update the weights to minimize errors, steadily improving the network's performance. To put this in simple terms, imagine navigating a complex maze without a clear path. Backpropagation acts as a guide, giving you step-by-step directions on how to adjust your course at each turn. While randomness plays a key role in deep learning, the goal is for models to eventually find optimal solutions. Mathematics provides the rules and methods that make this convergence happen systematically and efficiently.

1.2.3 Infusing Data with Meaning

In its natural form, data is just a collection of unprocessed facts. Mathematics helps us turn this raw data into meaningful information by representing it as mathematical objects like vectors, matrices, or higher-dimensional tensors. This mapping allows us to measure relationships, calculate distances, and find patterns, which are essential steps in deep learning. A key aspect of deep learning is how we represent data. Whether we're dealing with images, text, or sounds, real-world data is often converted into numerical arrays or tensors. This transformation uses the power of linear algebra, enabling us to manipulate and analyze data efficiently. By adding structure to what might be chaotic information, mathematics makes it easier to manage and understand complex data. Mathematics also helps us quantify relationships between different data points. This could involve calculating correlations, finding dependencies, or uncovering hidden patterns using techniques like dimensionality reduction. Additionally, mathematical tools allow us to measure distances or differences between data points. Methods like Euclidean distance or cosine similarity tell us how similar or dissimilar items are. This ability is vital in many deep learning applications, such as grouping similar items (clustering), detecting unusual data points (anomaly detection), and finding the closest matches in data (nearest-neighbor searches). By mapping data onto mathematical structures, we can identify patterns. For example, Fourier analysis can detect repeating patterns in time-based data, while convolution operations in convolutional neural networks (CNNs) find spatial patterns in images. Recognizing these patterns is fundamental to learning from data and making accurate predictions. To bring this idea to life, imagine grouping data points based on their similarities. Without mathematical techniques to measure these similarities, effectively clustering data would be nearly impossible.

Figure 1.1 illustrates how various mathematical constructs transform raw data into meaningful insights. The first subplot demonstrates the use of principal component analysis (PCA) to reduce the dimensionality of data, making it more interpretable and easier to visualize. In the second subplot, a heatmap reveals the Euclidean distances between data points, quantifying their dissimilarities and helping us understand the spatial relationships within the dataset. The third subplot displays a heatmap of cosine similarities, highlighting the degree of similarity between data points by focusing on their angular relationships. Finally, the K-means clustering subplot shows how data points are grouped into distinct clusters, revealing underlying patterns and structures in the dataset.

1.2.4 Designing and Interpreting Algorithms

Mathematical principles act like a compass, helping us choose the right functions, understand the effects of our choices, and keep improving existing algorithms. For example, consider how math is crucial when selecting activation functions for neural networks. Why are sigmoid functions often used? The answer lies in their unique mathematical properties. With their S-shaped curve, sigmoid functions accept any real number as input but always output a value between 0 and 1. This makes them perfect for situations where predictions, like probabilities, need to stay within a specific range. Understanding the math behind an algorithm also sheds light on how it behaves. For instance, knowing that sigmoid functions squeeze extreme input values into a narrow range between 0 and 1 helps us understand why neural networks using them might face vanishing gradients. This is a phenomenon where the gradients become very small, slowing down learning and affecting training efficiency. Mathematics is not just foundational for enhancing and creating new algorithms but also for understanding current ones.

1.2.5 Improving Models

Mathematics is essential for designing and understanding deep learning models, and it plays a crucial role in improving their performance. By learning mathematical concepts like overfitting, underfitting, and regularization, we can identify issues in our models, optimize their performance, and enhance their reliability. Overfitting and underfitting are common challenges in machine learning. Understanding these problems is key to fixing weaknesses in our models. Mathematics provides us with tools to tackle these issues directly. For example, regularization is a mathematical technique that improves a model's ability to generalize by reducing its complexity, helping to prevent overfitting. Techniques like $\mathbf{L}_1$ and $\mathbf{L}_2$ regularization, each with their own benefits, are crucial for simplifying models and making them more effective. Knowing these techniques empowers us to choose the best one for our specific needs, further enhancing our model's performance. Mathematics also plays a vital role in optimizing model parameters. Methods like gradient descent adjust model parameters step by step to minimize errors efficiently. Additionally, understanding the bias-variance trade-off, a key concept in statistics, helps us balance a model's accuracy on training data (bias) with its ability to perform well on new data (variance). Finding the right balance between bias and variance is crucial for building accurate and robust models. By applying mathematical insights, we can fine-tune our models to ensure they perform well both in theory and in real-world situations.

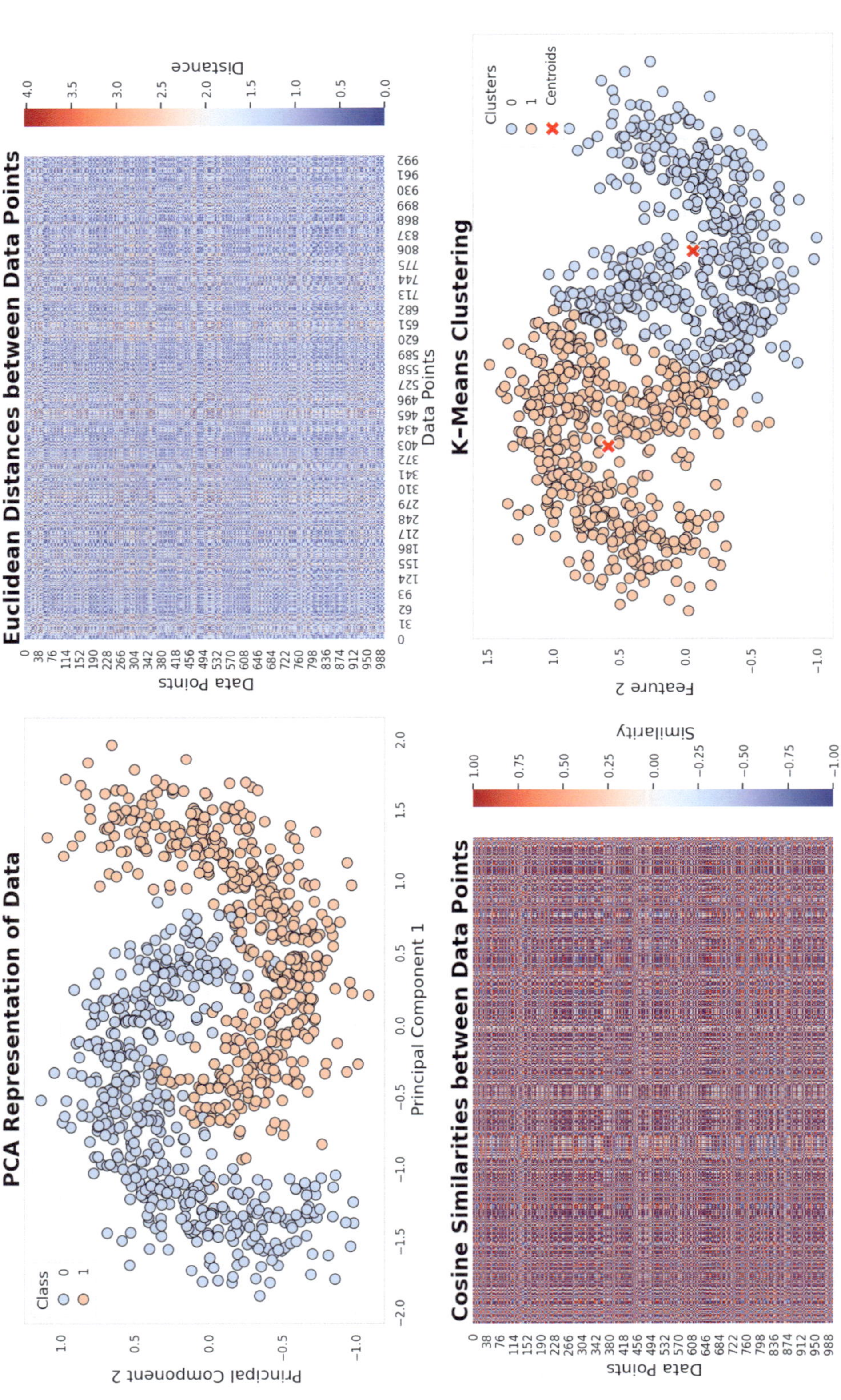

FIGURE 1.1 Transforming data and identifying patterns.

1.3 BRIEF OVERVIEW OF DEEP LEARNING

1.3.1 Simulating Human-Like Learning

Deep learning takes inspiration from how humans learn, especially through experience. Just like a child recognizes shapes, sounds, or faces by seeing them repeatedly, deep learning models process vast amounts of data through layers of artificial neural networks (ANNs). These models gradually learn to detect patterns, make connections, and develop a deeper understanding of the input data. At the core of deep learning is the idea of learning from data. Unlike traditional programming, where machines follow specific instructions for every task, deep learning models learn by example. They start with raw, unprocessed data and improve their understanding as they process it. This enables them to make decisions or predictions without being explicitly programmed for each specific task.

Much like humans, deep learning models improve with experience. They update their parameters with each training cycle to reduce prediction errors. The more data they are exposed to, the better they perform their tasks. This ongoing process mirrors human learning, where continuous practice leads to gradual improvement and mastery. By imitating the way humans learn, deep learning has become invaluable in many areas, from image recognition and natural language processing (NLP) to self-driving cars and medical diagnoses. Deep learning models can tackle complex problems that were once thought too difficult for artificial intelligence (AI).

1.3.2 Artificial Neural Networks

ANNs are the foundation of deep learning and are inspired by the neural networks in the human brain. These computational models imitate how the brain processes information but in a much simpler way. **ANNs** consist of interconnected layers of nodes, or "neurons", that work together to process information and learn patterns from data. An **ANN** is typically organized into three main layers:

- *Input Layer:* This layer receives the raw data that the network will process. Each neuron here represents a feature or input variable from your dataset.
- *Hidden Layers:* These are the intermediate layers that perform complex transformations on the input data. An **ANN** can have one or more hidden layers, each containing many neurons. These neurons are connected to neurons in the previous and next layers through weighted connections, similar to synapses in the brain. When a network has multiple hidden layers, it's called a deep neural network.
- *Output Layer:* This final layer produces the network's output, the result of processing the data. The number of neurons in this layer depends on how many output variables you need.

Each neuron in an **ANN** performs a simple operation:

1. *Receive Inputs:* The neuron gets inputs from the neurons in the previous layer. Each input is multiplied by a specific weight.
2. *Calculate Weighted Sum:* The neuron sums up all these weighted inputs and adds a bias term.
3. *Apply Activation Function:* The result is passed through an activation function, like the sigmoid function or ReLU (rectified linear unit). This function introduces nonlinearity into the model, allowing the network to learn complex patterns that aren't just straight lines.

The real strength of **ANNs** comes from the way neurons and layers are interconnected. During training, the network adjusts the weights and biases of each neuron to minimize the difference between its predictions and the actual values. This adjustment is done through a process called backpropagation, which uses optimization algorithms like gradient descent. Backpropagation calculates the gradient of the loss function (which measures the error) with respect to each weight.

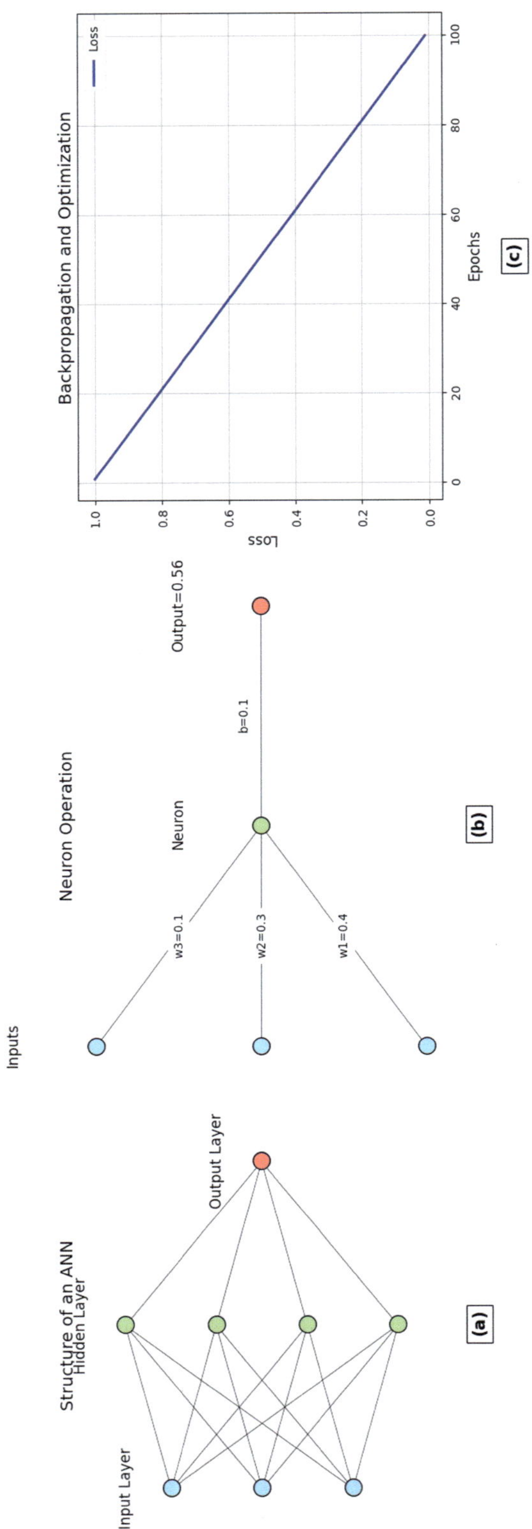

FIGURE 1.2 (a) Structure of an ANN, (b) neuron operation, and (c) backpropagation and optimization.

It then updates the weights to reduce this loss. **ANNs** are powerful because of their layered and interconnected structure, and they can model complex patterns and structures in data. They excel at finding hidden patterns and subtle relationships that traditional machine learning models might miss. This ability has led to major breakthroughs in areas such as computer vision, NLP, and speech recognition.

Figure 1.2 illustrates the key components of ANNs. Figure 1.2a depicts the basic structure of an ANN, showcasing the input layer, one or more hidden layers, and the output layer. Figure 1.2b demonstrates the operation of a single neuron, showing how it processes inputs, applies weights and biases, and utilizes an activation function to produce an output. Figure 1.2c visualizes the backpropagation process, highlighting how the loss decreases over epochs as the model's parameters are optimized through iterative weight adjustments.

1.3.3 Applications of Deep Learning

Deep learning, a major part of AI, has transformed many fields in technology, research, and business through its wide-ranging applications. Its ability to process and learn from large amounts of data has led to breakthroughs once thought impossible. In computer vision, deep learning excels at image recognition, enabling models to quickly and accurately identify objects, people, and scenes in images. This power is behind facial recognition systems, self-driving cars, and medical imaging. For example, deep learning algorithms can detect tumors in medical scans with accuracy similar to experienced doctors. In NLP, deep learning has made significant strides. Algorithms can now understand and generate human language, allowing for sentiment analysis, language translation, text summarization, and question-answering. These advancements power everyday tools such as voice assistants, customer service chatbots, and real-time translation services, improving communication and accessibility worldwide. Beyond recognition and understanding, deep learning fosters creativity through generative models like generative adversarial networks (GANs). **GANs** can produce new data that closely resembles their training data, creating realistic images, music, and even art. For instance, **GANs** can generate lifelike human faces of people who do not exist, which has exciting applications in entertainment, fashion, and virtual reality. In predictive analytics, deep learning is a powerful tool that can analyze vast amounts of data to forecast future events like stock prices, customer behavior, disease outbreaks, and natural disasters. Industries such as finance, marketing, healthcare, and disaster management use this predictive power to make informed, data-driven decisions with greater confidence. Additionally, reinforcement learning, a subset of deep learning, involves models that learn to make decisions by interacting with their environment. This approach has achieved remarkable success in game-playing AI, surpassing human champions in games like Go, Chess, and Poker. Reinforcement learning is also applied in robotics, helping robots learn to navigate environments and manipulate objects on their own, paving the way for more advanced automation and intelligent systems. These applications demonstrate the significant impact deep learning has across various fields, driving innovation and enhancing capabilities in many aspects of modern life. As deep learning continues to evolve, it holds the promise of unlocking new possibilities and addressing complex challenges that were once beyond our reach.

1.4 BOOK FEATURES AND STRUCTURE

1.4.1 Book Features

Math can often feel overwhelming, filled with strange symbols and complex rules, especially when you're new to deep learning. Concepts like vectors and matrices from linear algebra, the tricky operations of calculus, and the rules of probability are all deeply connected to deep learning. At first glance, these ideas might seem scary, like a wall that's hard to climb. That's where this book comes in. It aims to guide you through the maze of math by breaking down these abstract concepts and making them easy to grasp. One of our main goals is to uncover the complexity of mathematical ideas. Terms like complex numbers, high-dimensional vectors, and abstract spaces can seem intimidating. The abstract nature of math often makes it hard to see how it applies to real-life problems, leaving many wondering how these theories are useful. This book is dedicated to bridging that gap by directly linking mathematical theories to their uses in deep learning. By doing this, abstract ideas become practical tools you can use in AI. It's not just about understanding math; it's about seeing these theories come alive in deep learning. You might think of math and deep learning as completely separate fields. Math, with its abstract symbols and strict proofs, often feels very different from the hands-on, algorithm-focused world of deep learning. However, the connection between them is much deeper than it might seem. This book aims to connect these seemingly different areas by creating clear and practical links between mathematical concepts and deep learning techniques. Each math concept is paired with its real-world application in a deep learning scenario. This approach not only applies theoretical knowledge but also shows how to turn abstract formulas into useful algorithms. By exploring these connections, you'll gain deeper insights into deep learning, turning it from a mysterious "black box" into a system you can understand and explain. This foundational knowledge will help you see how algorithms work, spot potential issues, and find ways to improve them. Mastering the math behind deep learning isn't just about knowing how things work; it's about improving, adapting, and innovating based on a deep understanding of the core principles. Understanding deep learning goes beyond just using existing models and tools. It's about diving deep into the algorithms, figuring out how they work, and understanding why they succeed or fail in certain situations. Moreover, it's more than just understanding; it's about innovation, using your new knowledge to build on existing methods and create unique solutions to your challenges. That's why a key goal of this book is to empower you, the reader.

1.4.2 Overview of Chapters

- **Chapter 1: Introduction:** The introductory chapter sets the stage for the book, providing an overview of the core topics that will be covered. It emphasizes how these mathematical concepts contribute to the development and optimization of deep learning models, preparing you for the in-depth exploration ahead.
- **Chapter 2: Linear Algebra:** This chapter delves into the essential elements of linear algebra, which form the backbone of many deep learning algorithms. Topics include vectors, matrices, matrix operations, eigenvalues, and eigenvectors. Understanding these concepts is vital for grasping more advanced techniques, such as matrix factorization and singular value decomposition, which are critical for neural network computations and data transformations.
- **Chapter 3: Multivariate Calculus:** Multivariate calculus is integral to the optimization processes in deep learning. This chapter explores derivatives, gradients, partial derivatives, and the Hessian matrix. We focus on their applications in neural networks, especially during the training phase, where gradient-based optimization techniques like backpropagation are employed to minimize loss functions.
- **Chapter 4: Probability Theory and Statistics:** Deep learning models operate under uncertainty, making probability and statistics fundamental to their success. This chapter covers probability distributions, Bayesian inference, hypothesis testing, and statistical methods that help model uncertainty. These tools enable effective learning from data in environments where randomness and variability are significant factors.
- **Chapter 5: Optimization Theory:** At the heart of deep learning is the challenge of optimizing model parameters. This chapter introduces optimization theory, covering both convex and nonconvex optimization techniques, gradient descent methods, and advanced optimization strategies like stochastic gradient descent, Adam, and root mean squared propagation

(RMSprop). These methods ensure that models efficiently converge to the best possible performance.

- **Chapter 6: Information Theory:** Information theory provides tools for quantifying the information processed by neural networks. This chapter discusses key concepts such as entropy, mutual information, and Kullback–Leibler divergence (KL divergence), highlighting their relevance in model regularization, compression, and understanding information flow through networks. Applications in areas like variational autoencoders and information bottleneck methods are explored.
- **Chapter 7: Graph Theory:** Graph theory has gained importance in deep learning, particularly with the development of graph neural networks (GNNs). This chapter introduces the fundamentals of graphs, including nodes, edges, adjacency matrices, and graph Laplacians. It explores their applications in representing complex relationships in structured data such as social networks, molecular structures, and knowledge graphs.
- **Chapter 8: Differential Geometry:** Studying high-dimensional spaces is crucial for deep learning models, which often operate in complex, nonlinear environments. This chapter introduces the mathematical structures of manifolds and tangent spaces geometry. We demonstrate how differential geometry enhances our understanding of optimization landscapes and generalization properties in deep learning.
- **Chapter 9: Topology in Deep Learning:** Topology provides a way to understand the shape and structure of data. This chapter introduces topological data analysis (TDA), focusing on concepts like persistent homology and Betti numbers. We discuss how these tools are applied in deep learning to uncover hidden patterns and structures in data, leading to improved feature extraction and data representation techniques.
- **Chapter 10: Harmonic Analysis for CNNs:** Harmonic analysis focuses on the frequency components of signals and has significant applications in CNNs. This chapter explores Fourier transforms, wavelets, and spectral analysis, illustrating how these mathematical tools improve feature extraction, signal processing, and understanding of convolution operations in **CNNs**.
- **Chapter 11: Dynamical Systems and Differential Equations for RNNs:** Recurrent neural networks (RNNs) are designed to handle sequential data, where time-dependent behavior plays a crucial role. This chapter examines the theory of dynamical systems and differential equations, explaining how they are used to model temporal dependencies in **RNNs**.
- **Chapter 12: Quantum Computing:** Quantum computing represents the next frontier in computational power, with the potential to revolutionize deep learning. This chapter introduces quantum principles and quantum algorithms, discussing how quantum computing might be leveraged to accelerate and enhance neural network training and inference.

1.4.3 Chapter's Structure

Each chapter in this book is thoroughly crafted to provide a complete understanding of fundamental mathematical concepts and their relevance to deep learning. To facilitate your learning journey, every chapter includes the following sections:

- **Preface:** Each chapter begins with a preface that introduces the core concepts to be covered.
- **Chapter Contents: Definitions, Formulas, Examples, and Real-World Applications:** The main content delves into essential definitions, formulas, and mathematical frameworks. Concepts are explained step by step, accompanied by practical examples to ensure clarity. Real-world applications demonstrate how these principles are applied in deep learning models and related fields, making the transition from theory to practice seamless.

- **Hands-On Examples:** To bridge the gap between theory and practice, each chapter includes a hands-on section with programming examples. These examples allow you to implement the mathematical concepts in code, gaining practical experience and reinforcing your understanding of how these techniques are applied in deep learning frameworks.
- **Common Mistakes and Troubleshooting Tips:** Learning complex mathematical concepts often involves overcoming challenges. This section highlights frequent mistakes that students and practitioners make when applying the material, along with tips for troubleshooting and avoiding errors in both theoretical understanding and practical implementation.
- **Review Questions:** Review questions are provided to test your understanding of the chapter material. These questions challenge your understanding of definitions, formulas, and key concepts, allowing you to self-assess your learning progress.
- **Programming Exercises:** To deepen your programming skills, each chapter includes three programming problems categorized as easy, medium, and challenging. These exercises are directly related to the chapter's content, providing an opportunity to apply what you've learned in a practical coding environment. By solving these problems, you will enhance your problem-solving abilities and strengthen your grasp of how mathematics drives deep learning algorithms.

2 Linear Algebra

2.1 INTRODUCTION

Linear algebra provides the essential framework for describing vectors, matrices, and tensor operations that form the core of neural network computations. Beyond a set of mathematical tools, linear algebra is the thread weaving through the fabric of deep learning, enabling us to frame problems and devise solutions that are both computationally efficient and conceptually profound. In this chapter, we will delve into critical concepts such as vector spaces, matrix decompositions, and the operations that empower neural networks to process and learn from vast amounts of data. These concepts are not only theoretical foundations but also practical instruments that bring deep learning models to life. As you journey through this chapter, you will discover how the principles of linear algebra are applied to enable neural networks to learn from data, optimize models, and ultimately drive innovation in artificial intelligence.

2.2 VECTORS, MATRICES, AND TENSOR OPERATIONS

This section will dive into the fundamental concepts of vectors, matrices, tensor operations, linear transformations, eigenvalues, eigenvectors, and singular value decomposition (SVD).

2.2.1 Understanding Vectors

2.2.1.1 What Is a Vector?

A vector is a fundamental mathematical entity that possesses both magnitude (size) and direction. You can visualize a vector as an arrow: the length of the arrow represents its magnitude, and the arrow's orientation indicates its direction. This concept is not just theoretical; vectors are crucial in numerous real-world applications, including deep learning, where they are used to represent data points, model weights, and activations within neural networks (NNs). Imagine you're playing a video game where your character's movement is determined by a two-dimensional (2D) vector. This vector tells the character how much to move in the **x** (horizontal) and **y** (vertical) directions. For example, consider the vector:

$$v = \begin{bmatrix} 3 \\ 4 \end{bmatrix}.$$

This vector instructs the character to move three units to the right (along the x-axis) and four units upward (along the y-axis). By representing movement in this way, vectors provide a concise and

DOI: 10.1201/9781032690742-2

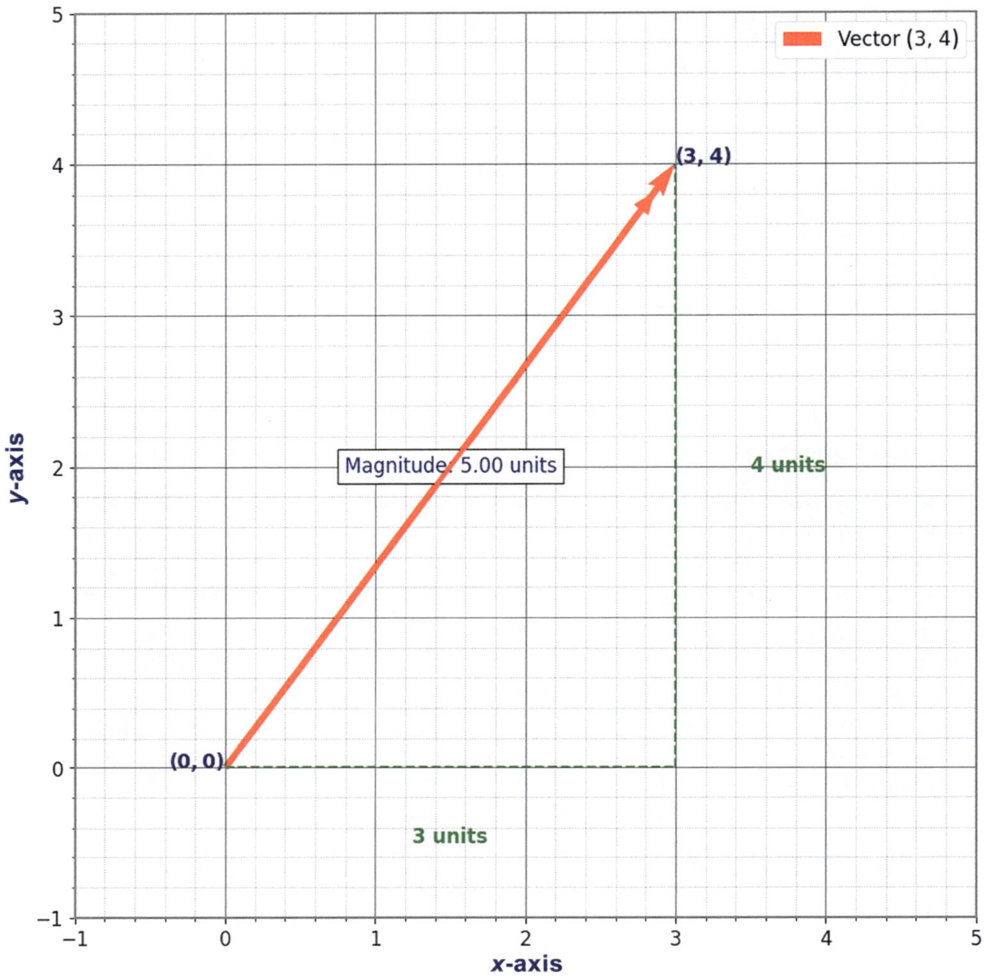

FIGURE 2.1 Vector representation [3, 4].

powerful way to encode directional information, which is essential in both physics and computational fields like deep learning.

Figure 2.1 illustrates a vector represented by a red arrow that starts from the origin (0, 0) and points toward the coordinates (3, 4). The length of the arrow indicates the vector's magnitude, while its orientation shows its direction. The magnitude of a vector can be thought of as its "strength" or "length." For the vector **v** given by:

$$v = \begin{bmatrix} 3 \\ 4 \end{bmatrix},$$

you can calculate the magnitude using the Pythagorean theorem:

$$\|\mathbf{v}\| = \sqrt{v_x^2 + v_y^2} = \sqrt{3^2 + 4^2} = \sqrt{9 + 16} = \sqrt{25} = 5.$$

Here, $\|\mathbf{v}\| = 5$ units. This value represents the direct distance from the origin to the point (3, 4) on a 2D plane.

2.2.1.2 Vector Operations

Vectors are fundamental in various fields, and understanding their operations is crucial. Here are some essential vector operations with examples.

(a) *Addition:* Adding two vectors involves adding their respective components. Geometrically, you place the tail of one vector at the head of the other and then draw a vector from the tail of the first to the head of the second. This new vector is the resultant or sum of the two vectors.

Example: Let **a** and **b** be defined as follows:

$$\boldsymbol{a} = [1, 2] \text{ } and \text{ } \boldsymbol{b} = [3, 4].$$

To add these vectors:

$$\boldsymbol{c} = \boldsymbol{a} + \boldsymbol{b} = [1, 2] + [3, 4] = [1 + 3, 2 + 4] = [4, 6].$$

In Figure 2.2, the red arrow represents **a**, the blue arrow represents **b**, and the green arrow **c** represents the resultant vector **c** = **a** + **b**.

(b) *Scalar Multiplication:* Multiplying a vector by a scalar changes its magnitude without changing its direction unless the scalar is negative, in which case the direction is reversed. To scale a vector by a scalar **k**, multiply each component of the vector by **k**.

Example: Let **v** and scalar **k** be defined as follows:

$$\boldsymbol{v} = \begin{bmatrix} 3 \\ 4 \end{bmatrix}, \quad k = 2 \text{ The scaled vector is:}$$

$$\boldsymbol{v} = 2\begin{bmatrix} 3 \\ 4 \end{bmatrix} = \begin{bmatrix} 6 \\ 8 \end{bmatrix}$$

In Figure 2.3, the red arrow represents **v** as **(a)**, and the blue arrow represents the scaled vector **kv** as **(3a)**. It illustrates the concept of vector scaling on a 2D plane. It shows two vectors, **a** and **3a**, originating from the origin (0, 0) and extending in the same direction. The vector **a**, with coordinates (2, 1), is marked in red, and its magnitude (or length) is labeled as approximately 2.24, calculated as $\sqrt{2^2 + 1^2} = \sqrt{5}$. The second vector, **3a**, represents a scaled version of **a** by a factor of 3, extending further along the same direction. Its length is shown as 6.71, three times the magnitude of **a** (since $|3a| = 3 \times |a|$). This blue vector demonstrates the effect of scalar multiplication on a vector's length, keeping the direction unchanged while proportionally increasing the magnitude.

(c) *Dot Product:* The dot product is an operation that multiplies two vectors to obtain a scalar. It helps determine the angle between two vectors and is calculated as the sum of the products of their respective components.

Example: Let **a** and **b** be defined as follows:

$$\boldsymbol{a} = \begin{bmatrix} 1 \\ 2 \end{bmatrix}, \quad \boldsymbol{b} = \begin{bmatrix} 3 \\ 4 \end{bmatrix}$$

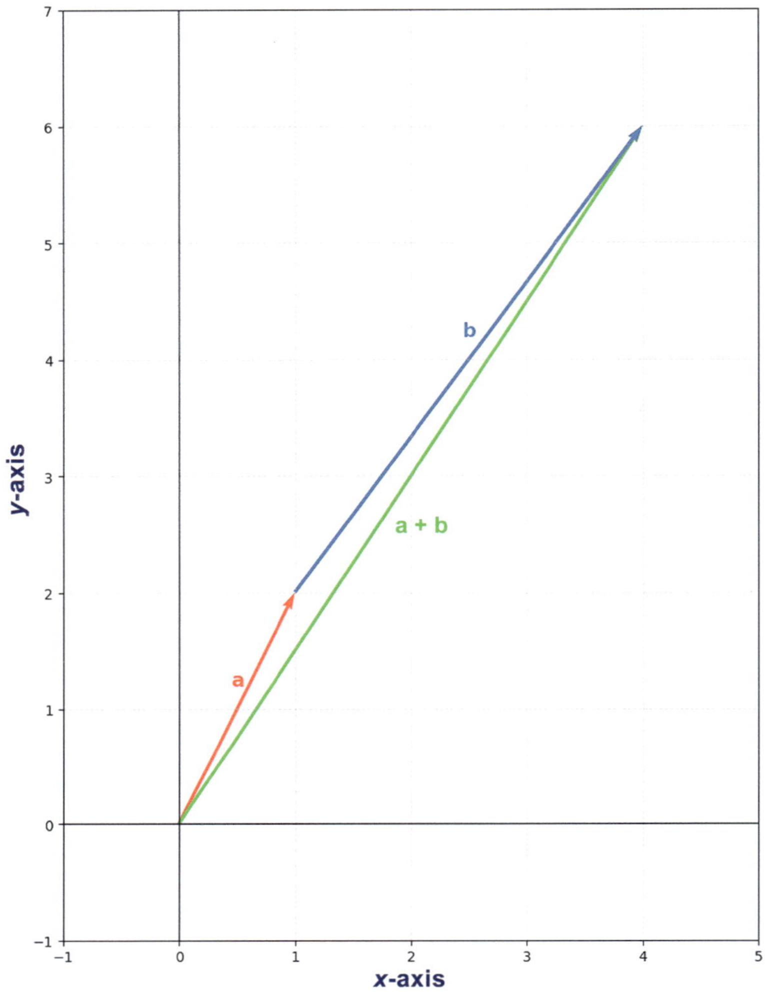

FIGURE 2.2 Vector addition: $a + b$.

The dot product of **a** and **b** is:

$$\boldsymbol{a} \cdot \boldsymbol{b} = (1)(3) + (2)(4) = 3 + 8 = 11.$$

The dot product is 11, and this scalar value is crucial in various applications, such as calculating the cosine of the angle between the two vectors:

$$\cos\theta = \frac{\boldsymbol{a} \cdot \boldsymbol{b}}{\|\boldsymbol{a}\| \ \|\boldsymbol{b}\|}.$$

Figure 2.4 shows two vectors, **a** (red) and **b** (blue), with the green vector representing the projection of **a** onto **b**. This projection shows how much of **a** aligns with **b**. The projection length depends on the cosine of the angle between **a** and **b**, indicating the part of **a** that points in **b**'s direction. The "Dot Product: 11" label represents the dot product, a measure of this alignment.

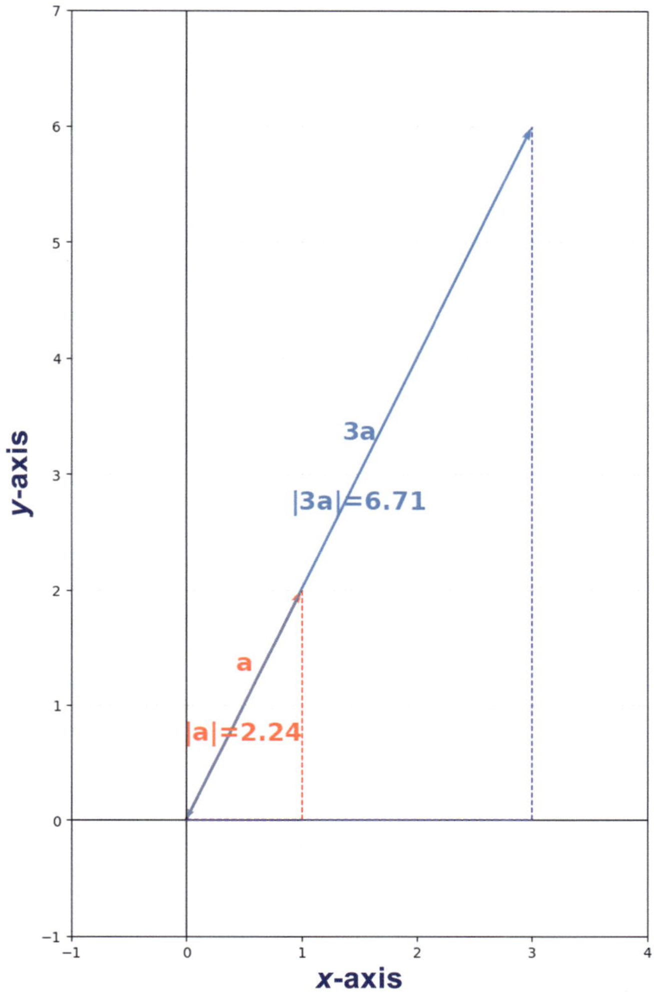

FIGURE 2.3 Scalar multiplication: **3a**.

(d) *Cross-Product:* The cross-product between two vectors in three-dimensional space produces another vector that is perpendicular to the plane formed by the two original vectors. The cross-product of vectors **a** and **b** produces a third vector, **c**, that is perpendicular to both **a** and **b**.

Example: Let **a** and **b** be defined as follows:

$$\boldsymbol{a} = \begin{bmatrix} 1 \\ 0 \\ 0 \end{bmatrix}, \quad \boldsymbol{b} = \begin{bmatrix} 0 \\ 1 \\ 0 \end{bmatrix}.$$

The cross-product is calculated using the determinant of a matrix composed of unit vectors and the components of **a** and **b**:

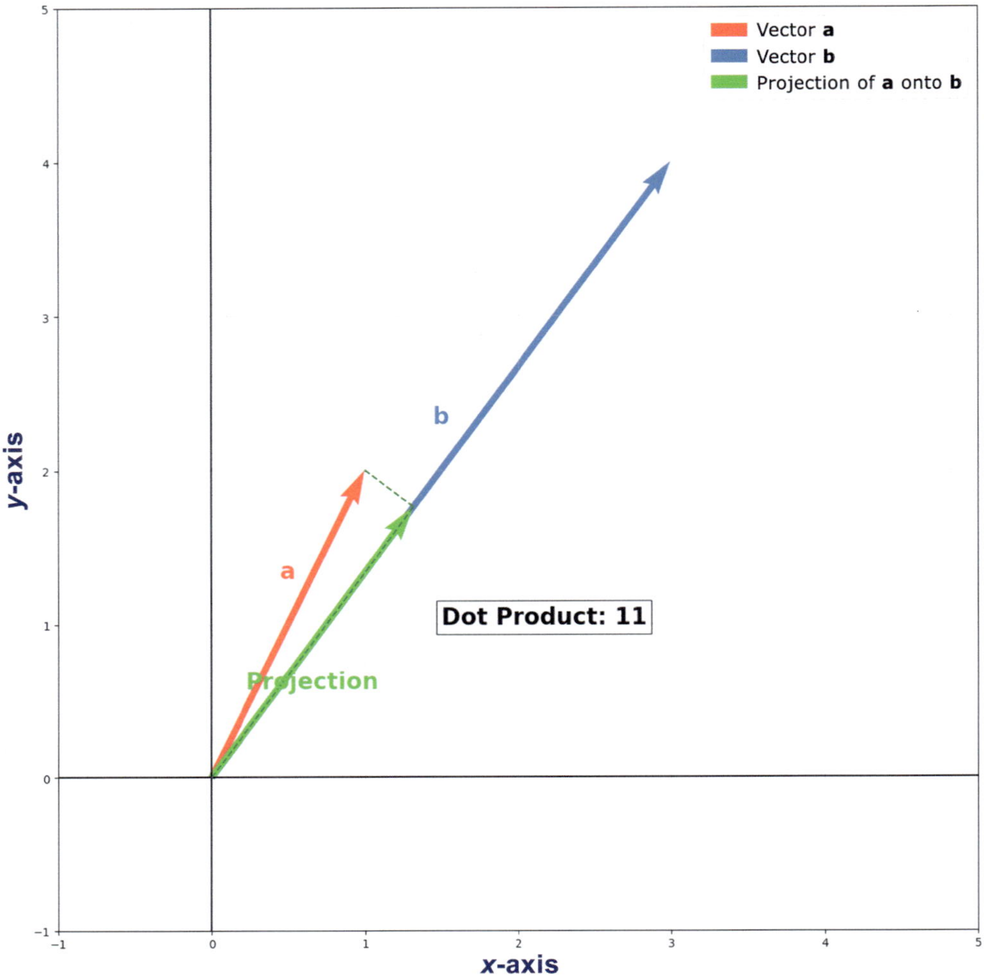

FIGURE 2.4 Dot product interpretation.

$$\boldsymbol{a}\times\boldsymbol{b}=\begin{bmatrix} x & y & z \\ 1 & 0 & 0 \\ 0 & 1 & 0 \end{bmatrix}, \text{ then } \quad c=\boldsymbol{a}\times\boldsymbol{b}=x\begin{vmatrix} 0 & 0 \\ 1 & 0 \end{vmatrix}-y\begin{vmatrix} 1 & 0 \\ 0 & 0 \end{vmatrix}+z\begin{vmatrix} 1 & 0 \\ 0 & 1 \end{vmatrix} \text{or}$$

$$\boldsymbol{c}=\boldsymbol{a}\times\boldsymbol{b}=\begin{bmatrix} a_y b_z - a_z b_y \\ a_z b_x - a_x b_z \\ a_x b_y - a_y b_x \end{bmatrix}$$

Substituting the components:

$$\boldsymbol{c}=\begin{bmatrix} (0)(0)-(0)(1) \\ (0)(0)-(1)(0) \\ (1)(1)-(0)(0) \end{bmatrix}=\begin{bmatrix} 0 \\ 0 \\ 1 \end{bmatrix}$$

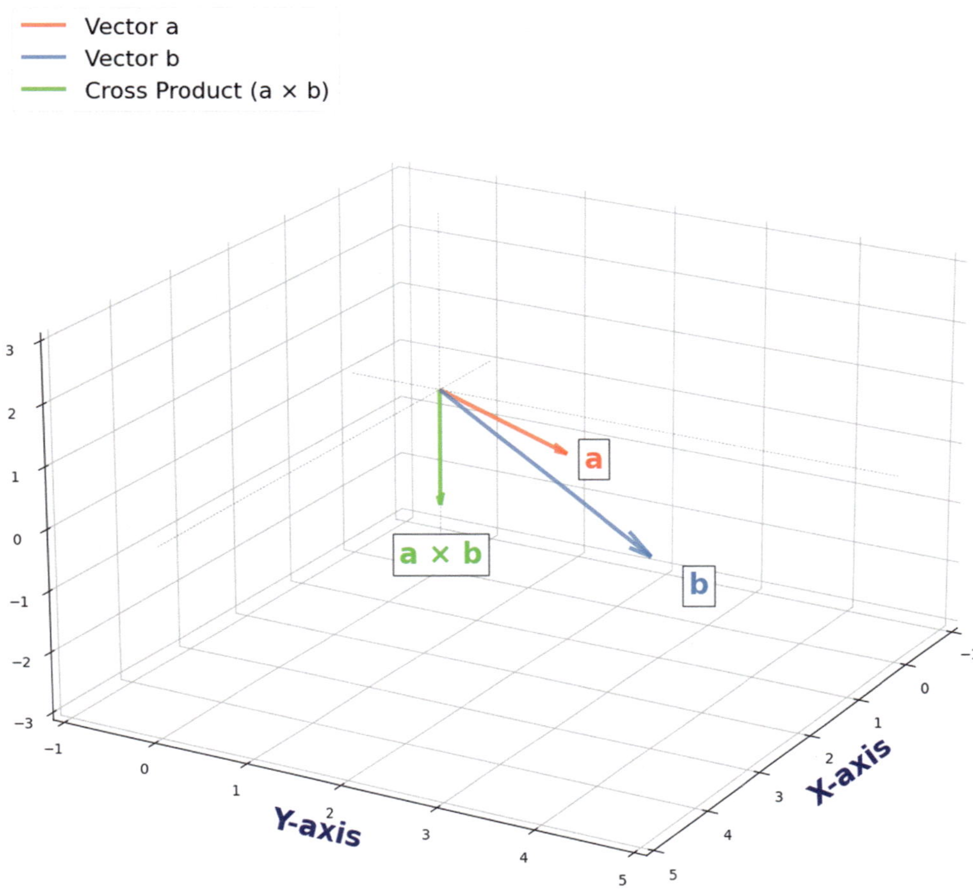

FIGURE 2.5 The cross-product of vectors **a** and **b**.

Figure 2.5 demonstrates the concept of the cross-product between two vectors, **a** (red) and **b** (blue). The cross-product, shown as the green vector labeled **a × b**, represents a new vector that is perpendicular to both **a** and **b**. This perpendicular direction follows the right-hand rule, meaning if you point your right-hand's fingers along **a** and curl them toward **b**, your thumb points in the direction of **a × b**. The length of the green vector reflects the magnitude of the cross-product, which corresponds to the area of the parallelogram covered by **a** and **b**.

2.2.1.3 Vector Components

In a 2D space, any vector can be decomposed into its components along the axes of the coordinate system.

Example: Consider a vector that terminates at the point (3, 4) on a 2D Cartesian plane. This vector can be represented by its components along the **x** and **y** axes: $\vec{A} = A_x\hat{i} + A_y\hat{j}$

Figure 2.6 illustrates the components of the vector $\vec{A}$ on a 2D Cartesian plane, where $\mathbf{A_x}$ and $\mathbf{A_y}$ are unit vectors along the **x** and **y** directions, respectively. For vector, $\vec{A}$, $\mathbf{A_x}$ is 3, and $\mathbf{A_y}$ is 4. The

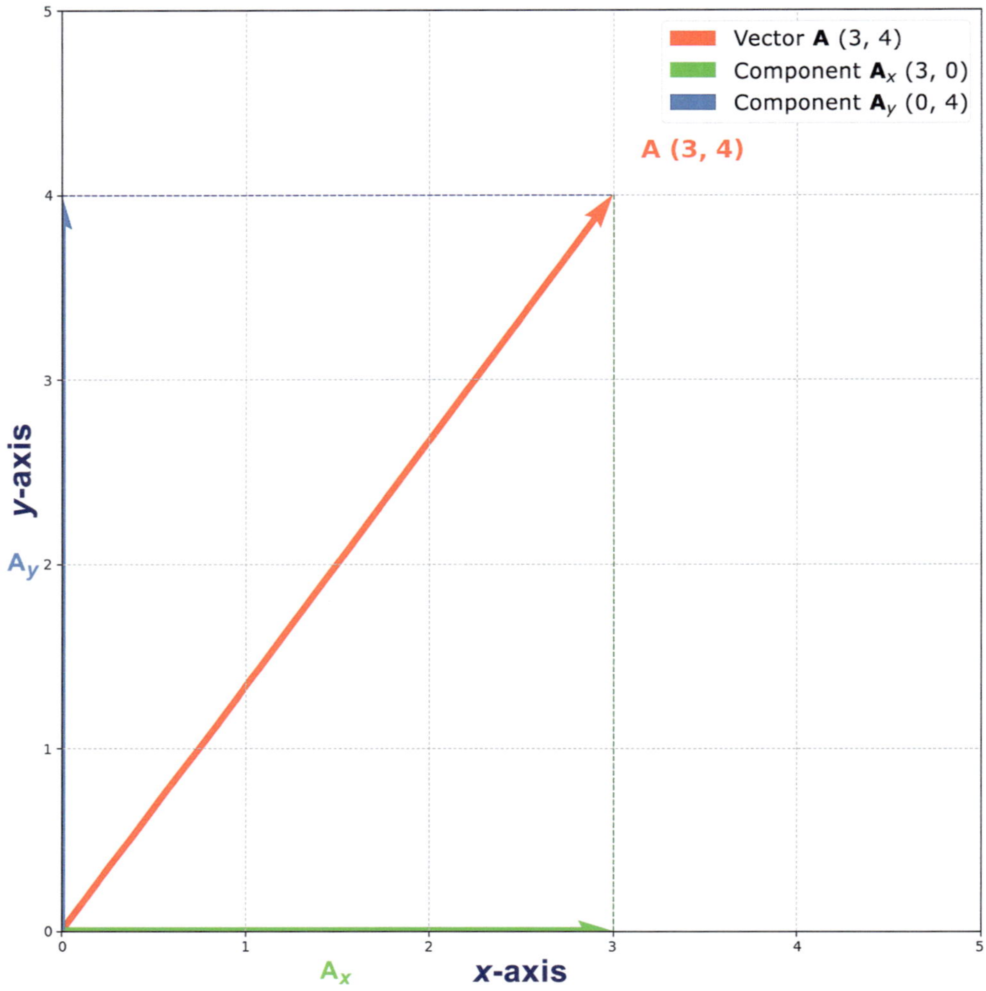

FIGURE 2.6 Vector components of **A**.

red arrow shows the vector $\vec{A}$ ending at point (3, 4). The green arrow represents the x-component extending along the x-axis to 3, and the blue arrow indicates the y-component extending along the y-axis to 4.

2.2.1.4 Magnitude and Direction

The magnitude of a vector $\vec{A}$ in 2D space is given by: $|\vec{A}| = \sqrt{A_x^2 + A_y^2}$. The direction can be found using trigonometry, typically with the tangent function:

$$\theta = \arctan\left(\frac{A_y}{A_x}\right).$$

Example: Consider a vector that terminates at the point (3, 4) on a 2D Cartesian plane. This vector can be represented by its components along the **x** and **y** axes: $\vec{A} = A_x\hat{i} + A_y\hat{j}$, where $A_x = 3$ and $A_y = 4$ The magnitude is: $|\vec{A}| = \sqrt{3^2 + 4^2} = \sqrt{9+16} = \sqrt{25} = 5$ and the direction is: $\theta = \arctan\left(\frac{4}{3}\right) = 53.13°$.

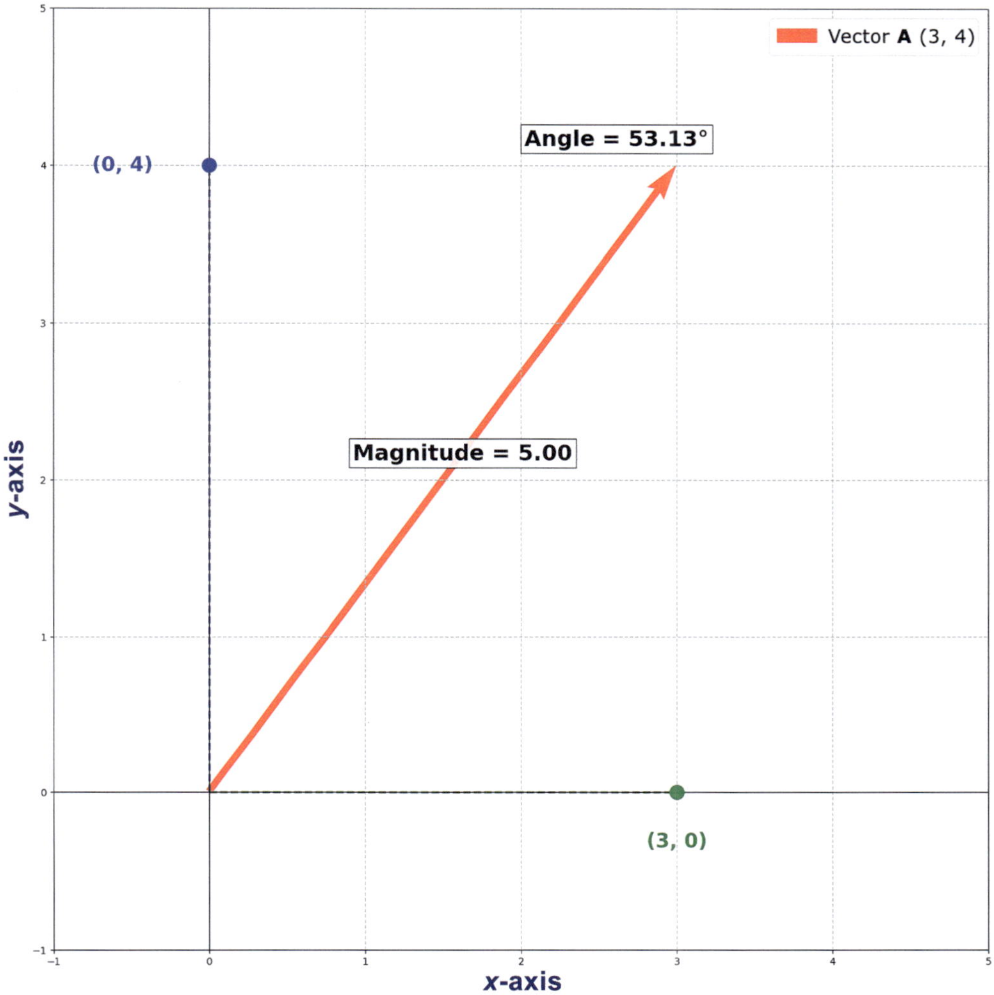

FIGURE 2.7 Vector magnitude and direction.

Figure 2.7 shows vector **A** = (3, 4) on a 2D plane. The red arrow represents vector **A**, starting from the origin (0, 0) and pointing to (3, 4). The length, or magnitude, of **A,** is labeled as 5.00, calculated from $\sqrt{3^2+4^2}=5$. The angle between **A** and the x-axis is marked as 53.13°, which is found using $\tan^{-1}\left(\frac{4}{3}\right)$. The blue and green dashed lines show the projections of **A** onto the y-axis at (0, 4) and the x-axis at (3, 0), highlighting its horizontal and vertical components.

2.2.1.5 Vector Spaces

A vector space is a set of vectors combined with two operations (vector addition and scalar multiplication) that satisfy specific properties. The space itself can be in any dimension, and the vectors do not necessarily have to be geometric; they can be functions, polynomials, or other mathematical entities as long as they follow the defined rules of a vector space. A vector space **V** over a field **F** (often the field of real numbers) must satisfy the following properties:

(a) *Vector Addition*
  1. **Commutativity:** $\mathbf{u}+\mathbf{v}=\mathbf{v}+\mathbf{u}$ for all $\mathbf{u},\ \mathbf{v} \in Vs$
  2. **Associativity:** $\mathbf{u}+(\mathbf{v}+\mathbf{w})=(\mathbf{u}+\mathbf{v})+\mathbf{w}$ for all $\mathbf{u},\ \mathbf{v},\mathbf{w} \in V$
  3. **Identity:** There exists a vector $0 \in V$ such that $\text{v}+0=\text{v}$ for all $\text{v} \in \text{V}$
  4. **Additive Inverse:** For every $v \in V$, there exists a vector $-v \in V$ such that $v+(-v)=0$

(b) *Scalar Multiplication*
  1. **Distributivity over Vector Addition:**

$$a(\mathbf{u}+\mathbf{v})=a\mathbf{u}+a\mathbf{v} \text{ for all scalars } a \text{ and vectors } \mathbf{u},\mathbf{v} \in V$$

  2. **Distributivity over Scalar Addition:**

$$(a+b)\mathbf{u}=a\mathbf{u}+a\mathbf{v} \text{ for all scalars } a,b \text{ and vectors } \mathbf{u} \in V$$

  3. **Associativity:** $a(b\mathbf{u})=(ab)\mathbf{u}$ for all scalars a,b and vectors $\mathbf{u} \in V$
  4. **Identity:** $1 \cdot \text{v}=\text{v}$ for all $v \in V$, where 1 is the multiplicative identity in the field F

Figure 2.8 shows different operations on two vectors, $v_1 = (1, 2)$ and $v_2 = (2, 3)$, on a 2D plane. The red vector represents v_1, starting from the origin and pointing to (1, 2), while the blue vector represents v_2, pointing to (2, 3). The pink vector, $3 \times v_1 = (3, 6)$, is a scaled version of v_1 by a factor of 3, extending in the same direction but with three times the length. The green vector, $v_1 + v_2 = (3, 5)$, shows the result of adding v_1 and v_2 together, combining their directions to reach the point (3, 5). Finally, the cyan vector, $-v_1 = (-1, -2)$, is v_1 reversed in direction, pointing to (−1, −2).

2.2.3 Vector Relevance in the Context of Deep Learning

2.2.3.1 Data Representation

(a) *Feature Vectors:* In machine learning and deep learning, data is typically represented as vectors, where each component (or dimension) of the vector corresponds to a specific feature of the data. For instance, consider an image from the MNIST dataset, which contains images of handwritten digits, each with a resolution of 28 × 28 pixels. This image can be flattened into a feature vector with 784 dimensions (since 28 × 28 = 784). Also, each dimension in this vector represents the intensity of a specific pixel, ranging from 0 (black) to 255 (white). For example, a pixel value of 0 represents a completely black pixel, and a value of 255 represents a fully white pixel. This representation transforms the image into a manageable vector format for machine learning models to process. Suppose we have a 3 × 3 image for simplicity, with the following pixel intensity values:

$$\begin{bmatrix} 255 & 128 & 0 \\ 64 & 255 & 32 \\ 0 & 128 & 64 \end{bmatrix}$$

Flattening this 3 × 3 image into a vector results in a nine-dimensional feature vector:

$$[255, 128, 0, 64, 255, 32, 0, 128, 64]$$

This flattened vector can now be used as input to a machine learning model.

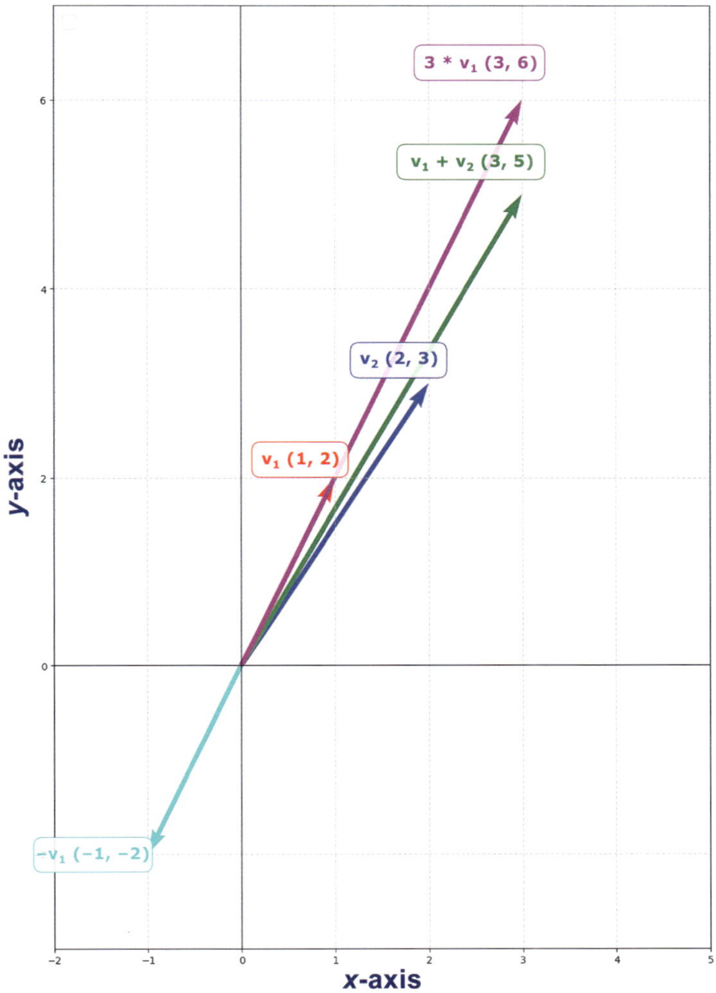

FIGURE 2.8 Vector operations in $\mathbb{R}^2$.

(b) *Word Embeddings:* In natural language processing (NLP), words or phrases are mapped to vectors of real numbers, known as word embeddings. These embeddings place words in a high-dimensional vector space, where the distance between words reflects their semantic relationships. For example, word embeddings might position "king" and "queen" close to each other because of their similar meanings, while "apple" and "banana" might be positioned close due to their shared category as fruits. Consider two words, "king" and "queen," represented by three-dimensional embeddings as follows: "king" = [0.8, 0.2, 0.7] and "queen" = [0.75, 0.3, 0.65]. The values for "king" and "queen" are just examples of values that illustrate how word embeddings work. For example, "king" and "queen" have similar vectors because they share common features like royalty, but small differences in their numbers capture distinctions like gender. In practice, the actual numbers would come from a trained model like Word2Vec, based on analyzing large text corpora. To measure their similarity, we can compute the cosine similarity between the two vectors:

$$\text{cosine similarity} = \frac{\text{king} \cdot \text{queen}}{|\text{king}||\text{queen}|} = \frac{(0.8 \times 0.75) + (0.2 \times 0.3) + (0.7 \times 0.65)}{\sqrt{0.8^2 + 0.2^2 + 0.7^2} \times \sqrt{0.75^2 + 0.3^2 + 0.65^2}}$$

$$\text{cosine similarity} = \frac{1.115}{1.1215} \approx 0.9942.$$

The resulting value (0.994 and close to 1) indicates a strong similarity between "king" and "queen," reflecting their semantic connection.

2.2.3.2 Neural Network Parameters

The parameters of NNs, specifically the weights and biases, are organized as vectors, matrices, or higher-dimensional tensors depending on the structure of the network. These parameters are essential in controlling how the network processes and transforms input data. When the network receives an input in the form of a feature vector, it performs matrix–vector multiplications using the weights and biases at each layer. The result is then passed through activation functions, such as the ReLU (Rectified Linear Unit) or sigmoid function, to introduce non-linearity. This step-by-step process transforms the input data, enabling the network to learn complex patterns and generate the desired output. Let's consider a simple NN layer with 3 input features and 2 output neurons. The input feature vector is:

$$x = [1.0, 0.5, -0.2]$$

The weight matrix for this layer is a 2 × 3 matrix, where each row corresponds to the weights connected to a single output neuron:

$$\boldsymbol{W} = \begin{bmatrix} 0.2 & -0.5 & 1.0 \\ -0.3 & 0.8 & 0.5 \end{bmatrix}.$$

The bias vector for the two output neurons is:

$$\mathbf{b} = [0.1, -0.1]$$

To calculate the output, we perform matrix–vector multiplication and add the bias:

$$z = \boldsymbol{W}\boldsymbol{x} + \boldsymbol{b} = \begin{bmatrix} 0.2 & -0.5 & 1.0 \\ -0.3 & 0.8 & 0.5 \end{bmatrix} \begin{bmatrix} 1.0 \\ 0.5 \\ -0.2 \end{bmatrix} + \begin{bmatrix} 0.1 \\ -0.1 \end{bmatrix}.$$

Performing the matrix multiplication:

$$z = \begin{bmatrix} (0.2 \times 1.0) + (-0.5 \times 0.5) + (1.0 \times -0.2) \\ (-0.3 \times 1.0) + (0.8 \times 0.5) + (0.5 \times -0.2) \end{bmatrix} + \begin{bmatrix} 0.1 \\ -0.1 \end{bmatrix} = \begin{bmatrix} 0.2 - 0.25 - 0.2 \\ -0.3 + 0.4 - 0.1 \end{bmatrix} + \begin{bmatrix} 0.1 \\ -0.1 \end{bmatrix}$$
$$= \begin{bmatrix} -0.25 + 0.1 \\ 0.0 - 0.1 \end{bmatrix} = \begin{bmatrix} -0.15 \\ -0.1 \end{bmatrix}.$$

Thus, the output of this layer before applying the activation function is [−0.15, −0.1]. If we apply a ReLU activation function (which outputs 0 for negative values), the final output will be [0, 0].

2.2.3.3 Activation Functions and Layers

In an NN, the output from each layer, known as the activations, is represented as a vector. These activations are computed by first multiplying the input vector by the layer's weights, adding the biases, and then applying an activation function to introduce non-linearity. The activation function transforms the output in such a way that the network can model complex relationships within the data. The resulting activation vector from one layer serves as the input to the next layer, allowing the network to progressively learn more intricate data representations. Consider an NN layer with a single input value $\mathbf{x} = 1.0$, a weight $\mathbf{w} = 0.5$, and a bias $\mathbf{b} = 0.2$. The layer applies the ReLU activation function, which outputs the input value if it is positive and returns 0 otherwise.

Step 1: Compute the linear combination of input, weight, and bias:

$$z = wx + b = (0.5 \times 1.0) + 0.2 = 0.5 + 0.2 = 0.7$$

Step 2: Apply the ReLU activation function:

$$\text{ReLU}(z) = \max(0, z) = \max(0, 0.7) = 0.7.$$

In this example, the activation output is 0.7, which becomes the input to the next layer. If we used a different activation function, like the sigmoid function, the output would have been:

$$\text{Sigmoid}(z) = \frac{1}{1 + e^{-z}} = \frac{1}{1 + e^{-0.7}} \approx 0.668.$$

By applying different activation functions, the network can handle various kinds of data patterns, enabling it to learn more complex representations as it passes data through multiple layers.

2.2.3.4 Dot Product in Neural Network

The dot product is a core mathematical operation in NNs, playing a critical role in how information is processed and propagated through layers. When an input vector is passed into a layer, it is multiplied by the weight matrix of that layer. This multiplication involves calculating the dot product between the input vector and each column (or row, depending on the network's architecture) of the weight matrix. The result is a transformed output that is passed on to the next layer, making the dot product essential for updating the network's activations and learning patterns from the data. Let us take a simple example of an NN with an input vector $\mathbf{x} = [2, 3]$ and a weight matrix $\mathbf{W}$ for a layer with two neurons:

$$\boldsymbol{W} = \begin{bmatrix} 0.1 & 0.4 \\ 0.2 & 0.5 \end{bmatrix}.$$

To calculate the dot product, we multiply the input vector by the weight matrix:

$$z = \boldsymbol{W} \cdot \boldsymbol{x} = \begin{bmatrix} 0.1 & 0.4 \\ 0.2 & 0.5 \end{bmatrix} \cdot \begin{bmatrix} 2 \\ 3 \end{bmatrix}.$$

This results in:

$$z = \begin{bmatrix} (0.1 \times 2) + (0.4 \times 3) \\ (0.2 \times 2) + (0.5 \times 3) \end{bmatrix} = \begin{bmatrix} 0.2 + 1.2 \\ 0.4 + 1.5 \end{bmatrix} = \begin{bmatrix} 1.4 \\ 1.9 \end{bmatrix}.$$

In this example, the dot product produces a new vector [1.4, 1.9], which serves as the output of this layer before applying any activation functions. This process enables the network to combine the input data with the learned weights, transforming the input into a new representation that is then passed to the next layer for further processing.

2.2.3.5 Gradient Descent and Backpropagation

A gradient is a vector that contains the partial derivatives of the loss function with respect to each parameter of the network (weights and biases). During training, the gradient of the loss function is computed with respect to these parameters, providing insights into how the loss changes as the parameters are adjusted. The parameters are then updated by moving in the opposite direction of the gradient, a method known as gradient descent. This iterative process helps the network reduce loss and improves its performance over time. The backpropagation algorithm is used to efficiently compute the gradients. It works by applying the chain rule of calculus to propagate the error backward through the network, layer by layer, calculating the gradient of the loss function with respect to each parameter. By updating the parameters step-by-step, backpropagation enables the network to learn from data and improve its accuracy. Suppose we have a simple NN with one parameter (weight) $\mathbf{w} = 2.0$ and a loss function $L(w) = (w-5)^2$. The goal is to minimize the loss using gradient descent.

Step 1: Compute the gradient of the loss function with respect to $\mathbf{w}$:

$$\frac{\partial L(\boldsymbol{w})}{\partial \boldsymbol{w}} = 2(\boldsymbol{w}-5).$$

Step 2: Update the parameter using gradient descent. Let the learning rate be $\boldsymbol{\alpha} = 0.1$:

$$w_{\text{new}} = w_{\text{old}} - \alpha \times \frac{\partial L(w)}{\partial w}.$$

$$w_{\text{new}} = 2.0 - 0.1 \times 2(2.0-5) = 2.0 - 0.1 \times 2(-3) = 2.0 + 0.6 = 2.6.$$

Step 3: Repeat the process. After the first update, the new weight is $\mathbf{w} = 2.6$, which brings the loss function closer to its minimum. Over multiple iterations, the weight will converge toward the optimal value of $\mathbf{w} = 5$, minimizing the loss.

2.2.3.6 Batch Processing

Learning models, particularly in machine learning and deep learning, often process data in batches to accelerate training and make more efficient use of computational resources. When handling a batch of data, each input, activation, gradient, and other intermediate values are represented as matrices or higher-dimensional tensors. This extends the concept of single input vectors to multiple dimensions, allowing the model to process multiple examples simultaneously. By doing so, the model can leverage efficient computation and parallel processing, speeding up training and making better use of available hardware, such as GPUs. Suppose we are training an NN with a batch size of 3, and each input is a vector of 4 features. Instead of processing each input separately, the model processes the batch as a whole by organizing the data into a matrix:

$$\boldsymbol{X} = \begin{bmatrix} x_{1,1} & x_{1,2} & x_{1,3} & x_{1,4} \\ x_{2,1} & x_{2,2} & x_{2,3} & x_{2,4} \\ x_{3,1} & x_{3,2} & x_{3,3} & x_{3,4} \end{bmatrix} = \begin{bmatrix} 1.0 & 2.0 & 3.0 & 4.0 \\ 5.0 & 6.0 & 7.0 & 8.0 \\ 9.0 & 10.0 & 11.0 & 12.0 \end{bmatrix}$$

Here, each row represents a single input vector, and the network processes all three input vectors in parallel. The same applies to activations and gradients, which are now computed for the entire batch in one step, making the training process much faster compared to handling one input at a time. In addition to speeding up training, batch processing helps stabilize learning by averaging gradients across multiple examples, which reduces the noise in parameter updates. This is particularly useful in stochastic gradient descent (SGD), where updating the model based on a single example can lead to noisy and inefficient convergence.

2.2.3.7 Convolutional Neural Networks

In convolutional neural networks (CNNs), filters (also known as kernels) are small-weight matrices that slide over the input data, such as images, to extract important local features. The convolution process involves taking the dot product between the filter and small patches of the input data at each position. This operation produces a feature map, which highlights the presence of specific patterns in the data, such as edges, textures, or corners. By detecting local patterns, CNNs are highly effective for tasks like image recognition, where spatial relationships between pixels are crucial for understanding the content. Consider a 3 × 3 filter applied to a 5 × 5 grayscale image. The filter is:

$$\boldsymbol{F} = \begin{bmatrix} 1 & 0 & -1 \\ 1 & 0 & -1 \\ 1 & 0 & -1 \end{bmatrix}.$$

The input image is:

$$\boldsymbol{I} = \begin{bmatrix} 255 & 255 & 255 & 0 & 0 \\ 255 & 255 & 255 & 0 & 0 \\ 255 & 255 & 255 & 0 & 0 \\ 0 & 0 & 0 & 0 & 0 \\ 0 & 0 & 0 & 0 & 0 \end{bmatrix}.$$

Step 1: Apply the filter at the top-left corner of the image. The dot product between the filter and the corresponding 3 × 3 patch of the image is:

$$(1\times255)+(0\times255)+(-1\times255)+(1\times255)+(0\times255)+(-1\times255)$$
$$+(1\times255)+(0\times255)+(-1\times255)=0$$

Slide the filter one step to the right (position (1, 2)):

$$(1\times255)+(0\times0)+(-1\times0)+(1\times255)+(0\times0)+(-1\times0)$$
$$+(1\times255)+(0\times0)+(-1\times0)=765$$

Slide the filter one step to the right again (position (1, 3)):

$$(1\times255)+(0\times0)+(-1\times0)+(1\times255)+(0\times0)+(-1\times0)+(1\times255)$$
$$+(0\times0)+(-1\times0)=765.$$

Move down to the second row (position (2, 1)):

$$(1\times255)+(0\times255)+(-1\times255)+(1\times255)+(0\times255)$$
$$+(-1\times255)+(1\times0)+(0\times0)+(-1\times0)=0$$

Step 2: Slide the filter to the right by one pixel and repeat the process. After sliding the filter over all positions of the image, we get a 3 × 3 feature map:

$$\boldsymbol{F}_{\text{map}} = \begin{bmatrix} 0 & 765 & 765 \\ 0 & 765 & 765 \\ 0 & 0 & 0 \end{bmatrix}.$$

This feature map highlights the edges or transitions in the original image where the filter detects changes, specifically horizontal edges, due to the structure of the filter.

2.3 MATRICES

2.3.1 What Is a Matrix?

A matrix is a rectangular array of numbers, symbols, or expressions arranged in rows and columns. Matrices are denoted by capital letters (e.g., **A**, **B**, **C**) and are often used in systems of linear equations, computer graphics, statistics, and many other areas.

Example: Let us delve into the basics of matrices. Imagine you have a table of numbers where each cell contains a value. This table can be thought of as a matrix. For instance:

$$A = \begin{pmatrix} 1 & 2 & 3 \\ 4 & 5 & 6 \\ 7 & 8 & 9 \end{pmatrix}.$$

2.3.2 Dimensions

The size or dimension of a matrix is defined by the number of rows and columns it contains. A matrix with **m** rows and **n** columns is called a matrix, often read as "**m** by **n**."

Example: Consider a matrix **B**:

$$B = \begin{pmatrix} 1 & 2 \\ 3 & 4 \\ 5 & 6 \end{pmatrix},$$

where **B** is a matrix which has three rows and two columns.

2.3.3 Types of Matrices

2.3.3.1 Row Matrix

A matrix with only one row.

Example: $R = \begin{pmatrix} 1 & 2 & 3 \end{pmatrix}$.

2.3.3.2 Column Matrix

A matrix with only one column.

Example: $C = \begin{pmatrix} 1 \\ 2 \\ 3 \end{pmatrix}$

2.3.3.3 Square Matrix

A matrix with the same number of rows and columns.

Example: $S = \begin{pmatrix} 1 & 2 \\ 3 & 4 \end{pmatrix}$

2.3.3.4 Diagonal Matrix

A square matrix where all elements outside the main diagonal are zero.

Example: $D = \begin{pmatrix} 1 & 0 & 0 \\ 0 & 2 & 0 \\ 0 & 0 & 3 \end{pmatrix}$

2.3.3.5 Identity Matrix (or Unit Matrix)

A diagonal matrix where all diagonal elements are 1.

Example: $I = \begin{pmatrix} 1 & 0 & 0 \\ 0 & 1 & 0 \\ 0 & 0 & 1 \end{pmatrix}$

2.3.3.6 Zero Matrix

All elements are zero.

Example: $O = \begin{pmatrix} 0 & 0 \\ 0 & 0 \end{pmatrix}$

2.3.3.7 Symmetric Matrix

$\mathbf{A}^{\mathrm{T}} = \mathbf{A}$ where $\mathbf{A}^{\mathrm{T}}$ is the transpose of $\mathbf{A}$.

Example: $A = \begin{pmatrix} 1 & 2 & 3 \\ 2 & 4 & 5 \\ 3 & 5 & 6 \end{pmatrix}, \quad A^T = \begin{pmatrix} 1 & 2 & 3 \\ 2 & 4 & 5 \\ 3 & 5 & 6 \end{pmatrix}$

2.3.3.8 Skew-Symmetric Matrix

A skew-symmetric matrix is a square matrix **A** that satisfies $A^T = -A$, meaning each element $a_{ij} = -a_{ij}$ and all diagonal elements are zero.

Example: $A = \begin{pmatrix} 0 & -2 & -3 \\ 2 & 0 & -5 \\ 3 & 5 & 0 \end{pmatrix}, \quad A^T = \begin{pmatrix} 0 & 2 & 3 \\ -2 & 0 & 5 \\ -3 & -5 & 0 \end{pmatrix}$

2.3.4 Matrix Operations

2.3.4.1 Addition and Subtraction

Matrices can be added or subtracted element-wise if they have the exact dimensions.

Example:

$$\text{Let}: A = \begin{pmatrix} 1 & 2 \\ 3 & 4 \end{pmatrix} \quad \text{and} \quad B = \begin{pmatrix} 2 & 0 \\ 1 & 3 \end{pmatrix}$$

$$\text{Addition}: A + B = \begin{pmatrix} 1+2 & 2+0 \\ 3+1 & 4+3 \end{pmatrix} = \begin{pmatrix} 3 & 2 \\ 4 & 7 \end{pmatrix}$$

$$\text{Subtraction}: A - B = \begin{pmatrix} 1-2 & 2-0 \\ 3-1 & 4-3 \end{pmatrix} = \begin{pmatrix} -1 & 2 \\ 2 & 1 \end{pmatrix}.$$

2.3.4.2 Scalar Multiplication

Every matrix element is multiplied by the scalar.

Example:

$$\text{Let}: A = \begin{pmatrix} 1 & 2 \\ 3 & 4 \end{pmatrix}. \text{ If } k = 3, \quad \text{then } k \cdot A = 3 \cdot \begin{pmatrix} 1 & 2 \\ 3 & 4 \end{pmatrix} = \begin{pmatrix} 3 \cdot 1 & 3 \cdot 2 \\ 3 \cdot 3 & 3 \cdot 4 \end{pmatrix} = \begin{pmatrix} 3 & 6 \\ 9 & 12 \end{pmatrix}.$$

2.3.4.3 Matrix Multiplication

For two matrices to be multiplied, the number of columns of the first matrix must equal the number of rows of the second matrix. The resulting matrix has the number of rows of the first matrix and the number of columns of the second matrix.

Example:

$$\text{Let}: A = \begin{pmatrix} 1 & 2 \\ 3 & 4 \end{pmatrix} \quad \text{and} \quad B = \begin{pmatrix} 2 & 0 \\ 1 & 3 \end{pmatrix},$$

$$C = A \cdot B = \begin{pmatrix} 1 & 2 \\ 3 & 4 \end{pmatrix} \cdot \begin{pmatrix} 2 & 0 \\ 1 & 3 \end{pmatrix} = \begin{pmatrix} (1 \cdot 2 + 2 \cdot 1) & (1 \cdot 0 + 2 \cdot 3) \\ (3 \cdot 2 + 4 \cdot 1) & (3 \cdot 0 + 4 \cdot 3) \end{pmatrix} = \begin{pmatrix} 2+2 & 0+6 \\ 6+4 & 0+12 \end{pmatrix} = \begin{pmatrix} 4 & 6 \\ 10 & 12 \end{pmatrix}.$$

2.3.4.4 Determinant

For square matrices, the determinant is a scalar value representing the volume scaling factor of the transformation the matrix represents.

Example:

$$\text{Let}: A = \begin{pmatrix} 1 & 2 \\ 3 & 4 \end{pmatrix}, \quad \det(A) = \begin{vmatrix} 1 & 2 \\ 3 & 4 \end{vmatrix} = (1 \cdot 4) - (2 \cdot 3) = 4 - 6 = -2.$$

2.3.4.5 Transpose

Reflects the matrix over its main diagonal.

Example:

$$\text{Let}: A=\begin{pmatrix}1&2&3\\4&5&6\end{pmatrix},\quad A^T=\begin{pmatrix}1&4\\2&5\\3&6\end{pmatrix}.$$

2.3.4.6 Inverse

For some square matrices, the inverse exists such that when the matrix is multiplied by its inverse, the result is the identity matrix.

Example:

$$\text{Let}: A=\begin{pmatrix}1&2\\3&4\end{pmatrix}, A^{-1}=\frac{1}{\det(A)}\begin{pmatrix}d&-b\\-c&a\end{pmatrix}, \det(A)=(1\cdot4)-(2\cdot3)=4-6=-2,$$

$$A^{-1}=\frac{1}{-2}\begin{pmatrix}4&-2\\-3&1\end{pmatrix}=\begin{pmatrix}-2&1\\1.5&-0.5\end{pmatrix} \text{ and } A\cdot A^{-1}=\begin{pmatrix}1&2\\3&4\end{pmatrix}\cdot\begin{pmatrix}-2&1\\1.5&-0.5\end{pmatrix}=\begin{pmatrix}1&0\\0&1\end{pmatrix}.$$

2.3.5 Applications in Linear Algebra

2.3.5.1 System of Linear Equations

Matrices can be used to represent and solve systems of linear equations. In a system of linear equations, the coefficients of the variables can be arranged in a matrix, and the equations can be solved using matrix operations.

Example:

$$\text{Consider}\begin{cases}\boldsymbol{x}+2\boldsymbol{y}=5\\3\boldsymbol{x}+4\boldsymbol{y}=6\end{cases},$$

and it can be written in this form

$$\begin{pmatrix}1&2\\3&4\end{pmatrix}\begin{pmatrix}\boldsymbol{x}\\\boldsymbol{y}\end{pmatrix}=\begin{pmatrix}5\\6\end{pmatrix}.$$

$$\text{Let}: A=\begin{pmatrix}1&2\\3&4\end{pmatrix},\quad X=\begin{pmatrix}x\\y\end{pmatrix},\quad \text{and}\quad B=\begin{pmatrix}5\\6\end{pmatrix}$$

Then, we have:

$$\boldsymbol{A}\cdot\boldsymbol{X}=\boldsymbol{B}.$$

Now for calculating X, we have:

$$\boldsymbol{X}=\boldsymbol{A}^{-1}\cdot\boldsymbol{B},$$

and for calculating $\boldsymbol{A}^{-1}$:

$$\det(A) = (1 \cdot 4) - (2 \cdot 3) = 4 - 6 = -2, \text{ and } A^{-1} = \frac{1}{-2}\begin{pmatrix} 4 & -2 \\ -3 & 1 \end{pmatrix} = \begin{pmatrix} -2 & 1 \\ 1.5 & -0.5 \end{pmatrix}$$

and now we have $\mathbf{A}^{-1}$ and $\mathbf{B}$, then we can calculate the $\mathbf{X}$:

$$X = \begin{pmatrix} -2 & 1 \\ 1.5 & -0.5 \end{pmatrix}\begin{pmatrix} 5 \\ 6 \end{pmatrix} = \begin{pmatrix} (-2 \cdot 5) + (1 \cdot 6) \\ (1.5 \cdot 5) + (-0.5 \cdot 6) \end{pmatrix} = \begin{pmatrix} -10 + 6 \\ 7.5 - 3 \end{pmatrix} = \begin{pmatrix} -4 \\ 4.5 \end{pmatrix},$$

then the solution is:

$$x = -4, \quad y = 4.5.$$

2.3.5.2 Eigenvalues and Eigenvectors

Eigenvalues and eigenvectors are fundamental in understanding the properties of linear transformations. For a square matrix $\mathbf{A}$, an eigenvector $\mathbf{v}$ and eigenvalue λ satisfy the equation:

$$\mathbf{A}\mathbf{v} = \lambda\mathbf{v}$$

2.3.5.3 Transformations

Matrices perform linear transformations in geometry and computer graphics, such as scaling, rotation, and translation.

Example:

- *Scaling:* $\mathbf{S} = \begin{pmatrix} s_x & 0 \\ 0 & s_y \end{pmatrix}$
- *Rotation:* $\mathrm{R} = \begin{pmatrix} \cos\theta & -\sin\theta \\ \sin\theta & \cos\theta \end{pmatrix}$
- *Translation:* This is typically handled in homogeneous coordinates, involving an additional dimension for translation vectors. In homogeneous coordinates, a 2D point (x, y) becomes $(x, y, 1)$, allowing translation by a vector(t_x, t_y) using the matrix $\begin{pmatrix} 1 & 0 & t_x \\ 0 & 1 & t_y \\ 0 & 0 & 1 \end{pmatrix}$. This matrix is applied to the point $(x, y, 1)$ to produce the translated point (x', y'):

$$\begin{bmatrix} x' \\ y' \\ 1 \end{bmatrix} = \begin{bmatrix} 1 & 0 & t_x \\ 0 & 1 & t_y \\ 0 & 0 & 1 \end{bmatrix}\begin{bmatrix} x \\ y \\ 1 \end{bmatrix},$$

  where $\mathbf{x}' = \mathbf{x} + \mathbf{t}_x$ and $\mathbf{y}' = \mathbf{y} + \mathbf{t}_y$.

2.3.6 Matrices in Deep Learning

Matrices are fundamental in deep learning, serving as the backbone for data representation, network parameters, and essential operations. Let us dive into their importance in various aspects of deep learning.

2.3.6.1 Data Representation

(a) *Input Data:* In deep learning, data is often structured and represented in matrix form, particularly for tasks like image processing. For example, a grayscale image is typically represented as a 2D matrix where each element corresponds to the pixel intensity. A 28 × 28 pixel grayscale image from the MNIST dataset can be represented as a 28 × 28 matrix, with each entry holding a value between 0 (black) and 255 (white). This matrix representation allows the model to easily interpret and process the image data by performing matrix operations like convolutions or transformations. Consider a 3 × 3 grayscale image with the following pixel intensities:

$$\boldsymbol{I} = \begin{bmatrix} 100 & 150 & 200 \\ 50 & 100 & 150 \\ 0 & 50 & 100 \end{bmatrix}$$

Here, each element of the matrix represents the intensity of a pixel. This matrix can be fed into the deep learning model for further processing.

(b) *Batch Processing*: When processing multiple samples at once, deep learning models stack the input data into a larger matrix or tensor. Each row in this matrix represents a different sample from the batch. For example, if we are processing 5 grayscale images of 28 × 28 pixels, the data would be represented as a 3D tensor of shape (5, 28, 28), where 5 represents the batch size, and 28 × 28 represents the dimensions of each image. Batch processing allows for more efficient computation and faster training, as multiple samples are processed in parallel. Suppose we have a batch of 3 images, each represented by a 2 × 2 matrix:

$$\boldsymbol{I}_1 = \begin{bmatrix} 100 & 150 \\ 50 & 100 \end{bmatrix}, \quad \boldsymbol{I}_2 = \begin{bmatrix} 200 & 250 \\ 150 & 200 \end{bmatrix}, \quad \boldsymbol{I}_3 = \begin{bmatrix} 50 & 75 \\ 25 & 50 \end{bmatrix}$$

Stacking these images forms a 3D tensor of shape (3, 2, 2):

$$\textbf{Batch} = \begin{bmatrix} \boldsymbol{I}_1 \\ \boldsymbol{I}_2 \\ \boldsymbol{I}_3 \end{bmatrix} = \left[\begin{bmatrix} 100 & 150 \\ 50 & 100 \end{bmatrix}, \begin{bmatrix} 200 & 250 \\ 150 & 200 \end{bmatrix}, \begin{bmatrix} 50 & 75 \\ 25 & 50 \end{bmatrix} \right]$$

This format allows deep learning models to process multiple images simultaneously, improving computational efficiency and reducing training time.

2.3.6.2 Parameters of the Network

(a) *Weights:* In an NN, the weights connecting two layers are represented as a matrix. Each element of this matrix corresponds to a connection between a neuron in the first layer and a neuron in the subsequent layer. For example, if the first layer has **n** neurons and the second layer has **m** neurons, the weight matrix will have dimensions **m × n**. These weights are learned during training, as they define how the input data is transformed and passed through the network. Consider an NN with 3 input neurons and 2 output neurons. The weights between the input and output layers are represented by a 2 × 3 matrix:

$$\boldsymbol{W} = \begin{bmatrix} 0.1 & 0.2 & 0.3 \\ 0.4 & 0.5 & 0.6 \end{bmatrix}$$

Here, the value 0.1 connects the first input neuron to the first output neuron, 0.2 connects the second input neuron to the first output neuron, and so on. These weights determine how much influence each input has on the outputs.

(b) *Biases:* Biases are typically represented as a vector, with each element corresponding to a neuron in the output layer. Bias values allow the network to shift the activation function and provide additional flexibility in learning. In the case of batch processing, bias values are often broadcast across the entire batch during computation, meaning the same bias is applied to each example in the batch. The bias vector is added to the result of the matrix multiplication between the input data and the weight matrix, ensuring that each output neuron has a corresponding bias term. For the previous example with 2 output neurons, the bias vector might look like this:

$$\boldsymbol{b} = \begin{bmatrix} 0.1 \\ 0.2 \end{bmatrix}.$$

If we are processing a batch of three input samples, each input will be multiplied by the weight matrix, and the same bias vector will be added to the result. For instance, given an input vector $\mathbf{x} = [1, 0, -1]$, the output calculation would be:

$$z = \boldsymbol{W} \cdot \boldsymbol{x} + \boldsymbol{b} = \begin{bmatrix} 0.1 & 0.2 & 0.3 \\ 0.4 & 0.5 & 0.6 \end{bmatrix} \cdot \begin{bmatrix} 1 \\ 0 \\ -1 \end{bmatrix} + \begin{bmatrix} 0.1 \\ 0.2 \end{bmatrix} = \begin{bmatrix} 0.1 - 0.3 + 0.1 \\ 0.4 - 0.6 + 0.2 \end{bmatrix} = \begin{bmatrix} -0.1 \\ 0.0 \end{bmatrix}$$

2.3.6.3 Operations in Layers

(a) *Matrix Multiplication*: In fully connected (dense) layers of an NN, matrix multiplication is the fundamental operation. The input data, represented as a matrix, is multiplied by the weight matrix to produce the output, which is then shifted by adding the bias vector. This operation transforms the input data and propagates it to the next layer. The matrix multiplication allows the network to combine features from the input data based on the learned weights. Suppose we have an input vector $\mathbf{x} = [1, 2]$, a weight matrix $\boldsymbol{W} = \begin{bmatrix} 0.5 & 0.1 \\ 0.3 & 0.7 \end{bmatrix}$, and a bias vector $\mathbf{b} = [0.2, 0.3]$. The output of this fully connected layer is computed as:

$$z = \boldsymbol{W} \cdot \boldsymbol{x} + \boldsymbol{b} = \begin{bmatrix} 0.5 & 0.1 \\ 0.3 & 0.7 \end{bmatrix} \cdot \begin{bmatrix} 1 \\ 2 \end{bmatrix} + \begin{bmatrix} 0.2 \\ 0.3 \end{bmatrix} = \begin{bmatrix} 0.5 \times 1 + 0.1 \times 2 \\ 0.3 \times 1 + 0.7 \times 2 \end{bmatrix} + \begin{bmatrix} 0.2 \\ 0.3 \end{bmatrix} = \begin{bmatrix} 0.7 \\ 1.7 \end{bmatrix} + \begin{bmatrix} 0.2 \\ 0.3 \end{bmatrix} = \begin{bmatrix} 0.9 \\ 2.0 \end{bmatrix}$$

(b) *Convolution*: In convolutional layers, commonly used in CNNs, the input matrix (or tensor, for color images) is convolved with a filter (or kernel) matrix to produce feature maps. The filter slides across the input, and at each position, a dot product is computed between the filter and the input patch. This operation allows CNNs to detect local patterns, such as edges or textures, in the input data. Consider a 3 × 3 filter applied to a 4 × 4 input image:

$$\boldsymbol{I} = \begin{bmatrix} 1 & 0 & 2 & 1 \\ 0 & 1 & 0 & 2 \\ 1 & 2 & 1 & 0 \\ 0 & 1 & 0 & 1 \end{bmatrix}, \quad \boldsymbol{F} = \begin{bmatrix} 1 & 0 & -1 \\ 1 & 0 & -1 \\ 1 & 0 & -1 \end{bmatrix}$$

The convolution operation slides the filter over the input, performing a dot product at each position. For example, when the filter is applied to the top-left corner (the first 3 × 3 patch), the result is:

$$(1\times1)+(0\times0)+(2\times-1)+(0\times1)+(1\times0)+(0\times-1)+(1\times1)+(2\times0)+(1\times-1)$$
$$=1+0-2+0+0+0+1+0-1=-1$$

Sliding the filter over the rest of the input produces a feature map that highlights specific patterns in the image. This convolution process enables CNNs to focus on local features in the data and is crucial for tasks like image recognition.

2.3.6.4 Activation Functions

After matrix multiplication in a layer, the resulting matrix is often passed through an activation function element-wise, such as ReLU, sigmoid, or tanh.

(a) *ReLU:* Returns $\mathbf{x}$ if $\mathbf{x} > 0$; otherwise, it returns 0. It is commonly used in CNNs and deep feedforward networks.
(b) *Sigmoid:* Squeezes the output between 0 and 1. It is helpful in early NNs but less common due to limitations like vanishing gradients.
(c) *Tanh (Hyperbolic Tangent):* Output values between –1 and 1. Similar to sigmoid but zero-centered.

2.3.6.5 Backpropagation

(a) *Gradient Matrices:* In the process of training an NN, backpropagation computes the gradients (partial derivatives) of the loss function with respect to each network parameter, including weights and biases. These gradients are represented as matrices, where each element corresponds to the partial derivative of the loss with respect to a specific weight. The gradient matrix indicates how much the loss function would change if a small change were made to the corresponding weights. These gradients are crucial for updating the parameters to minimize the loss and improve model performance. Suppose we have a weight matrix $\boldsymbol{W}=\begin{bmatrix}0.5 & 0.1\\0.3 & 0.7\end{bmatrix}$, and during backpropagation, the gradient matrix with respect to the loss is computed as:$\boldsymbol{G}=\begin{bmatrix}0.02 & -0.01\\0.03 & 0.04\end{bmatrix}$. This gradient matrix informs us how the loss will change based on small adjustments to each corresponding weight.
(b) *Weight Update:* Once the gradient matrix is calculated, the weights are updated to reduce the loss by moving in the opposite direction of the gradient. This process involves subtracting a fraction of the gradient matrix from the current weight matrix. The fraction is determined by the learning rate, a hyperparameter that controls how large the weight updates should be. Let's assume a learning rate $\boldsymbol{\alpha} = 0.1$. The updated weight matrix $\boldsymbol{W}_{\text{new}}$ is computed as:

$$\boldsymbol{W}_{\text{new}}=\boldsymbol{W}-\alpha\cdot\boldsymbol{G}=\begin{bmatrix}0.5 & 0.1\\0.3 & 0.7\end{bmatrix}-0.1\cdot\begin{bmatrix}0.02 & -0.01\\0.03 & 0.04\end{bmatrix}$$
$$=\begin{bmatrix}0.5-0.002 & 0.1+0.001\\0.3-0.003 & 0.7-0.004\end{bmatrix}=\begin{bmatrix}0.498 & 0.101\\0.297 & 0.696\end{bmatrix}$$

After updating, the weight matrix is slightly adjusted in a way that reduces the loss. This iterative process of calculating gradients and updating weights continues until the model converges to an optimal set of weights, minimizing the loss function. Backpropagation ensures that each weight and bias in the network is gradually adjusted to improve performance, allowing the network to learn complex patterns from the data.

2.3.6.6 Regularization

(a) *L_2 Regularization:* It, also known as weight decay, adds a penalty to the loss function for large weights, discouraging the model from relying too heavily on any particular parameter. This penalty helps prevent overfitting by encouraging the network to keep the weights small, making the model simpler and more generalizable. The loss function with $\mathbf{L_2}$ regularization is given by:

$$\text{Loss}_{\text{L2}} = \text{Loss}_{\text{original}} + \frac{\lambda}{2} \sum w^2$$

where $\text{Loss}_{\text{original}}$ is the original loss (e.g., mean squared error (MSE)), $\mathbf{w}$ represents the weights, and $\boldsymbol{\lambda}$ is the regularization coefficient. A larger $\boldsymbol{\lambda}$ increases the penalty for larger weights, thus enforcing stronger regularization. Consider a simple network with two weights, $\mathbf{w_1}$=0.5 and $\mathbf{w_2}$=0.8, and an original loss of 1.0. If the regularization coefficient $\boldsymbol{\lambda} = 0.1$, the $\mathbf{L_2}$ regularized loss is:

$$\begin{aligned}\text{Loss}_{\text{L2}} &= 1.0 + \frac{0.1}{2} \times (0.5^2 + 0.8^2) = 1.0 + 0.05 \times (0.25 + 0.64) = 1.0 + 0.05 \times 0.89 \\ &= 1.0 + 0.0445 = 1.0445\end{aligned}$$

The additional term penalizes large weights, encouraging the model to learn simpler patterns.

(b) *Dropout:* Dropout is a regularization technique used to prevent overfitting by randomly "dropping out" or turning off neurons during each forward and backward pass with a probability $\mathbf{p}$. During training, at each iteration, a fraction of neurons (typically $\mathbf{p} = 0.5$) are set to zero, which forces the network to rely on multiple pathways for learning. This helps make the network more robust and prevents neurons from co-adapting too strongly to specific features. Suppose we have a layer with four neurons and the following activations during training:

$$\boldsymbol{a} = [0.9, 0.8, 0.4, 0.7]$$

With a dropout probability of $\mathbf{p} = 0.5$, two neurons might randomly be dropped, resulting in:

$$\boldsymbol{a}_{\text{dropout}} = [0.9, 0, 0.4, 0].$$

The network then continues training with these modified activations. At test time, dropout is turned off, but the output is scaled by a factor of $1 - \mathbf{p}$ to account for the dropped neurons during training, making the network's predictions more robust. Both $\mathbf{L_2}$ regularization and dropout are essential techniques for preventing overfitting and ensuring that NNs generalize well to unseen data.

2.3.6.7 Optimizers

(a) *RMSprop:* RMSprop (Root Mean Square Propagation) is an optimizer that adjusts the learning rate for each parameter by using a moving average of the squared gradients. This helps prevent the learning rate from decaying too quickly, as seen in Adagrad, making RMSprop well-suited for training deep NNs. The update rule for RMSprop is as follows:

$$v_t = \beta v_{t-1} + (1-\beta) g_t^2 \quad \text{and} \quad w_t = w_{t-1} - \frac{\eta}{\sqrt{v_t + \epsilon}} g_t.$$

Here, $\mathbf{v_t}$ is the moving average of the squared gradients, $\mathbf{g_t}$ is the current gradient, $\boldsymbol{\beta}$ is the decay rate, $\boldsymbol{\eta}$ is the learning rate, and $\boldsymbol{\epsilon}$ is a small constant for numerical stability. RMSprop helps normalize the gradients, ensuring that each parameter has a more stable and balanced learning rate. Suppose $\boldsymbol{\eta} = 0.01$, $\boldsymbol{\beta} = 0.9$, and the gradient $\mathbf{g_t} = 0.5$. Initially, $\mathbf{v_0} = 0$, and the updated $\mathbf{v_t}$ and $\mathbf{w_t}$ values can be computed as:

$$v_1 = 0.9 \times 0 + 0.1 \times 0.5^2 = 0.1 \times 0.25 = 0.025 \quad \text{and} \quad w_1 = w_0 - \frac{0.01}{\sqrt{0.025 + \epsilon}} \times 0.5.$$

For small $\boldsymbol{\epsilon}$, the update normalizes the gradient, leading to a more balanced update.

(b) *Adam:* Adam (Adaptive Moment Estimation) combines the benefits of RMSprop and momentum. It maintains two moving averages: one for the gradient (momentum) and another for the squared gradient. This dual mechanism allows Adam to adapt the learning rate based on the gradient's magnitude and direction. The update rules are:

$$m_t = \beta_1 m_{t-1} + (1-\beta_1) g_t, \quad v_t = \beta_2 v_{t-1} + (1-\beta_2) g_t^2, \quad \hat{m}_t = \frac{m_t}{1-\beta_1^t}, \quad \hat{v}_t = \frac{v_t}{1-\beta_2^t}$$

$$w_t = w_{t-1} - \frac{\eta}{\sqrt{\hat{v}_t} + \epsilon} \hat{m}_t.$$

Here, $\mathbf{m_t}$ is the moving average of the gradient, $\mathbf{v_t}$ is the moving average of the squared gradient, $\boldsymbol{\beta_1}$ and $\boldsymbol{\beta_2}$ are decay rates, $\boldsymbol{\eta}$ is the learning rate, and $\boldsymbol{\epsilon}$ is a small constant. Suppose $\boldsymbol{\eta} = 0.01$, $\boldsymbol{\beta_1} = 0.9$, $\boldsymbol{\beta_2} = 0.999$, and $\mathbf{g_t} = 0.5$. Initially, $\mathbf{m_0} = 0$ and $\mathbf{v_0} = 0$. The updated values for $\mathbf{m_t}$, $\mathbf{v_t}$, and the weight update $\mathbf{w_t}$ are:

$$m_1 = 0.9 \times 0 + 0.1 \times 0.5 = 0.05 \quad \text{and} \quad v_1 = 0.999 \times 0 + 0.001 \times 0.5^2 = 0.00025.$$

The bias-corrected estimates are:

$$\hat{m}_1 = \frac{0.05}{1-0.9} = 0.5, \quad \hat{v}_1 = \frac{0.00025}{1-0.999} = 0.25.$$

The weight update becomes:

$$w_1 = w_0 - \frac{0.01}{\sqrt{0.25} + \epsilon} \times 0.5.$$

Adam's ability to adapt learning rates based on both the gradient's momentum and its squared magnitude makes it one of the most effective and widely used optimizers in deep learning. RMSprop and Adam are both adaptive learning rate optimizers widely used in training deep NNs, but they differ in their mechanisms and resulting updates. RMSprop adjusts the learning rate for each parameter by maintaining a moving average of the squared gradients, effectively normalizing the gradient and stabilizing updates. Its single moving average (using a decay rate like $\beta = 0.9$) ensures that the learning rate adapts based solely on the magnitude of recent gradients. In contrast, Adam extends RMSprop by also incorporating a moving average of the gradients themselves (momentum), using separate decay rates for the first and second moments (e.g., $\beta_1 = 0.9$ and $\beta_2 = 0.999$). Additionally, Adam applies bias correction to these moving averages, especially during the initial training steps, to produce unbiased estimates. As a result, Adam typically provides more balanced and efficient updates by considering both the direction and magnitude of gradients, often leading to faster convergence and better performance in practice. Consequently, while both optimizers may appear to produce similar updates in simplified examples, especially in early iterations, their outputs diverge as training progresses due to Adam's additional momentum and bias correction, making Adam generally more robust and effective for a wider range of NN architectures.

2.3.6.8 Batch Normalization

Batch normalization is a technique used to normalize the activations of a layer in an NN. This helps maintain a consistent mean and variance across different batches during training, improving the model's stability and performance. Batch normalization works in the following steps:

1. Calculate the mean ($\boldsymbol{\mu}$) and variance ($\boldsymbol{\sigma}^2$) of the input **X** over the batch:

$$\mu_B = \frac{1}{m}\sum_{i=1}^{m} X_i \quad \text{and} \quad \sigma_B^2 = \frac{1}{m}\sum_{i=1}^{m}(X_i - \mu_B)^2,$$

   where **m** is the number of samples in the batch, and $\mathbf{X_i}$ represents each input in the batch.
2. Normalize the input **X** by subtracting the mean and dividing by the standard deviation:

$$\hat{X}_i = \frac{X_i - \mu_B}{\sqrt{\sigma_B^2 + \epsilon}}.$$

   Here, $\boldsymbol{\epsilon}$ is a small constant added for numerical stability to avoid division by zero.
3. Scale and shift the normalized input using two learnable parameters, $\boldsymbol{\gamma}$ (scale) and $\boldsymbol{\beta}$ (shift):

$$Y_i = \gamma \hat{X}_i + \beta$$

   The parameters $\boldsymbol{\gamma}$ and $\boldsymbol{\beta}$ are learned during training and allow the network to restore the representational power that might be lost due to normalization.

Figure 2.9 provides a visualization of the forward and backward pass in an NN layer, represented through a series of matrices. The input data matrix illustrates the initial data fed into the NN, with each element representing feature values across rows and columns. Following this, the Batch Data (Sample 1) matrix shows how a single sample from the batch is processed, demonstrating the model's handling of data during training in smaller chunks, known as mini-batches. Moving to the Weights Matrix, this highlights the learned weights, which connect the input data to the neurons in the layer. Complementing the weights, the Biases Vector is shown, which contains bias values added to the weighted sum of inputs for each neuron, providing additional flexibility in shifting the neuron's output. The Output Matrix shows the raw result after multiplying the input data by the weights and adding the biases, yet before any activation function is applied. The ReLU Activation Output matrix then demonstrates the effect of applying the ReLU activation function, which zeros out any negative values while keeping positive values the same. This step introduces non-linearity into the model, making it more capable of

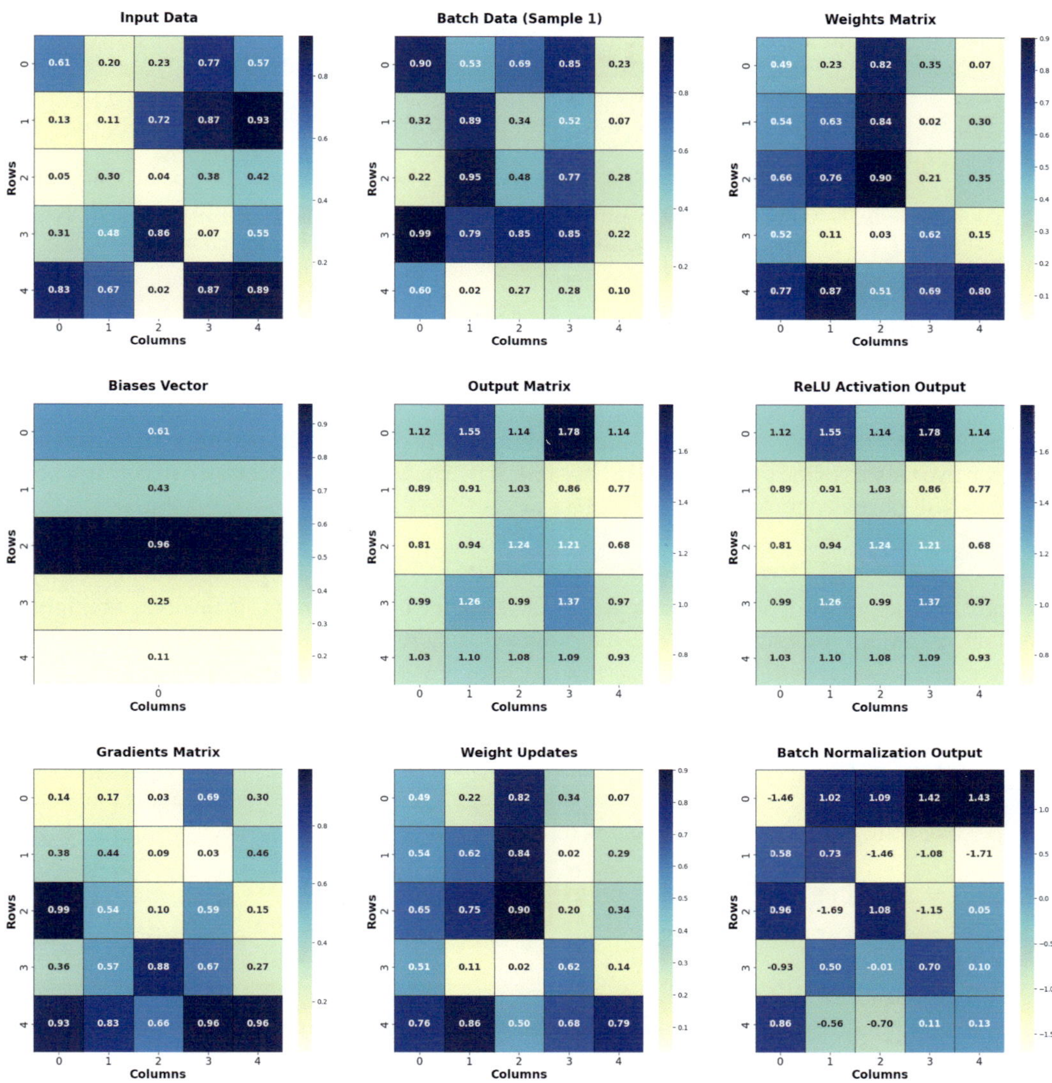

FIGURE 2.9 Visual representation of matrix concepts in deep learning.

learning complex patterns. Following the forward pass, the Gradients Matrix represents the calculated gradients during backpropagation, where the network computes how much the weights should be adjusted to minimize the loss function. These gradients are then used to adjust the model's parameters, as illustrated in the weight updates matrix. Finally, the Batch Normalization Output matrix presents the normalized output values after applying batch normalization.

2.4 TENSOR AND ITS OPERATIONS

2.4.1 Tensors

A tensor is a multi-dimensional array of numerical values and can be seen as a generalization of scalars, vectors, and matrices. In deep learning and many applications in physics, tensors are used to represent many data structures.

- *Scalar* (zero-dimensional tensor): A single number. For example, **c** = 5.
- *Vector* (one-dimensional tensor): An array of numbers. For example, **v** = [1, 2, 3].
- *Matrix* (two-dimensional tensor): A 2D array of numbers. For example, $\boldsymbol{M} = \begin{bmatrix} 1 & 2 \\ 3 & 4 \end{bmatrix}$.
- *Higher-dimensional Tensor:* Tensors can have three or more dimensions. For instance, color images can be represented as a three-dimensional tensor (height, width, color channels), such as **T** for an image with dimensions 100 × 100 pixels and 3 color channels (RGB): $\mathbf{T}_{\mathbf{ijk},}$ where $i = 1,\ldots,100;\ j = 1,\ldots,\ 100; k = 1,2,3$.

2.4.2 Tensor Operations

2.4.2.1 Element-Wise Operations

Operations like addition, subtraction, multiplication, and division are performed element-wise between two tensors of the same shape. For example, if two matrices (2D tensors) exist, corresponding elements will be added to **A** and **B**.

2.4.2.2 Tensor Dot Product (or Tensor Contraction)

Extends the idea of the dot product in vectors and matrix multiplication. It involves multiplying elements of tensors and summing the result, often reducing the tensor's dimensionality.

(a) *Tensor Reshaping:* It changes the shape of a tensor while preserving its data. It is beneficial in deep learning when preparing data for different layers.
(b) *Reduction Operations:* This refers to operations like sum, mean, max, or min, where you reduce a tensor to a more minor rank. For example, you might take the sum along one dimension, resulting in a tensor of one less dimension.

2.4.2.3 Matrix-Specific Operations on 2D Tensors

There are several operations including:

(a) *Transpose:* Swap rows with columns.
(b) *Inverse:* Find the matrix that gives the identity matrix when multiplied with the original.
(c) *Determinant:* A scalar value representing specific properties of the matrix.
(d) *Tensor Slicing and Indexing:* Extracting specific portions of a tensor is analogous to slicing lists or arrays.
(e) *Broadcasting:* They make element-wise binary operations compatible with tensors of different shapes. For instance, adding a vector to a matrix by duplicating the vector along the rows of the matrix.
(f) *Outer Product:* Produces a higher-rank tensor from two lower-rank tensors. For vectors, it produces a matrix.
(g) *Tensor Decompositions:* Techniques like **SVD** or **QR** decomposition can be extended to tensors.

In Figure 2.10, various tensor operations are performed on sample tensors using TensorFlow and visualized through heatmaps. The first row of subplots shows the results of element-wise operations. The element-wise addition plot displays the sum of corresponding elements from tensors **A** and **B**. The element-wise subtraction plot illustrates the difference between the corresponding elements of **A** and **B**. The element-wise multiplication plot highlights the product of corresponding elements from **A** and **B**, while the element-wise division plot shows the quotient of corresponding elements of **A** divided by **B**. The second row of subplots continues with more complex operations. The tensor

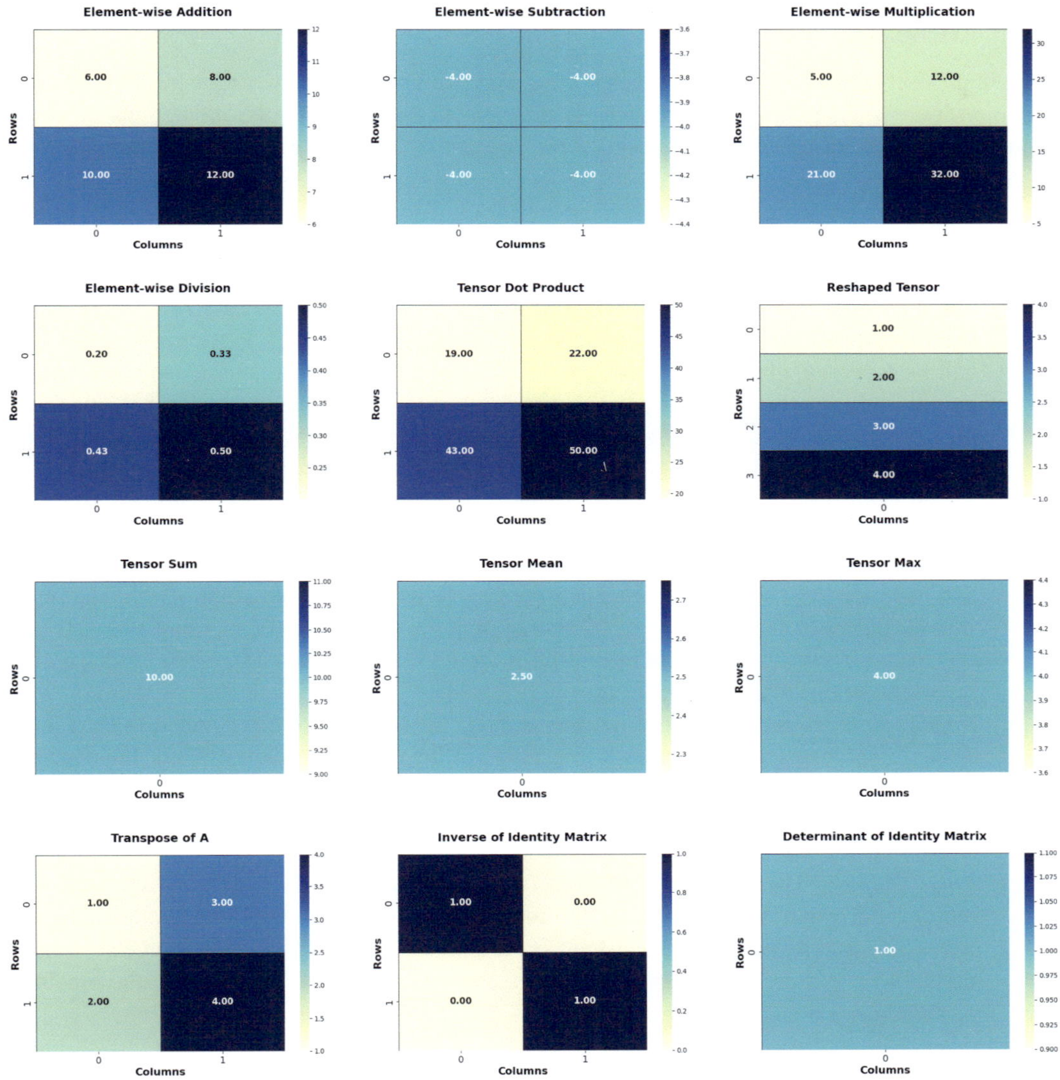

FIGURE 2.10 Visualization of tensor operations.

dot product plot presents the result of a dot product operation between tensors **A** and **B**, which involves summing the products of corresponding elements along the specified axes. The reshaped tensor plot shows tensor **A** rearranged into a column vector with a shape of (4, 1). The third row includes summary statistics and transformations of tensor **A**.

The tensor sum plot displays the sum of all elements in **A** as a single value. Similarly, the tensor mean plot presents the mean of all elements in **A**, and the tensor max plot shows the maximum value among all elements in **A**. The fourth row focuses on matrix transformations and properties. The transpose of **A** plot shows the result of swapping the rows and columns of tensor **A**. The inverse of the identity matrix plot displays the inverse of a 2×2 identity matrix, which remains the identity matrix itself. Finally, the determinant of the identity matrix plot shows that the determinant of a 2×2 identity matrix is 1. Additionally, the **SVD** results are printed. The singular values (**s**), left singular vectors (**u**), and right singular vectors (**v**) of tensor **A** are provided. These values and vectors are crucial in many applications, such as data compression and noise reduction.

2.4.3 Tensors in Deep Learning

At a high level, deep learning involves building and training NNs, which can be thought of as a composition of functions. These functions are applied to data to make predictions or create representations. In this paradigm, tensors are the primary data structure for various operations and transformations.

(a) *Input Data:* In deep learning, input data is typically multi-dimensional. For example, in Image Data, Images are represented as 3D tensors with dimensions corresponding to height, width, and channels (RGB). For instance, an RGB image of size 32 × 32 pixels can be represented as a tensor of shape, and for Sequential Data, Text or time series data can be represented as 2D tensors, where one dimension represents the sequence length, and the other represents features or embedding dimensions. For example, a sequence of 50 words, each represented by a 300-dimensional embedding, would be a tensor of shape (50, 300).

(b) *Weights and Biases:* The parameters of NNs, including weights and biases, are stored as tensors. The architecture of the network and the connections between neurons dictate the shape of these tensors. For example, for a convolutional layer in an NN with 16 filters of size 3 × 3 applied to an input with 3 channels (such as RGB images), the weights would be stored in a 4D tensor of shape (16, 3, 3, 3), representing the number of filters, input channels, and the spatial dimensions of each filter. Biases would be stored in a 1D tensor of shape (16), representing one bias per filter.

(c) *Intermediate Values:* As data progresses through an NN, it gets transformed at each layer. These transformations yield new tensors, representing the intermediate outputs. For example, applying a convolutional layer to input images of shape (32, 32, 3) with 16 filters can produce intermediate tensors of shape (32, 32, 16), where each filter produces a separate feature map of the same spatial dimensions as the input image.

(d) *Final Outputs:* The predictions or classifications made by an NN are also represented as tensors. For example, after passing the intermediate tensor through an activation function like ReLU and a final dense layer with 10 output neurons (for a classification task with 10 classes), we get the final outputs as a 1D tensor of shape (10), representing the predicted probabilities for each class.

2.4.4 Tensor Operations in Deep Learning

(a) *Linear Transformations:* In fully connected layers, the primary operation is matrix multiplication, a tensor operation. For example, given an input tensor $\mathbf{X} = [1.0, 2.0]$ and a weight tensor $\boldsymbol{W} = \begin{bmatrix} 0.5 & 0.3 \\ 0.7 & 0.9 \end{bmatrix}$, the output is calculated as:

$$\boldsymbol{Z} = \boldsymbol{W} \cdot \boldsymbol{X} + \boldsymbol{b} = \begin{bmatrix} 0.5 & 0.3 \\ 0.7 & 0.9 \end{bmatrix} \cdot \begin{bmatrix} 1.0 \\ 2.0 \end{bmatrix} + \begin{bmatrix} 0.1 \\ 0.2 \end{bmatrix} = \begin{bmatrix} 1.1 \\ 2.5 \end{bmatrix},$$

where **b** is the bias tensor.

(b) *Activation Functions:* After a linear transformation, activation functions are applied element-wise to tensors. For instance, applying the ReLU activation to the previous result Z = [1.1, 2.5] results in:

$$\text{ReLU}(\mathbf{Z}) = [\max(0,1.1), \max(0,2.5)] = [1.1, 2.5]$$

(c) *Convolutions:* In CNNs, convolution operations involve sliding a filter tensor over an input tensor to produce a feature map. For instance, convolving a 3 × 3 filter $\boldsymbol{F} = \begin{bmatrix} 1 & 0 & -1 \\ 1 & 0 & -1 \\ 1 & 0 & -1 \end{bmatrix}$ over a 4 × 4 input tensor, **I** produce a feature map after calculating dot products at each position.

(d) *Pooling*: Pooling layers reduce the spatial dimensions of feature maps. For example, max-pooling with a 2 × 2 window on a tensor $\boldsymbol{I} = \begin{bmatrix} 1 & 3 \\ 2 & 4 \end{bmatrix}$ would result in:

$$\text{Max-Pooling}(\mathbf{I}) = \max(1, 2, 3, 4) = 4$$

(e) *Batch Operations:* When processing data in batches, multiple samples are stacked into a single tensor. For example, processing a batch of two input tensors $\mathbf{X}_1 = [1.0, 2.0]$ and $\mathbf{X}_2 = [0.5, 1.5]$ results in a batch tensor:

$$\boldsymbol{X}_{\text{batch}} = \begin{bmatrix} 1.0 & 2.0 \\ 0.5 & 1.5 \end{bmatrix}$$

Tensor operations, such as matrix multiplication, are then performed simultaneously on this entire batch.

(f) *Backpropagation:* During training, gradients are computed as tensors with the same shape as the weights. For instance, if the weight tensor $\boldsymbol{W} = \begin{bmatrix} 0.5 & 0.3 \\ 0.7 & 0.9 \end{bmatrix}$ has a gradient tensor $\boldsymbol{G} = \begin{bmatrix} 0.02 & -0.01 \\ 0.03 & 0.04 \end{bmatrix}$, the weight update is:

$$\boldsymbol{W}_{\text{new}} = \boldsymbol{W} - \eta \cdot \boldsymbol{G} = \begin{bmatrix} 0.5 - 0.02 & 0.3 + 0.01 \\ 0.7 - 0.03 & 0.9 - 0.04 \end{bmatrix}$$

(g) *Sequence Processing:* In RNNs and transformers, sequences are processed using tensor operations like the dot product. For instance, calculating the dot product between two sequences (tensors) determines how much one sequence element influences another.

(h) *Broadcasting:* Broadcasting allows tensors of different shapes to be combined. For example, adding a vector [1.0, 2.0] to each row of a matrix $\boldsymbol{M} = \begin{bmatrix} 3 & 4 \\ 5 & 6 \end{bmatrix}$ results in:

$$\boldsymbol{M}_{\text{new}} = \begin{bmatrix} 3 + 1.0 & 4 + 2.0 \\ 5 + 1.0 & 6 + 2.0 \end{bmatrix} = \begin{bmatrix} 4 & 6 \\ 6 & 8 \end{bmatrix}.$$

(i) *Regularization and Normalization:* Techniques like dropout randomly drop units during training, represented by setting some elements of the activation tensor to zero. Batch normalization normalizes activations using tensor operations to maintain a consistent mean and variance.

(j) *Loss Computation:* The loss between predictions and actual targets is computed using tensor operations. For instance, if the predicted tensor is $\mathbf{Y}_{\text{pred}} = [0.8, 0.4]$ and the actual target tensor is $\mathbf{Y}_{\text{true}} = [1.0, 0.0]$, the MSE loss is:

$$\text{Loss} = \frac{1}{2}((0.8 - 1.0)^2 + (0.4 - 0.0)^2) = \frac{1}{2}(0.04 + 0.16) = 0.1.$$

In Figure 2.11, various tensor operations related to a CNN are performed on a sample input image and visualized through heatmaps. The first subplot, "Input Image (Channel 1)," displays one channel of the RGB input image, which is an 8 × 8 pixel grid with intensity values ranging from 0 to 255. This visualization provides a grayscale representation of the first color channel of the image. The second subplot, "Convolution Filter 1," shows the weights of the first convolutional filter applied to the input image. This 3 × 3 filter is used to detect specific features within the input image. The heatmap illustrates the values of the filter weights, which will be convolved with the input image. The third subplot, "Intermediate Tensor (Filter 1)," presents the result of applying the convolutional filter to the input image. This intermediate tensor captures the feature map produced by the convolution operation, highlighting areas of the input image that match the filter's pattern. The fourth subplot, "Activated Tensor (ReLU)," shows the result of applying the ReLU activation function to the intermediate tensor. The ReLU function sets all negative values to zero, introducing non-linearity to the model. This heatmap represents the activated feature map, emphasizing the features detected by the filter. The fifth subplot, "Flattened Tensor (Segment)," illustrates a segment of the flattened tensor obtained by reshaping the activated tensor into a 1D array. This transformation prepares the tensor for input into a dense (fully connected) layer. The heatmap visualizes a segment of this flattened tensor as an 8 × 8 grid for better interpretation. The sixth subplot, "Final Output Tensor," displays the final output tensor after passing the flattened tensor through a dense layer. This layer performs a linear transformation using learned weights and biases to produce the final output. The heatmap represents the output tensor as a row vector, showing the resulting values for each of the 10 output units.

2.5 LINEAR TRANSFORMATIONS

A linear transformation, often called a linear map, is a function between two vector spaces that preserves the operations of vector addition and scalar multiplication. In simpler terms, it is a transformation that does not "bend" or "twist" the space in a non-linear way. Mathematically, a function **T** from a vector space **V** to a vector space **W** is called a linear transformation if the following two properties hold for all vectors **u** and **v** in **V** and any scalar **c**:

- *Additivity:* $T(u + v) = T(u) + T(v)$
- *Homogeneity:* $T(\mathrm{cu}) = \mathrm{cT}(u)$

2.5.1 Matrix Representation of Linear Transformations

A matrix can represent every linear transformation. Given a vector space **V** and a basis for that space, a linear transformation can be associated with a matrix **A**. When you multiply this matrix by a column vector (representing a point in space), you get a new column vector representing the transformed point. When you have a linear transformation defined by a matrix and want to apply this transformation to a vector **V**, you perform the matrix–vector multiplication to obtain the transformed vector.

2.5.2 Examples of Linear Transformations

2.5.2.1 Scaling

Scaling involves multiplying vectors by a scalar, stretching or shrinking them but not changing their direction unless the scalar is negative. A vector can be scaled using a diagonal matrix where each diagonal entry corresponds to a scaling factor along each dimension.

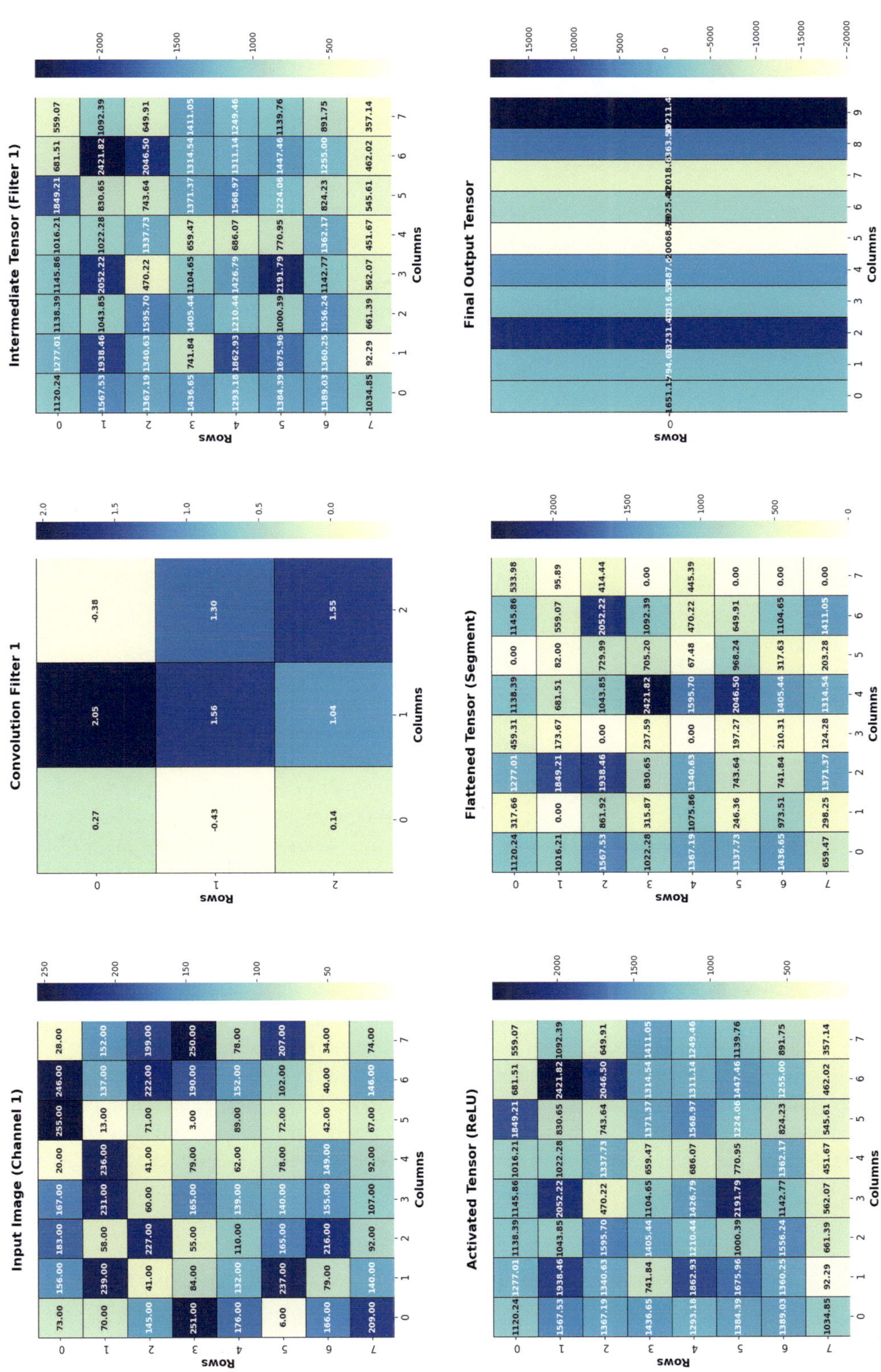

FIGURE 2.11 Visual representation of key Tensor Operations in deep learning.

2.5.2.2 Rotating

Rotating vectors around the origin involves using a rotation matrix. To rotate a 2D vector by a specific angle:

$$A_{\text{rotation}} = \begin{pmatrix} \cos\theta & -\sin\theta \\ \sin\theta & \cos\theta \end{pmatrix}$$

2.5.2.3 Reflecting

Reflecting vectors across a line (in 2D) or a plane (in 3D) can be represented using a reflection matrix. Reflecting a 2D vector across the x-axis uses the matrix:

$$A_{\text{reflection}} = \begin{pmatrix} 1 & 0 \\ 0 & -1 \end{pmatrix}$$

2.5.2.4 Shearing

Shearing involves "sliding" vectors along a fixed line or plane. A shear transformation can be represented by:

$$A_{\text{shear}} = \begin{pmatrix} 1 & k \\ 0 & 1 \end{pmatrix},$$

where **k** is the shear factor.

Figure 2.12 visualizes various linear transformations applied to an original vector. The original vector, shown in blue, represents the vector before any transformation. The scaled vector, depicted in red, is the result of scaling the vector by different factors along the **x** and **y** directions, which stretches or shrinks the vector. The rotated vector, shown in green, demonstrates the effect of rotating the vector by 45° around the origin. The reflected vector, represented in magenta, illustrates the transformation of reflecting the vector across the x-axis, flipping it over the x-axis. Lastly, the sheared vector, shown in yellow, represents the vector after being sheared along the x-direction, sliding the vector along the x-axis while changing its shape without altering its area.

2.5.3 Linear Transformations in Deep Learning

Linear transformations play a crucial role in deep learning due to their modeling power, efficiency, and interpretability. While individual linear transformations are limited in their ability to capture complex relationships, when stacked and combined with non-linear activation functions, they enable NNs to model intricate, non-linear patterns in data. Additionally, linear operations in matrix form are highly optimized on modern hardware, particularly GPUs, making the training and inference processes in deep learning more efficient. Furthermore, linear transformations are often more interpretable than their non-linear counterparts due to their simplicity, which can be valuable in applications where understanding the model's decision-making process is essential.

2.5.3.1 Neural Network Layers

In an NN, a dense (fully connected) layer is a linear transformation of the input. Given an input vector $\boldsymbol{x} = \left[x_1, x_2, \dots, x_n\right]$, the dense layer transforms it using a weight matrix **W** and a bias vector **b**. Mathematically, the transformation is expressed as:

$$\mathbf{z} = \mathbf{W}\mathbf{x} + \mathbf{b}.$$

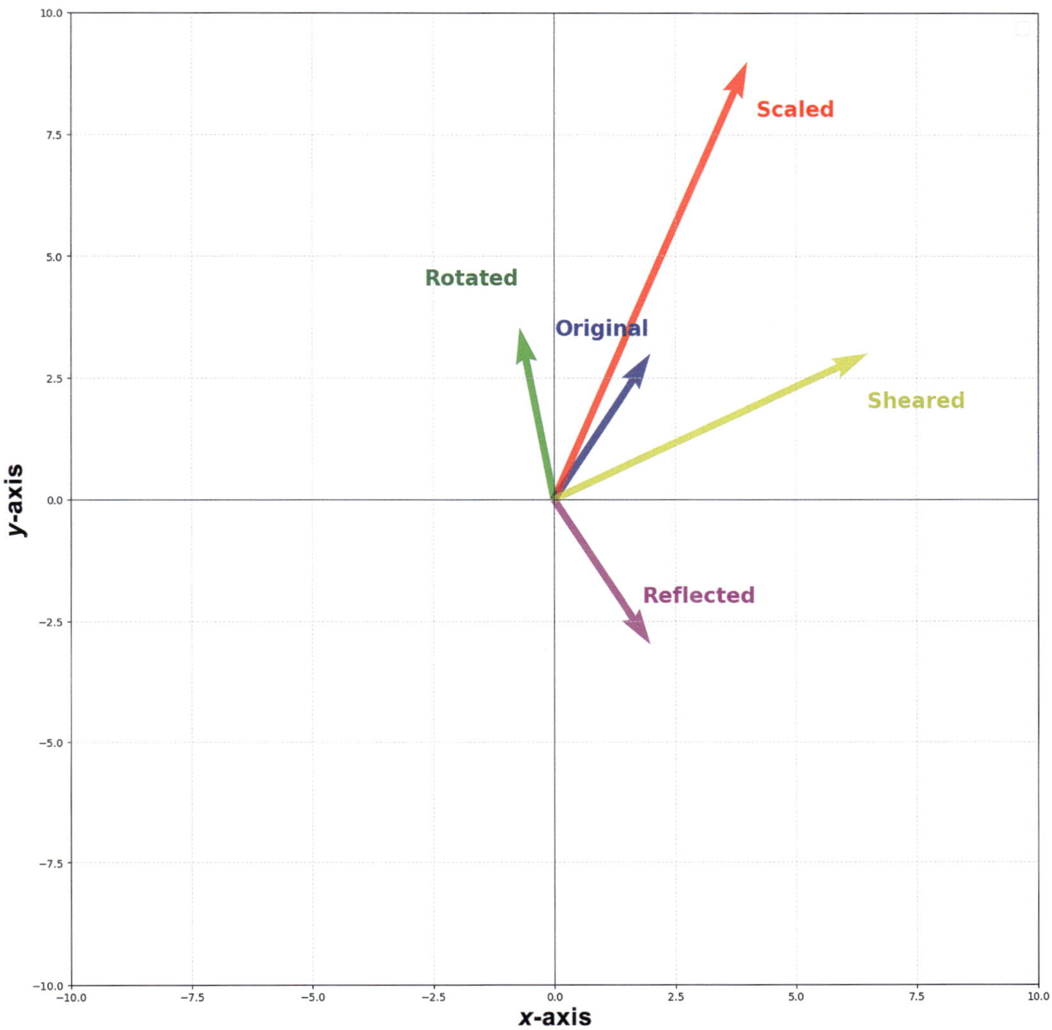

FIGURE 2.12 Visual representation of linear transformations.

Here, **W** is the weight matrix, and **b** is the bias vector. This linear transformation maps the input vector to an output vector, and the bias vector shifts the output. After this linear transformation, an activation function (e.g., ReLU or sigmoid) is typically applied to introduce non-linear capabilities to the model. Suppose we have an input vector **x** = [1, 2], a weight matrix $\boldsymbol{W} = \begin{bmatrix} 0.5 & 0.3 \\ 0.7 & 0.9 \end{bmatrix}$ and a bias vector **b** = [0.1, 0.2]. The output is computed as:

$$z = \boldsymbol{W}\boldsymbol{x} + \boldsymbol{b} = \begin{bmatrix} 0.5 & 0.3 \\ 0.7 & 0.9 \end{bmatrix} \cdot \begin{bmatrix} 1 \\ 2 \end{bmatrix} + \begin{bmatrix} 0.1 \\ 0.2 \end{bmatrix} = \begin{bmatrix} 1.1 \\ 2.7 \end{bmatrix}.$$

2.5.3.2 Convolutional Neural Networks

In CNNs, the convolution operation can be viewed as a series of small, local linear transformations. A convolutional filter (or kernel) slides over different local regions of the input image, applying a linear transformation to produce a feature map. By stacking multiple convolution layers, CNNs detect complex patterns such as edges, textures, and shapes.

2.5.3.3 Embeddings

In NLP, embedding layers map discrete words or tokens to continuous vector representations in a high-dimensional space. The transformation from a token to its corresponding vector is a linear operation. Embedding lookup involves selecting a specific row (vector) from the embedding matrix, which corresponds to the input token. For an embedding matrix **E** of size 10 × 4, where each row represents a word vector of dimension 4, input token 3 corresponds to the third row of **E**, resulting in a vector $\mathbf{v}_3$.

2.5.3.4 Regularization Techniques

While normalization techniques like batch normalization are not linear by nature, they include learnable parameters (gamma and beta) that apply a linear transformation to the normalized output. Specifically, after normalization, the output is scaled by γ and shifted by $\boldsymbol{\beta}$, introducing a linear transformation that adjusts the normalization effect based on the data.

2.5.3.5 Initialization

When initializing the weights of an NN, the initial linear transformations applied to the data must be carefully chosen. Proper initialization helps ensure that the transformations do not excessively shrink or expand the data, preventing issues like vanishing or exploding gradients during training.

2.5.3.6 Loss Functions and Optimization

Many loss functions, including MSE, involve linear components. For example, the MSE loss function, which computes the squared difference between predictions and actual values, includes a summation of squared errors, a linear operation. During optimization, gradients of the loss function with respect to the weights are computed, often involving linear operations, and used to update the model's parameters. For a simple linear regression model with predicted values $\hat{y} = [2,3]$ and actual values $\mathbf{y} = [1, 4]$, the **MSE** is: $\frac{1}{2}((2-1)^2 + (3-4)^2) = 0.5$

The gradient of this loss function with respect to the weights leads to a weight update that is a linear transformation of the error.

In Figure 2.13, the linear transformation process applied to a 2D input vector using a dense layer is visualized. This figure illustrates the transformation of the input vector by applying weights and biases, resulting in an output vector. The input vector is represented in blue, starting from the origin (0, 0) and pointing to the coordinates (2, 3). The transformed vector is depicted in red, also originating from (0, 0) but pointing to the coordinates determined by the linear transformation. In this case, the output vector is calculated using the weights and biases defined for the dense layer. Additionally, dashed lines indicate the path of each vector from the origin to their respective endpoints, with blue representing the input vector and red for the transformed vector.

2.6 MATRIX FACTORIZATIONS

Matrix factorization techniques allow us to decompose matrices into products of simpler matrices. These techniques are beneficial in various applications, from solving systems of equations to data compression and dimensionality reduction. Here are essential matrix factorization methods.

2.6.1 LU Decomposition

LU Decomposition factorizes a matrix as the product of a lower **L** and upper triangular matrix **U**. This is particularly useful for solving systems of linear equations, inverting matrices, and computing determinants.

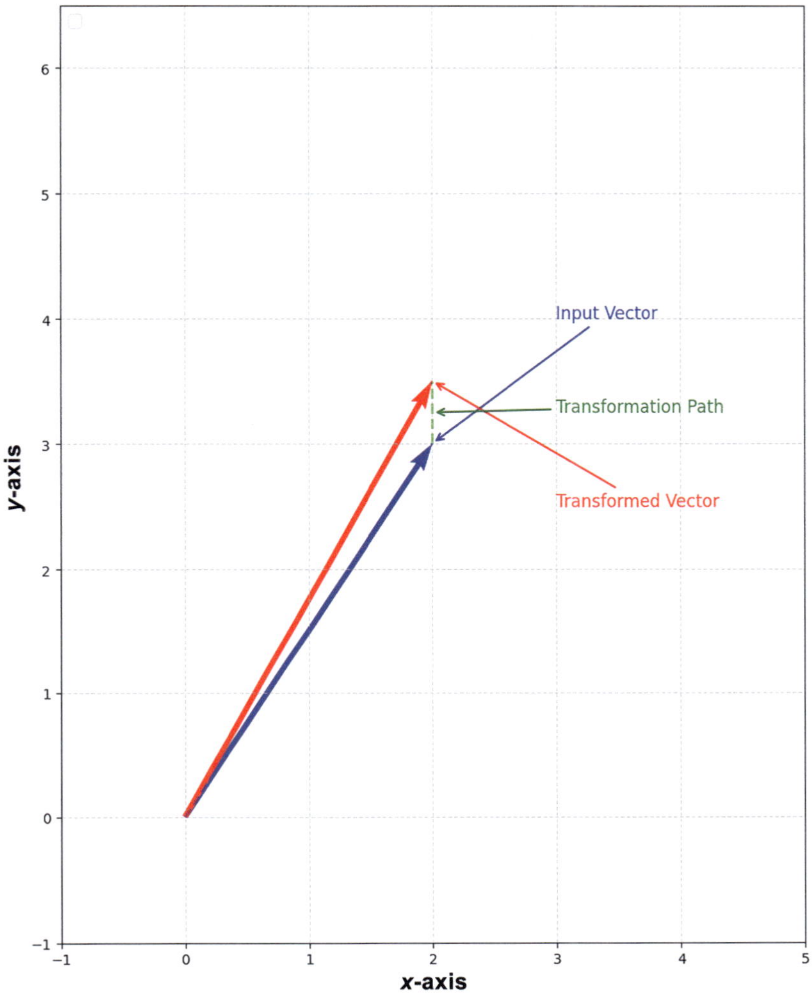

FIGURE 2.13 Visual representation of linear transformation in a dense layer.

Example: Consider a matrix A : $A = \begin{pmatrix} 4 & 3 \\ 6 & 3 \end{pmatrix}$. The **LU** Decomposition of **A** expresses it as the product of a lower triangular matrix **L** and an upper triangular matrix **U**:

$$\boldsymbol{A} = \boldsymbol{L} \cdot \boldsymbol{U} \textbf{ where } \boldsymbol{L} = \begin{pmatrix} 1 & 0 \\ 1.5 & 1 \end{pmatrix} \text{ and } \boldsymbol{U} = \begin{pmatrix} 4 & 3 \\ 0 & -1.5 \end{pmatrix}.$$

$$\textbf{Then } A = \begin{pmatrix} 1 & 0 \\ 1.5 & 1 \end{pmatrix} \begin{pmatrix} 4 & 3 \\ 0 & -1.5 \end{pmatrix} = \begin{pmatrix} 4 & 3 \\ 6 & 3 \end{pmatrix}.$$

This decomposition is useful for solving equations, inverting **A**, and finding determinants efficiently

(a) ***L** Matrix:* The lower triangular matrix from the **LU** decomposition of the matrix **A**. This matrix has non-zero elements only on and below the main diagonal:

$$L = \begin{pmatrix} l_{11} & 0 & 0 & \cdots & 0 \\ l_{21} & l_{22} & 0 & \cdots & 0 \\ l_{31} & l_{32} & l_{33} & \cdots & 0 \\ \vdots & \vdots & \vdots & \ddots & 0 \\ l_{n1} & l_{n2} & l_{n3} & \cdots & l_{nn} \end{pmatrix}$$

The lower triangular matrix from the **LU** decomposition of the matrix **A**. This matrix has non-zero elements only on and below the main diagonal.

$$\boldsymbol{L} = \begin{pmatrix} \mathbf{1} & \mathbf{0} \\ \mathbf{1.5} & \mathbf{1} \end{pmatrix}$$

(b) ***U** Matrix:* The upper triangular matrix from the **LU** decomposition. This matrix has non-zero elements only on and above the main diagonal.

$$U = \begin{pmatrix} u_{11} & u_{12} & u_{13} & \cdots & u_{1n} \\ 0 & u_{22} & u_{23} & \cdots & u_{2n} \\ 0 & 0 & u_{33} & \cdots & u_{3n} \\ \vdots & \vdots & \vdots & \ddots & \vdots \\ 0 & 0 & 0 & \cdots & u_{nn} \end{pmatrix}.$$

The upper triangular matrix from the **LU** decomposition. This matrix has non-zero elements only on and above the main diagonal.

$$\boldsymbol{U} = \begin{pmatrix} 4 & 3 \\ 0 & -1.5 \end{pmatrix}.$$

(c) *Reconstructed Matrix* ***A**:* The original matrix **A** was reconstructed by multiplying **L** and **U**. This step verifies the correctness of the decomposition. The original matrix was reconstructed by multiplying **L** and **U**. This step verifies the correctness of the decomposition.

$$\boldsymbol{A} = \boldsymbol{L} \cdot \boldsymbol{U} = \begin{pmatrix} 1 & 0 \\ 1.5 & 1 \end{pmatrix} \begin{pmatrix} 4 & 3 \\ 0 & -1.5 \end{pmatrix} = \begin{pmatrix} 4 & 3 \\ 6 & 3 \end{pmatrix}.$$

2.6.2 QR Decomposition

QR decomposition decomposes a matrix into an orthogonal **Q** and upper triangular matrix **R**. This decomposition is widely used in numerical linear algebra to solve most minor square problems and in eigenvalue algorithms.

Example:

$$A = Q \cdot R,\ A = \begin{pmatrix} 4 & -2 & 1 \\ 6 & 3 & -8 \\ -5 & 7 & 10 \end{pmatrix}, Q = \begin{pmatrix} 0.36 & -0.48 & 0.8 \\ 0.54 & 0.87 & 0.0 \\ -0.72 & 0.12 & 0.68 \end{pmatrix}, R = \begin{pmatrix} 11.18 & 2.15 & -5.95 \\ 0 & 7.68 & 8.02 \\ 0 & 0 & 3.87 \end{pmatrix}.$$

(a) ***Q** Matrix (**QR** Decomposition)*: The orthogonal matrix **Q** from the **QR** decomposition of the matrix **A**. Orthogonal matrices have the property that their columns are orthonormal vectors.

Example:

$$Q = \begin{pmatrix} 0.36 & -0.48 & 0.8 \\ 0.54 & 0.87 & 0.0 \\ -0.72 & 0.12 & 0.68 \end{pmatrix}.$$

(b) ***R** Matrix (**QR** Decomposition):* The upper triangular matrix from the **QR** decomposition. This matrix has non-zero elements only on and above the main diagonal.

Example:

$$R = \begin{pmatrix} 11.18 & 2.15 & -5.95 \\ 0 & 7.68 & 8.02 \\ 0 & 0 & 3.87 \end{pmatrix}.$$

(c) *Reconstructed:* **A**(Q *R) the original matrix was reconstructed by multiplying **Q** and **R**. This step confirms that the decomposition accurately represents the original matrix.

Example:

$$A = Q \cdot R = \begin{pmatrix} 0.36 & -0.48 & 0.8 \\ 0.54 & 0.87 & 0.0 \\ -0.72 & 0.12 & 0.68 \end{pmatrix} \begin{pmatrix} 11.18 & 2.15 & -5.95 \\ 0 & 7.68 & 8.02 \\ 0 & 0 & 3.87 \end{pmatrix} = \begin{pmatrix} 4 & -2 & 1 \\ 6 & 3 & -8 \\ -5 & 7 & 10 \end{pmatrix}.$$

2.6.3 Singular Value Decomposition

SVD represents a matrix as the product of three matrices: an orthogonal matrix **U**, a diagonal matrix Σ, and the transpose of an orthogonal matrix $\mathbf{V}^{\mathrm{T}}$

Example: If the matrix **A** in the image is something like this:

$$A = \begin{pmatrix} 1 & 0 & 0 \\ 0 & 4 & 0 \\ 0 & 0 & 0 \end{pmatrix}.$$

Its SVD would be: $\boldsymbol{A} = \boldsymbol{U\Sigma V}^{\top}$,

where

- **A** is the original matrix,
- **U** is an orthogonal matrix (with orthonormal columns),
- Σ is a diagonal matrix containing the singular values of **A**,
- $V^{\top}$ is the transpose of an orthogonal matrix **V**,

where

$$U = \begin{pmatrix} 0 & 1 & 0 \\ 1 & 0 & 0 \\ 0 & 0 & 1 \end{pmatrix}, \quad \Sigma = \begin{pmatrix} 4 & 0 & 0 \\ 0 & 1 & 0 \\ 0 & 0 & 0 \end{pmatrix}, \quad V^{\top} = \begin{pmatrix} 0 & 1 & 0 \\ 1 & 0 & 0 \\ 0 & 0 & 1 \end{pmatrix}.$$

Each decomposition technique has specific applications and provides unique insights into the matrix structure. **LU** decomposition is particularly useful for solving systems of linear equations, **QR** decomposition is widely used in numerical linear algebra for solving most minor squares problems, and **SVD** is fundamental in applications like principal component analysis (PCA), data compression, and signal processing.

In Figure 2.14, the results of three different matrix decompositions (**LU**, **QR**, and **SVD**) are visualized. For the **LU** decomposition, the **L** matrix shows the lower triangular matrix obtained from the decomposition of matrix **A**, featuring non-zero elements below the main diagonal and zeros above it. The **U** matrix displays the upper triangular matrix, which has non-zero elements on and above the main diagonal, with zeros below it. The third heatmap shows the reconstructed matrix **A** obtained by multiplying the **L** and **U** matrices, confirming the accuracy of the decomposition. For the **QR** decomposition, the **Q** matrix is an orthogonal matrix with orthonormal columns, obtained from the decomposition of matrix **A**. The **R** matrix is an upper triangular matrix with non-zero elements on and above the main diagonal and zeros below it. The reconstructed matrix **A**, shown in the third heatmap, is obtained by multiplying the **Q** and **R** matrices, demonstrating the correctness of the **QR** decomposition. Finally, for the **SVD**, the **U** matrix represents the left singular vectors of matrix **A**, forming an orthogonal matrix. The Sigma matrix is a diagonal matrix of singular values, which are non-negative and sorted in descending order. The last heatmap shows the reconstructed matrix **A**, obtained by multiplying the **U**, Sigma, and $\mathbf{V}^t$ (transpose of **V**) matrices, validating the **SVD** process.

2.6.4 Eigenvalues and Eigenvectors

Eigenvalues and eigenvectors are fundamental concepts in linear algebra with widespread applications in various fields, including physics, engineering, and data science. Given a square matrix **A**, if there exists a non-zero vector **v** and a scalar λ such that the following equation holds:

$$Av = \lambda v.$$

Then is an eigenvector of **A** and is the corresponding eigenvalue.

In Figure 2.15, the visual representation of eigenvalues and eigenvectors for a given square matrix **A** is displayed. The matrix **A** is defined as: $A = \begin{pmatrix} 4 & 2 \\ 1 & 3 \end{pmatrix}$. The plot begins with the original eigenvectors, which are represented in green. Each eigenvector is plotted starting from the origin (0, 0) and pointing to the coordinates defined by its components. Labels, such as **v1** and **v2**, are placed at the endpoints of these eigenvectors to clearly identify them. Next, the transformed eigenvectors are shown in red. These vectors are obtained by multiplying the original eigenvectors by matrix **A**, demonstrating how **A** transforms its eigenvectors.

In Figure 2.16, the SVD of a matrix **M** is visualized by transforming a grid of points. The matrix **M** is defined as:

$$M = \begin{pmatrix} 1 & 2 & 3 \\ 4 & 5 & 6 \\ 7 & 8 & 9 \end{pmatrix}$$

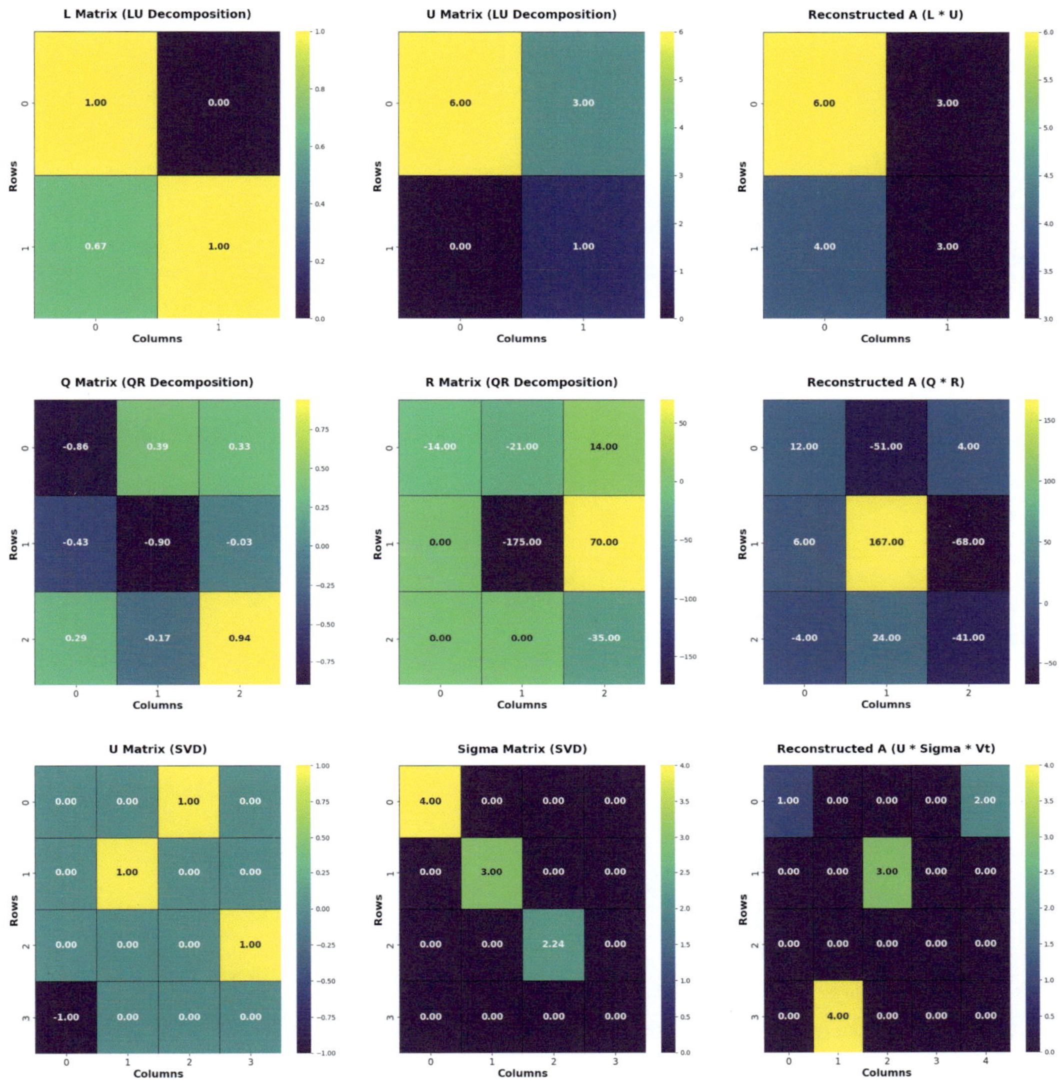

FIGURE 2.14 Visual representation of LU decomposition, QR decomposition, and singular value decomposition.

The SVD decomposes **M** into three matrices: **U**, **Σ**, and $\mathbf{V}^T$, where **U** contains the left singular vectors, **Σ** is a diagonal matrix of singular values, and $\mathbf{V}^T$ contains the right singular vectors. It presents the effects of these matrices on a grid of points, allowing for a visual understanding of the SVD components. The original grid is plotted in blue, representing the initial arrangement of points before any transformation. These points form a regular grid centered around the origin. The **U** matrix (left singular vectors) is visualized by transforming the original grid using the **U** matrix. This transformation primarily represents a rotation. The green lines and points show how the left singular vectors of **M** rotate the grid. The Sigma matrix (singular values) illustrates the scaling effect of the singular values. The red lines and points depict the grid after being scaled by the diagonal matrix **Σ**. This scaling adjusts the lengths of the vectors but does not change their directions. The $\mathbf{V}^T$ matrix (right singular vectors) is visualized by transforming the original grid using the $\mathbf{V}^T$ matrix. The orange lines and points show the effect of applying the right singular vectors to the grid, which typically involves another rotation.

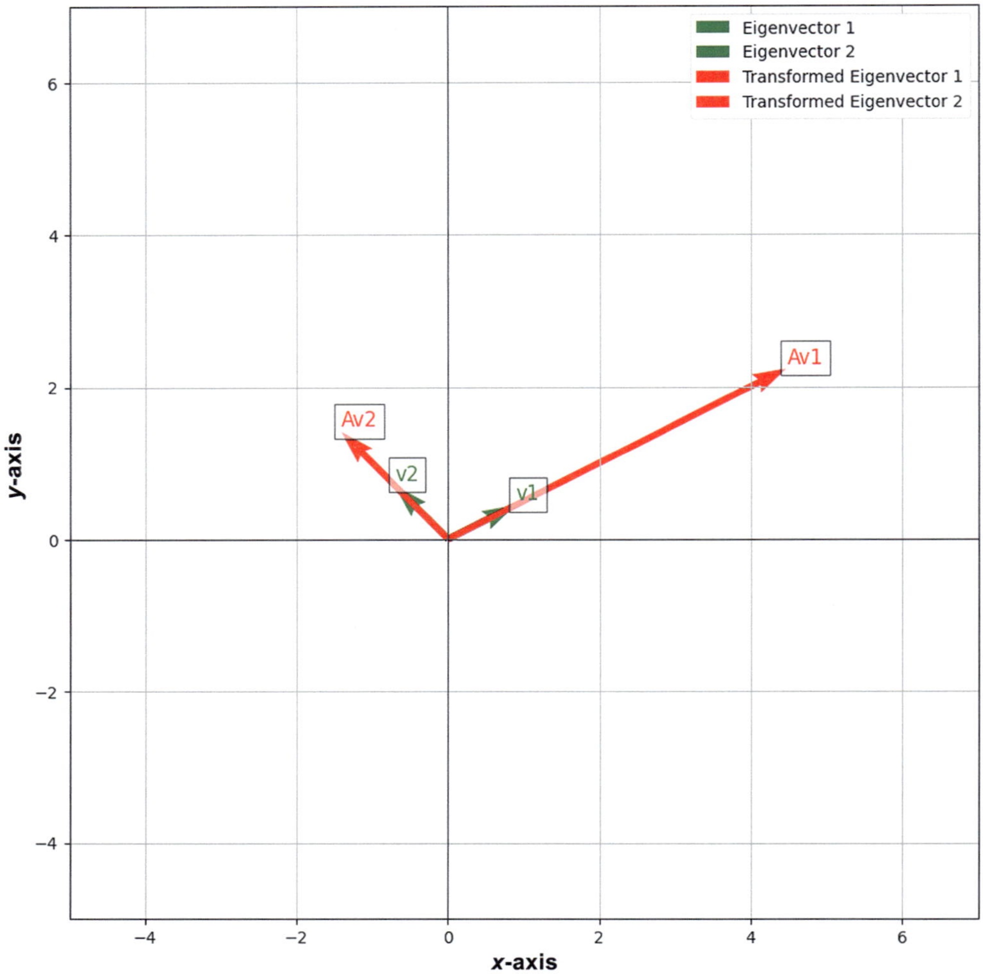

FIGURE 2.15 Visual representation of eigenvalues and eigenvector.

In Figure 2.17, the SVD of a matrix **M** is visualized by transforming a grid of points. The first subplot, titled "Original Grid," shows the original grid in blue. This grid represents the initial arrangement of points before any transformation, forming a regular grid centered around the origin. The second subplot, titled "Left Singular Vectors (**U**)," illustrates the effect of the **U** matrix on the grid. The **U** matrix primarily represents a rotation. The green lines and points show how the left singular vectors of **M** rotate the grid. The third subplot, titled "Singular Values (Sigma)," visualizes the effect of the Sigma matrix on the grid. The Sigma matrix scales the grid. The red lines and points depict the grid after being scaled by the diagonal matrix $\mathbf{\Sigma}$, which adjusts the lengths of the vectors while maintaining their directions. The fourth subplot, titled "Right Singular Vectors ($\mathbf{V}^T$)," shows the effect of the $\mathbf{V}^T$ matrix on the grid. The $\mathbf{V}^T$ matrix typically involves another rotation. The orange lines and points illustrate the transformed grid, highlighting the rotation applied by the right singular vectors.

2.6.5 Eigenvalues and Eigenvectors in Deep Learning

The eigenvalues of the Hessian matrix (a matrix of second-order partial derivatives) can indicate the curvature of the loss landscape around a point. A large eigenvalue indicates a steep curvature

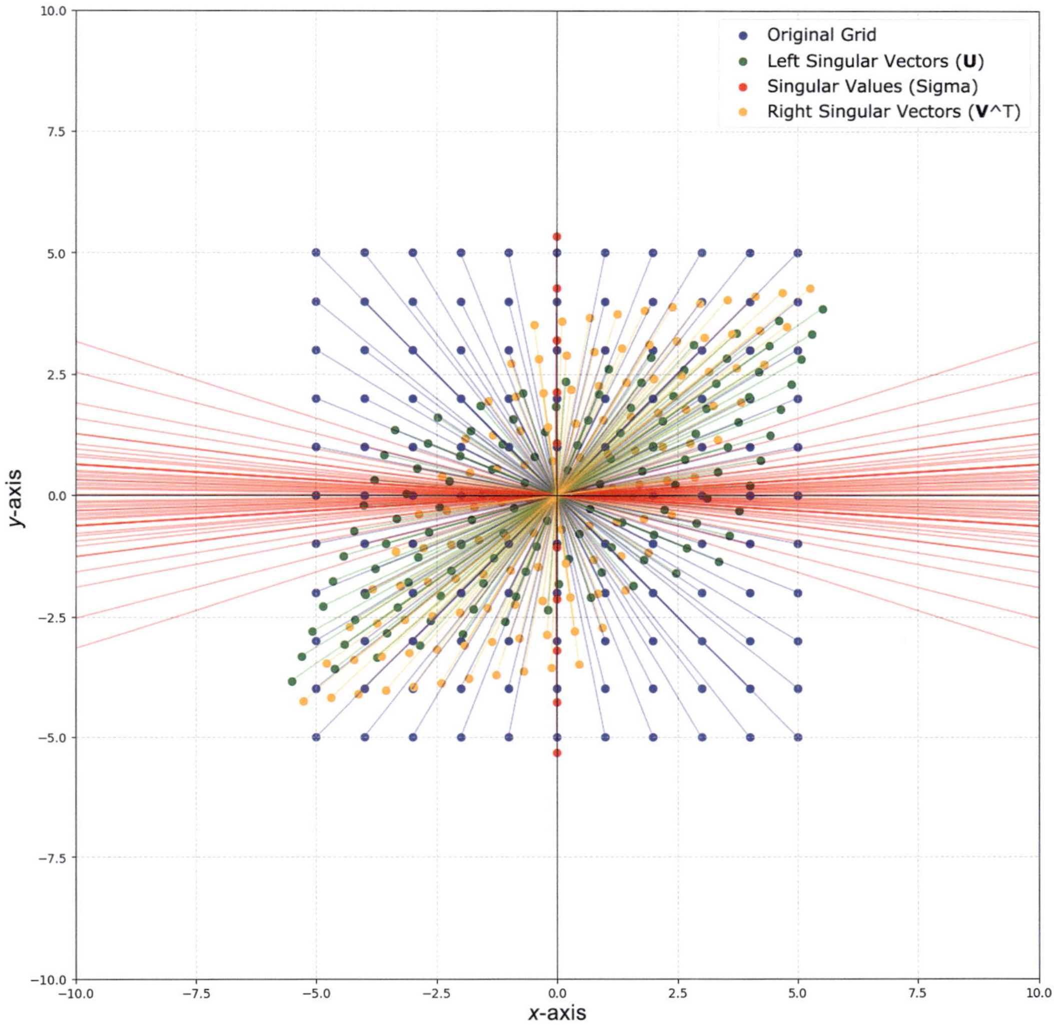

FIGURE 2.16 Visual representation of singular value decomposition.

(possibly a sharp minimum or a narrow ravine), making optimization challenging. If the largest eigenvalue of the weight matrices in recurrent neural networks (RNNs) is much larger or much smaller than 1, it can lead to exploding or vanishing gradients, respectively. A fundamental property in deep learning optimization is the curvature of the loss landscape, which can influence training dynamics.

In Figure 2.18, the contours of a quadratic function $f(x,y) = ax^2 + by^2$ are visualized along with the eigenvectors of its Hessian matrix. The function parameters are defined as **a** = 2 and **b** = 5, resulting in the quadratic function:

$$f(x, y) = 2x^2 + 5y^2.$$

The Hessian matrix **H** of this function is calculated as:

$$H = \begin{pmatrix} 4 & 0 \\ 0 & 10 \end{pmatrix}.$$

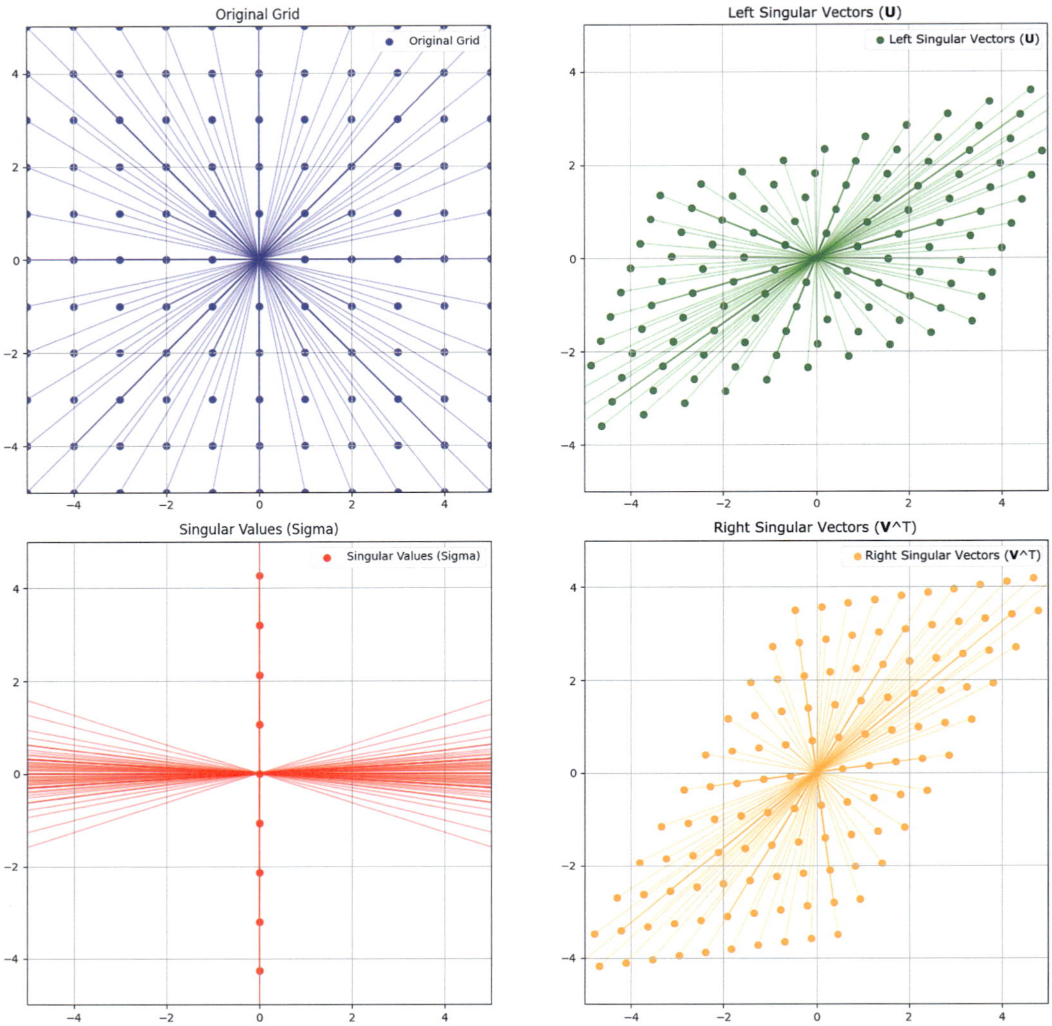

FIGURE 2.17 Visual representation of singular value decomposition.

The eigenvalues represent the principal curvatures, while the eigenvectors indicate the directions of these curvatures. The plot includes several key elements. First, it features the contours of the quadratic function, where the contour lines represent the levels of the function **f**(**x**, **y**). These lines help visualize how the function values change across the **x** and **y** planes and are plotted using a color gradient from the colormap with 50 contour levels. Additionally, the plot displays the eigenvectors of the Hessian matrix, represented as arrows originating from the origin (0, 0). The first eigenvector is shown in red, and the second in blue, indicating the principal directions of the quadratic function's curvature. Each eigenvector is labeled with its corresponding eigenvalue, and the labels are positioned near the tips of the arrows, with a semi-transparent white background to enhance readability. Using the top eigenvectors (principal components) corresponding to the largest eigenvalues of the data's covariance matrix, we can project data into a lower-dimensional space while retaining most of the variance.

Figure 2.19 consists of two subplots: one showing the original 3D data and the other displaying the data projected onto the first two principal components. The original 3D data is generated using a multivariate normal distribution with a specified mean and covariance matrix. This data is then reduced to 2D using PCA, a technique that identifies the directions (principal components) along

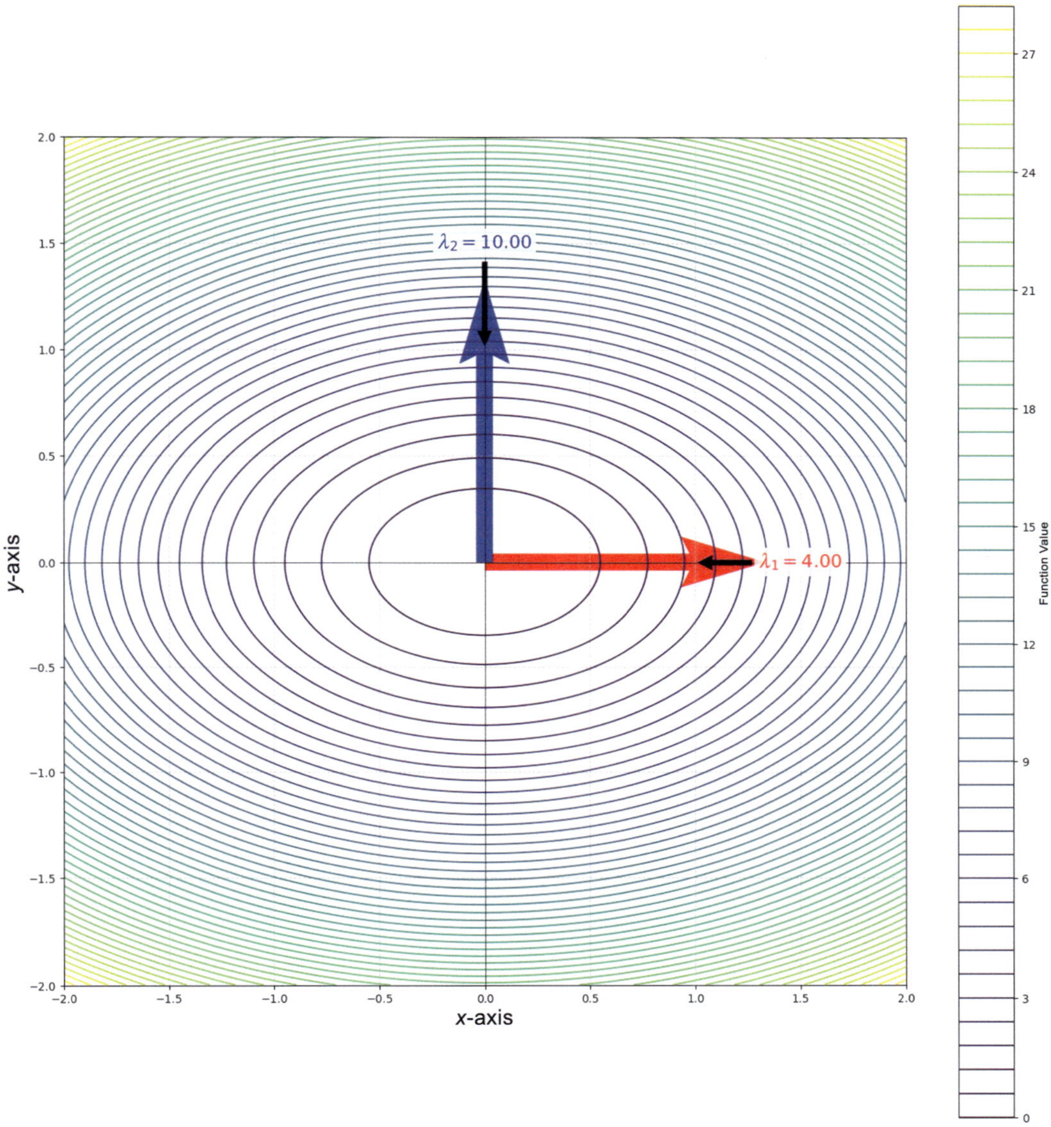

FIGURE 2.18 Contours of the quadratic function with eigenvalues of the Hessian matrix.

which the variance in the data is maximized. The left subplot, titled "Original 3D Data," presents a scatter plot of the original data points in three-dimensional space. Each data point is represented by a dot, with its coordinates corresponding to the values in the **X**, **Y**, and **Z** dimensions. The right subplot, titled "Data Projected onto the First Two Principal Components," shows the result of projecting the 3D data onto the first two principal components identified by PCA. The scatter plot in this subplot displays the data points in the reduced 2D space, with the *X*-axis representing the first principal component and the *Y*-axis representing the second principal component. The axes are labeled with the percentage of variance explained by each principal component, indicating their significance. The points are colored based on their values along the second principal component, and a colorbar is added to provide a reference for the color scale.

$\mathbf{L_2}$ regularization adds a penalty to the loss function proportional to the sum of the squares of the model's weights. This encourages the model to have smaller weights, leading to a smoother model.

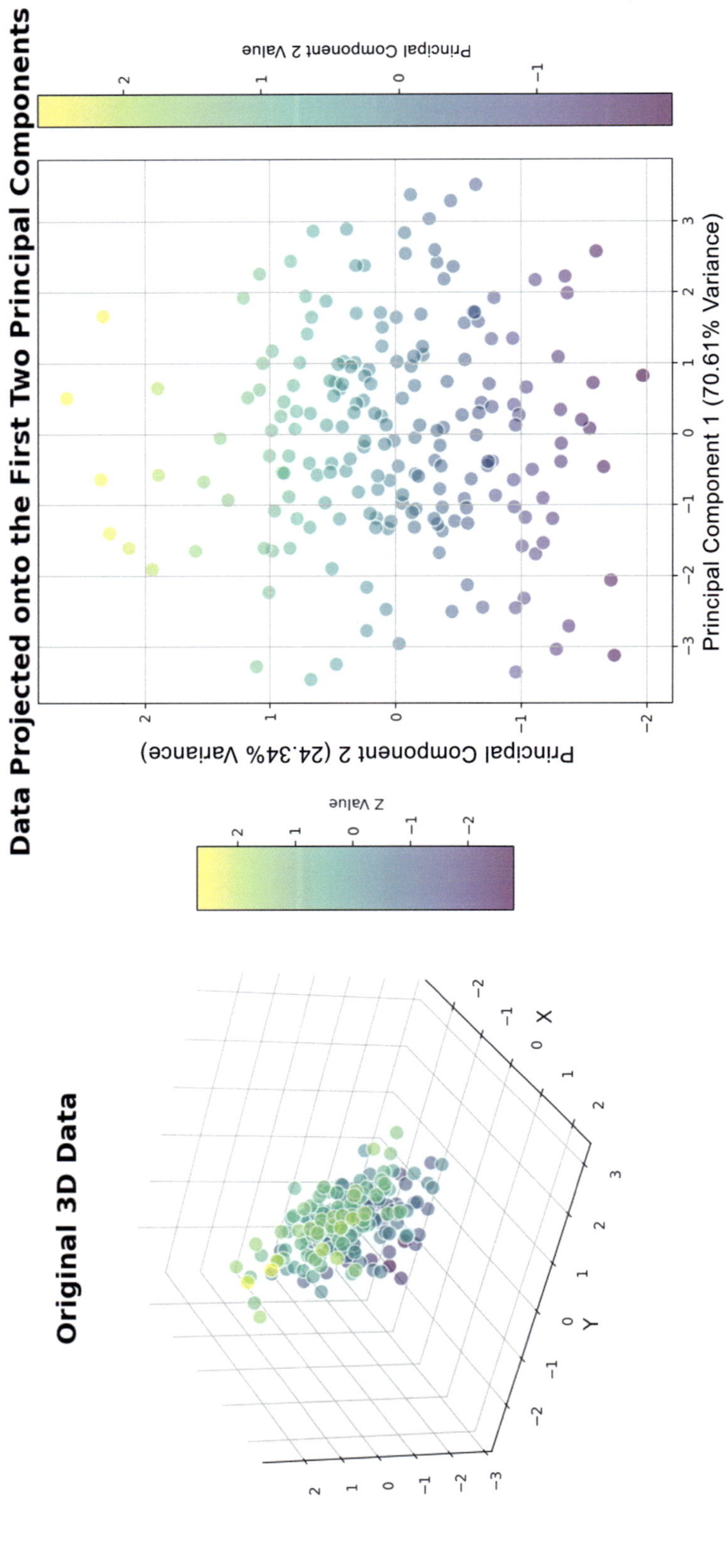

FIGURE 2.19 Visualization of PCA—original 3D data and projected 2D Data. 2D, two-dimensional; 3D, three-dimensional; PCA, principal component analysis.

In Figure 2.20, the impact of $\mathbf{L}_2$ regularization on polynomial regression is illustrated using a set of sample data points generated from a noisy cubic function. The figure contrasts the fitting results of polynomial regression with and without $\mathbf{L}_2$ regularization. The sample data consists of 10 random points in the interval [−5, 5], with the corresponding *y*-values generated from a cubic function with added Gaussian noise:

$$y = 0.5x^3 - 20x + 90 + \text{noise}.$$

Two polynomial regression models of degree 9 are fitted to the data. The first model is trained without regularization, whereas the polynomial regression model has no penalty applied (alpha = 0). The second model incorporates $\mathbf{L}_2$ regularization, specifically Ridge regression, where the regularization strength is set to alpha = 10. Both models are trained on the same dataset, allowing a comparison between the effects of no regularization and $\mathbf{L}_2$ regularization on the polynomial fit.

2.6.6 Singular Value Decomposition in Deep Learning

One of the uses of **SVD** in deep learning is in model compression. It can be used to compress fully connected layers in NNs. We can reduce the number of parameters by approximating the weight matrix using a low-rank approximation (using the top **k** singular values and the corresponding singular vectors), leading to a more compact model with faster inference times and potentially reduced overfitting. It also can be used in visualization and analysis. **SVD**, mainly in techniques like LSA (Latent Semantic Analysis), can help visualize word embeddings or document embeddings in a lower-dimensional space. Another application is noise reduction in data. By keeping only the top **k** singular values (and discarding smaller ones), **SVD** can filter out noise from data. This is commonly used in image processing but can be extended to any data preprocessing in deep learning tasks.

In Figure 2.21, the effect of **SVD** on denoising an image is demonstrated through a step-by-step process involving the addition of noise and subsequent reconstruction. The original image of an astronaut, converted to grayscale format, is shown in the first subplot. This image serves as the reference for comparison and is displayed without any modifications to highlight its clarity and baseline quality. The second subplot presents the image after introducing salt-and-pepper noise, which randomly replaces some pixels with either black or white. The noise amount is set to 10% of the total pixels, significantly degrading the image quality and making the random black and white pixels visibly scattered throughout. The third subplot illustrates the reconstructed image using **SVD**, where the image is decomposed into three matrices: **U**, **S**, and $\mathbf{V}^\mathrm{T}$. The reconstruction leverages only the top 50 singular values, effectively reducing the noise while retaining essential features of the image. This selective reconstruction results in a smoother image with more defined features compared to the noisy version.

2.7 REAL-WORLD APPLICATIONS AND EXAMPLES

2.7.1 Image Processing and Computer Vision

In computer vision, linear algebra is fundamental to the processing and analysis of images. Images are often represented as matrices where each element corresponds to a pixel's intensity. Operations such as image rotation, scaling, and translation can be performed using matrix transformations. For example, a rotation matrix can be applied to an image matrix to rotate the image around a specific point. **CNNs**, which are widely used in tasks like facial recognition and object detection, rely heavily on matrix operations to filter and process images, enabling the network to learn features such as edges, textures, and shapes.

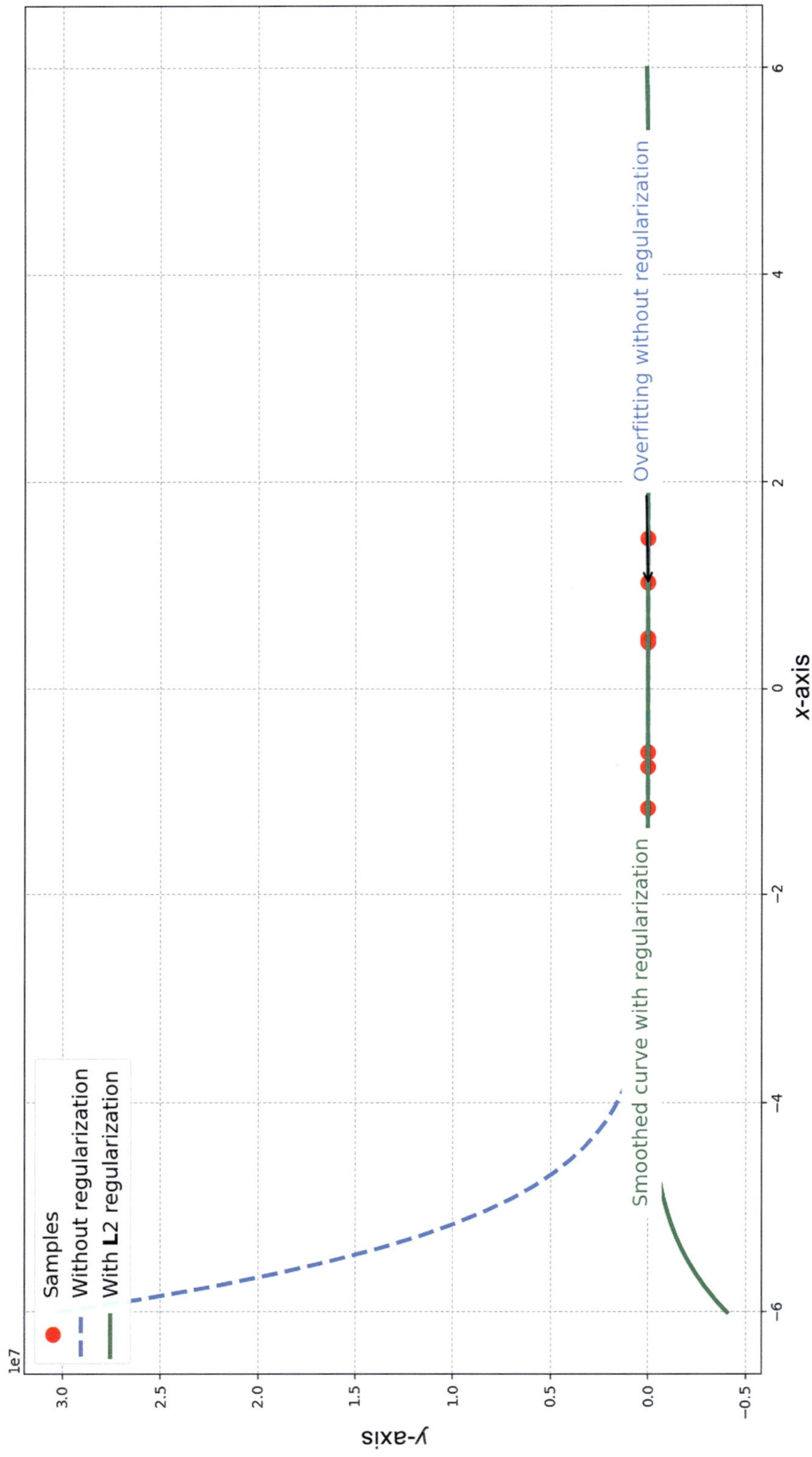

FIGURE 2.20 Effect of $\mathbf{L}_2$ regularization on polynomial regression.

FIGURE 2.21 SVD-based noise reduction in images. SVD, singular value decomposition.

2.7.2 Natural Language Processing

In **NLP**, vectors and matrices are used to represent words and sentences in a way that machines can understand. Word embeddings, such as Word2Vec, map words to vectors in a high-dimensional space where semantically similar words are positioned closely together. These embeddings are created using matrix operations and linear transformations. Additionally, RNNs and transformers utilize tensor operations to handle text sequences, allowing for the efficient processing of sentences and paragraphs in tasks like translation, summarization, and sentiment analysis.

2.7.3 Robotics and Autonomous Systems

In robotics, linear algebra is critical for modeling the motion and control of robots. The position and orientation of a robot in space can be represented using vectors and matrices, and transformations between different coordinate systems are performed using matrix operations. For example, a robot's arm movement is often modeled as a series of matrix multiplications that describe the rotation and translation of each joint. Autonomous systems, such as self-driving cars, use these principles to navigate and interact with their environment, calculating paths and making real-time adjustments based on sensor data.

2.8 HANDS-ON EXAMPLE

In this example, we will generate random input data, define an NN layer with weight and bias matrices, and perform the forward pass using matrix multiplications and transformations.

Step 1: Setting Up the Environment

In this step, we are generating random input data that will serve as a small dataset for demonstration purposes. By using np.random.seed(0), we ensure that the random values generated will be the same every time we run the code, which is important for reproducibility. The function np.random.randn(10, 3) generates a 10×3 matrix filled with random numbers sampled from a standard normal distribution (mean 0, standard deviation 1). This matrix represents 10 samples, each containing 3 features. Printing the matrix helps us inspect the input data before using it in further computations or visualizations.

```
import numpy as np
import matplotlib.pyplot as plt
# Seed for reproducibility
```

```
np.random.seed(0)
# Generate some random input data (10 samples, 3 features)
input_data = np.random.randn(10, 3)
print("Input Data:")
print(input_data)
```

Step 2: Define the NN Layer
In this step, we are defining a weight matrix and a bias vector that will be used for a simple transformation of the input data, simulating a linear layer of an NN. The weight matrix, weights, is initialized with random values using np.random.randn (3, 2), which means it has 3 rows (corresponding to the 3 features of the input data) and 2 columns (corresponding to the 2 output units). The bias vector is also initialized randomly with 2 values, representing the biases for each of the 2 output units. By printing both the weights and biases, we can examine the initialized values before applying them to the input data.

```
# Define weight matrix (3 features to 2 outputs)
weights = np.random.randn(3, 2)
# Define bias vector (2 outputs)
biases = np.random.randn(2)
print("\nWeights:")
print(weights)
print("\nBiases:")
print(biases)
```

Step 3: Perform the Forward Pass
In this step, we are performing a forward pass through a simple linear layer, which is a fundamental operation in NNs. The formula $Z = \mathbf{WX} + \mathbf{B}$ represents the transformation of the input data $\mathbf{X}$ using the weight matrix $\mathbf{W}$ and the bias vector $\mathbf{B}$. The operation np.dot(input_data, weights) computes the dot product between the input data and the weight matrix, effectively combining the features and transforming them to match the 2 output units. We then add the vector biases to adjust the output values. The result, stored in output_data, represents the output of the linear transformation, which we print to inspect the result of this forward pass.

```
# Perform the forward pass Z = W * X + B
output_data = np.dot(input_data, weights) + biases
print("\nOutput Data:")
print(output_data)
```

Step 4: Visualize the Transformations
In this final step, we are visualizing the transformation of the data through a plot that compares the input data and output data. The code uses matplotlib to create a figure with two subplots: one for the input data and another for the output data. In the input data plot, the first two features of the input matrix are visualized as a scatter plot with blue points and labeled as "Feature 1" and "Feature 2." Similarly, in the output data plot, the transformed output from the forward pass is visualized with red points, representing "Output 1" and "Output 2." This comparison helps in understanding how the linear transformation affects the original data.

```
# Visualize the input data and output data
fig, ax = plt.subplots(1, 2, figsize=(14, 7))
# Plot input data
ax[0].scatter(input_data[:, 0], input_data[:, 1], c='blue',
s=50, edgecolors='w', label='Input Data')
ax[0].set_title('Input Data')
ax[0].set_xlabel('Feature 1')
ax[0].set_ylabel('Feature 2')
ax[0].legend()
ax[0].grid(True)
# Plot output data
ax[1].scatter(output_data[:, 0], output_data[:, 1], c='red',
s=50, edgecolors='w', label='Output Data')
ax[1].set_title('Output Data')
ax[1].set_xlabel('Output 1')
ax[1].set_ylabel('Output 2')
ax[1].legend()
ax[1].grid(True)
plt.tight_layout()
plt.show()
```

Figure 2.22 displays the original randomly generated data points in a 2D feature space. Each point represents a sample with two features plotted on the *x*-axis (Feature 1) and *y*-axis (Feature 2). The points are colored blue and labeled as "Input Data." The right plot illustrates the "Output Data" after an NN layer has transformed it.

2.9 COMMON MISTAKES AND TROUBLESHOOTING TIPS

2.9.1 Understanding Vectors

- *Mistake:* Confusing vectors and scalars.
- *Tip:* Remember that vectors have both magnitude and direction, whereas scalars only have magnitude.
- *Mistake:* Incorrect calculation of vector magnitude.
- *Tip:* Use the Pythagorean theorem: for a vector $\mathbf{v} = (x, y)$, the magnitude is.$\sqrt{x^2 + y^2}$.

2.9.2 Vector Operations

- *Mistake:* Adding vectors incorrectly by not matching corresponding components.
- *Tip:* Ensure you add vectors component-wise. For $\mathbf{u} = (u_1, u_2)\, and\, \mathbf{v} = (v_1, v_2)$, the sum is $\mathbf{u} + \mathbf{v} = (u_1 + v_1, u_2 + v_2)$.
- *Mistake:* Misunderstanding the dot product operation.
- *Tip:* The dot product of **u** and **v** is calculated as $\mathbf{u} \cdot \mathbf{v} = \mathbf{u_1 v_1} + \mathbf{u_2 v_2}$. It results in a scalar, not a vector.
- *Mistake:* Incorrectly calculating the cross-product.
- *Tip:* For $\mathbf{u} = u_1 + u_2 + u_3$ and $\mathbf{v} = v_1 + v_2 + v_3$, the cross-product is $\mathbf{a} \times \mathbf{v} = (u_2\ v_3 - u_3 v_2, u_3 v_1 - u_1 v_3, u_1 v_2 - u_2 v_1)$. Ensure the resulting vector is perpendicular to both **u** and **v.**

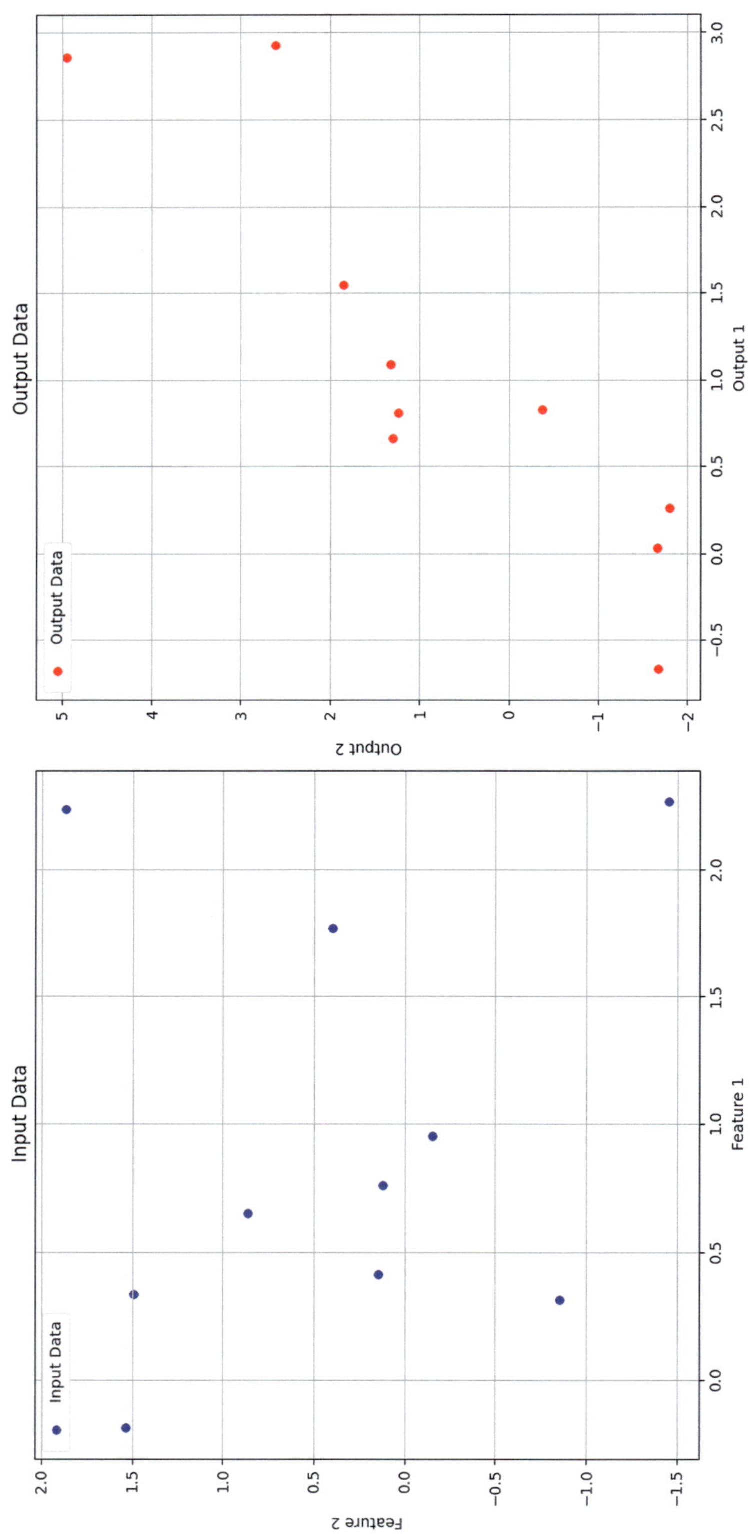

FIGURE 2.22 Comparison of input and output data in neural network transformations.

2.9.3 Matrix Operations

- *Mistake:* Mismatching matrix dimensions for multiplication.
- *Tip:* Ensure the number of columns in the first matrix matches the number of rows in the second matrix. For matrices **A** of size $n \times p$, the product is defined and results in a matrix of size $m \times p$.
- *Mistake:* Forgetting the properties of matrix addition and multiplication.
- *Tip:* Remember that matrix multiplication is not commutative$\left(A + B \neq B + A\right)$. However, matrix addition is commutative $\left(A + B = B + A\right)$.
- *Mistake:* Incorrectly computing the determinant.
- *Tip:* For a matrix $\begin{pmatrix} a & b \\ c & d \end{pmatrix}$, the determinant is $ad - bc$. Practice finding determinants for larger matrices using cofactor expansion.

2.9.4 Eigenvalues and Eigenvectors

- *Mistake:* Misidentifying eigenvectors and eigenvalues.
- *Tip:* For a matrix **A**, an eigenvector and eigenvalue satisfy $Av = \lambda v$. Ensure it is non-zero and is a scalar.
- *Mistake:* Overlooking the geometric interpretation of eigenvalues and eigenvectors.
- *Tip:* Understand that eigenvectors indicate directions that remain invariant under the transformation represented by the matrix, while eigenvalues indicate the scaling factor along those directions.

2.9.5 Singular Value Decomposition

- *Mistake:* Confusing **SVD** with eigenvalue decomposition.
- *Tip:* SVD applies to any $\mathbf{m} \times \mathbf{n}$ matrix, decomposing it into $\mathbf{A} = \mathbf{U}\boldsymbol{\Sigma}\mathbf{V}^{\mathrm{T}}$, where **U** and **V** are orthogonal matrices, and $\boldsymbol{\Sigma}$ is a diagonal matrix of singular values. Eigenvalue decomposition applies only to square matrices.

2.9.6 Practical Applications and Troubleshooting

- *Mistake:* Neglecting the importance of normalization in deep learning.
- *Tip:* Normalize vectors and matrices to improve numerical stability and convergence during training. This can prevent issues like vanishing or exploding gradients.
- *Mistake:* Misapplying regularization techniques.
- *Tip:* Use techniques like $\mathbf{L}_2$ regularization (adding a penalty proportional to the square of the magnitude of coefficients) to prevent overfitting. Ensure you understand the implications of each regularization method on your model.

2.10 REVIEW QUESTIONS

1. Describe the various roles that vectors play in NNs. How do they contribute to data representation, weight storage, and activation functions?
2. Explain how matrices are utilized within deep learning models. Provide an example of how matrix operations are critical in the functioning of dense layers or convolutional layers.
3. Discuss why tensors are essential in deep learning. What advantages do they offer over vectors and matrices, particularly in handling multi-dimensional data?
4. How do tensor operations, such as reshaping and broadcasting, benefit deep learning computations?

5. Why are linear transformations fundamental in NNs? Discuss how these transformations are applied in different layers of an NN.
6. How do eigenvalues and eigenvectors influence model analysis and training? Consider their role in understanding the curvature of the loss landscape and the impact on gradient-based optimization.
7. How does PCA, based on eigen-decomposition, assist in deep learning? Provide an example of its application in reducing dimensionality while preserving data variance.
8. Discuss the significance of **SVD** in deep learning. How does it enhance the efficiency and performance of NNs?
9. Why are tensors preferred over matrices for handling image data in deep learning? Explore the advantages of using tensors in CNNs.
10. How does **SVD** support orthogonal initialization in deep NNs? Explain why this is important for maintaining numerical stability during training.

2.11 PROGRAMMING QUESTIONS

2.11.1 Easy

In this exercise, you will generate random input data and apply a transformation using a weight matrix and a bias vector, mimicking the forward pass of a single layer in an NN.

1. Generate random input data
2. Define weights and biases
3. Forward pass with linear transformation

2.11.2 Medium

In this exercise, you will compare the effects of different weight matrices on the same input data.

1. Generate random input data
2. Define two different weight matrices
3. Forward pass with first weight matrix
4. Forward pass with a second weight matrix

2.11.3 Hard

In this exercise, you will generate random input data, apply a linear transformation using a weight matrix and bias vector, and then apply the **ReLU** activation function.

1. Generate random input data
2. Define weights and biases
3. Forward pass with linear transformation
4. Apply ReLU activation

3 Multivariate Calculus

3.1 INTRODUCTION

In this chapter, we delve into the calculus that underpins key processes such as backpropagation and optimization. We begin by introducing derivatives and gradients, which are crucial for optimizing neural networks through methods like gradient descent. Each mathematical concept discussed here directly impacts how models are built, trained, and optimized, making this chapter an important step in your journey through deep learning.

3.2 PARTIAL DERIVATIVES

Partial derivatives help us understand how a multivariable function changes to each independent variable, assuming all other variables remain constant. This concept is fundamental in multivariable calculus and finds applications across various fields. When dealing with functions of multiple variables, the rate at which the function changes concerning one variable while keeping other variables constant is known as the partial derivative. Consider the function $f(x,y)$, when taking the partial derivative of **f** to **x**, treat **y** as a constant and usually differentiate it to **x**. The same logic applies when taking the partial derivative to **y**. Let: $f(x,y) = x^2y + xy^2$, the partial derivative of **f** to **x** is $\frac{\partial f}{\partial x} = 2xy + y^2$, here, treat **y** as a constant and differentiate to **x**. The partial derivative of **f** for **y** is $\frac{\partial f}{\partial y} = x^2 + 2xy$ here, treat **x** as a constant and differentiate concerning **y**.

3.2.1 Geometric Interpretation

Imagine a three-dimensional surface described by $z = f(x,y)$. The partial derivative $\frac{\partial f}{\partial x}$ represents the slope of the tangent line to the curve obtained by slicing the surface along the plane where **y** is constant. Similarly, $\frac{\partial f}{\partial y}$ provides the slope of the tangent line to the slice given by **x** = constant. Consider the surface $z = x^2 + y^2$, which represents a paraboloid. At the point **(x, y)** = (1, 2), the partial derivatives are as follows:

$$\frac{\partial z}{\partial x} = 2x = 2(1) = 2, \quad \frac{\partial z}{\partial y} = 2y = 2(2) = 4$$

DOI: 10.1201/9781032690742-3

This means that at point (1, 2), the slope of the tangent line in the x-direction is 2, and in the y-direction, it is 4.

3.2.2 Higher-Order Partial Derivatives

Like single-variable functions, we can have higher-order partial derivatives for multivariable functions. For example, the second partial derivative of **f** concerning **x** is denoted $\frac{\partial^2 f}{\partial x^2}$ and represents the rate of change $\frac{\partial f}{\partial x}$ about **x**. We can also have mixed partial derivatives, denoting the change rate relating to **y**. Consider the function $f(x, y) = x^2 y + 3xy^2$. The first partial derivatives are as follows:

$$\frac{\partial f}{\partial x} = 2xy + 3y^2, \quad \frac{\partial f}{\partial y} = x^2 + 6xy$$

The second-order partial derivatives are as follows:

$$\frac{\partial^2 f}{\partial x^2} = 2y, \quad \frac{\partial^2 f}{\partial y^2} = 6x, \quad \frac{\partial^2 f}{\partial x \partial y} = 2x + 6y$$

The mixed partial derivative $\frac{\partial^2 f}{\partial x \partial y} = 2x + 6y$ tells us how the change in **f** with respect to **x** is affected by changes in **y**.

Figure 3.1 shows the visualization of the mathematical function and its partial derivatives through a combination of 3D surface and 2D contour plots. The left subplot features a 3D surface plot, illustrating how the function $\frac{\partial f}{\partial x} = 2xy + y^2$ values change over a grid of **x** and **y** values ranging from –2 to 2. The surface, colored using the colormap, transitions from blue to yellow to represent different function values, with black edges highlighting the surface $\frac{\partial f}{\partial y} = x^2 + 2xy$ for better visual distinction. The right subplot is a 2D contour plot showing the partial derivatives of the function concerning **x** and **y**. The filled contours represent the partial derivative with respect to **x**, using the colormap, which spans from cool (blue) to warm (red) colors, indicating the magnitude of the derivative. The black contour lines represent the partial derivative with respect to **y**. The color bar beside the filled contours provides a reference for the values of the partial derivative with respect to **x**. The plot includes three sample points marked in red with their coordinates labeled for additional reference: (–1, 1), (1, –1), and (1, 1).

3.3 PARTIAL DERIVATIVES IN DEEP LEARNING

In deep learning, we typically work with deep neural networks comprising multiple layers, each containing numerous weights and biases. The primary goal during training is to adjust these weights and biases to optimize the network's performance on a given task. Partial derivatives guide this optimization process. Every deep learning model utilizes a loss function (or cost function) that measures the discrepancy between the model's predictions and the targets. The objective of training is to minimize this loss. We need to understand how the loss changes with tiny adjustments to these parameters to adjust each weight and bias effectively to reduce the loss. This is where partial

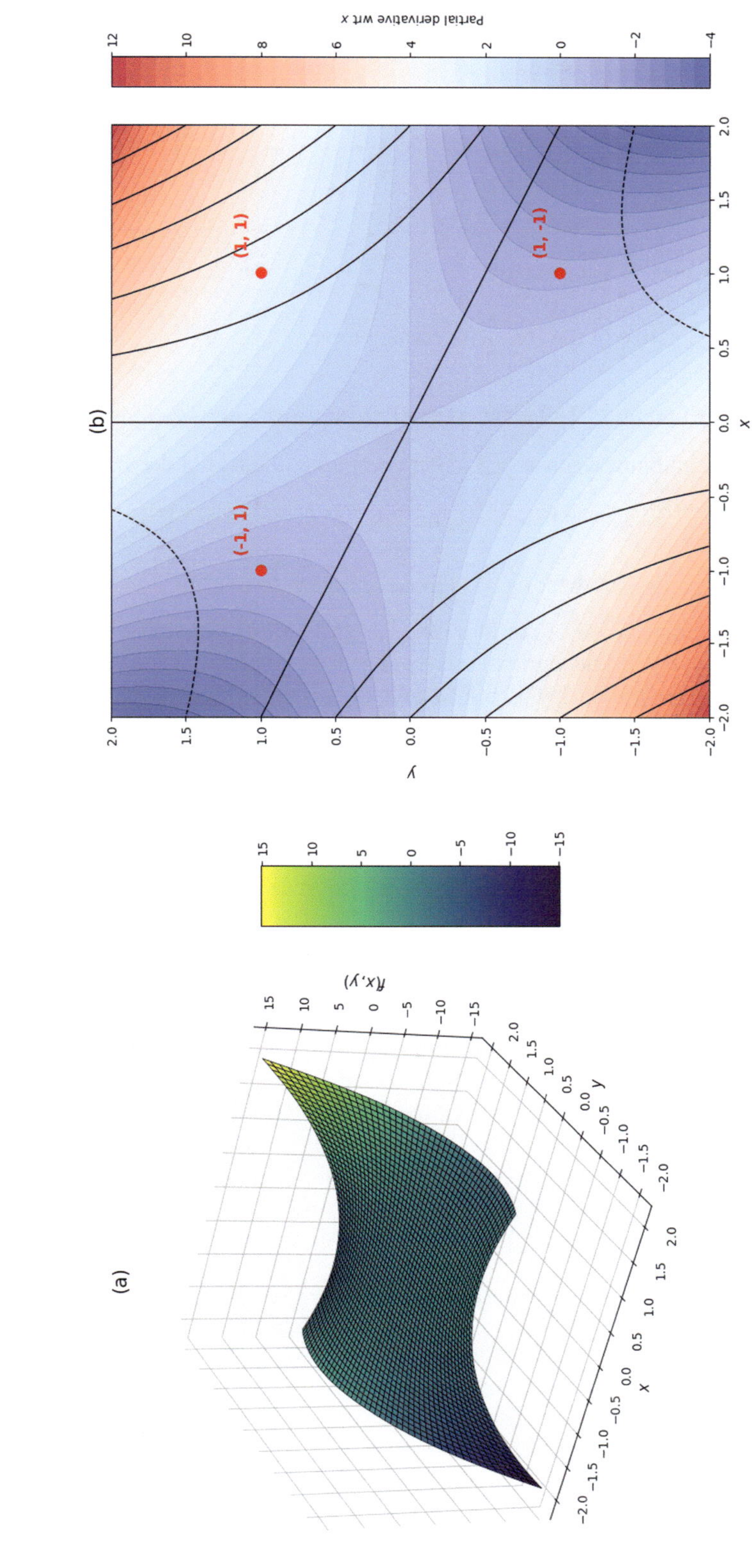

FIGURE 3.1 (a) Surface plot of $f(x,y) = x^2 y + xy^2$ and (b) contours of partial derivatives.

derivatives come into play. It has two features of changes: direction and magnitude. In gradient descent, the weight update rule is as follows:

$$w^{(t+1)} = w^{(t)} - \eta \frac{dL}{dw}$$

where:

- $\mathbf{w^{(t)}}$ is the weight at iteration **t**,
- $\boldsymbol{\eta}$ is the learning rate,
- $\frac{dL}{dw}$ is the partial derivative of the loss with respect to **w**.

Consider a quadratic loss function $L(w) = (w-3)^2$, where **w** represents a weight in the model. This loss function reaches its minimum when **w** = 3. After computing the partial derivative $\frac{dL}{dw} = 2(w-3)$ the sign of this derivative tells us whether to increase or decrease **w**. If **w** = 4, the derivative $\frac{dL}{dw} = 2(4-3)$, which is positive, indicating we should reduce **w** to minimize the loss.

Figure 3.2 illustrates the behavior of the loss function and how its derivative indicates the direction of the steepest descent toward the optimal weight. The first subplot on the left displays the quadratic loss function $L(w) = (w-3)^2 + 2$, which measures the error or "loss" associated with different values of the weight **w**. This plot shows a U-shaped curve where the minimum point represents the optimal weight. A vertical dashed gray line at **w** = 3 indicates the optimal weight, where the loss function reaches its minimum value. The second subplot on the right displays the derivative of the loss function, which is given by indicating the slope of the loss function. This derivative is crucial for optimization algorithms like gradient descent. It shows a straight line that crosses the x-axis at **w** = 3, where the derivative is zero, corresponding to the optimal weight. A horizontal dashed gray line at **y** = 0 indicates where the derivative is zero.

3.4 GRADIENTS

In multivariable calculus, the gradient is a central concept, especially when dealing with functions of several variables. The gradient provides a way to encapsulate the rates of change of a function in every direction in a single vector. By pointing in the direction of the steepest ascent and having a magnitude representing the maximum increase rate, the gradient offers an understanding of how a function changes in different directions. Given a scalar-valued function **f** of several variables, such as **f(x, y)** or **f(x, y, z)**, the gradient of **f**, denoted by ∇f or "grad **f**," is a vector whose components are the partial derivatives of **f** concerning each of its variables. For a function **f(x, y)**, the gradient is $\nabla f = \left(\frac{\partial f}{\partial x}, \frac{\partial f}{\partial y}\right)$. For a function **f(x, y, z)**, the gradient is $\nabla f = \left(\frac{\partial f}{\partial x}, \frac{\partial f}{\partial y}, \frac{\partial f}{\partial z}\right)$, and so on for functions with more variables. Its geometric interpretation includes the following:

(a) *Direction:* The gradient of a function indicates the direction of the steepest ascent. If you imagine standing on a hill represented by the function **f** and walking toward the gradient, you will climb the mountain as steeply as possible.
(b) *Magnitude:* The gradient vector's magnitude (or length) represents the function's maximum increase rate. If the gradient is zero at a point, then that point is a local maximum, local minimum, or a saddle point.

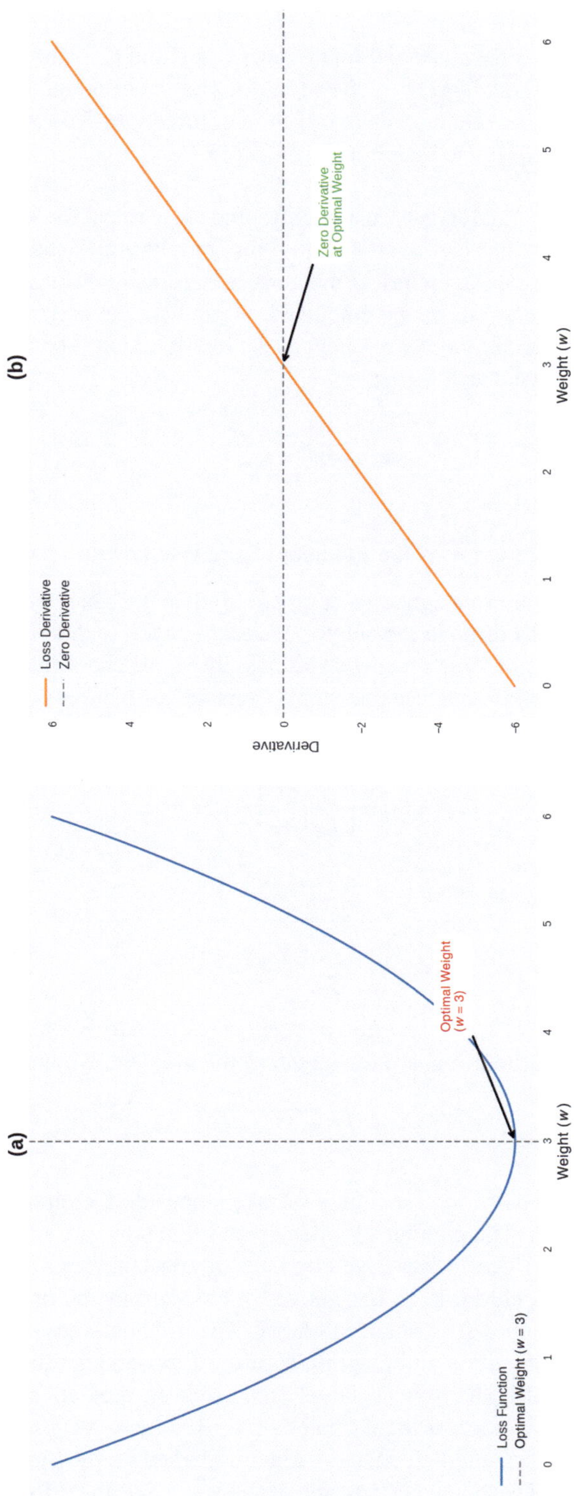

FIGURE 3.2 (a) Loss function and (b) derivative of the loss function.

Gradient has two main properties:

1. *Linearity:* For scalar constants **a** and **b** and functions **F** and **G**: $\nabla(af + bg) = a\nabla f + b\nabla g$
2. *Dot Product with Directional Derivative:* The dot product of the gradient of **f** at a point and a unit vector **u** gives the rate of change of **f** in your direction. This is also known as the directional derivative of **F** in the direction $D_{\vec{u}} f = \nabla f \cdot \vec{u}$.

Gradients can be computed analytically using differentiation rules for each partial derivative. However, analytical computations can be challenging for complicated functions, especially in high-dimensional spaces. In such cases, numerical methods or automatic differentiation tools (standard in machine learning frameworks) might be employed. In gradient descent, the weights are updated iteratively in the opposite direction of the gradient of the loss function. The update rule for a weight **w** at iteration **t** can be expressed as follows:

$$w_{t+1} = w_t - \eta \frac{\partial L}{\partial w}$$

where $\boldsymbol{\eta}$ is the learning rate and $\frac{\partial L}{\partial w}$ is the gradient of the loss **L** with respect to **w**. Additionally, backpropagation is an algorithm designed to compute gradients efficiently in neural networks. It propagates the error backward through the network, layer by layer, to determine the gradient of the loss with respect to each weight. This process relies heavily on the chain rule of calculus to decompose the gradient calculation into manageable steps. Consider a function $f(x, y) = x^2 + y^2$, the gradient of **f** is:

$$\nabla f(x, y) = \left(\frac{\partial f}{\partial x}, \frac{\partial f}{\partial y} \right) = (2x, 2y)$$

At the point **(x, y)** = (1, 2), the gradient is:

$$\nabla f(1,2) = (2 \times 1, 2 \times 2) = (2,4)$$

This tells us that at the point (1, 2), the function increases most rapidly in the direction (2, 4), and the steepness of the increase is proportional to the length of the gradient, which is:

$$\nabla f(1,2) = \sqrt{2^2 + 4^2} = \sqrt{20} \approx 4.47$$

Figure 3.3 illustrates the shape of the function's surface alongside the direction and magnitude of its gradient at various points. The contour plot depicts the function $f(x, y) = x^2 + y^2$, representing a symmetrical paraboloid surface characterized by lines of constant function value. These contour lines indicate the function's elevation as one moves outward from the origin, with labels specifying their corresponding values for enhanced clarity. The gradient vectors are visualized as red arrows, which embody the direction and magnitude of the function's gradient ∇f(x, y) = (2x, 2y). The length and orientation of each arrow accurately represent the gradient's strength and direction, effectively demonstrating the steepest ascent paths of the function at those locations. Key points such as (1, 1), (−1, −1), (1, −1), and (−1, 1) are distinctly marked on the plot with blue annotations and corresponding arrows. These annotations are strategically positioned to prevent overlap with other plot elements, ensuring that each key point is easily identifiable and their associated gradient vectors are clearly understood. Additionally, black lines are drawn along the **y** = 0 and **x** = 0 axes to

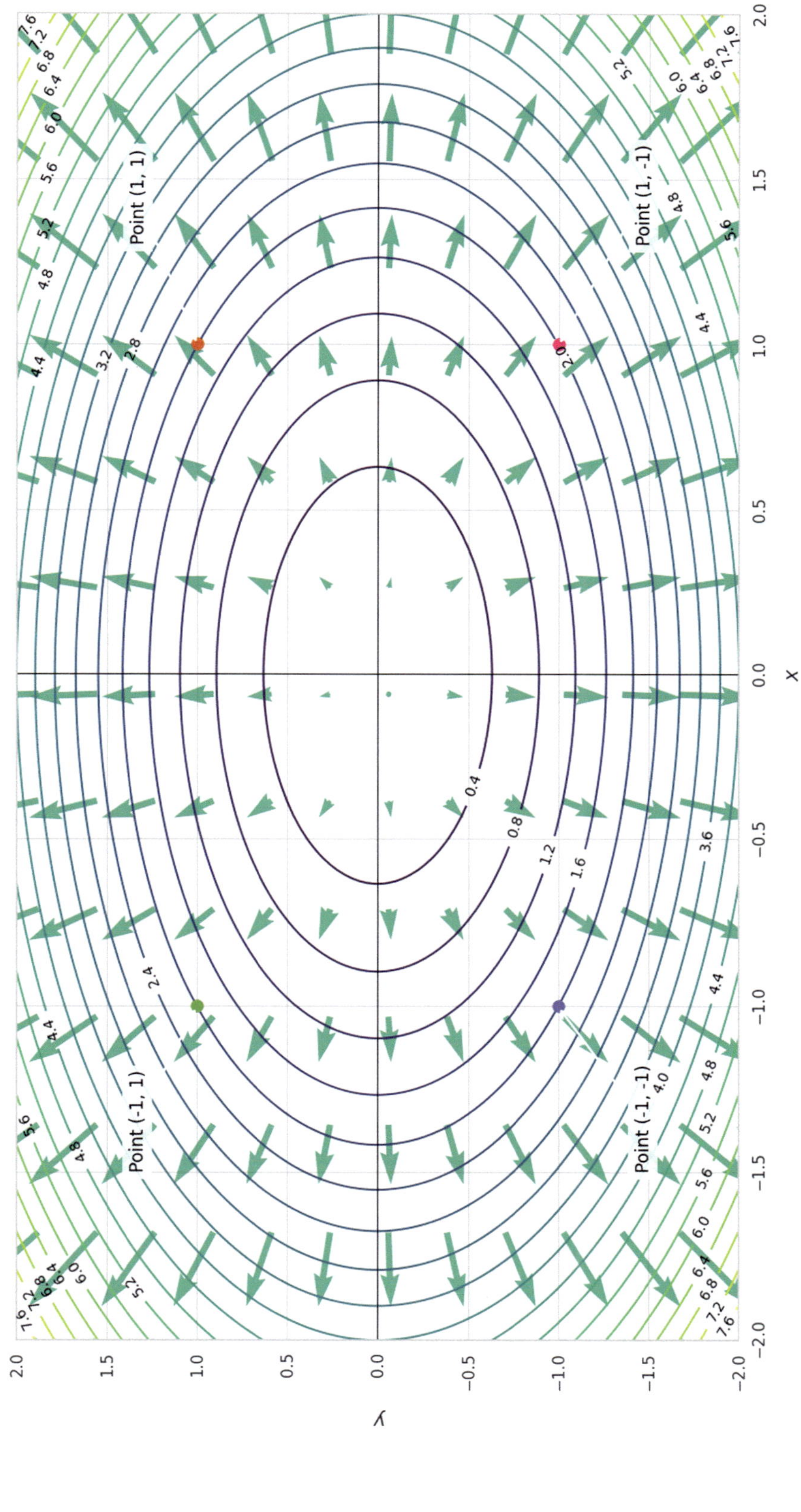

FIGURE 3.3 Contour plot $f(x,y) = x^2 + y^2$ with gradient vectors.

denote the origin, aiding in spatial orientation and providing a reference framework for interpreting the plot's elements.

3.5 GRADIENT IN DEEP LEARNING

In deep learning, neural networks are trained for classification and regression by fine-tuning their weights and biases to minimize a specified loss (or cost) function. Gradients are the driving force behind the training of deep neural networks. By understanding how the loss changes concerning model parameters, we can iteratively adjust these parameters to improve the model's performance. Despite the challenges associated with gradient calculations, the proper techniques and tools make it possible to harness the power of gradients effectively, maintaining their status as a cornerstone concept in deep learning. The gradient plays several critical roles in deep learning. One key role is in training a neural network, where the goal is to minimize the loss function by updating the network's weights and biases to reduce the loss. The gradient of the loss function with respect to each weight and bias indicates both the direction and magnitude of the adjustments needed to decrease the loss. Before delving into gradients, it is crucial to understand the role of the loss function in neural networks. The loss function quantifies how well the neural network's predictions align with the data. The objective of training a neural network is to minimize this loss function. To achieve this, we must understand how the loss changes for the model's parameters (weights and biases). This is where the gradient comes into play. The gradient of the loss function concerning the model's parameters shows how the loss function changes in direction and magnitude when these parameters are adjusted. Specifically, for a neural network with thousands (or even millions) of parameters, the gradient is a high-dimensional vector where each component represents the partial derivative of the loss function with respect to one of these parameters. Each partial derivative indicates how much the loss will increase or decrease if the corresponding parameter is adjusted by a small amount. There are several challenges for gradient in deep learning, here we review some of the most common and their possible solution.

(a) *Vanishing Gradients:* In deep networks, gradients can become very small as they propagate through the layers, leading to slow learning. Techniques like normalization, appropriate activation functions, and advanced architectures like long short-term memory networks help mitigate this issue. Using ReLU (Rectified Linear Unit) or its variants can mitigate the vanishing gradient problem.
(b) *Exploding Gradients*: Conversely, gradients can become excessively large, causing unstable updates. Gradient clipping is widely used to tackle this issue by limiting the gradients to a maximum value. Gradient clipping involves setting a threshold value and scaling down gradients that exceed this threshold, preventing the exploding gradient problem.
(c) *Saddle Points*: Points where the gradient is zero but is neither a minimum nor a maximum of the loss function can cause neural networks to get stuck, slowing down the training process. Algorithms like Adam, RMSprop, and Adagrad adjust the learning rate dynamically and consider past gradients to provide more stable and faster convergence than vanilla gradient descent.

3.5.1 Gradient Descent

Gradient descent is an optimization algorithm that minimizes the loss function in neural networks. The idea is simple, iteratively adjust each parameter in the opposite direction of its corresponding gradient component, aiming to reduce the loss function. The update rule for gradient descent is:

$$\theta^{(t+1)} = \theta^{(t)} - \eta \nabla_{\theta} J\left(\theta^{(t)}\right)$$

where

- $\boldsymbol{\theta}$ is the parameter vector,
- $\boldsymbol{\eta}$ is the learning rate,
- $\nabla_{\theta} J\left(\theta^{(t)}\right)$ is the gradient of the loss function with respect to $\boldsymbol{\theta}$ at iteration **t**.

The variations differ in how much data is used to compute the gradient, affecting the speed and efficiency of the optimization process. Suppose you are training a neural network with 100,000 parameters, and the goal is to minimize the loss function. The dataset contains 1,000 training samples. Variations of gradient descent include:

(a) *Stochastic Gradient Descent (SGD)*: In this case, after processing each individual data sample, the network updates its parameters. So, if you process the first sample, calculate the loss, and compute the gradient, the parameters are updated immediately, and then you proceed to the next sample. For example, if the current parameter value is $\boldsymbol{\theta}^{(t)} = 0.5$, the learning rate is $\boldsymbol{\eta} = 0.01$, and the gradient for a particular parameter is $\boldsymbol{\nabla\theta} = 0.2$, the update would be: $\theta^{(t+1)} = 0.5 - 0.01 \times 0.2 = 0.498$.

(b) *Mini-Batch Gradient Descent*: In mini-batch gradient descent, the model updates its parameters after processing small, manageable batches of data. For example, if the batch size is 32 samples, the model processes these 32 samples, computes the average gradient for the batch, and updates the parameters accordingly. This approach balances the trade-off between the efficiency of SGD and the stability of batch gradient descent, leading to faster convergence and smoother updates.

(c) *Batch Gradient Descent:* In batch gradient descent, the model processes the entire dataset (e.g., all 1,000 samples) before updating the parameters. After computing the gradient for the entire dataset, the parameters are updated in one large step. This method provides a more stable gradient estimate, but it is computationally expensive, especially for large datasets, as it requires going through all data points before making an update.

3.5.2 Backpropagation

The backpropagation algorithm is essential for determining the gradient of the loss function relative to each network parameter. It operates through two primary phases:

1. *Forward Propagation:* Input data is processed layer by layer to produce the network's output.
2. *Reverse Propagation:* By applying the chain rule, the gradient of the loss function concerning each parameter is calculated, starting from the output layer and working backward to the input layer.

The gradient of the loss function **L** with respect to a weight **w** in layer **l** is computed using the chain rule:

$$\frac{\partial L}{\partial w_l} = \frac{\partial L}{\partial a_l} \cdot \frac{\partial a_l}{\partial z_l} \cdot \frac{\partial z_l}{\partial w_l}$$

where:

- $\mathbf{a}_\mathbf{l}$ is the activation at layer **l**,
- $\mathbf{z}_\mathbf{l}$ is the pre-activation (linear combination of weights and inputs) at layer **l**.

Backpropagation efficiently computes gradients for all parameters in the network, making it fundamental for training deep neural networks. Suppose you're training a neural network with three layers to classify images. During forward propagation, a sample image passes through the network, and the model predicts a class based on its current parameters. For example, if the true class label is 2 and the network predicts class 3, a loss function (such as cross-entropy) is used to measure the error between the predicted class and the true class. During reverse (backward) propagation, the algorithm calculates the gradient of the loss with respect to the parameters (weights and biases), starting from the output layer and propagating back through the three layers to the input. Using the chain rule of calculus, backpropagation computes how each weight and bias contributes to the loss. These gradients are then used to update the parameters in the opposite direction of the gradient (through gradient descent), reducing the loss and improving the model's predictions. This process is repeated for multiple samples in the training set, gradually fine-tuning the network's parameters.

Figure 3.4 illustrates the behavior of a quadratic loss function and the corresponding gradient descent optimization process used to minimize it. In the subplot (a), the loss function $L(w) = (w - 3)^2 + 2$ is displayed as a smooth blue curve. This parabolic curve depicts how the loss varies with different weight values **w**, reaching its minimum at the optimal weight **w** = 3. The Gradient Descent Path is illustrated with a dashed orange line and highlighted with large red markers, tracing the iterative steps taken from an initial weight of **w** = 0 toward the minimum. Each red marker represents an updated weight after applying the gradient descent update rule with a learning rate of 0.1, showcasing the step-by-step approach toward minimizing the loss. A vertical dashed green line marks the optimal weight (**w** = 3), providing a clear reference point for the goal of the optimization. The subplot (b) focuses on the gradient of the loss function $\nabla L(w) = 2(w - 3)$ depicted as a robust orange line. This linear function illustrates how the gradient changes with different values of **w**, crossing zero precisely at the optimal weight of 3. The gradient steps are marked with purple scatter points, corresponding to the gradient values at each step of the gradient descent process. These points visually demonstrate how the gradient diminishes as the weight approaches the optimal value, guiding the descent toward the minimum loss. A horizontal dashed gray line indicates the zero gradient level, providing a reference for where the gradient equals zero, signifying the optimal point.

3.6 JACOBIANS

Jacobian is a fundamental concept in various branches of mathematics, including calculus, differential equations, and geometry. It is essential in multivariate calculus and has numerous applications in theoretical and applied mathematics. The Jacobian matrix of a system of transformation equations is a matrix of the first-order partial derivatives of the functions involved. For a vector-valued function/mapping it can be $\mathbf{f} : \mathbb{R}^n \to \mathbb{R}^m$, $\mathbf{f}(x) = \begin{bmatrix} f_1(x) \\ f_2(x) \\ \vdots \\ f_m(x) \end{bmatrix}$ $\mathbb{R}^n$ *to* $\mathbb{R}^m$, the Jacobian matrix is $m \times n$, defined as follows:

$$J(f) = \begin{pmatrix} \frac{\partial f_1}{\partial x_1} & \frac{\partial f_1}{\partial x_2} & \cdots & \frac{\partial f_1}{\partial x_n} \\ \frac{\partial f_2}{\partial x_1} & \frac{\partial f_2}{\partial x_2} & \cdots & \frac{\partial f_2}{\partial x_n} \\ \vdots & \vdots & \ddots & \vdots \\ \frac{\partial f_m}{\partial x_1} & \frac{\partial f_m}{\partial x_2} & \cdots & \frac{\partial f_m}{\partial x_n} \end{pmatrix}$$

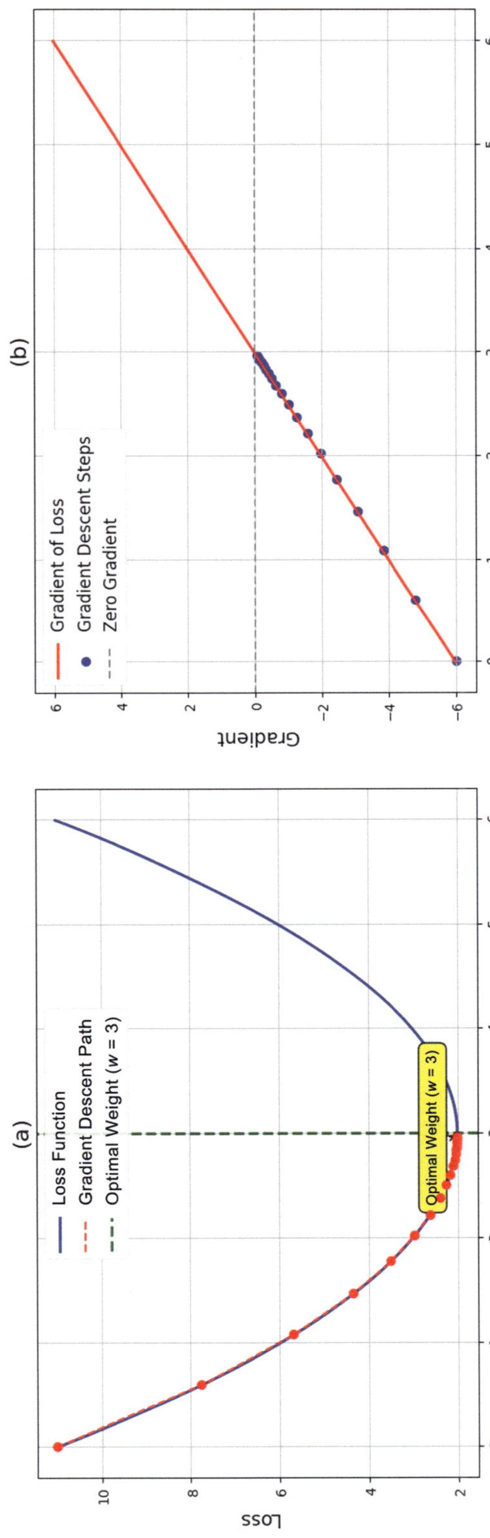

FIGURE 3.4 (a) Loss function and gradient descent path, (b) gradient of the loss function.

Here, F_i represents the $\mathbf{i^{th}}$ component of the vector-valued function F. The Jacobian matrix has several critical applications. In optimization, the Jacobian provides information about the gradient and direction to move toward or away from a local optimum, aiding in the optimization process.

3.6.1 Jacobian Determinant

The determinant of a Jacobian matrix is called the Jacobian determinant or simply the Jacobian. For a function that maps from $\mathbb{R}^n \to \mathbb{R}^n$, the Jacobian determinant indicates how the function scales areas or volumes in the neighborhood of a point. If the Jacobian determinant at a point is positive, the function preserves orientation near that point; if it is negative, the function reverses orientation. For a transformation function $f : \mathbb{R}^n \to \mathbb{R}^n$, the Jacobian determinant is:

$$\det(J) = \det\left(\frac{\partial(f_1, f_2, \ldots, f_n)}{\partial(x_1, x_2, \ldots, x_n)}\right)$$

The Jacobian determinant informs how the function scales areas or volumes locally and whether it preserves or reverses orientation. Consider a transformation function that maps from $\mathbb{R}^2 to\, \mathbb{R}^2$. The function transforms the coordinates **(x, y)** to new coordinates **(u, v)** using the equations:

$$u = 2x + y, \quad v = x - y$$

The Jacobian matrix is:

$$J = \begin{pmatrix} 2 & 1 \\ 1 & -1 \end{pmatrix}$$

The Jacobian determinant is calculated as follows:

$$\det(J) = (2 \times -1) - (1 \times 1) = -2 - 1 = -3$$

As the determinant is negative, the transformation reverses orientation near the point.

3.6.2 Relationship with the Chain Rule

The Jacobian matrix is closely related to the chain rule in calculus. The chain rule allows us to find the derivative of a composite function, and when working with tasks from $\mathbb{R}^n$to $\mathbb{R}^m$, it involves the Jacobian matrix. Specifically, if **f** and **g** are functions such that $f : \mathbb{R}^n \to \mathbb{R}^p$ and $\mathbb{R}^p \to \mathbb{R}^m$, then the Jacobian matrix of the composite function $h = g^\circ f$ is the product of the Jacobian matrices of **g** and **f**. If $f : \mathbb{R}^n \to \mathbb{R}^p$ and $g : \mathbb{R}^p \to \mathbb{R}^m$, the Jacobian of the composite function $h = g(f(x))$ is:

$$J_h(x) = J_g(f(x)) \cdot J_f(x)$$

The chain rule in multi-dimensional calculus uses this product of Jacobians to compute the derivative of composite functions. Suppose you have two functions:

1. f(x, y) = (2x + y, x − y)
2. g(u, v) (where **u** and **v** are the outputs of f(x, y))

To find the derivative of the composite function g(f(x, y)), you need to use the chain rule for multivariable functions, which involves multiplying the Jacobian matrices of **g** and **f**. Let us do the process step by step:

1. *Compute the Jacobian of **f(x, y)***: The Jacobian matrix of f(x, y) is the matrix of partial derivatives of each component of **f** with respect to **x** and **y**:

$$J_f(x,y) = \begin{pmatrix} \frac{\partial(2x+y)}{\partial x} & \frac{\partial(2x+y)}{\partial y} \\ \frac{\partial(x-y)}{\partial x} & \frac{\partial(x-y)}{\partial y} \end{pmatrix} = \begin{pmatrix} 2 & 1 \\ 1 & -1 \end{pmatrix}$$

2. *Compute the Jacobian of **g(u, v)**:* Suppose g(u, v) is a function of **u** and **v**, you need to compute its Jacobian with respect to **u** and **v**:

$$J_g(u,v) = \begin{pmatrix} \frac{\partial g_1}{\partial u} & \frac{\partial g_1}{\partial v} \\ \frac{\partial g_2}{\partial u} & \frac{\partial g_2}{\partial v} \end{pmatrix}$$

3. *Apply the Chain Rule:* The chain rule for multivariable functions states that the Jacobian of the composite function g(f(x, y)) is the product of the Jacobian matrices of **g** and **f**:

$$J_{g(f)}(x,y) = J_g(u,v) \cdot J_f(x,y)$$

3.6.3 Computational Aspects

In computational fields, the Jacobian is often used in numerical methods, such as Newton's Method. This technique finds the roots of equations with multiple variables, utilizing the Jacobian matrix to approach the solution iteratively. The Jacobian assists in linearizing complex systems, making numerical approximation methods more efficient. Newton's Method for systems of equations uses the Jacobian matrix to update guesses iteratively:

$$\boldsymbol{x}^{(n+1)} = \boldsymbol{x}^{(n)} - J^{-1}\left(\boldsymbol{x}^{(n)}\right) \cdot \boldsymbol{f}\left(\boldsymbol{x}^{(n)}\right)$$

where

- $\mathbf{x}^{(n)}$is the current guess,
- **J** is the Jacobian matrix of the system,
- **f** represents the system of equations.

This iterative process converges to the roots, making it effective for solving non-linear systems. Suppose you want to find the roots of the system of equations:

$$f_1(x,y) = x^2 + y^2 - 4 = 0, \quad f_2(x,y) = x - y = 0$$

Using Newton's Method, you start with an initial guess, say $(\mathbf{x_0}, \mathbf{y_0}) = (1, 1)$, and iteratively apply the method to get closer to the solution. The Jacobian matrix for this system is:

$$J = \begin{pmatrix} 2x & 2y \\ 1 & -1 \end{pmatrix}$$

This matrix is used in the iterative formula to update the guesses for **x** and **y**.

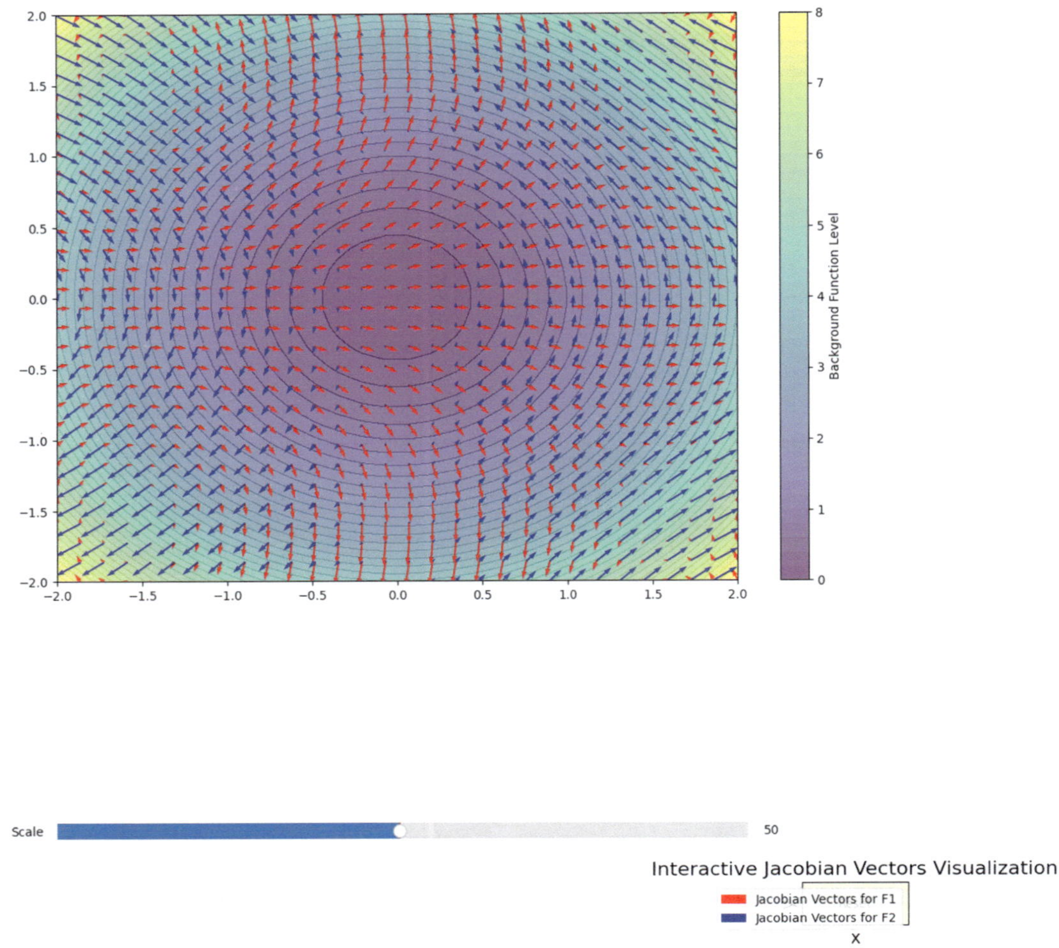

FIGURE 3.5 Visualization of Jacobian vectors for the function $F(x,y) = [xcos(y), ysin(x)]$.

Figure 3.5 shows the visualization of the Jacobian vectors of a vector-valued function **F(x, y)**. The function $F(x,y) = [xcos(y), ysin(x)]^{T}$ is depicted along with its Jacobian matrix, which is calculated as $J_F(x,y)$. The figure is divided into two main components: a contour plot and Jacobian vectors. The background contour plot displays a quadratic function $Z = x^2 + y^2$, providing a visual reference for the domain over which the Jacobian vectors are computed, with filled contours using the colormap and an accompanying color bar indicating the function levels. Superimposed on the contour plot are red and blue arrows representing the Jacobian vectors corresponding to the components of **F**. The red arrows depict the partial derivatives associated with them, while the blue arrows depict those associated with them.

3.7 JACOBIANS IN DEEP LEARNING

In the context of deep learning, Jacobians play a crucial role in understanding how small changes in input can affect the output of a neural network model. This is especially important in scenarios where the sensitivity and responsiveness of a model to inputs are critical. In deep learning, a neural network functions as a mapping between input vectors and output vectors. The Jacobian matrix details the rate at which each output changes in response to variations in each input. Specifically, for a network that has **m** outputs and **n** inputs, the Jacobian is a matrix. Formally, if $y = \mathbf{F}(x)$ where $x \in \mathbb{R}^n$ and $y \in \mathbb{R}^m$, the Jacobian matrix **J** is given by

$$J(f) = \begin{pmatrix} \frac{\partial y_1}{\partial x_1} & \frac{\partial y_1}{\partial x_2} & \cdots & \frac{\partial y_1}{\partial x_n} \\ \frac{\partial y_2}{\partial x_1} & \frac{\partial y_2}{\partial x_2} & \cdots & \frac{\partial y_2}{\partial x_n} \\ \vdots & \vdots & \ddots & \vdots \\ \frac{\partial y_m}{\partial x_1} & \frac{\partial y_m}{\partial x_2} & \cdots & \frac{\partial y_m}{\partial x_n} \end{pmatrix}$$

The Jacobian matrix is a valuable tool for analyzing the sensitivity of a neural network's outputs to input variations. This sensitivity analysis is crucial in adversarial attacks, where minor input alterations can cause substantial output changes. By understanding this sensitivity, we can enhance the robustness and reliability of the model. While the Jacobian is not directly used in backpropagation, the concept is closely related. Backpropagation relies on the gradient to adjust weights and biases in a network. This gradient quantifies how the loss function changes as you adjust the input or weights, like how the Jacobian quantifies the output change concerning inputs.

In adversarial machine learning, attackers often use the Jacobian matrix to generate adversarial examples. By understanding how changes in input affect output, attackers can make slight modifications to an input sample to fool a neural network into misclassifying it. This understanding is crucial for developing defenses against such attacks. The Jacobian can be useful for visualizing how changes in specific inputs affect outputs. This visualization aids in interpreting how a deep learning model functions and helps debug or improve model behavior. By examining the Jacobian, researchers can gain insights into the model's internal workings and response to varying inputs. Recent research in deep learning involves regularizing the Jacobian to ensure that the model is not overly sensitive to input perturbations. Regularizing the Jacobian can help make the model more robust, improving its performance on real-world data where slight variations are common. The Jacobian matrix can be massive for large neural networks, making computation and storage computationally challenging. In many practical scenarios, approximations or sampling methods might be used to manage these challenges effectively. Despite these limitations, understanding and utilizing the Jacobian remains a critical aspect of advanced neural network training and analysis. Consider a simple neural network with two inputs $\mathbf{x}_1$, $\mathbf{x}_2$ and two outputs $\mathbf{y}_1$, $\mathbf{y}_2$, where the output is calculated as follows:

$$y_1 = 2x_1 + 3x_2, \quad y_2 = x_1^2 + 4x_2$$

The Jacobian matrix for this network is:

$$\boldsymbol{J} = \begin{bmatrix} \frac{\partial y_1}{\partial x_1} & \frac{\partial y_1}{\partial x_2} \\ \frac{\partial y_2}{\partial x_1} & \frac{\partial y_2}{\partial x_2} \end{bmatrix} = \begin{bmatrix} 2 & 3 \\ 2x_1 & 4 \end{bmatrix}$$

At the point $(\mathbf{x}_1, \mathbf{x}_2) = (1, 2)$, the Jacobian becomes:

$$\boldsymbol{J}(1,2) = \begin{bmatrix} 2 & 3 \\ 2\times 1 & 4 \end{bmatrix} = \begin{bmatrix} 2 & 3 \\ 2 & 4 \end{bmatrix}$$

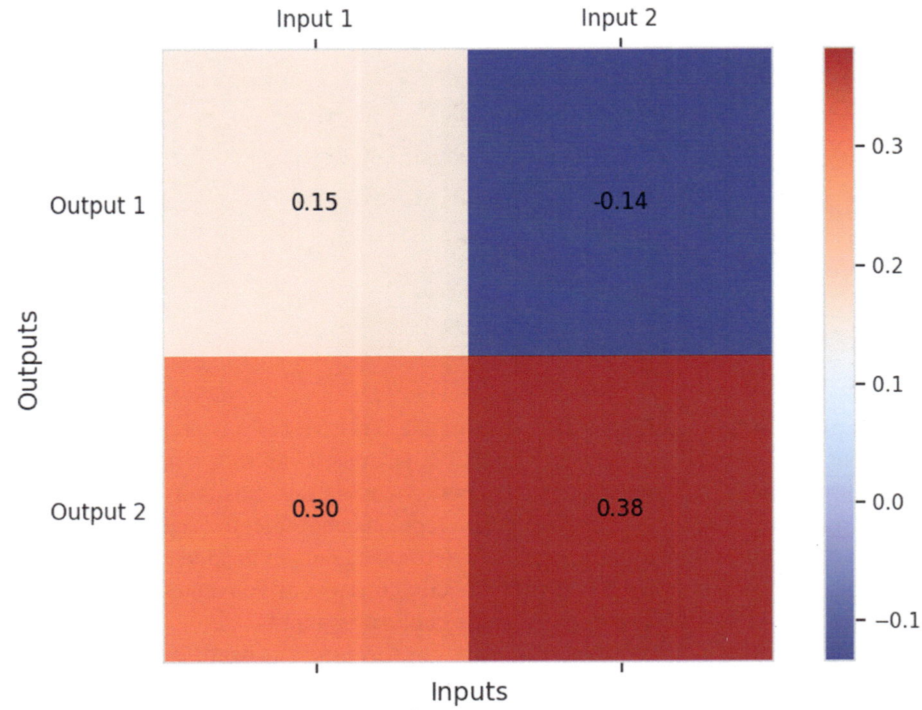

FIGURE 3.6 Visualization of the Jacobian matrix for a simple neural network.

This matrix shows how small changes in $\mathbf{x}_1$ and $\mathbf{x}_2$ affect $\mathbf{y}_1$ and $\mathbf{y}_2$. For instance, increasing $\mathbf{x}_1$ by 1 unit increases $\mathbf{y}_1$ by 2 units and $\mathbf{y}_2$ by 2 units, while increasing $\mathbf{x}_2$ by 1 unit increases $\mathbf{y}_1$ by 3 units and $\mathbf{y}_2$ by 4 units.

Figure 3.6 presents a visualization of the Jacobian matrix of a simple neural network. The Jacobian matrix, in this context, represents the partial derivatives of the network's outputs with respect to its inputs. Each cell in the matrix indicates the sensitivity of an output to a change in a specific input. The color gradient in the matrix helps to visually differentiate the magnitude and direction of the sensitivities. Warmer colors (reds) indicate higher positive values, while cooler colors (blues) represent higher negative values. Neutral values are closer to white. Each cell is annotated with the exact value of the partial derivative it represents. The x-axis is labeled with the inputs (Input 1 and Input 2), and the y-axis is labeled with the outputs (Output 1 and Output 2). By examining the Jacobian matrix, you can understand how changes in each input variable will affect each output variable. For example, a high positive value in a cell indicates that a small increase in the corresponding input will result in a significant increase in the corresponding output.

3.8 HESSIAN MATRICES

While the Hessian matrix provides valuable insights into the curvature of a function and can influence optimization strategies, its direct application in deep learning is limited by practical computational challenges. However, methods that leverage Hessian information, either directly or indirectly, can be crucial in specific deep learning scenarios. The Hessian matrix, a square matrix, showcases the second-order partial derivatives of a scalar function. For a function$\mathbb{R}^n \to \mathbb{R}$, the Hessian $H(f)$ is given by:

$$H(f)=\begin{pmatrix} \frac{\partial^2 f}{\partial x_1^2} & \frac{\partial^2 f}{\partial x_1 \partial x_2} & \cdots & \frac{\partial^2 f}{\partial x_1 \partial x_n} \\ \frac{\partial^2 f}{\partial x_2 \partial x_1} & \frac{\partial^2 f}{\partial x_2^2} & \cdots & \frac{\partial^2 f}{\partial x_2 \partial x_n} \\ \vdots & \vdots & \ddots & \vdots \\ \frac{\partial^2 f}{\partial x_n \partial x_1} & \frac{\partial^2 f}{\partial x_n \partial x_2} & \cdots & \frac{\partial^2 f}{\partial x_n^2} \end{pmatrix}$$

The importance of Hessian matrices include:

(a) *Convexity and Concavity:* A function is locally convex if the Hessian is positive definite at a point. If the Hessian is negative definite, the function is locally concave. This property is critical in optimization, as it helps identify the nature of critical points, distinguishing between local minima, maxima, or saddle points.
(b) *Second-order Optimization Methods:* The Hessian is used in second-order optimization methods, such as Newton's Method, which can achieve faster convergence than first-order methods like gradient descent under certain conditions. However, in deep learning, computing the Hessian is often computationally expensive due to the large number of parameters.
(c) *Approximations:* Quasi-Newton methods approximate the Hessian. These methods are used in optimization scenarios where computing the actual Hessian might be too costly.

Consider a simple function $f(x,y)=x^2+y^2$. The Hessian matrix of this function, representing the second-order partial derivatives, is:

$$H=\begin{pmatrix} \frac{\partial^2 f}{\partial x^2} & \frac{\partial^2 f}{\partial x \partial y} \\ \frac{\partial^2 f}{\partial y \partial x} & \frac{\partial^2 f}{\partial y^2} \end{pmatrix}=\begin{pmatrix} 2 & 0 \\ 0 & 2 \end{pmatrix}$$

As the Hessian matrix is positive definite (both eigenvalues are positive), the function is locally convex, and any critical point is a local minimum. The Hessian helps determine the curvature of the function, which influences the optimization strategy by providing insights into local minima, maxima, and saddle points.

Figure 3.7 shows the behavior of the function $f(x,y)=x^2+y^2+2xy$ and its Hessian matrix. The left subplot features a 3D surface plot of the function, displaying its values over a grid of **x** and **y** ranging from –2 to 2. The axes are labeled **x**, **y**, and **f(x, y)**, and a color bar accompanies the plot to show the corresponding function values. The right subplot presents the Hessian matrix of the function at the specific point **(x, y)** = (1, 1). For $f(x,y)=x^2+y^2+2xy$, the Hessian matrix at this point is $\begin{pmatrix} 2 & 2 \\ 2 & 2 \end{pmatrix}$. The heatmap visualizes this matrix using the "Blues" colormap, with each cell's color reflecting the value of the corresponding second-order partial derivative.

3.9 HESSIAN IN DEEP LEARNING

While direct computation and use of the Hessian in deep learning training are rare due to its high computational and storage costs, its properties and associated concepts are precious. They provide

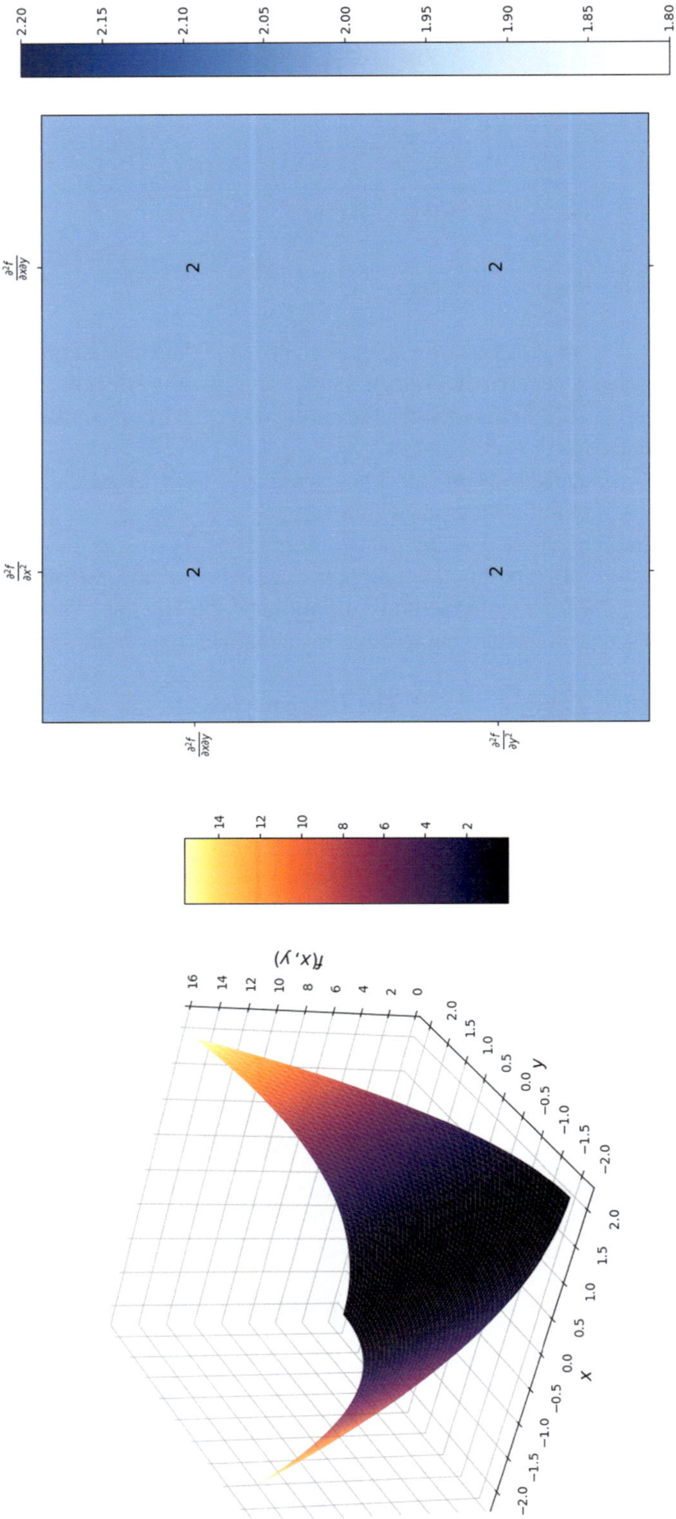

FIGURE 3.7 Visualization of the Hessian matrix for the function $f(x,y) = x^2 + y^2 + 2xy$.

insights into model optimization, robustness, and generalization and influence the development of advanced optimization algorithms tailored for deep learning. Research suggests that the eigenvalues of the Hessian can provide insights into model generalization. Flatter minima (where the Hessian has smaller eigenvalues) might correspond to better generalization, as they indicate less sensitivity to small perturbations in the input data. Regularizing the Hessian can help prevent overfitting, leading to models that better generalize unseen data. The Hessian's properties can be leveraged to understand model robustness, especially in adversarial attacks. The curvature information can provide insights into the model's susceptibility to perturbations in the input space. By studying the Hessian, researchers can develop better defense mechanisms against adversarial attacks, enhancing the robustness and reliability of deep learning models. Given the high computational cost of the Hessian in deep networks, Hessian-free methods aim to leverage its second-order information without explicitly computing it. Here are some of its features:

(a) *High-Dimensional, Non-Convex Landscapes:* Deep learning involves training models on high-dimensional, non-convex optimization landscapes. This makes the training process susceptible to challenges like getting stuck in saddle points.
(b) *Curvature Information:* The Hessian matrix provides second-order information about the curvature of the loss surface. The eigenvalues of the Hessian can indicate whether a point is a minimum, maximum, or saddle point. Specifically, if all eigenvalues are positive, it is a local minimum; if all are negative, it is a local maximum; and if there is a mix, it is a saddle point.
(c) *Saddle Points:* In deep learning, saddle points are more common than local minima, making it crucial to understand the role of the Hessian in navigating such points. Understanding the Hessian can help develop strategies to escape saddle points and improve training efficiency.

For example, consider that we are training a neural network with a loss function $\mathbf{L(w)}$, where $\mathbf{w}$ represents the weight parameters of the network. Suppose at a certain point in the weight space, the Hessian matrix has the following eigenvalues:

$$\lambda_1 = 0.01, \quad \lambda_2 = 0.1, \quad \lambda_3 = 0.5, \quad \lambda_4 = -0.2$$

The negative eigenvalue $\boldsymbol{\lambda}_4 = -0.2$ indicates the presence of a saddle point, meaning the training process might slow down or get stuck at this point. The small positive eigenvalues $\boldsymbol{\lambda}_1 = 0.01$, $\boldsymbol{\lambda}_2 = 0.1$ suggest a flat region in the loss surface, meaning progress may be slow along these directions. The larger positive eigenvalue $\boldsymbol{\lambda}_3 = 0.5$ indicates that in this direction, the curvature is steeper, meaning changes in this direction will more rapidly impact the loss.

In Figure 3.8, the Hessian matrix of a simple neural network is visualized to provide insights into the second-order derivatives of the network's loss function with respect to its inputs. The neural network consists of a single fully connected layer that maps a two-dimensional input to a two-dimensional output using ReLU activation. The sample input tensor [1.0, 2.0] is used for this visualization, and the forward pass through the network produces an output. The loss, defined as the sum of the outputs, is computed, and the gradients are calculated. The Hessian matrix, representing the second-order partial derivatives, is then computed using these gradients. The Hessian matrix is visualized as a heatmap using the colormap, which provides a clear distinction between positive and negative values. The **x** and **y** axes of the heatmap are labeled with x_1 and x_2, representing the inputs to the neural network.

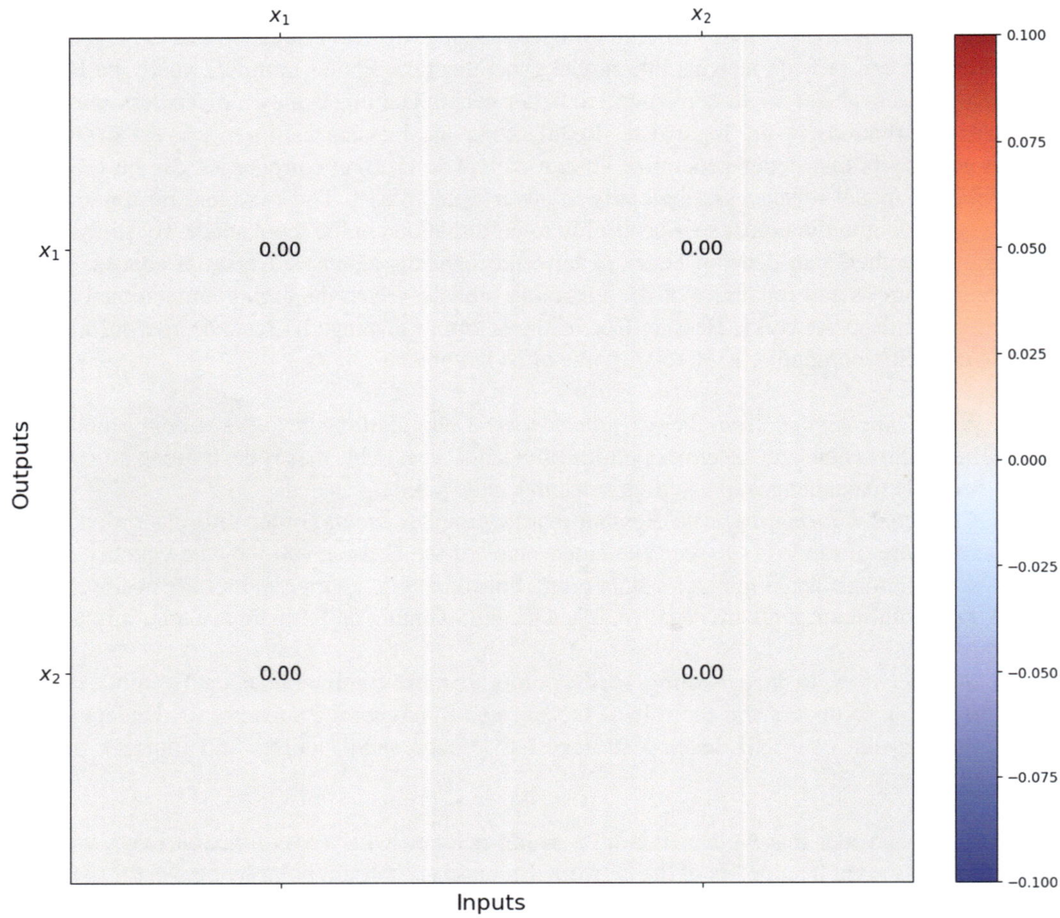

FIGURE 3.8 Visualization of the Hessian matrix for a neural network at $(x_1,x_2) = (1,2)$.

The challenges in using the Hessian in deep learning include the following:

(a) *Computational Complexity:* The Hessian is a square matrix of size **n** × **n**, where **n** is the number of parameters in the model. In modern deep networks, which can have millions or even billions of parameters, computing the full Hessian becomes computationally prohibitive. The time and resources required to calculate second-order derivatives for such high-dimensional models are often beyond the capabilities of standard hardware.

(b) *Storage Costs:* Storing the full Hessian matrix for large-scale models is impractical, as its memory requirements scale quadratically with the number of parameters. Even using techniques like sparse or low-rank approximations, the storage costs remain substantial, creating a bottleneck for efficient computation and optimization.

(c) *Non-convexity:* The loss surfaces in deep learning models are highly non-convex, meaning they contain numerous saddle points and local minima. While the Hessian provides information about the curvature of the loss surface, the presence of these saddle points complicates the optimization process. At these points, the Hessian can have both positive and negative eigenvalues, making it difficult to determine a clear direction for convergence and potentially leading to slow or unstable training.

3.10 REAL-WORLD APPLICATIONS

3.10.1 Autonomous Vehicles

Calculus, particularly gradient descent, is pivotal in training the deep learning models that power autonomous vehicles. These models rely on calculus to optimize their decision-making processes in real time. For instance, during navigation, the vehicle must continually adjust its steering and speed based on the environment, which is captured and processed by sensors. The optimization process involves minimizing a loss function that represents the difference between the predicted path and the ideal path. By using gradients, the model updates its parameters to improve accuracy and safety, ensuring that the vehicle can adapt to changing road conditions and unexpected obstacles.

3.10.2 Healthcare and Medical Imaging

In healthcare, calculus plays a critical role in enhancing the accuracy of diagnostic tools, particularly in medical imaging. Techniques like backpropagation, which relies on the chain rule of calculus, are used to train deep learning models for tasks such as detecting tumors in radiographic images. By optimizing the loss function, which quantifies the discrepancy between the predicted diagnosis and the actual condition, these models can learn to identify subtle patterns in the images that may be indicative of early-stage diseases. This application not only improves diagnostic accuracy but also facilitates earlier detection and treatment, ultimately saving lives.

3.10.3 Robotics and Control Systems

In robotics, calculus is integral to designing control systems that allow robots to interact with their environment in real time. Calculus-based algorithms help optimize the robot's movements and actions to perform tasks with precision. For instance, in robotic surgery, the robot's movements must be incredibly accurate to perform delicate procedures. The optimization of these movements relies on minimizing a loss function that measures the deviation from the desired motion. By calculating gradients, the control system can continuously adjust the robot's actions, ensuring safe and effective operation.

3.10.4 Natural Language Processing (NLP)

In NLP, calculus underlies the training of models that understand and generate human language. These models, such as transformers used in machine translation or sentiment analysis, require extensive optimization during training. The gradients of the loss function guide the updates to the model's parameters, improving its ability to understand context, syntax, and semantics. For example, in machine translation, the model must minimize the error between the translated output and the correct translation, which is achieved through calculus-driven optimization techniques like backpropagation.

3.10.5 Image and Video Processing

Calculus is fundamental in processing and enhancing images and videos through deep learning models. These models are trained to perform tasks such as image recognition, video analysis, and facial recognition by optimizing a loss function that measures how well the model's predictions match the actual content. For instance, in facial recognition, the model learns to distinguish between different faces by minimizing the difference between the predicted identity and the true identity. This process relies on gradient calculations to adjust the model parameters, improving accuracy and reliability in real-world applications.

3.11 HANDS-ON EXAMPLE

This section will walk through a programming example to demonstrate how multivariate calculus is used in deep learning, specifically focusing on gradient descent optimization.

Step 1: Import necessary libraries

In this step, we import essential libraries that will help with data manipulation and visualization. The NumPy library is imported as np, which is widely used for numerical computations and handling arrays or matrices. The matplotlib.pyplot module, imported as plt, is used for creating static, interactive, and animated visualizations in Python. Additionally, Axes3D from mpl_toolkits.mplot3d is imported to enable the creation of 3D plots, which will allow us to visualize data in three dimensions, providing deeper insight into relationships between variables.

```
import numpy as np
import matplotlib.pyplot as plt
from mpl_toolkits.mplot3d import Axes3D
```

Step 2: Define the multivariate function

We'll use a simple quadratic function of two variables: $f(x, y) = x^2 + y^2$

```
def f(x, y):
  return x**2 + y**2
```

Step 3: Compute the gradient

The gradient of the function is given by the partial derivatives with respect to each variable: $\nabla f = \left(\frac{\partial f}{\partial x}, \frac{\partial f}{\partial y}\right)$

```
def gradient(x, y):
  df_dx = 2 * x
  df_dy = 2 * y
  return np.array([df_dx, df_dy])
```

Step 4: Implement gradient descent

In this function, we are implementing a basic version of the gradient descent optimization algorithm. The gradient_descent function takes in three parameters: starting_point, which is the initial point where the optimization process begins; learning_rate, which controls the size of the steps taken in the direction of the gradient; and iterations, which specifies how many times the process should update the point. The function starts at the given starting_point and iteratively updates the position based on the negative of the gradient (calculated using the gradient function). The learning rate controls how much the point moves in the direction of the negative gradient. The new positions (points) are stored in a list and returned, giving a trace of the steps taken during the optimization process.

```
def gradient_descent(starting_point, learning_rate, iterations):
  points = [starting_point]
  point = starting_point
  for _ in range(iterations):
    grad = gradient(point[0], point[1])
    point = point - learning_rate * grad
    points.append(point)
  return points
```

Step 5: Run gradient descent

In this part of the code, we are setting up the initial parameters to run the gradient descent algorithm. The starting_point is initialized as a NumPy array [5.0, 5.0], which represents the initial point in a 2D space where the optimization begins. The learning_rate is set to 0.1, which defines the step size in the direction of the gradient at each iteration, controlling how quickly the algorithm converges. The iterations parameter is set to 50, meaning that the gradient descent will run for 50 steps. The gradient_descent function is then called with these parameters, and the points generated throughout the optimization process are stored in the points variable. Finally, the list of points is converted into a NumPy array for easier manipulation or visualization in the following steps. This setup is crucial for tracking how the algorithm converges towards the minimum of the function being optimized.

```
starting_point = np.array([5.0, 5.0])
learning_rate = 0.1
iterations = 50
points = gradient_descent(starting_point, learning_rate,
iterations)
points = np.array(points)
```

Step 6: Visualize the function and Gradient Descent Path

We will create a 3D plot of the function and the path taken by gradient descent.

```
x = np.linspace(-6, 6, 400)
y = np.linspace(-6, 6, 400)
X, Y = np.meshgrid(x, y)
Z = f(X, Y)
fig = plt.figure(figsize=(12, 8))
ax = fig.add_subplot(111, projection='3d')
ax.plot_surface(X, Y, Z, cmap='viridis', alpha=0.6)
ax.plot(points[:, 0], points[:, 1], f(points[:, 0], points[:, 1]),
color='r', marker='o')
ax.set_xlabel('X')
ax.set_ylabel('Y')
ax.set_zlabel('f(X, Y)')
ax.set_title('Gradient Descent on f(X, Y) = X^2 + Y^2')
plt.show()
```

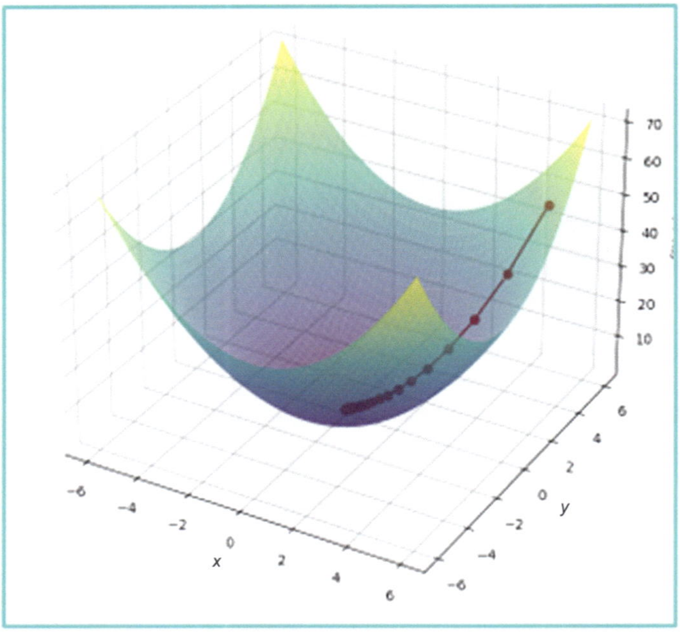

FIGURE 3.9 Gradient descent on $f(x, y) = x^2 + y^2$.

Figure 3.9 is the output of the program, which is a 3D plot that shows the surface of the function and the path taken by gradient descent.

3.12 COMMON MISTAKES AND TROUBLESHOOTING TIPS

3.12.1 Understanding Partial Derivatives

- *Mistake:* Confusing partial derivatives with total derivatives.
- *Tip:* Remember that partial derivatives measure how a function changes with respect to one variable while keeping other variables constant. In contrast, total derivatives consider the change with respect to all variables simultaneously.
- *Mistake:* Incorrectly treating constants while taking partial derivatives.
- *Tip:* Always treat other variables as constants when computing the partial derivative with respect to a specific variable. For example, when finding $\frac{\partial f}{\partial x}$ for $f(x, y)$, treat it as a constant.

3.12.2 Higher-Order Partial Derivatives

- *Mistake:* Misinterpreting mixed partial derivatives.
- *Tip:* Understand that mixed partial derivatives involve taking the derivative with respect to different variables in succession. For example, it means taking the partial derivative of **f** with respect to **y** first, then with respect to **x** .
- *Mistake:* Assuming mixed partial derivatives are always equal.
- *Tip:* Clairaut's theorem states that mixed partial derivatives are equal if the function and its partial derivatives are continuous. Verify continuity before assuming equality.

3.12.3 Gradients

- *Mistake:* Misunderstanding the direction of the gradient vector.
- *Tip:* Remember that the gradient vector points in the direction of the steepest ascent. For minimization problems, move in the opposite direction of the gradient (steepest descent).
- *Mistake:* Incorrectly computing the gradient for functions with multiple variables.
- *Tip:* The gradient is a vector of partial derivatives. For a function $f(x, y)$, the gradient is $\nabla f = \left(\frac{\partial f}{\partial x}, \frac{\partial f}{\partial y}\right)$.

3.12.4 Optimization and Gradient Descent

- *Mistake:* Using a fixed learning rate that is too high or too low.
- *Tip:* Adjust the learning rate dynamically or use adaptive learning rate algorithms like Adam or RMSprop to improve convergence.
- *Mistake:* Ignoring the potential for vanishing or exploding gradients.
- *Tip:* To mitigate vanishing gradients, use techniques like gradient clipping to handle exploding gradients and appropriate activation functions (e.g., ReLU).

3.12.5 Jacobians

- *Mistake:* Confusing the Jacobian matrix with the Hessian matrix.
- *Tip:* The Jacobian matrix is used for vector-valued functions and contains first-order partial derivatives. The Hessian matrix is used for scalar-valued functions and contains second-order partial derivatives.
- *Mistake:* Misinterpreting the role of the Jacobian in transformations.
- *Tip:* Understand that the Jacobian matrix represents the rate of change of each output with respect to each input. It is crucial for sensitivity analysis and understanding how input changes affect outputs.

3.12.6 Hessian Matrices

- *Mistake:* Overlooking the computational complexity of the Hessian.
- *Tip:* Be aware that the Hessian matrix can be computationally expensive for large models. Use approximations or Hessian-free optimization methods when necessary.
- *Mistake:* Misinterpreting the eigenvalues of the Hessian.
- *Tip:* Positive eigenvalues indicate a local minimum, negative eigenvalues indicate a local maximum, and mixed eigenvalues indicate a saddle point.

3.13 REVIEW QUESTIONS

1. How is matrix multiplication used during the forward propagation of a neural network?
2. What roles do vectors and matrices play in representing data and weights within the network?
3. Define a partial derivative and discuss its importance in the context of multivariate functions in deep learning.
4. How does a partial derivative differ when calculated with respect to different variables?
5. What is a gradient in multivariable calculus? Explain its significance in optimizing neural network parameters during training.
6. Compare and contrast stochastic, mini-batch, and batch gradient descent. Under what circumstances might each be preferred?

7. How do Jacobians and Hessians contribute to deep learning? Provide examples of how these higher-order derivatives can be applied in model optimization and analysis.
8. How does the learning rate influence the convergence of neural network training? What potential issues can arise if the learning rate is set too high or too low?
9. Explain the backpropagation algorithm.
10. How does backpropagation use the chain rule of calculus to update the weights and biases in a neural network?

3.14 PROGRAMMING QUESTIONS

3.14.1 EASY

Implement gradient descent to find the minimum of the function $g(x, y) = x^2 + 3y^2$.

1. Define the function g(x, y).
2. Compute the gradient of the function.
3. Implement the gradient descent algorithm.
4. Choose a starting point, learning rate, and number of iterations.

3.14.2 MEDIUM

Implement gradient descent with momentum to minimize the function $h(x, y) = (x-2)^2 + (y+3)^2$.

1. Define the function h (x, y).
2. Compute the gradient of the function.
3. Implement the gradient descent algorithm with momentum.
4. Choose a starting point, learning rate, momentum factor, and number of iterations.
5. Visualize the function and the path taken by gradient descent with momentum.

3.14.3 HARD

Implement gradient descent to minimize the function $k(x, y) = \sin(x) + \cos(y)$.

1. Define the function k(x, y).
2. Compute the gradient of the function.
3. Implement the gradient descent algorithm.
4. Choose a starting point, learning rate, and number of iterations.
5. Visualize the function and the path taken by gradient descent.

4 Probability Theory and Statistics

4.1 INTRODUCTION

In this chapter, we begin by exploring the fundamental role of probability distributions in understanding and modeling the randomness inherent in real-world data. From the discrete simplicity of a dice roll to the continuous variations in financial markets, probability distributions help us frame our predictive models in terms that align closely with observed phenomena. As we navigate this chapter, we introduce you to various probability distributions tailored to specific data and analysis requirements. We also discuss how these distributions interact with neural network (NN) architectures, particularly Bayesian Neural Networks (BNNs), to provide a robust framework for handling uncertainty and making informed predictions.

4.2 PROBABILITY DISTRIBUTIONS

Probability distributions are fundamental statistical tools, offering mathematical models that predict the likelihood of various outcomes. These outcomes could be as simple as a dice roll or as complex as forecasting investment returns. Each type of probability distribution is tailored to specific kinds of data and assumptions, making them necessary for analyzing patterns, making predictions, and drawing meaningful conclusions about data.

4.2.1 Discrete Probability Distributions

These types of distributions apply discrete random variables with many possible outcomes. For example, when you roll a dice, the outcomes are discrete numbers ranging from 1 to 6. Another instance is counting the number of heads when you flip three coins. Here are a few widely used discrete probability distributions.

1. *Binomial Distribution:* It measures the number of successes in a fixed set of independent yes/no trials, each with the same probability of success. For instance, it can predict the likelihood of flipping heads 10 times out of 20 coins' tosses. Here is the mathematical representation for binomial distribution:

$$P(X = k) = \binom{n}{k} p^k (1-p)^{n-k}$$

DOI: 10.1201/9781032690742-4

where:

- **n** is the number of trials,
- **p** is the probability of success, and
- **k** is the number of successes.

For binomial distribution, suppose you flip a coin 20 times, and the probability of heads is 0.5. The binomial distribution can predict the likelihood of getting exactly 10 heads. Using the binomial formula:

$$P(X = 10) = \binom{20}{10}(0.5)^{10}(0.5)^{10} = 0.176$$

2. *Poisson Distribution:* It is ideal for estimating the probability of a particular number of events occurring within a fixed interval; this distribution is often used for rare events, such as the frequency of earthquakes in a specific area over a year. Here is the mathematical representation for Poisson distribution:

$$P(X = k) = \frac{\lambda^k e^{-\lambda}}{k!}$$

where:

- **λ** is the average rate of occurrence,
- **k** is the number of events.

For the Poisson distribution, if, on average, two earthquakes occur per year in a region, the Poisson distribution can estimate the probability of having exactly three earthquakes in a year, with **λ** = 2:

$$P(X = 3) = \frac{2^3 e^{-2}}{3!} = 0.18$$

3. *Geometric Distribution:* It shows the probability of achieving the first success on the **n**th trial. Here is the general formula for geometric distribution:

$$P(X = k) = (1 - p)^{k-1} \cdot p$$

where:

- **X** is the random variable representing the number of trials until the first success,
- **p** is the probability of success on any given trial,
- **k** is the trial number where the first success occurs, and
- **(1 – p)** is the probability of failure on a given trial.

The formula calculates the probability that the first success happens on the **k**th trial. The term $(1 - p)^{k-1}$ represents the probability of failing **k – 1** times before achieving success on the **k**th trial. Suppose you are flipping a coin, and the probability of getting heads (success) is 0.5. You want to calculate the probability that the first heads appear on the fifth flip. Using the formula:

$$P(X=5)=(1-0.5)^{5-1}\cdot 0.5=(0.5)^4\cdot 0.5=0.03125$$

The probability of getting the first heads on the fifth flip is 0.03125, or 3.125%.

4. *Uniform Distribution:* In uniform distribution (discrete), all outcomes have an equal probability of occurring, such as rolling any number from 1 to 6 on a fair die. Here is the general formula for discrete uniform distribution:

$$P(X=k)=\frac{1}{n}$$

where:

- **X** is the random variable representing the outcome,
- **k** is a specific outcome, and
- **n** is the number of equally likely outcomes.

The formula is used to calculate the probability of any specific outcome **k** when all outcomes are equally probable. For example, rolling any number on a fair six-sided die would have the same probability because the die is fair, and each face is equally likely. Suppose you roll a fair six-sided die. The probability of rolling any specific number (like rolling a 3) is equal for all outcomes. As there are six possible outcomes, the probability is:

$$P(X=3)=\frac{1}{6}\approx 0.1667$$

The probability of rolling a 3, or any other specific number, is 0.1667, or 16.67%.

4.2.2 Continuous Probability Distributions

They are used for continuous random variables, assuming any value within a specified range. A typical example is the height of individuals in a population, which can vary continuously. Some of the standard continuous probability distributions are as follows.

4.2.2.1 Uniform Distribution (Continuous)

It is applied when any value within a specified range is equally likely. This distribution is typically used to model situations where every outcome in an interval is equally probable, such as randomly selecting a number between two values. Here is its general formula:

$$f(x)=\frac{1}{b-a},\quad \text{for } a\le x\le b$$

where:

- **a** is the lower bound of the interval,
- **b** is the upper bound of the interval,
- **f(x)** is the probability density function (PDF).

This distribution is "flat," meaning every outcome between **a** and **b** has the same probability density. The total area under the PDF equals 1, ensuring it is a valid distribution. Suppose you want to model the probability of selecting a random number between 0 and 1. The PDF would be:

$$f(x)=\frac{1}{1-0}=1,\quad \text{for } 0\le x\le 1$$

Thus, the probability of any specific range (e.g., between 0.2 and 0.8) can be calculated as:

$$P(0.2 \le X \le 0.8) = (0.8 - 0.2) \cdot 1 = 0.6$$

This means the probability of selecting a number between 0.2 and 0.8 is 60%.

4.2.2.2 Normal (Gaussian) Distribution

The normal distribution is one of the most widely used continuous distributions, known for its bell-shaped curve. It is determined by two parameters: the mean (**μ**) and the standard deviation (**σ**). Here is its general formula:

$$f(x) = \frac{1}{\sigma\sqrt{2\pi}} e^{-\frac{(x-\mu)^2}{2\sigma^2}}$$

where:

- **μ** is the mean,
- **σ** is the standard deviation,
- **f(x)** is the PDF.

The mean **μ** represents the center of the distribution, while the standard deviation **σ** determines the spread of the data. The total area under the curve equals 1. The further a value is from the mean, the lower the probability density. Suppose the height of individuals in a population is normally distributed with a mean of 170 cm and a standard deviation of 10 cm. The probability of an individual being between 160 and 180 cm can be found using the cumulative distribution function (CDF). Approximately 68% of individuals fall within 1 standard deviation of the mean:

$$P(160 \le X \le 180) \approx 0.68$$

This means there is a 68% probability that an individual's height will be between 160 and 180 cm.

4.2.2.3 Exponential Distribution

It is used to model the time between continuous, independent events that occur at a constant average rate. It is frequently used in scenarios like waiting times, such as the time between incoming calls at a call center. Here is its general formula:

$$f(x) = \lambda e^{-\lambda x}, \quad x \ge 0$$

where:

- **λ** is the rate parameter (the inverse of the mean time between events), and **f(x)** is the PDF.

The rate parameter **λ** describes how frequently events occur. The exponential distribution has the "memoryless" property, meaning the probability of an event occurring in the future is independent of how much time has already passed. If the average time between incoming calls at a call center is 2 minutes, the rate parameter is $\lambda = \frac{1}{2} = 0.5$ per minute. The probability of waiting less than 1 minute for the next call is:

$$P(X < 1) = 1 - e^{-\lambda \cdot 1} = 1 - e^{-0.5} \approx 0.3935$$

Thus, there is a 39.35% probability of receiving the next call within 1 minute.

4.2.2.4 Beta and Gamma Distributions

The beta distribution is useful for modeling probabilities and proportions for values bounded between 0 and 1, while the gamma distribution is often used for modeling waiting times or life durations and can be seen as a generalization of the exponential distribution. The general formula for beta distribution is:

$$f(x) = \frac{x^{\alpha-1}(1-x)^{\beta-1}}{B(\alpha,\beta)}, \quad 0 \le x \le 1$$

where:

- **α** and **β** are shape parameters,
- **$B(\alpha, \beta)$** is the beta function.

The beta distribution is highly flexible and can take on various shapes depending on **α** and **β**. It is particularly useful when modeling events that have an inherent probability or proportion. Suppose you want to model the distribution of success probabilities in a series of experiments with **α** = 2 and **β** = 3, the beta distribution models the probabilities, for example, by getting a success rate between 0.2 and 0.8, using the CDF. The general formula for gamma distribution is:

$$f(x) = \frac{\lambda^{\alpha} x^{\alpha-1} e^{-\lambda x}}{\Gamma(\alpha)}, \quad x \ge 0$$

where:

α is the shape parameter,
λ is the rate parameter,
$\Gamma(\alpha)$ is the Gamma function.

The gamma distribution is a generalization of the exponential distribution (when **α** = 1) and is used for modeling waiting times when there is more than one event occurring. If the average lifespan of a device is modeled by a Gamma distribution with shape parameter **α** = 2 and rate parameter **λ** = 1/3 (representing 3 units of time on average), you can calculate the probability of a device lasting less than 5 units of time.

4.2.3 Characteristics of Probability Distributions

4.2.3.1 Mean (Expected Value)

The mean (or expected value) represents the average or central value of the distribution. It is calculated as the weighted average of all possible values, with the weights being the probabilities associated with those values. The mathematical representation is as follows:

$$\mu = E(X) = \sum_i x_i \cdot P(x_i)$$

For a continuous distribution:

$$\mu = E(X) = \int_{-\infty}^{\infty} x \cdot f(x) dx$$

where:

μ is the mean,
x$_i$ represents the possible values,
f(x) is the PDF (for continuous distributions).

For a normal distribution with a mean of **μ** = 50, the mean represents the center of the distribution. Most of the data will be centered around 50, with equal probabilities on either side of the mean.

4.2.3.2 Variance and Standard Deviation

Variance and standard deviation measure how much the values of the distribution deviate from the mean. Variance is the average of the squared differences from the mean, while the standard deviation is the square root of the variance. The mathematical representation of variance is as follows:

$$\sigma^2 = \text{Var}(X) = \sum_i (x_i - \mu)^2 \cdot P(x_i)$$

For a continuous distribution:

$$\sigma^2 = \int_{-\infty}^{\infty} (x - \mu)^2 \cdot f(x) dx$$

The mathematical representation of standard deviation is as follows:

$$\sigma = \sqrt{\text{Var}(X)}$$

For a normal distribution with **μ** = 50 and a standard deviation **σ** = 5, the variance is $\sigma^2 = 25$. This means that most values fall within 1 standard deviation of the mean, that is, between 45 and 55.

4.2.3.3 Skewness

Skewness measures the asymmetry of the distribution. A skewness value of zero indicates a symmetric distribution. Positive skewness means the distribution has a longer tail on the right, and negative skewness means it has a longer tail on the left. The mathematical representation is as follows:

$$\text{Skewness} = \frac{E[(X - \mu)^3]}{\sigma^3} = \frac{\sum_i (x_i - \mu)^3 \cdot P(x_i)}{\sigma^3}$$

For a right-skewed distribution with a mean of **μ** = 40, if the skewness is positive, it indicates that there are more extreme values on the right side of the distribution. This often happens in income distributions where there are a few very high values.

4.2.3.4 Kurtosis

Kurtosis measures the heaviness of the tails of the distribution relative to a normal distribution. A higher kurtosis value indicates heavier tails, meaning more outliers. A kurtosis value of 3 is typical for a normal distribution (this is referred to as mesokurtic). The mathematical representation is as follows:

$$\text{Kurtosis} = \frac{E[(X-\mu)^4]}{\sigma^4} - 3$$

Subtracting 3 from the formula centers the kurtosis at zero for a normal distribution (called excess kurtosis). Positive excess kurtosis indicates a heavy-tailed distribution (leptokurtic), while negative kurtosis suggests a light-tailed distribution (platykurtic). If a distribution has a kurtosis value greater than 3, it indicates that it has heavy tails and is prone to producing extreme values or outliers.

4.2.3.5 Mode

The mode is the value that appears most frequently in the distribution. In a unimodal distribution (like the normal distribution), the mode coincides with the mean. However, for multimodal distributions, there may be more than one mode. The mode is useful for understanding the most frequent outcomes in a distribution, particularly for discrete data or skewed distributions. In a normal distribution, the mode is equal to the mean (**μ** = 50), and the most frequent value is 50. In a right-skewed distribution, the mode would be less than the mean and closer to the peak of the distribution.

4.2.4 Function Representations

4.2.4.1 Probability Mass Function

The probability mass function (PMF) represents the probability that a discrete random variable is exactly equal to a specific value. It is used for discrete distributions, where the random variable takes on a finite or countably infinite number of values. Its general formula is as follows:

$$P(X = x) = p(x)$$

where:

- **P(X = x)** is the probability that the random variable **X** takes on the value **x**,
- **p(x)** represents the PMF for the discrete variable **X**.

Consider a discrete random variable representing the number of heads when flipping two coins. The possible outcomes for the number of heads are 0, 1, or 2. The PMF would give the following probabilities:

$$P(X=0)=\frac{1}{4},\quad P(X=1)=\frac{2}{4},\quad P(X=2)=\frac{1}{4}$$

This shows that there is a 25% probability of getting 0 heads, a 50% probability of getting 1 head, and a 25% probability of getting 2 heads.

4.2.4.2 Probability Density Function

It is used for continuous random variables and describes the relative likelihood of the random variable taking on a particular value. Unlike the PMF, the PDF does not give the probability of the random variable being exactly equal to a value (because for continuous variables, that probability is zero), but rather the likelihood of it falling within a small interval around that value. The general formula is as follows:

$$f(x) = \lim_{\Delta x \to 0} \frac{P(x \leq X \leq x + \Delta x)}{\Delta x}$$

where:

f(x) is the PDF for the continuous variable **X**,
X is the continuous random variable.

For a normal distribution with a mean **μ** = 0 and standard deviation **σ** = 1, the PDF is given by:

$$f(x) = \frac{1}{\sqrt{2\pi}} e^{-\frac{x^2}{2}}$$

This PDF describes the likelihood of values around the mean **μ** = 0. The value of **f(x)** is highest at the mean and decreases symmetrically as **x** moves away from the mean. For instance, around **x** = 0, the PDF will have its peak, and as you move toward **x** = ±1, the likelihood of these values decreases.

4.2.4.3 Cumulative Distribution Function

It gives the probability that a random variable **X** is less than or equal to a specific value **x**. The CDF is used for both discrete and continuous random variables and is defined as the integral (for continuous) or sum (for discrete) of the PMF or PDF. The general formula for discrete random variables is as follows:

$$F(x) = P(X \leq x) = \sum_{t \leq x} P(X = t)$$

The general formula for continuous random variables is as follows:

$$F(x) = P(X \leq x) = \int_{-\infty}^{x} f(t)dt$$

where:

- **F(x)** is the CDF,
- **f(x)** is the PDF for continuous variables, or **P(X = t)** for discrete variables.

For a normal distribution with **μ** = 0 and **σ** = 1, the CDF gives the probability that the random variable is less than or equal to a particular value. For example:

$$P(X \leq 1.96) \approx 0.975$$

This means that approximately 97.5% of the values in this distribution are less than or equal to 1.96.

Figure 4.1 shows four common probability distributions: binomial, Poisson, normal, and uniform. In subplot Figure 4.1a (binomial distribution), represented by blue bars, the probability of achieving a specific number of successes in a fixed number of trials is clearly illustrated. An annotation labeled "Peak at n*p" points to the most probable number of successes, emphasizing the distribution's central tendency around the expected value n × p. Adjacent to it, Figure 4.1b (Poisson distribution) showcases orange bars depicting the probability of a given number of events occurring within a fixed interval. The annotation "Peak at λ" highlights the mode of the distribution, where the expected number of events λ occurs with the highest probability. Moving to Figure 4.1c (normal distribution), a smooth green curve with a shaded area beneath it represents the continuous PDF. An annotation labeled "Mean (μ)" points to the center of the distribution, underscoring the concept of the mean as the peak of the bell curve. Finally, Figure 4.1d (uniform distribution) is depicted with a solid purple line and a filled area, illustrating a constant probability density across the interval from 0 to 1. The *x*-axis covers the range of possible values, and the *y*-axis shows the uniform probability density. The annotation "Constant Probability" succinctly conveys the essence of the uniform distribution, where every outcome within the specified interval is equally likely.

4.2.5 Connection Between Overfitting and Underfitting to Probability Distributions

From a probabilistic standpoint, overfitting can be seen as maximizing the likelihood of the parameters given the training data excessively to the point where the model is improbable to perform well on new data. This is similar to tuning a musical instrument so precisely to a specific environment that it sounds out of tune in any other setting. Regularization techniques, such as $\mathbf{L}_1$ and $\mathbf{L}_2$, address these issues by adding a penalty term to the complexity of the model. This approach can be conceptualized through Bayesian statistics: the penalty term acts as a prior distribution that inherently favors simpler models, reducing the likelihood of overfitting by penalizing complexity.

Figure 4.2 demonstrates the concepts of underfitting, good fit, and overfitting using polynomial regression models of varying complexity. The dataset consists of 30 data points generated using the function $y = x\sin(x)$ with some added Gaussian noise. These data points are represented as black dots on the plot in Figure 4.2, providing a visual reference for the underlying pattern in the data. Three models with varying degrees of complexity are fitted to the data. The first model, represented by the blue dashed line, uses simple linear regression. This model fails to capture the underlying pattern of the data, resulting in a poor fit. This underfitting is characterized by a high bias and low variance, indicating that the model is too simple to represent the data accurately. The second model, represented by the green solid line, uses a cubic polynomial (degree 3). This model accurately captures the underlying pattern of the data, providing a good balance between bias and variance. This good fit is characterized by an appropriate level of complexity that models the data well without overfitting. The third model, represented by the red dash-dot line, uses a polynomial of degree 10. This high-degree polynomial model fits the training data almost perfectly, capturing both the underlying pattern and the noise in the data. This overfitting is characterized by a low bias and high variance, indicating that the model is too complex and is likely to perform poorly on new, unseen data.

4.2.6 Connecting BNNs to Probability Distributions

BNNs incorporate principles of Bayesian probability into traditional neural network (NN) architectures. This integration allows BNNs to provide a measure of uncertainty in their predictions, enhancing decision-making processes in scenarios where certainty is crucial. Unlike traditional NNs that use fixed weights, BNNs treat weights as probability distributions. This approach means that a single fixed number does not represent each weight but a distribution that reflects our beliefs or uncertainties about the values of these weights. This probabilistic treatment allows BNNs to express and manage uncertainty more effectively. In Bayesian statistics, prior distributions represent our initial beliefs about the parameters before observing any data. As data is observed and processed

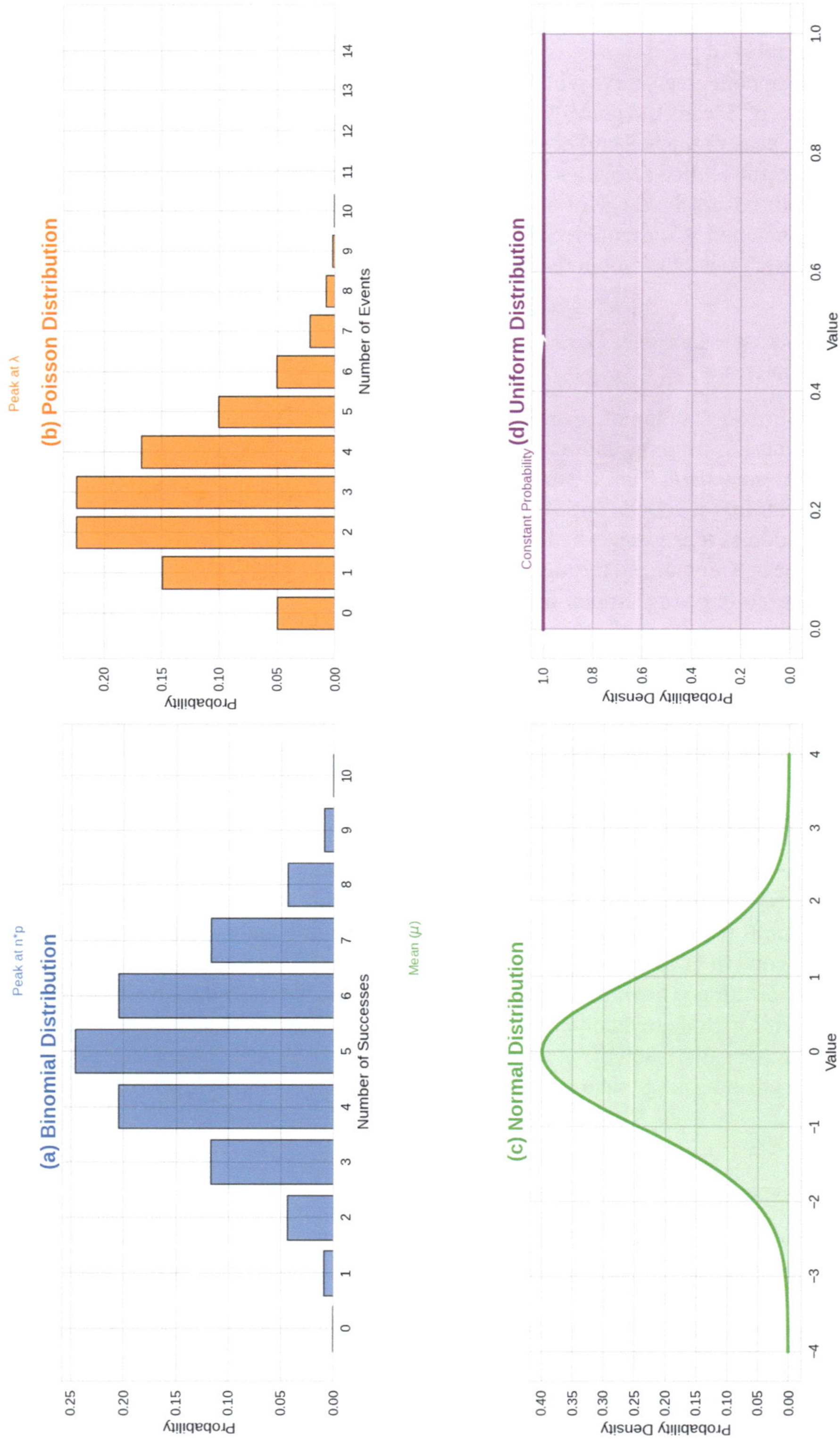

FIGURE 4.1 Common probability distributions.

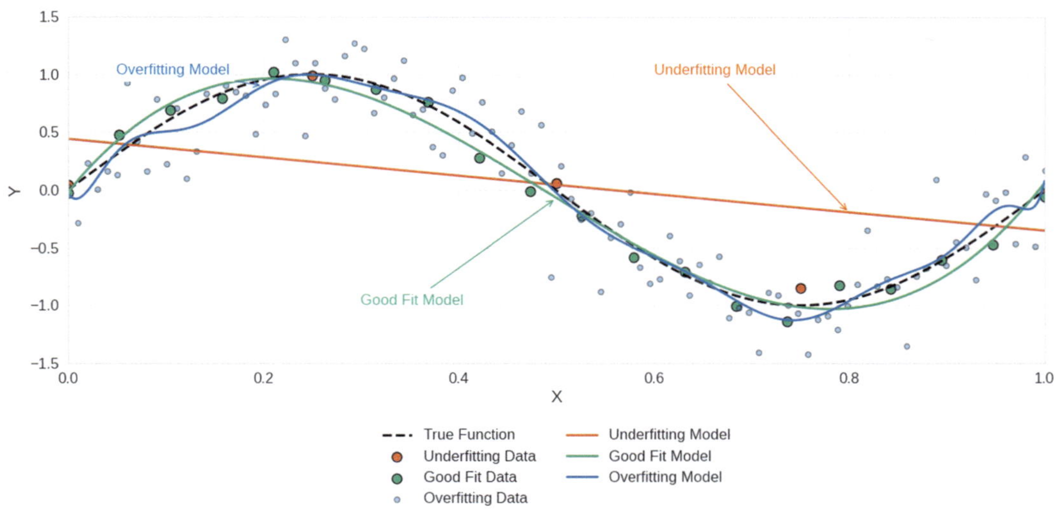

FIGURE 4.2 Demonstration of underfitting, good fit, and overfitting.

by a BNN, these prior beliefs are updated to form posterior distributions, which reflect new understandings gained from the data. This dynamic updating mechanism, absent in traditional NNs, allows BNNs to adapt their beliefs based on incoming information continuously. BNNs generate a distribution over possible outcomes when making predictions rather than producing a single-point estimate. This output provides a detailed spectrum of potential results and their associated probabilities. In Bayesian frameworks like BNNs, regularization is naturally incorporated through prior distributions. These priors can act as constraints or regularizes, penalizing deviations from established beliefs or biases. This regularization helps prevent overfitting by smoothing the learning process, ensuring that the model adheres to the training data at the expense of its generalization ability. The mathematical representations for the distributions are as follows.

4.2.6.1 Prior Distribution

It represents the distribution of the model's parameters (e.g., weights) before any data is observed. In Bayesian inference, this reflects our initial belief or assumptions about the model's parameters. The distribution of weights before observing the data:

$$P(w) \sim \mathcal{N}(0, \sigma^2)$$

where:

- **w** represents the weights of the NN,
- **P(w)** is the prior distribution of the weights, which could follow a certain distribution (e.g., normal distribution).

Suppose you are using a BNN to predict house prices, and you initially believe the weight (or impact) of the number of rooms on house price follows a normal distribution $N(0.5, 0.1^2)$. This means that before seeing any data, you assume the effect of the number of rooms on house price is centered around 0.5, but with some uncertainty (variance of 0.01).

4.2.6.2 Posterior Distribution

It represents the updated distribution of the model's parameters after observing the data. The posterior distribution combines the prior beliefs with the likelihood of the observed data to give an updated belief about the parameters. After observing the data, the posterior distribution of weights is:

$$P(w \mid \text{data}) = \frac{P(\text{data} \mid w)\, P(w)}{P(\text{data})}$$

where:

w are the weights,
data is the observed data,
P(data | w) is the likelihood of the data the weights,
P(w) is the prior distribution,
P(data) is the evidence (normalizing constant).

The posterior distribution adjusts the prior belief based on the data observed. The more data you have, the more the posterior distribution reflects the influence of the data rather than the prior. After observing data (e.g., historical house prices and the number of rooms), the BNN updates its belief about the weight of the number of rooms. Now, instead of assuming a fixed value for this weight (e.g., 0.5), the posterior distribution could be refined to something like $N(0.55, 0.05^2)$, which means the model now believes the number of rooms has a slightly higher impact on price, but with reduced uncertainty (variance of 0.0025).

4.2.6.3 Prediction

In BNNs, predictions are made by integrating the possible weights. This means the model considers the uncertainty in the weights rather than using a single fixed value. The model predicts a distribution over possible outcomes rather than a single-point estimate. The general formula for prediction is as follows:

$$P(y \mid x, D) = \int P(y \mid x, w) P(w \mid D) dw$$

where:

- **y** is the predicted output,
- **x** is the input (e.g., features like the number of rooms),
- **P(w | D)** is the posterior distribution of weights,
- **P(y | x, w)** is the likelihood of predicting **y** given input **x** and weights **w**.

The prediction integrates all possible weight values based on the posterior distribution, resulting in a prediction that captures the uncertainty in the model's parameters. This is a key feature of BNNs compared to traditional NNs, which make predictions based on fixed weight values. For predicting the price of a house, the BNN does not output a single price. Instead, it provides a range of possible prices, say from \$300,000 to \$350,000, with different probabilities assigned to each outcome. This prediction reflects the model's uncertainty about how strongly the number of rooms influences the price, as the weight for that feature is not fixed but distributed.

Figure 4.3 presents a visualization of Bayesian ridge regression applied to a synthetic dataset. Figure 4.3 illustrates several important elements of this regression technique. The data points (in blue) represent the observed values from the dataset. The predictive mean line (in dark green)

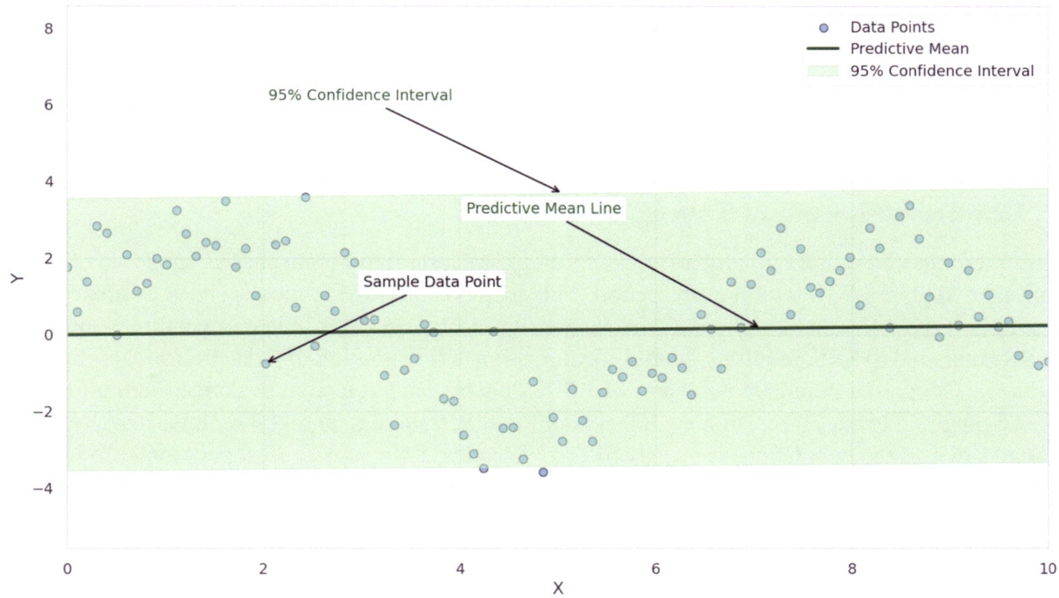

FIGURE 4.3 Bayesian ridge regression with uncertainty.

indicates the model's prediction for the mean outcome as a function of the input variable **X**. The shaded region around the predictive mean represents the 95% confidence interval (in light green), which provides a measure of uncertainty in the model's predictions. This interval shows where we expect the true values to lie with 95% confidence, allowing us to assess the model's reliability.

4.3 SAMPLING METHODS

Sampling methods are essential in statistics, especially in fields like Bayesian statistics, where they approximate distributions that are otherwise too complex or costly to compute directly. These methods are vital in various applications, including surveys and statistical model fitting, particularly in scenarios involving large or inaccessible full datasets due to constraints such as time, cost, or accessibility.

4.3.1 Simple Random Sampling

Simple random sampling (SRS) is a sampling method where each member of the population has an equal chance of being selected, ensuring no selection bias. This method is straightforward and easy to implement, making it a popular choice for many studies. However, its simplicity can be a drawback when dealing with heterogeneous populations, as it may not be efficient in such cases and could lead to the underrepresentation of certain subgroups. Suppose you have a population of 1,000 people, and you want to select a sample of 100 individuals. Using SRS, each person has a 1 in 10 chance of being selected, ensuring that everyone has an equal probability of inclusion in the sample. The probability of selecting any individual in SRS from a population **N** with a sample size **n** is:

$$P(\text{selection}) = \frac{n}{N}$$

If **N** = 1000 and **n** = 100, the probability of selecting any individual is:

$$P(\text{selection}) = \frac{100}{1{,}000} = 0.1$$

4.3.2 Stratified Random Sampling

Stratified random sampling involves dividing the population into homogeneous subgroups (strata) and then applying SRS within each section. This method can provide more precise estimates, particularly if the differences between divisions are significant. However, it is more complex to implement compared to SRS, requiring careful identification and separation of strata before sampling can proceed. Suppose a population consists of 1,000 students, and you want to divide them into strata based on grade level: 200 freshmen, 300 sophomores, 250 juniors, and 250 seniors. If you want to sample 10% of the population, you apply SRS by selecting 20 freshmen, 30 sophomores, 25 juniors, and 25 seniors. This ensures that each grade level is proportionally represented in the sample. For a population divided into **k** strata, the sample size for each stratum **h** is calculated as:

$$n_h = \frac{N_h}{N} \times n$$

where:

- $\mathbf{N_h}$ is the population size of stratum **h**,
- **N** is the total population size,
- **n** is the total sample size.

If **N** = 1000, $\mathbf{N_h}$ = 300 (sophomores), and **n** = 100, the sample size for sophomores would be:

$$n_{\text{sophomores}} = \frac{300}{1000} \times 100 = 30$$

4.3.3 Cluster Sampling

Cluster sampling divides the population into clusters, randomly selects certain clusters, and studies all members within these chosen clusters. This method is particularly efficient for geographically dispersed populations, reducing the costs and logistical challenges of data collection. However, it can be potentially less precise than SRS of an equivalent size, as the variability within clusters might not fully represent the entire population. Suppose a city has 50 schools, and you want to sample students. Using cluster sampling, you divide the population by schools (each school is a cluster). You randomly select 10 schools and study all the students in these selected schools. If each school has 200 students, you end up with a sample size of 2,000 students from the selected clusters. The probability of selecting a cluster is:

$$P(\text{cluster selected}) = \frac{\text{Number of clusters selected}}{\text{Total number of clusters}}$$

If 10 clusters are selected from 50 schools, the probability of selecting any specific school is:

$$P(\text{cluster selected}) = \frac{10}{50} = 0.2$$

4.3.4 Systematic Sampling

Systematic sampling selects a random starting point, and then every $\mathbf{k}^{\text{th}}$ member of the population is chosen. This method is more convenient than SRS due to its straightforward approach and ease of implementation. However, there is a risk of bias if the population has a periodic pattern, as this could result in a non-representative sample. Suppose you have a population of 1,000 people, and you want to sample 100 individuals. You randomly select a starting point, say 5, and then select every 10th person after that (e.g., 5th, 15th, 25th, and so on) until you have your sample of 100 people. If **N** is the population size, and **n** is the desired sample size, the sampling interval **k** is calculated as:

$$k = \frac{N}{n}$$

If **N** = 1000 and **n** = 100, the interval **k** is:

$$k = \frac{1,000}{100} = 10$$

4.3.5 Quota Sampling (Non-Probability Method)

Quota sampling is a non-probability method that divides the population into subgroups and selects individuals non-randomly. This approach is simple and cost-effective, making it attractive for studies with limited resources. However, the absence of random selection makes it less reliable, as it may introduce bias and limit the generalizability of the results. Suppose you want to survey 500 people about their favorite type of transportation. You decide to divide the population into subgroups by age: 200 people aged 18–29, 150 people aged 30–49, and 150 people aged 50+. Instead of randomly selecting individuals from each group, you select people conveniently or based on certain characteristics until you meet the quota for each age group. In quota sampling, the sample size for each subgroup is predefined based on the desired proportions. If the total sample is 500 and 40% of the population is aged 18-29, you need to select:

$$n = 0.40 \times 500 = 200 \text{ individuals from the 18 to 29 group.}$$

4.3.6 Monte Carlo Sampling

Monte Carlo sampling employs repeated random sampling to compute numerical results based on the law of large numbers. This method is widely used in numerical integration and probability computations, providing robust estimates for complex mathematical problems by simulating random variables and averaging the results. The Monte Carlo estimate of an integral **I** over a domain **D** is computed as:

$$I \approx \frac{1}{N}\sum_{i=1}^{N} f(x_i)$$

where:

- **N** is the number of random samples,
- $\mathbf{f(x_i)}$ is the function value at random points $\mathbf{x_i}$ drawn from a uniform distribution over **D**.

Suppose you want to estimate the value of π by using the Monte Carlo method. You randomly generate 100,000 points in a square with side length 2 and count how many fall inside a circle with radius 1 centered at the origin. If 78,539 points fall inside the circle, the estimate for π is:

$$\pi \approx 4 \times \frac{78,539}{100,000} = 3.14156.$$

4.3.7 Importance Sampling

Importance sampling focuses on sampling more frequently from important but rare regions of the sample space. This method is particularly effective in scenarios where certain outcomes are rare, as it allows for more efficient and accurate estimation of probabilities and expectations by emphasizing the critical regions of the distribution. The expected value $\mathbb{E}\left[f(X)\right]$ is estimated as:

$$\mathbb{E}\left[f(X)\right] \approx \frac{1}{N}\sum_{i=1}^{N}\frac{f(X_i)p(X_i)}{q(X_i)}$$

where:

- **p(X)** is the original probability distribution,
- **q(X)** is the importance sampling distribution, and
- **N** is the number of samples.

Suppose you want to estimate the probability of an event that occurs rarely, like a tail risk in finance. Instead of uniformly sampling from the entire space, you sample more frequently from the tail of the distribution. If you perform 10,000 simulations with a focus on the tail, and 200 events fall in the rare region, you can adjust the estimates based on the importance of this region, improving accuracy over random sampling.

4.3.8 Rejection Sampling

Rejection sampling involves sampling from a simple distribution and then accepting or rejecting each sample based on a predefined criterion. This method is particularly useful for sampling from complex distributions, as it allows for the generation of samples that meet specific criteria even when direct sampling from the desired distribution is challenging. Given a target distribution **p(x)** and a simpler proposal distribution **q(x)**, you accept a sample **x** if:

$$u \le \frac{p(x)}{M_q(x)}$$

where:

- **u** is a random uniform number between 0 and 1,
- **M** is a constant such that $\mathbf{M_q(x) \geq p(x)}$ for all **x**.

Suppose you want to sample from a distribution with a complex shape, like a Gaussian distribution, but instead, you sample from a uniform distribution between 0 and 1. You generate 10,000 points uniformly, and for each point, you compute its likelihood under the target distribution. If the ratio of the likelihood to the maximum is greater than a random number, you accept the point; otherwise, you reject it. After performing this process, you are left with 2,500 valid samples.

4.3.9 Markov Chain Monte Carlo

Markov Chain Monte Carlo (MCMC) constructs a Markov chain that reaches the desired distribution as its balance. This approach is widely used for sampling from complex probability distributions.

Key algorithms within MCMC include Metropolis–Hastings, Gibbs sampling, and Hamiltonian Monte Carlo, each offering different strategies for constructing and navigating the Markov chain to efficiently explore the sample space and achieve accurate results. In Metropolis–Hastings, the acceptance probability for moving from the current state **x** to a proposed state **x′** is:

$$\alpha(x \to x') = \min\left(1, \frac{p(x')q(x \mid x')}{p(x)q(x' \mid x)}\right)$$

where:

- **p(x)** is the target distribution,
- **q(x′ | x)** is the proposal distribution for transitioning between states.

Suppose you want to estimate the posterior distribution of a parameter in a Bayesian model. Using MCMC, specifically the Metropolis–Hastings algorithm, you generate a Markov chain of 100,000 samples. The first 10,000 samples are discarded as burn-in to ensure the chain has reached equilibrium, and the remaining 90,000 samples represent the parameter's posterior distribution.

Figure 4.4 demonstrates the results of a Metropolis–Hastings MCMC simulation targeting a standard normal distribution. It illustrates the dynamics of the MCMC algorithm and its ability to sample from complex probability distributions. Here, the trace plot illustrates the progression of sampled values over 1,000 iterations. The blue line represents the MCMC chain, showing how the sampled values evolve as the algorithm explores the distribution. A vertical red dashed line marks the end of the burn-in period at iteration 200, after which the samples are considered to have reached a stable distribution. An annotation highlights the burn-in period, emphasizing its importance in allowing the chain to converge toward the target distribution before collecting samples for analysis.

Figure 4.5 displays the posterior distribution of the samples collected after the burn-in period. The green histogram depicts the density of the sampled values, providing a visual approximation of the target distribution based on the MCMC samples. Overlaid on the histogram is a black dashed line representing the true standard normal distribution. The close alignment between the histogram and the target distribution indicates that the MCMC sampler has effectively approximated the standard normal distribution through the sampling process.

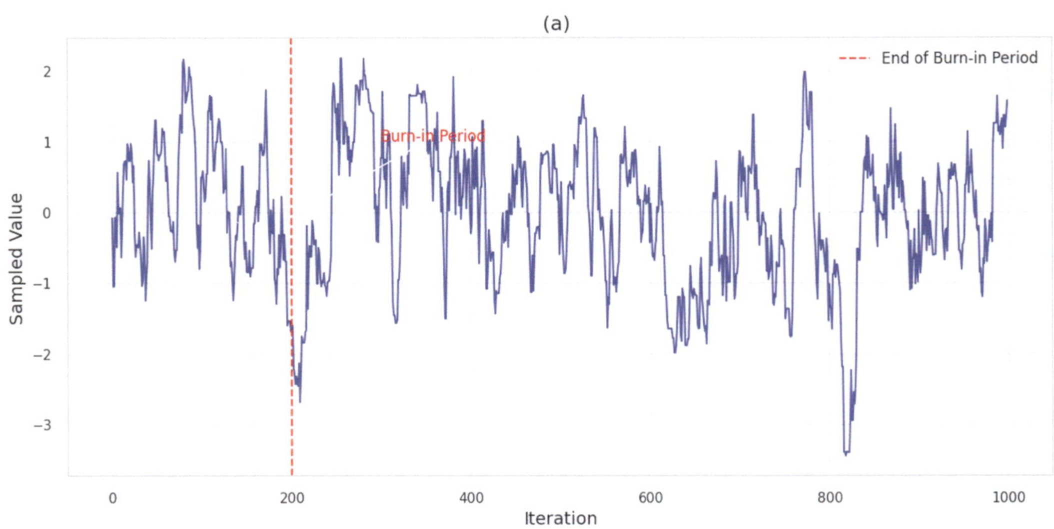

FIGURE 4.4 MCMC trace plot: convergence to target distribution.

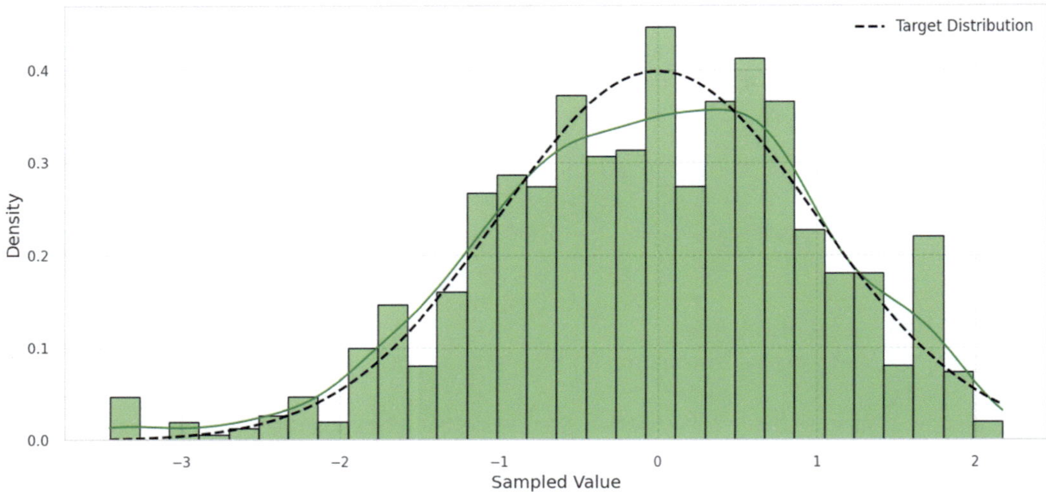

FIGURE 4.5 MCMC sample distribution after burn-in.

4.3.10 Gibbs Sampling

Gibbs sampling is a specific type of MCMC method that is particularly suitable for multivariate distributions. It is often used when the components of the distribution are conditionally independent. By iteratively sampling each variable conditional on the current values of the other variables, Gibbs sampling efficiently explores the sample space and can handle complex, high-dimensional distributions. For a joint distribution **p(X, Y)**, Gibbs sampling iteratively updates the variables by sampling from the conditional distributions:

$$X^{(t+1)} \sim p\left(X \mid Y^{(t)}\right)$$

$$Y^{(t+1)} \sim p\left(Y \mid X^{(t+1)}\right)$$

Suppose you want to sample from a bivariate distribution with variables **X** and **Y**. Using Gibbs sampling, you iteratively sample **X** conditional on **Y**, and **Y** conditional on **X**. After 10,000 iterations, the algorithm generates a sample that approximates the joint distribution of **X** and **Y**.

4.3.11 Latin Hypercube Sampling

Latin hypercube sampling (LHS) is a stratified sampling method that divides the sample space into equally probable subdivisions. This approach is particularly efficient for multi-dimensional sampling, ensuring that each variable is sampled across its entire range, which leads to more representative and coverage of the sample space compared to SRS. In LHS, for each variable $\mathbf{X_i}$ (where **i** indexes the dimensions), the range is divided into **N** equally probable intervals. Each sample is then selected such that:

$$x_i^{(j)} \in \left[\frac{j-1}{N}, \frac{j}{N}\right] \text{ for } j = 1, 2, \ldots, N.$$

Suppose you need to sample from a three-dimensional (3D) parameter space, with each dimension divided into 10 intervals. Using LHS, you randomly select one point from each interval for all three dimensions, ensuring that every part of the parameter space is represented. This results in 10 points per dimension, and the total sample size is 10 rather than the 1,000 points needed for full factorial sampling.

4.3.12 Resampling Methods

Techniques like bootstrapping estimate properties of estimators by sampling from an approximating distribution. These methods are commonly used to assess the variability or stability of statistical models, providing insights into the accuracy and reliability of the model's predictions by repeatedly sampling with replacement from the dataset to create numerous simulated samples. For each bootstrap sample **B**, drawn with replacement from the original dataset **D**:

$$\hat{\theta}_B^* = \text{statistic}(\text{resample})$$

The empirical distribution of $\hat{\theta}^*$ (e.g., the sample mean) across many resamples provides an estimate of the variability and confidence intervals for $\boldsymbol{\theta}$, the population parameter. Suppose you have a dataset with 500 observations and want to estimate the mean of the population. Using bootstrapping, you generate 1,000 resampled datasets by randomly selecting 500 observations with replacements from the original dataset. For each of the 1,000 resamples, you calculate the mean, and the distribution of these means provides an estimate of the variability and confidence intervals for the population mean.

Figure 4.6 illustrates three distinct sampling methods, SRS, stratified random sampling, and cluster sampling, through a series of subplots, each providing a visual representation of how samples are drawn from a population. In subplot Figure 4.6a, the SRS method is depicted, where 15 samples are randomly selected from the entire population without replacement. The population data is represented by light gray dots spread uniformly along the index axis, while the SRS samples are highlighted with red dots edged in black, scattered randomly across the index range. Figure 4.6b showcases stratified random sampling, where the population is divided into three distinct strata. Each stratum is sampled separately, with five samples drawn randomly from each. The sampled data points are colored differently, green, blue, and purple, to represent each stratum, and are also edged in black for emphasis. In Figure 4.6c, cluster sampling is presented, where the population is divided into five clusters, and two clusters (specifically clusters 2 and 5) are entirely selected for sampling. The sampled clusters are depicted with orange and purple dots, again edged in black, and an annotation points to one of the sampled clusters to illustrate that entire groups, rather than individual random samples, are included.

4.3.13 Sampling Methods and Overfitting/Underfitting

Sampling methods are pivotal in how well a statistical model can learn and generalize from data. Improper sampling can lead to overfitting and underfitting, each affecting model performance significantly.

4.3.13.1 Overfitting

Overfitting occurs when a model learns the specific details and noise within the training data, which can ultimately reduce its performance on new, unseen data. One contributing factor to overfitting is biased sampling, where non-representative sampling techniques lead to training on a skewed subset of the population. For example, if a model is trained primarily on data from a particular demographic due to sampling biases, it may struggle to generalize to data from other demographics. Another factor is the overuse of resampling techniques. While bootstrapping can be useful for estimating model accuracy, excessive use without proper cross-validation can result in overly optimistic

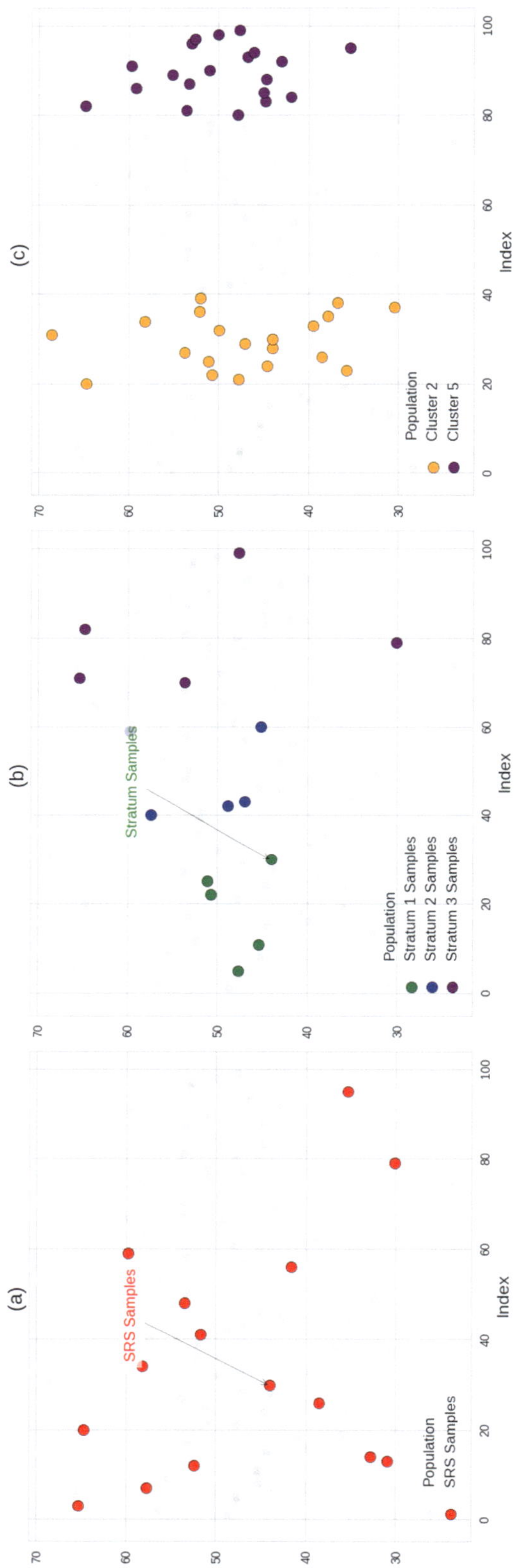

FIGURE 4.6 Comparison of sampling methods. (a) Simple random, (b) stratified random, (c) cluster sampling.

performance evaluations. This can create a false impression of a model's effectiveness, leading to overly complex models that are prone to overfitting the training data.

4.3.13.2 Underfitting

Underfitting occurs when a model is too simple to capture the underlying patterns in the data effectively or when it has not been exposed to a sufficient amount of data. One cause of underfitting is insufficient sampling, where the sampling method fails to gather enough data or fails to capture the complexity of the underlying population. For instance, in a large and diverse dataset, sparse sampling may result in a model that struggles to generalize beyond the limited training examples it has seen. Another issue that can lead to underfitting is the use of sparse sampling methods, which may provide too few data points or systematically overlook key aspects of the data distribution. This risk is common with systematic sampling techniques, where an overly large step size may skip important variations in the data, preventing the model from learning the full range of patterns needed for accurate predictions.

Figure 4.7 demonstrates the concepts of overfitting and underfitting through two illustrative plots. In subplot Figure 4.7a, the overfitting example showcases how an overly complex model, a high-degree polynomial, fits the sampled data perfectly within a limited range but fails to generalize to the broader dataset. The sampled data, represented by red dots, are collected from a biased portion of the dataset (specifically where $x < 4$), and the overfitted model is depicted by a dark red curve closely following these points. However, this model diverges significantly from the true underlying function, indicated by the green dashed line, when extended beyond the sampled region, highlighting the model's inability to generalize. In subplot Figure 4.7b, the underfitting example illustrates how a simplistic linear model fails to capture the complexity of the data due to insufficient model capacity and sparse sampling. The sampled data here are blue dots sparsely scattered across the entire range of **x**, and the underfit model is represented by a dark blue straight line that does not align well with the true function's quadratic and sinusoidal patterns.

4.4 BAYESIAN STATISTICS

Bayesian statistics is a branch of statistics that employs probability to represent all forms of uncertainty. After Thomas Bayes formulated the fundamental theorem underpinning this method, Bayesian statistics offered a robust framework for making inferences. At the core of Bayesian statistics lies Bayes' theorem, which provides a method for updating probabilities as new evidence is introduced. The theorem is expressed as:

$$P(A|B) = \frac{P(B|A)P(A)}{P(B)}$$

where:

- **P(A | B)** is the posterior probability,
- **P(B | A)** is the likelihood,
- **P(A)** is the prior probability, and
- **P(B)** is the evidence (the marginal likelihood).

The prior represents initial beliefs about an event or model parameters before any new evidence is considered. It sets the baseline from which Bayesian inference starts. Likelihood measures how probable the observed data are, given the model parameters. It plays a crucial role in updating the prior into the posterior. The posterior is the result of combining the prior and the likelihood of the observed data. It represents updated beliefs after considering new evidence. Often acting as a

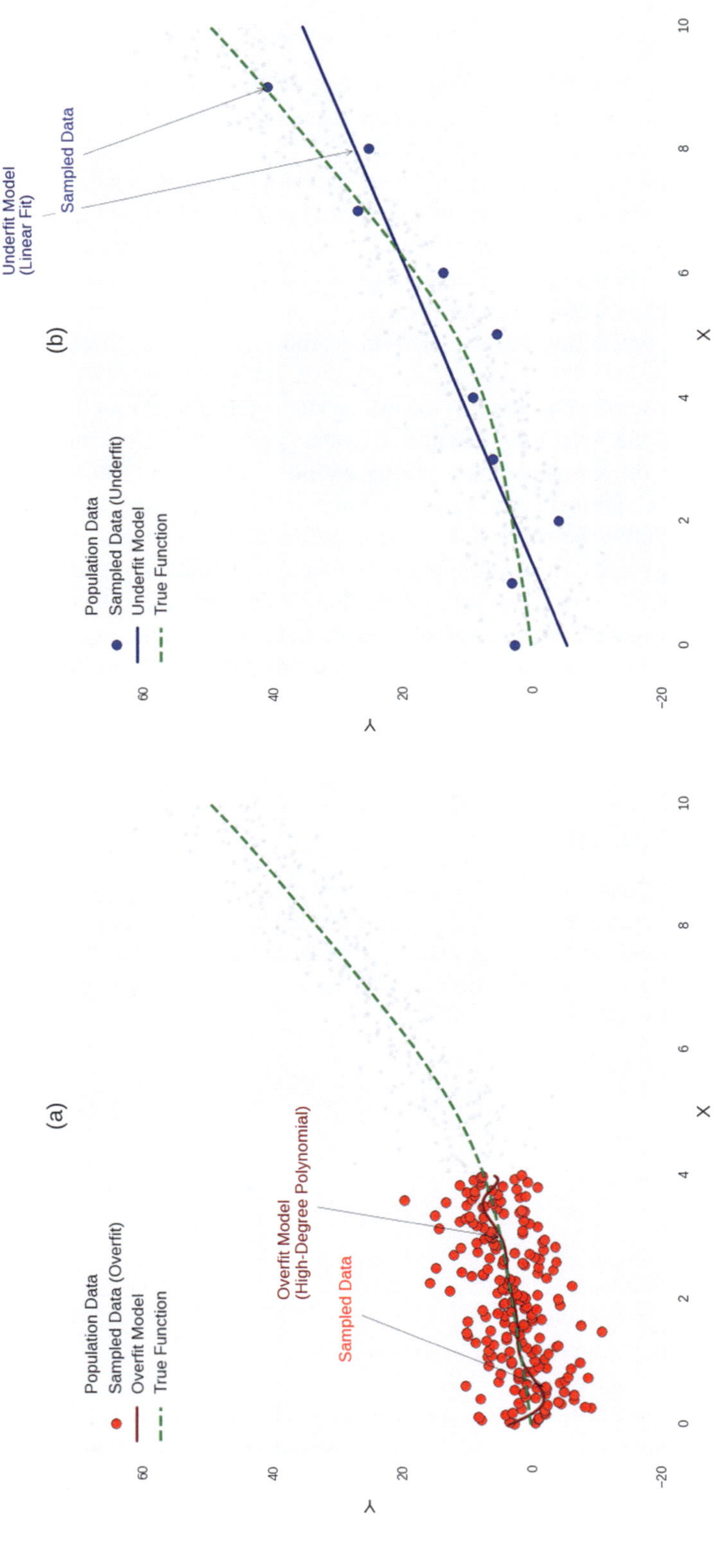

FIGURE 4.7 Impact of sampling methods on model. (a) Overfitting scenario and (b) underfitting.

normalization constant, the evidence ensures that the posterior probabilities sum to one. It integrates the likelihood over all possible values of the unknown parameters. The advantages of Bayesian statistics are:

1. *Flexibility:* Incorporates both new data and prior knowledge.
2. *Interpretability:* Results are expressed as probabilities, making them intuitively understandable.
3. *Full Distribution:* Provides a complete probabilistic description of model parameters, not just point estimates.

Given the complexity of deriving posterior distributions analytically, computational methods like MCMC and variational inference are crucial for approximating these distributions. Bayesian methods are employed across various fields, including machine learning, genetics, medicine, economics, and astronomy. There are still some challenges like:

1. *Choice of Prior:* The subjective nature of selecting a prior can lead to biases.
2. *Computational Intensity:* Bayesian methods can require significant computational resources, particularly for complex models.

Unlike frequentist statistics, which often focuses on likelihoods and provides point estimates and confidence intervals without incorporating prior beliefs, Bayesian statistics integrate prior knowledge with observed data to offer a probabilistic view of model parameters.

Figure 4.8 illustrates the Bayesian updating of beliefs regarding the probability of heads in a coin flip. It compares the prior belief with the posterior belief after observing the outcome of 50 coin's flips. Initially, the prior belief is represented by a beta distribution with parameters $\boldsymbol{\alpha} = 2$ and $\boldsymbol{\beta} = 2$, reflecting an assumption that the coin is fair. The blue dashed line shows this prior distribution, which is symmetric around 0.5. The prior mean, indicated by a vertical blue dashed line, is 0.5. After conducting 50 coin's flips with an observed true probability of heads being 0.7, the data yields 35 heads and 15 tails. This observed data updates the prior distribution to form the posterior distribution, depicted by the red solid line. The updated posterior parameters are $\boldsymbol{\alpha} = 37$ and $\boldsymbol{\beta} = 17$. The posterior mean, shown by a vertical red dashed line, shifts towards the observed probability of heads and is approximately 0.69, reflecting the new belief after incorporating the evidence from the coin flips.

4.4.1 Overfitting and Bayesian Statistics

Overfitting occurs when a statistical model captures noise or random fluctuations in the data instead of the true underlying patterns. This is often a result of model complexity exceeding what the data can support. Bayesian statistics offer several tools and concepts to mitigate this common problem. In Bayesian modeling, priors serve as a form of regularization. By incorporating prior beliefs about the parameters, Bayesian methods can constrain parameter estimates, pulling them toward more plausible values despite what the raw data might suggest. This is particularly useful when the likelihood derived from a small or noisy dataset might lead the model to overfit. A well-chosen strong prior can effectively shrink estimates, thus preventing the model from fitting too closely to the noise. Bayesian model comparison inherently includes a penalty for complexity. Tools like the Bayesian Information Criterion (BIC) or the Deviance Information Criterion (DIC) balance model fit with complexity, discouraging unnecessary complexity in model structure. This approach helps select complex models to capture essential data characteristics that are simple enough that they overfit the noise. The penalties ensure that a simpler, more generalizable model may be chosen over a more complex one that fits the training data slightly better. Unlike frequentist approaches that often yield point estimates,

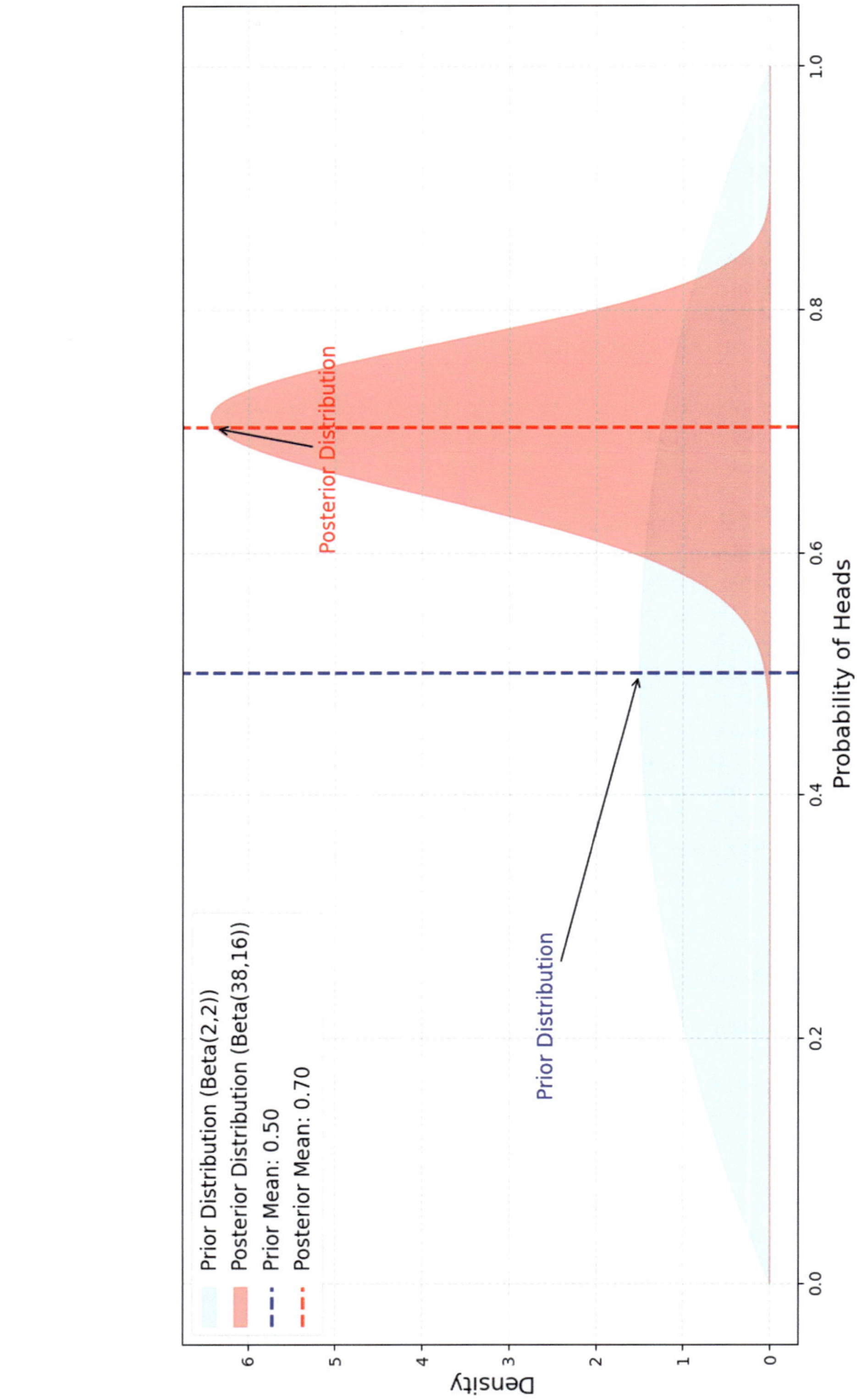

FIGURE 4.8 Bayesian updating of beliefs in coin toss.

Bayesian methods provide complete probability distributions over model parameters, offering a deeper insight into the uncertainty associated with each parameter. This feature of Bayesian analysis is particularly beneficial for identifying overfitting; an overfitted model may display implausibly narrow confidence intervals around predictions, indicating excessive confidence. In contrast, a well-fitted Bayesian model will show wider uncertainty intervals in areas where the data does not provide strong evidence, thereby reflecting a more realistic level of confidence.

Figure 4.9 compares two approaches to modeling noisy data: an overfitted model and a Bayesian model with a strong prior. The synthetic data is generated by adding Gaussian noise to a sine wave, creating a realistic scenario where noise is present. In Figure 4.9, the data points are shown as gray dots, representing the noisy observations of the sine function. The overfitted model, represented by the red line, is fitted using a high-degree polynomial (degree 15). This model captures the noise along with the underlying trend, resulting in a wavy and complex fit that closely follows the data points but is likely to generalize poorly to new data. The Bayesian model with a strong prior is depicted by the blue line. This model uses a simple linear fit with added Gaussian noise to simulate uncertainty. The simplicity of the model, enforced by the strong prior, prevents it from overfitting the noise in the data. The shaded blue region around the Bayesian model line represents the uncertainty interval, illustrating the confidence range of the predictions.

4.4.2 Underfitting and Bayesian Statistics

Underfitting in statistical modeling occurs when a model is too simplistic, failing to capture the data's complexities or underlying distribution. Bayesian statistics offer robust tools to tackle underfitting, enhancing model complexity appropriately without overcomplicating the model. Bayesian statistics allow for constructing hierarchical models, which is particularly useful for managing model complexity. One model's parameters can be modeled in hierarchical models, creating layers of parameters, each with its priors. By allowing parameters to vary across groups or contexts within a structured framework, hierarchical models can introduce complexity dynamically, adapting to the data's structure. This helps capture underlying patterns without needing an overly complex global model that could lead to overfitting. Hierarchical modeling facilitates borrowing strength across groups, improving estimation accuracy, especially in cases where data for certain groups might be sparse. One common reason for underfitting is overly restrictive priors that must adequately reflect the data's reality or variability. Bayesian statistics provide:

1. *Adaptability:* The flexibility to adjust and choose priors based on updated knowledge or new data. If initial assumptions lead to underfitting, priors can be revised to be less informative or adjusted to capture the data's distribution better.
2. *Incorporation of Expert Knowledge:* Bayesian methods allow the integration of expert knowledge through the selection of priors, which can be particularly beneficial in fields where prior research or domain expertise can inform model parameters.

Figure 4.10 illustrates the differences between an underfitted model, and a Bayesian model applied to a synthetic dataset. The orange line represents the underfitted model, a low-degree polynomial that is too simplistic to capture the true underlying function of the data (dashed green line). This model fails to represent the complexity of the data, leading to poor performance. In contrast, the blue lines represent samples from a Bayesian model, which introduces uncertainty around the mean prediction (solid blue line) by incorporating priors. This allows the model to balance complexity and uncertainty, improving its ability to generalize to new data. The true underlying function is shown by the dashed green line, and the observed noisy data points are plotted in gray. Figure 4.10 demonstrates how Bayesian regularization can avoid the issues of underfitting by dynamically adjusting model complexity to better capture the underlying patterns in the data.

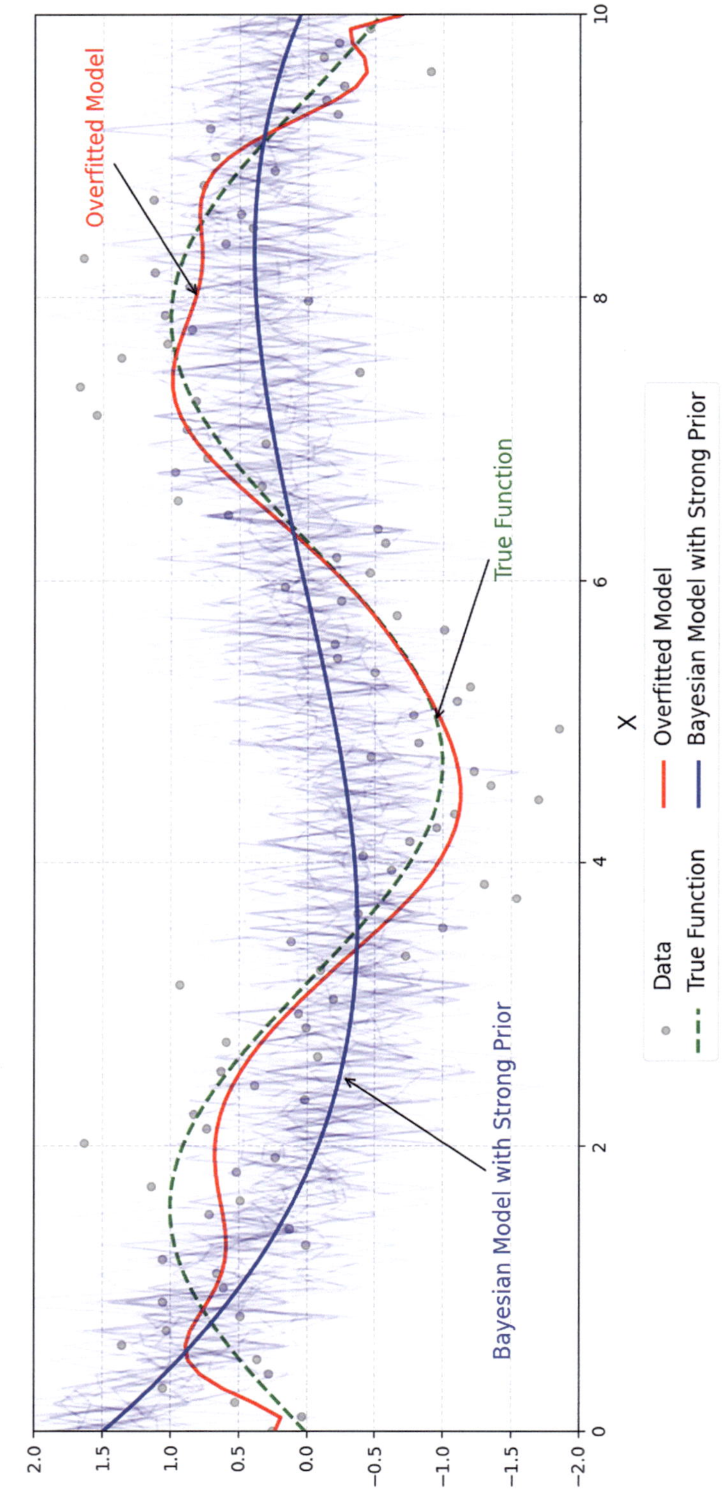

FIGURE 4.9 Comparison of overfitting vs. Bayesian regularization.

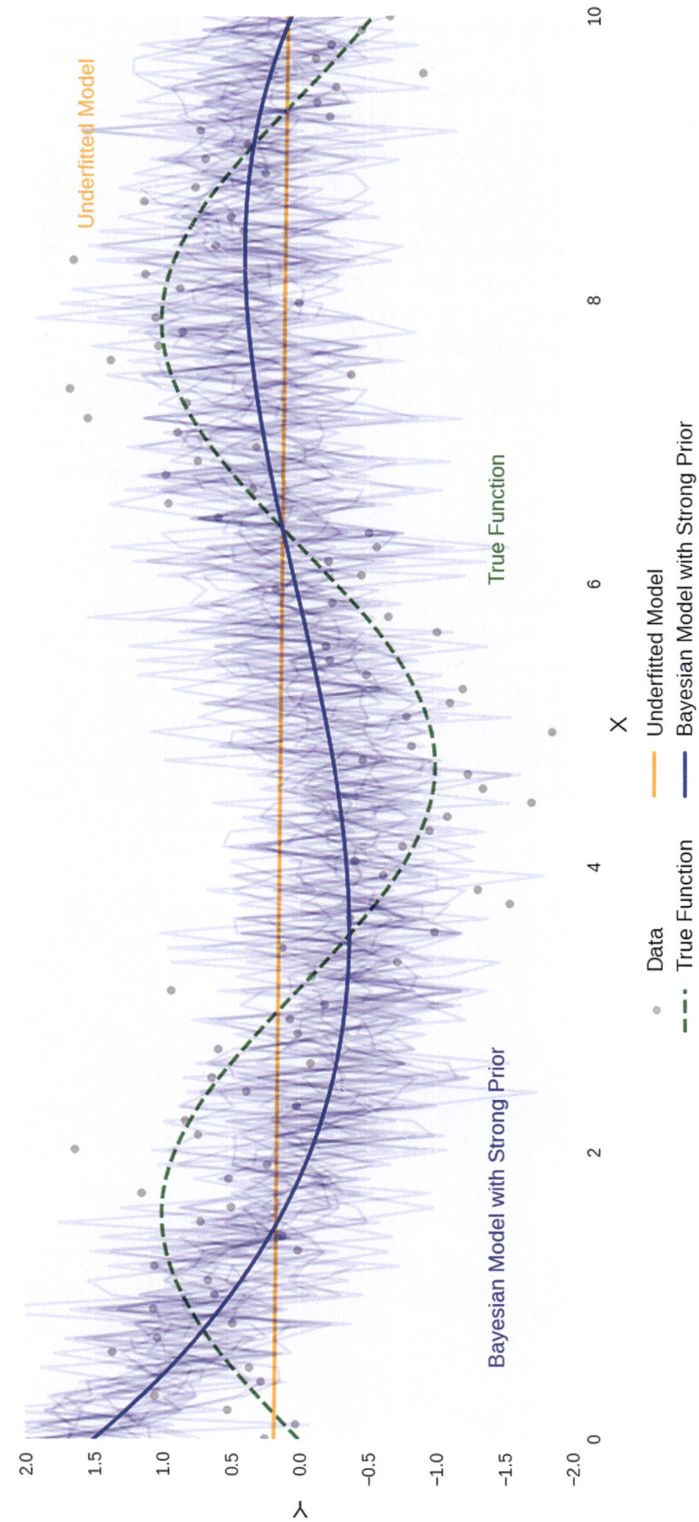

FIGURE 4.10 Comparison of underfitting and Bayesian regularization.

4.4.3 Bayesian Neural Networks

BNNs extend traditional NN architectures by incorporating probabilistic inference into their framework. This involves placing a prior distribution on the network weights, fundamentally changing how learning and prediction are approached. In BNNs, the use of priors acts as a natural form of regularization. By imposing prior beliefs about the distribution of weights, BNNs inherently reduce the risk of overfitting. These priors can penalize weights that stray too far from zero (or some other prior assumption), much like $\mathbf{L}_1$ or $\mathbf{L}_2$ regularization in traditional networks. Unlike traditional NNs that output single-point estimates, BNNs distribute possible outputs for each input. This probabilistic output is crucial for applications where understanding the uncertainty of predictions is important, such as in medical diagnosis, where it is essential to quantify confidence in diagnostic decisions. The probabilistic nature of BNNs contributes to their robustness, particularly in handling adversarial attacks or responding to shifts in data distribution. BNNs can maintain performance even when conditions change or when faced with deliberately misleading input data by considering a range of possible weight configurations rather than a single fixed set. Bayesian methods are known for their data efficiency. In the context of NNs, BNNs can often achieve comparable or superior performance to traditional NNs with less data. This is because they effectively leverage prior knowledge and the inherent uncertainty in their parameters to make more informed predictions. The major drawback of BNNs is their computational cost. The process of maintaining and updating distributions over weights is computationally intensive.

Figure 4.11 compares a traditional NN prediction and a BNN prediction for a synthetic dataset. The gray dots represent the original data points generated using a polynomial function with added noise. These points are the actual observations from which predictions are made. The red dashed line, marked with circles, represents the prediction made by a traditional NN. This line is fitted to the data using a polynomial regression as a proxy for the traditional NN. The red circles indicate specific prediction points along the red dashed line, making it easier to see the line's trajectory. The solid blue line represents the mean prediction made by the BNN. This line is also fitted using the same polynomial regression but incorporates Bayesian inference to estimate the mean

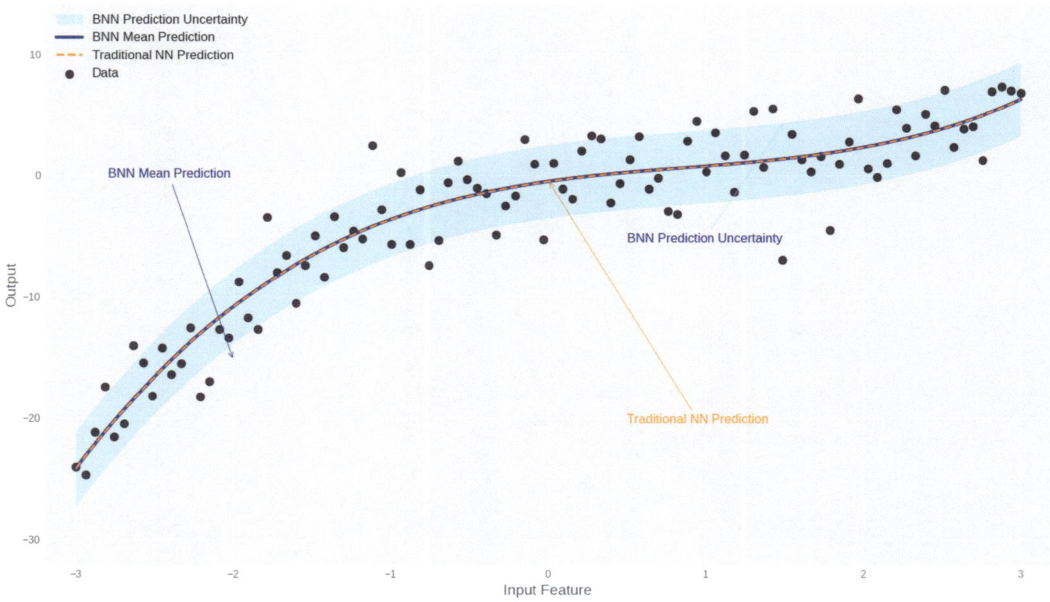

FIGURE 4.11 Comparison of traditional NN and BNN predictions.

prediction. The light blue shaded area around the blue line represents the uncertainty in the BNN predictions. This region is defined by an upper and lower bound, calculated as the mean prediction plus or minus a constant uncertainty value. The shaded area highlights the range within which the true values are expected to fall with a certain confidence level, reflecting the model's uncertainty in its predictions. The red arrow points to the red dashed line, representing the traditional NN prediction.

4.5 MOMENTS IN STATISTICS AND PROBABILITY THEORY

In statistics and probability theory, moments are fundamental metrics that describe the shape and characteristics of probability distributions. The **n^th^** moment of a random variable **X** about a value **c** is the expected value $(X-c)^n$. If this expectation exists and is finite, the **n^th^** moment of **X** is said to exist. The types of moments are as follows.

1. *Raw Moments (Crude Moments):* The raw moments of a distribution describe the expected values of powers of the random variable. When **c** = 0, these moments are referred to as raw or crude moments. They provide insights into the overall shape of the distribution. The general formula is as follows:

$$\mu_n' = E\left(X^n\right)$$

where:

- μ_n' is the $\mathbf{n^{th}}$ raw moment,
- $E\left(X^n\right)$ is the expected value of the nth power of the random variable **X**.

The raw moments capture the distribution's shape relative to the origin (0). The first raw moment is the mean, which provides the central tendency of the distribution. For a normal distribution with a mean of 0 and a standard deviation of 1, the first raw moment **E(X)** is 0, and the second raw moment $E\left(X^2\right)$ is 1, representing the variance.

2. *Central Moments:* The central moments are the expected values of powers of deviations from the mean. The $\mathbf{n^{th}}$ central moment measures the variability of the random variable around the mean. The general formula is as follows:

$$\mu_n = E[\left(X - E\left(X\right)\right)^n]$$

where:

- $\boldsymbol{\mu_n}$ is the $\mathbf{n^{th}}$ central moment,
- **E(X)** is the mean of the random variable **X**.

The first central moment is always zero because it represents the deviation of the data from the mean, and the second central moment is the variance, which quantifies the dispersion or spread of the distribution. Higher-order central moments help describe the shape of the distribution. For a normal distribution with a mean of 50 and standard deviation of 5, the first central moment (deviation from the mean) is zero, and the second central moment (variance) is $5^2 = 25$.

3. *Standardized Moments:* Standardized moments are the central moments divided by an appropriate power of the standard deviation. These moments provide shape-related characteristics of the distribution, including skewness and kurtosis. The general formula is as follows:

$$\text{Standardized moment} = \frac{\mu_n}{\sigma^n}$$

where:

- $\boldsymbol{\mu_n}$ is the $\mathbf{n^{th}}$ central moment,
- $\boldsymbol{\sigma}$ is the standard deviation.

4. *Skewness (Third Standardized Moment):* We explained skewness as it measures the asymmetry of the distribution. A skewness value of zero indicates a symmetric distribution. Positive skewness indicates a longer right tail, and negative skewness indicates a longer left tail:

$$\text{Skewness} = \frac{\mu_3}{\sigma^3}$$

4. *Kurtosis (Fourth standardized moment):* We explained kurtosis as it measures the "tailedness" of the distribution. A kurtosis of 3 corresponds to a normal distribution. Values greater than 3 indicate heavy tails (leptokurtic), while values less than 3 indicate light tails (platykurtic).

$$\text{Kurtosis} = \frac{\mu_4}{\sigma^4} - 3$$

For a right-skewed distribution with a positive skewness value of 1.5, this means that the distribution has a longer right tail compared to the left. If the kurtosis is greater than 3, it indicates the distribution has more extreme values or outliers compared to a normal distribution.

Figure 4.12 illustrates a comparative analysis of three different probability distributions using histograms: the normal distribution, the skewnorm (positive skew) distribution, and the Laplace (heavy tails) distribution. Figure 4.12, top-left, subplot displays the normal distribution in blue, showcasing a symmetric bell-shaped curve centered around the mean, highlighting its characteristic of light tails and symmetry, which indicates that data is evenly distributed around the central value. Figure 4.12, top-right, subplot presents the skewnorm distribution in green, demonstrating a positive skew where the tail on the right side is longer or fatter than the left. This indicates that a majority of the data values fall to the left of the mean, with fewer high-value outliers stretching the distribution to the right. Figure 4.12, bottom-left, subplot depicts the Laplace distribution in red, characterized by a sharper peak at the mean and heavier tails compared to the normal distribution, signifying a higher probability of extreme values occurring and more variability in the data. Figure 4.12, bottom-right, subplot overlays all three distributions using their respective colors, blue for normal, green for skewnorm, and red for Laplace, with semi-transparency, allowing for direct comparison of the distributions within the same scale and axes. This composite view highlights the differences in their shapes, central tendencies, and tail behaviors. Each subplot includes grid lines for better readability and axes labeled with "Value" and "Frequency" to indicate the data range and the count of occurrences within each bin, respectively. The normal distribution shows symmetry and light tails, suggesting data points are commonly close to the mean. The skewnorm distribution

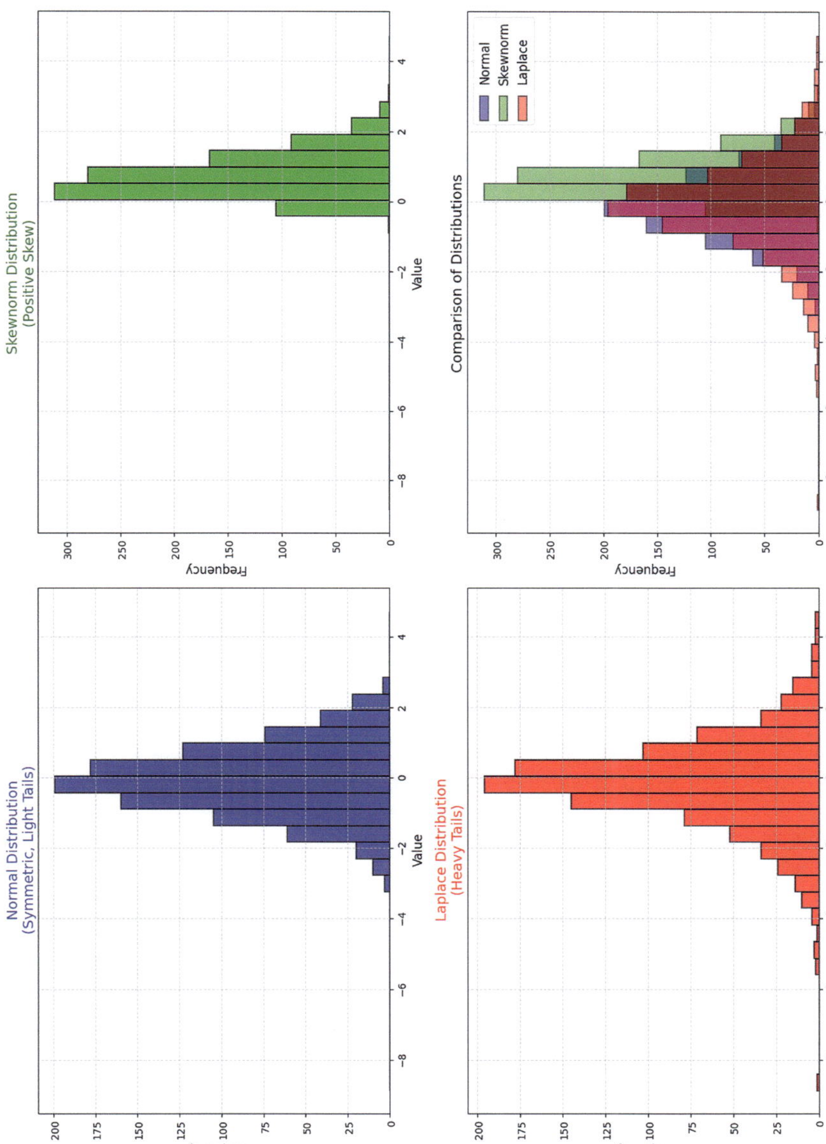

FIGURE 4.12 Comparison of distribution shapes.

highlights asymmetry due to its positive skew, indicating a tendency toward lower values with occasional higher outliers. The Laplace distribution emphasizes the presence of heavy tails, implying a higher likelihood of extreme deviations from the mean.

4.5.1 Moments and BNNs

BNNs fundamentally alter the traditional NN paradigm by directly incorporating uncertainty into the model's structure using probability distributions for weights. Let us explore how moments are critical in managing this uncertainty and influencing model performance. In BNNs, each weight is characterized not by a single fixed value but by a probability distribution. The first two moments of these distributions, the mean and variance, are particularly significant. The mean represents the expected value of the weight, essentially indicating the "average" strength of the connection the weight represents. The variance measures the uncertainty or reliability of this weight. A higher variance suggests less confidence in the weight's exactness, introducing a degree of flexibility or hesitance in the model's decisions.

The complexity of a BNN can have a profound impact on its performance. A highly complex BNN, with numerous parameters or low variance in weight distributions, might be overfitted by capturing not just the underlying data patterns but also the noise, including higher moments like skewness and kurtosis. Conversely, a too simplistic BNN might fail to capture sufficient moments of the data distribution, overlooking crucial information that could lead to underfitting. In BNNs, variance also plays a dual role by acting as a form of regularization. High variance on certain weights can indicate areas where the model should be less confident, preventing it from relying too heavily on data that may represent noise rather than signal. This mechanism helps to balance the model, ensuring it does not become overly confident based on the limited or noisy training data, thereby mitigating the risk of overfitting. Overfitting could occur if the model's weights had variances close to 0, leading to excessive certainty in potentially noisy data patterns. In BNNs, each weight $\mathbf{w}_i$ is modeled as a distribution characterized by moments (mean and variance):

$$w_i \sim \mathcal{N}\left(\mu_i, \sigma_i^2\right)$$

where:

- $\boldsymbol{\mu}_i$ is the mean (expected value) of the weight,
- $\boldsymbol{\sigma}_i$ is the variance (uncertainty) of the weight.

Suppose in a BNN a particular weight has a mean of 0.5 and a variance of 0.1. The model is confident in the connection strength but allows for some uncertainty. If another weight has a variance of 0.3, this indicates more uncertainty in that weight's value, meaning the model is less confident in the connection.

Figure 4.13 illustrates the effect of weight variance in BNNs on prediction stability and confidence. Figure 4.13 is divided into two subplots: In subplot Figure 4.13a (BNN with low weight variance), the blue prediction lines are closely clustered around the true function, represented by the green line. This clustering indicates that the model's predictions are consistent and closely aligned with the actual underlying relationship, demonstrating high confidence and stability. The light blue shaded area encompasses the range of these predictions, highlighting the model's narrow uncertainty band. Conversely, subplot Figure 4.13b (BNN with high weight variance) showcases a more dispersed set of orange prediction lines surrounding the same green true function. This dispersion reflects the model's increased uncertainty due to high variance in its weights, resulting in a broader prediction range illustrated by the light orange shaded area.

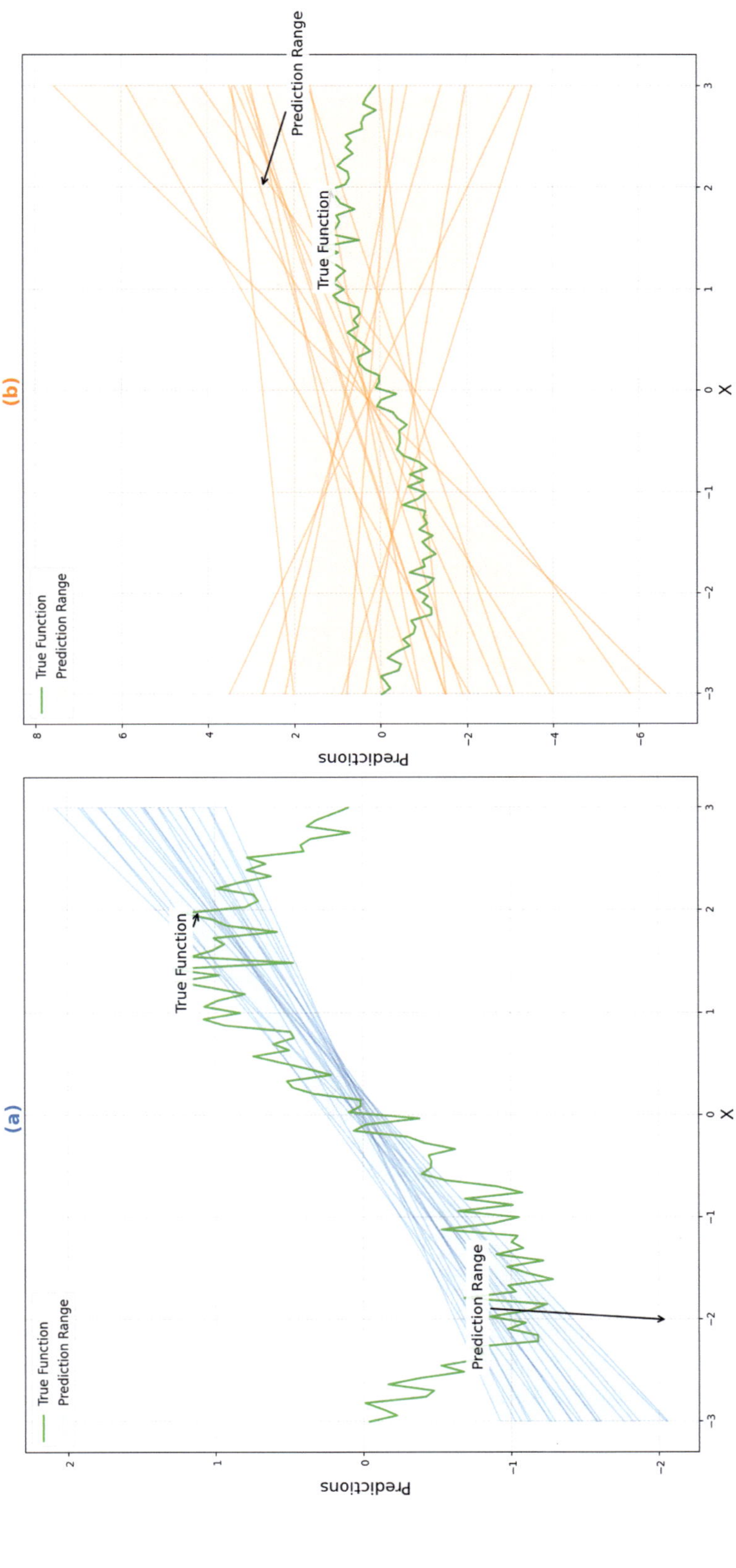

FIGURE 4.13 Effect of weight variance in BNN. (a) BNN with low variance, (b) BNN with high variance.

4.5.2 Connection Between Moments and Overfitting/Underfitting

Moments are crucial in shaping our understanding of data distributions and how they relate to model fitting in statistics and machine learning. Here, we explore how moments impact the assumptions behind model building and the common pitfalls of overfitting and underfitting. Statistical and machine learning models often rely on specific assumptions about their data distributions. For example, linear regression models assume that residuals are normally distributed, which pertains to having zero skewness (symmetry) and a certain kurtosis (peakedness). We can validate these assumptions by understanding moments such as mean, variance, skewness, and kurtosis. Significant deviations in these moments from expected values can indicate a potential misfit, suggesting that the model may not perform well on data that does not adhere to these assumed distributions. Overfitting occurs when a model too closely fits the training data, capturing noise and outliers as if they were true underlying patterns. This issue can arise if the model is excessively complex or needs to be correctly regularized. While mean and variance often receive most of the focus, higher moments like skewness and kurtosis are also critical. If these higher moments are influenced by noise, and the model attempts to accommodate them, it may overfit. Monitoring these moments can serve as a diagnostic tool; for instance, a model that captures extreme skewness or kurtosis in the training data may be overfitting, especially if these characteristics are not representative of the broader dataset. Conversely, underfitting happens when a model is too simplistic, often considering only the first moment (mean) and possibly the second (variance) but ignoring more complex moments such as skewness and kurtosis. This lack of complexity can prevent the model from capturing essential patterns in the data, leading to poor training and unseen data performance. Monitoring all moments provides insights into potentially overlooked data distribution aspects. For example, if the data exhibits high skewness or kurtosis and the model needs to account for these, it might underfit, failing to generalize effectively. The higher moments (skewness γ and kurtosis $\boldsymbol{\kappa}$) are computed as:

$$\gamma = \frac{\mathbb{E}[(X-\mu)^3]}{\sigma^3}, \quad \kappa = \frac{\mathbb{E}[(X-\mu)^4]}{\sigma^4}$$

where:

- $\boldsymbol{\mu}$ is the mean,
- $\boldsymbol{\sigma}$ is the standard deviation.

Suppose you are fitting a regression model to a dataset with a mean (first moment) of 0, variance (second moment) of 1, skewness (third moment) of 0.5, and kurtosis (fourth moment) of 3.5. If your model only accounts for the mean and variance, it may underfit by failing to capture the skewness and kurtosis of the data. Alternatively, if the model attempts to fit extreme skewness (e.g., skewness of 2) caused by noise in the training set, it may overfit. In overfitting, the model may overly conform to deviations in these higher moments (e.g., skewness and kurtosis), while in underfitting, it ignores these moments, leading to poor generalization. Monitoring these metrics helps diagnose and mitigate fitting issues.

Figure 4.14 presents a comparative analysis of how different data distributions, specifically positively skewed data and high kurtosis data, impact the performance of a simple linear regression model. Figure 4.14a displays positively skewed data generated using a skewed normal distribution. The scatter plot, depicted in blue, shows a concentration of data points on the left side with a tail extending to the right, characteristic of positive skewness. The orange line represents the linear fit applied to this skewed data. Due to the asymmetry in the data distribution, the linear model fails to capture the underlying trend effectively, resulting in a fit that does not align well with the majority of

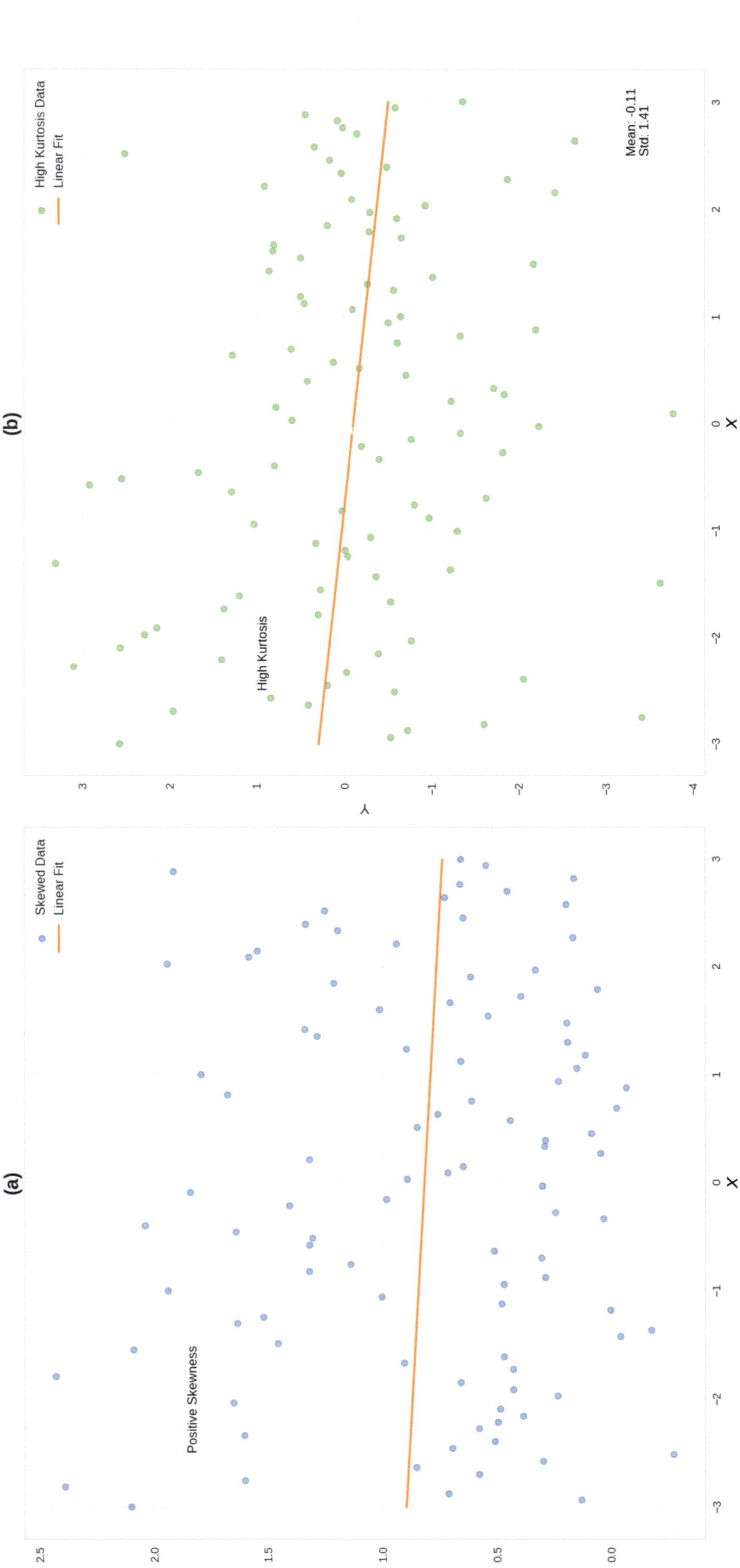

FIGURE 4.14 (a) Fit to positively skewed data, and (b) fit to high kurtosis data.

the data points. Conversely, Figure 4.14b on the right showcases high kurtosis data generated using the Laplace distribution. The scatter plot in green exhibits a sharp peak with heavy tails, indicative of high kurtosis. The same orange linear fit is applied to this dataset, revealing that the model struggles to accommodate the pronounced peak and the extreme values in the tails. This results in a linear fit that oversimplifies the data's variability, failing to accurately represent the distribution's heavy tails.

4.6 REAL-WORLD APPLICATIONS AND EXAMPLES

4.6.1 Image Recognition and Processing

Linear algebra is fundamental in image recognition and processing tasks. Images are typically represented as matrices, where each pixel is a value in a matrix. Operations like image filtering, transformation, and enhancement involve matrix manipulations such as convolution, eigenvalue decomposition, and singular value decomposition (SVD). convolutional neural networks (CNNs), which are widely used for image recognition tasks, rely on these linear algebra operations to extract features, recognize patterns, and classify images accurately. For example, in facial recognition systems, *CNNs* process the pixel values of images through multiple layers of convolutions, pooling, and fully connected layers to identify and verify individuals based on their facial features.

4.6.2 Natural Language Processing

Probability theory and statistics are essential in Natural Language Processing (*NLP*), where models often need to handle the uncertainty and variability inherent in human language. Language models, such as those used in machine translation or text generation, rely on probability distributions to predict the likelihood of a word or sequence of words given a context. Bayesian methods, which combine prior knowledge with observed data to update beliefs, are particularly useful in *NLP* tasks that involve uncertainty. For example, in a spam detection system, Bayesian classifiers use probability distributions to calculate the likelihood that an email is spam based on the presence of certain keywords or phrases, making decisions even when there is ambiguity in the data.

4.6.3 Robotics and Control Systems

In robotics, the principles of multivariate calculus and linear algebra are applied to control systems and motion planning. Robots must navigate and interact with their environment, which requires solving optimization problems in real time. For example, the control of a robotic arm involves computing the joint angles needed to move the end effector to a desired position. This computation requires solving a system of non-linear equations, which is achieved using techniques from multivariate calculus and linear algebra. Additionally, probability theory is used in robotics for state estimation and sensor fusion, where the robot must estimate its position and orientation based on noisy sensor data.

4.6.4 Healthcare and Medical Diagnostics

In healthcare, statistical methods and probability theory are applied to medical diagnostics and treatment planning. Bayesian networks, which model the probabilistic relationships between different variables, are used to predict the likelihood of diseases based on patient symptoms and medical history. For example, in a diagnostic system for cancer detection, a Bayesian network can integrate various risk factors, such as genetic predispositions and lifestyle choices, to estimate the probability of a patient having cancer. This probabilistic approach allows for more personalized and accurate diagnoses, enabling better treatment decisions.

4.7 HANDS-ON EXAMPLE

In this section, we will walk through a hands-on example that demonstrates the application of probabilistic models and the handling of uncertainty in NNs using BNNs. We will use a synthetic dataset to show how *BNNs* provide a measure of uncertainty in their predictions compared to traditional NNs.

4.7.1 Step 1. Setup and Import Libraries

In this section, we are installing and importing the necessary packages for our project. The command!pip install tensorflow tensorflow-probability matplotlib NumPy ensures that the latest versions of the TensorFlow, TensorFlow Probability, matplotlib, and NumPy libraries are installed. After installing, we import these libraries to make use of their functions and capabilities throughout the project.

```
# Install the required packages
!pip install tensorflow tensorflow-probability matplotlib numpy
import tensorflow as tf
import tensorflow_probability as tfp
import numpy as np
import matplotlib.pyplot as plt
```

4.7.2 Step 2. Generate Synthetic Data

In this step, we are generating synthetic data to use in our model or for visualization purposes. First, by setting np.random.seed(42), we ensure that the random number generation is consistent across runs, which is important for reproducibility. Then, x = np.linspace(−3, 3, 100) creates an array of 100 evenly spaced values between −3 and 3, which serves as the input data points. The corresponding y values are generated using the sine function (np.sin(x)) with added noise (0.3 * np.random.randn(100)) to simulate a real-world scenario where data is often noisy. Finally, a scatter plot of the data is created using matplotlib, with appropriate labels for the axes and a title to describe the plot.

```
# Generate synthetic data
np.random.seed(42)
x = np.linspace(-3, 3, 100)
y = np.sin(x) + 0.3 * np.random.randn(100)
plt.scatter(x, y, label='Data')
plt.xlabel('x')
plt.ylabel('y')
plt.legend()
plt.title('Synthetic Data')
plt.show()
```

4.7.3 Step 3. Define the BNN

In this step, we are defining a BNN using TensorFlow and TensorFlow Probability. First, the tf. keras.backend.set_floatx('float64') command ensures that all computations are done with double precision (64-bit floats) for better numerical stability. The create_bnn_model() function builds a sequential model with three layers using DenseFlipout from TensorFlow Probability. DenseFlipout layers allow the model to learn weight distributions, enabling it to estimate uncertainty in predictions. The model has two hidden layers, each with 64 units and Rectified Linear Unit (ReLU) activation, and one output layer with a single unit for regression. After creating the model, it is compiled with the Adam optimizer and mean squared error loss, which are commonly used for training NNs. This setup prepares the BNN for training on data, enabling it to handle uncertainty in predictions effectively.

```
# Define a Bayesian Neural Network
tf.keras.backend.set_floatx('float64')
def create_bnn_model():
  model = tf.keras.Sequential([
    tfp.layers.DenseFlipout(64, activation='relu', input_
    shape=(1,)),
    tfp.layers.DenseFlipout(64, activation='relu'),
    tfp.layers.DenseFlipout(1)
  ])
  return model
bnn_model = create_bnn_model()
bnn_model.compile(optimizer=tf.optimizers.Adam(learning_rate=
0.01),
    loss='mean_squared_error')
```

4.7.4 Step 4. Train the BNN

```
# Train the Bayesian Neural Network
bnn_model.fit(x, y, epochs=1000, verbose=0)
```

4.7.5 Step 5. Make Predictions and Plot Uncertainty

In this step, we are using the previously defined BNN to make predictions and visualize the results along with uncertainty estimates. The x_test array is created using np.linspace(−3, 3, 100) to generate 100 evenly spaced points within the range of −3 to 3, which serve as test inputs. The BNN model is run 100 times (for _ in range(100)) to obtain multiple predictions for each input point, capturing the model's uncertainty. These predictions are stored in y_pred, which is then converted into a NumPy array for easier manipulation. We calculate the mean (y_pred_mean) and standard deviation (y_pred_std) of the predictions at each test point, reflecting the model's predictive distribution. Finally, we plot the results: the original data points are shown as blue scatter points, the mean predictions as a red line, and the uncertainty (±2 standard deviations) as a shaded region around the mean.

```
# Make predictions with the BNN
x_test = np.linspace(-3, 3, 100)
y_pred = [bnn_model(x_test) for _ in range(100)]
y_pred = np.array(y_pred)
# Calculate mean and standard deviation of predictions
y_pred_mean = y_pred.mean(axis=0).flatten()
y_pred_std = y_pred.std(axis=0).flatten()
# Plot the results
plt.figure(figsize=(10, 6))
plt.scatter(x, y, label='Data', color='blue')
plt.plot(x_test, y_pred_mean, label='Predictive Mean', color='red')
plt.fill_between(x_test, y_pred_mean - 2 * y_pred_std, y_
pred_mean + 2 * y_pred_std, color='red', alpha=0.3, label=
'Uncertainty (±2 std)')
plt.xlabel('x')
plt.ylabel('y')
plt.legend()
plt.title('Bayesian Neural Network Predictions with
Uncertainty')
plt.show()
```

Figure 4.15 demonstrates the predictive capability and uncertainty quantification of a BNN on a synthetic dataset. Figure 4.15a shows the predictive mean (red line) and the actual data points (blue). Figure 4.15b highlights the uncertainty (±2 standard deviations), predictive mean, and actual data. The shaded area in Figure 4.15b represents the uncertainty, showing how the **BNN** captures the variability in its predictions.

4.8 COMMON MISTAKES AND TROUBLESHOOTING TIPS

4.8.1 Understanding Probability Distributions

- *Mistake:* Confusing discrete and continuous distributions.
- *Tip:* Remember that discrete distributions deal with countable outcomes (e.g., number of heads in coin flips), while continuous distributions handle outcomes over a range (e.g., height of individuals).

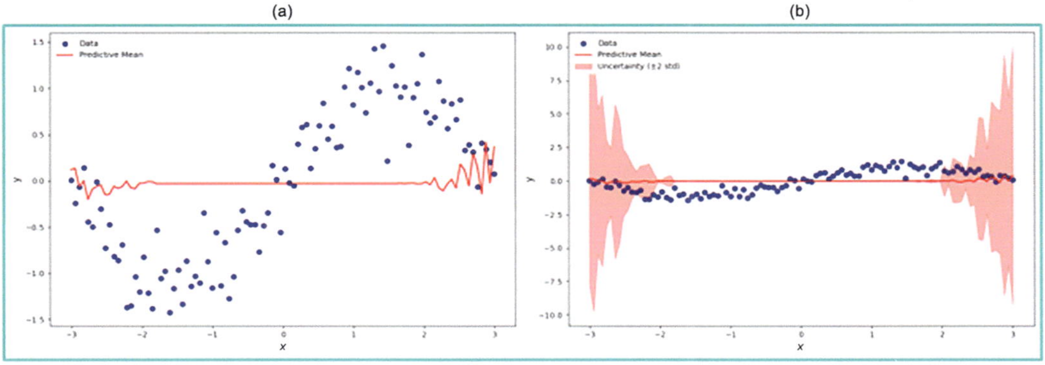

FIGURE 4.15 Predictive mean and actual data points of a BNN on synthetic data.

- *Mistake:* Misinterpreting the properties of distributions (e.g., thinking all data are normally distributed).
- *Tip:* Always check the data and select the appropriate distribution. For example, a Poisson distribution can model the number of events occurring in a fixed interval.

4.8.2 Applying Probability Distributions

- *Mistake:* Using inappropriate distributions for the data type.
- *Tip:* Match your data characteristics to the distribution. For instance, the binomial distribution can be used for binary outcomes, and the normal distribution can be used for continuous, symmetric data.

4.8.3 Overfitting and Underfitting

- *Mistake:* Ignoring signs of overfitting and underfitting in model performance.
- *Tip:* Monitor model performance on both training and validation sets. High training accuracy but low validation accuracy indicates overfitting. Low accuracy on both sets suggests underfitting.
- *Mistake:* Not using regularization techniques to combat overfitting.
- *Tip:* To penalize model complexity, Implement regularization methods like $\mathbf{L}_1/\mathbf{L}_2$ regularization, dropout, or Bayesian priors.

4.8.4 Bayesian Neural Networks

- *Mistake:* Treating BNNs like traditional NNs without considering uncertainty.
- *Tip:* Embrace the probabilistic nature of *BNNs*. Use the posterior distributions of weights to gauge uncertainty in predictions, which is crucial for making informed decisions.

4.8.5 Moments in Statistics

- *Mistake:* Misinterpreting the significance of higher-order moments like skewness and kurtosis.
- *Tip:* Use skewness to understand asymmetry in data and kurtosis to assess tail heaviness. These moments provide deeper insights into data distribution beyond mean and variance.
- *Mistake:* Relying solely on mean and variance for data analysis.
- *Tip:* Consider higher-order moments when data exhibits skewness or heavy tails. This can prevent misinterpretation of data characteristics and improve model fitting.

4.8.6 Sampling Methods

- *Mistake:* Not ensuring randomness in sampling.
- *Tip:* Implement truly random sampling methods to avoid bias. For systematic sampling, ensure that the population does not have periodic patterns that could bias results.
- *Mistake:* Ignoring the need for stratified sampling in heterogeneous populations.
- *Tip:* Use stratified sampling to ensure all population subgroups are adequately represented, improving the precision of estimates.

4.8.7 Bayesian Statistics

- *Mistake:* Choosing inappropriate priors in Bayesian models.
- *Tip:* Select priors that reflect prior knowledge or are non-informative when little prior information is available. Sensitivity analysis can help understand the impact of different priors on posterior distributions.

4.9 REVIEW QUESTIONS

1. How do BNNs handle uncertainty differently from traditional NNs? Discuss the role of probability distributions in *BNNs*.
2. Explain the concepts of overfitting and underfitting in the context of statistical modeling. What are some methods used to prevent these issues in machine learning models?
3. Define and explain the importance of the following terms in the context of probability distributions: mean, variance, skewness, kurtosis.
4. What is Bayes' theorem, and how is it applied to Bayesian statistics? Provide a simple example to illustrate its use.
5. Discuss how model complexity affects the performance of a machine learning model. How do Bayesian statistics help manage model complexity?
6. Compare and contrast the three sampling methods discussed in the chapter. Which method would you choose for a given scenario involving a large, heterogeneous population, and why?
7. Provide an example of a real-world application where the statistical principles discussed in this chapter can be applied. How do these principles improve the prediction accuracy or reliability of the model?
8. What are regularization techniques, and how do they relate to Bayesian statistics? Explain how these techniques help in handling overfitting.
9. How does understanding the uncertainty in model predictions benefit the deployment of machine learning models in sensitive areas like healthcare or autonomous driving?
10. How do prior and posterior distributions differ in Bayesian statistics? Discuss their roles in updating beliefs as new data becomes available and provide an example of how this concept is applied in machine learning.

4.10 PROGRAMMING QUESTIONS

4.10.1 EASY

Implement a program that generates a set of data points following a normal distribution.

1. Use a random number generator to create a set of data points that follow a normal distribution with a specified mean and standard deviation.
2. Calculate the theoretical *PDF* of the normal distribution using the specified mean and standard deviation.

4.10.2 MEDIUM

Implement Bayesian linear regression to predict a dependent variable based on an independent variable.

1. Create a dataset with an independent variable, **x**, and a dependent variable, **y**, with added Gaussian noise.
2. Specify the Bayesian linear regression model with priors for the weights and noise.
3. Use a Bayesian inference method to estimate the posterior distributions of the weights.
4. Generate predictions for new data points and compute credible intervals for the predictions.

4.10.3 Hard

Implement variational inference for a BNN to perform classification on a synthetic dataset.

1. Create a synthetic dataset for a binary classification problem with clear decision boundaries.
2. Specify the architecture of the *BNN* using variational inference methods (e.g., Bayes by Backprop or Flipout layers in TensorFlow Probability).
3. Train the BNN on the synthetic dataset using an appropriate optimizer and loss function.
4. Perform variational inference to estimate the posterior distributions of the weights.
5. Generate predictions for a grid of points covering the input space. Compute the uncertainty in the predictions for each point in the grid.

5 Optimization Theory

5.1 INTRODUCTION

Optimization theory is a key part of mathematics and applied sciences, and it takes a deep look into how we make decisions. At its core, optimization is about finding the best solution from many possible options. It's focused on accuracy, blending logic and function to achieve the most effective results. Whenever we seek knowledge or try to solve a problem, we're quietly pursuing perfection, and optimization guides us on this journey toward excellence. From dividing up resources in an economy to improving the algorithms that run our digital world, the search for the best choice is always ongoing. Optimization theory provides the tools and viewpoints we need to navigate this area, offering solutions and a deeper understanding of how we reach them. This chapter explores some of the most popular optimization methods in deep learning.

5.2 OPTIMIZATION THEORY

Optimization theory is basically about making choices and picking the best option from a set of available alternatives based on specific, measurable criteria. At its core, it sheds light on the principle of selection: when faced with many options, how do we identify the most suitable or advantageous one? This field provides systematic and analytical methods to determine the "best" choice, where "best" could mean minimizing costs, maximizing efficiency, reducing waste, or achieving any other measurable goal. For example, imagine a business evaluating different marketing strategies to find the most cost-effective one, an engineer choosing the most durable material for a new design, or a traveler planning the quickest route to their destination. Constraints are an essential part of optimization problems. They define the feasible solutions and ensure that these solutions are practical and applicable to real-world scenarios. Constraints limit the decision variables and set the boundaries for the optimal solution. They are restrictions or limitations placed on decision variables and can be equality constraints (e.g., $x + y = 10$) or inequality constraints (e.g., $x + y \leq 10$). These constraints shape the feasible region in the decision variable space, within which the optimal solution must be found. There are different types of constraints:

- **Equality constraints** must be satisfied exactly, such as the total weight of items in a knapsack problem.
- **Inequality constraints** set an upper or lower limit, like a budget constraint where total costs must not exceed a specific value.
- **Bound constraints** directly limit the range of decision variables, such as a percentage that must be between 0 and 100.

DOI: 10.1201/9781032690742-5

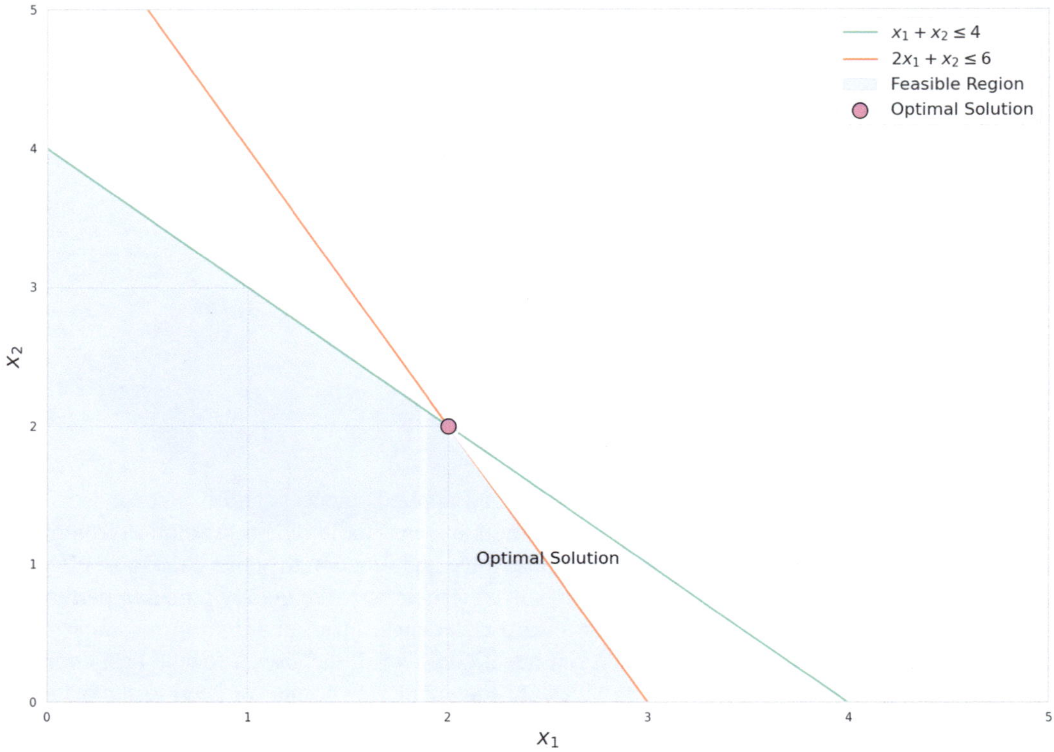

FIGURE 5.1 Feasible region and constraints in linear programming.

Feasible solutions are those that meet all the constraints, while infeasible solutions violate one or more constraints. Active constraints are those that, if slightly changed, would alter the optimal solution. Binding constraints are specific inequality constraints that are exactly met at the optimal solution and directly influence it. Many optimization algorithms are initially designed for problems without constraints. However, there are several effective methods to handle constraints. Penalty methods convert a constrained problem into an unconstrained one by adding penalties to the objective function for any violations of the constraints.

Figure 5.1 features two primary constraint lines: the green line represents the inequality $x_1 + x_2 \leq 4$, while the orange line corresponds to $2x_1 + x_2 \leq 6$. These lines present the boundaries within which the variables $\mathbf{x}_1$ and $\mathbf{x}_2$ must lie to satisfy both constraints simultaneously. The area where these constraints overlap is a shaded area, highlighting the feasible region where potential solutions exist. At the core of this feasible region lies the optimal solution, marked prominently with a large purple scatter point. This point, situated at the coordinates ($\mathbf{x}_1$, $\mathbf{x}_2$), represents the values of $\mathbf{x}_1$ and $\mathbf{x}_2$ that maximize the objective function $3\mathbf{x}_1 + 2\mathbf{x}_2$ while adhering to the defined constraints.

5.3 TYPES OF OPTIMIZATIONS

5.3.1 Linear Optimization

Linear optimization, also known as linear programming, focuses on finding the best outcome, like maximizing profit or minimizing cost, in models where all relationships are linear. The main component is the objective function that you want to optimize. For example, if you're trying to maximize profit from selling two products, **A** and **B**, the objective function could be written as $P = aA + bB$, where **a** and **b** represent the profit per unit of products **A** and **B**, respectively. In linear optimization, you have limits called constraints, which are expressed as linear equations or inequalities. Using the example above, constraints might include limitations on available resources like raw materials or labor, represented in linear terms. These constraints define the set of all feasible solutions that meet the requirements. The optimal solution is the one that maximizes or minimizes the objective function within this feasible set. In two-dimensional cases, this optimal point is often found at a vertex of the feasible polygon, the corner points of the area defined by the constraints. The simplex method is the most recognized technique for solving linear optimization problems, but other methods, like interior point methods, are also effective, especially in specific situations. However, linear optimization assumes that both the objective function and the constraints are linear, which might not be true in more complex, real-world scenarios. Nonetheless, some non-linear problems can be approximated as linear to make use of linear optimization methods, broadening their applicability. The general form of a linear optimization problem is:

$$\text{Maximize or minimize } Z = c_1x_1 + c_2x_2 + \ldots + c_nx_n$$

Subject to:

$$a_{11}x_1 + a_{12}x_2 + \ldots + a_{1n}x_n \leq b_1$$

$$a_{21}x_1 + a_{22}x_2 + \ldots + a_{2n}x_n \leq b_2$$

$$x_1, x_2, \ldots, x_n \geq 0$$

Suppose a company produces two products, **A** and **B**, with profit margins of \$3 per unit for product **A** and \$4 per unit for product **B**. The objective function for maximizing profit is:

$$\text{Maximize } P = 3A + 4B$$

Subject to the following constraints: **A** + 2**B** ≤ 14 (raw material constraint), **3A** + **B** ≤ 18 (labor constraint), and **A, B** ≥ 0.

Figure 5.2 features two primary constraint lines: the light blue line represents the inequality $A + 2B \leq 14$, while the blue line corresponds to $A + B \leq 18$. These lines present the boundaries within which the variables **A** and **B** must reside to satisfy both constraints simultaneously. The intersection of these constraints forms the vertices of the feasible region, which is shaded to visually demarcate the area where potential solutions lie. At the core of this feasible region is the optimal solution, marked prominently with a large green point. This point, situated at approximately (4, 3), represents the values of **A** and **B** which maximize the objective function **P** = 3**A** + 4**B** while adhering to the defined constraints. Additionally, the red dashed line represents the objective function at its optimal value, **P** = 24, illustrating the level curve of the objective function that intersects the feasible region at the optimal point. This line serves as a visual cue, demonstrating how the objective function interacts with the constraints to determine the optimal solution.

5.3.2 Non-linear Optimization

Non-linear optimization deals with situations where the objective function, the constraints, or both involve non-linear relationships. Unlike linear relationships, which are represented by straight lines or flat surfaces, non-linear interactions add complexity, making these problems more challenging and more reflective of real-world dynamics. This flexibility is crucial in many scientific and industrial applications because it models complex relationships. In non-linear optimization, the objective function includes non-linear expressions of the decision variables. For example, consider the

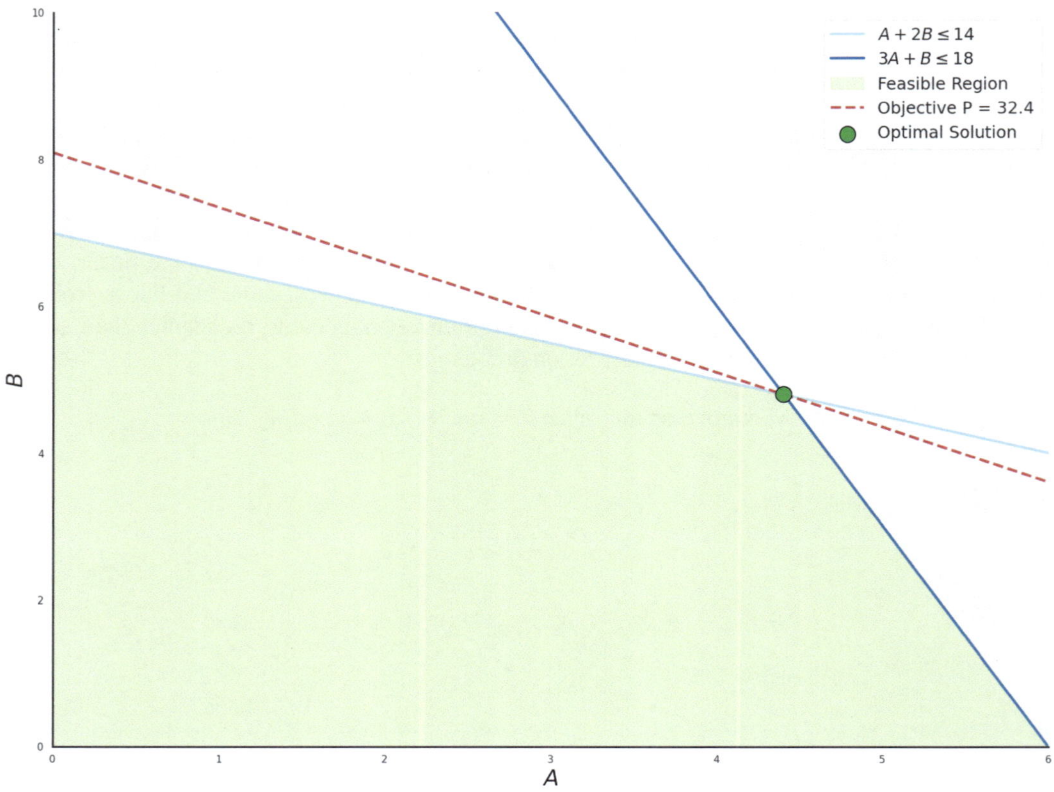

FIGURE 5.2 Linear optimization: feasible region and objective function.

function $P = A^2 + 3B^3 - 4AB$, where the goal is to maximize or minimize **P**. This creates a dynamic landscape of possible solutions, making the optimization process more complicated. Constraints can also be non-linear, such as $A^2 + B^2 \leq 9$, which represents a circular boundary in a two-dimensional space for variables **A** and **B**. This geometric constraint shapes the feasible solution space into forms that are not just flat or polyhedral. The solution techniques include:

(a) *Gradient-Based Methods:* These include gradient descent, Newton's method, and quasi-Newton methods, which utilize derivatives to guide the optimization process.
(b) *Direct Search Methods:* For instance, the Nelder–Mead method operates without derivative information and is suitable for problems where derivatives are infeasible to calculate.
(c) *Evolutionary and Genetic Algorithms:* Inspired by natural selection, these methods use heuristic approaches to explore the solution space, which is especially useful in complex landscapes.

A general non-linear optimization problem can be formulated as:

$$\text{Maximize or minimize } f\left(x_1, x_2, \ldots, x_n\right)$$

Subject to:

$$g_1\left(x_1, x_2, \ldots, x_n\right) \leq b_1, \quad g_2\left(x_1, x_2, \ldots, x_n\right) \leq b_2, \ldots, \quad g_m\left(x_1, x_2, \ldots, x_n\right) \leq b_m$$

In this case, $\mathbf{f(x_1, x_2)}$ represents the non-linear objective function, and $\mathbf{g(x_1, x_2)}$ represents the non-linear constraints, shaping the feasible region. Techniques like gradient descent would be used to find optimal solutions. Suppose you aim to maximize the non-linear objective function $P(A,B) = A^2 + B^2$, subject to the non-linear constraint $A^2 + B^2 \leq 9$, which represents a circular boundary with a radius of 3. The feasible solution space is the interior of this circle, and the maximum value of P occurs at the boundary where $P = A^2 + 3B^3 - 4AB$, producing a maximum value of P(3, 0)=9.

Figure 5.3 features a vibrant contour map of the objective function $P = A^2 + 3B^3 - 4AB$, rendered using the perceptually uniform colormap. Central to the visualization is the constraint $A^2 + B^2 \leq 9$, depicted as a prominent dashed black circle with a radius of 3. This boundary defines the feasible region in which optimization must occur. The area inside the circle, where both constraints are satisfied, is subtly highlighted with a semitransparent overlay of the same colormap, reinforcing the feasible region's significance without overwhelming the objective function's contours. Additionally, a specific point within the feasible region, such as (1, 1), is marked with a large red scatter point.

5.3.3 Integer Optimization

Integer optimization (IO), or integer programming (IP), focuses on optimization problems where some or all decision variables must assume integer values. This requirement arises when fractional solutions are impractical, such as determining the number of items to produce, personnel assignments, or scheduling tasks. IO is critical for problems demanding discrete solutions, presenting unique challenges distinct from those of continuous optimization. The types of IO are:

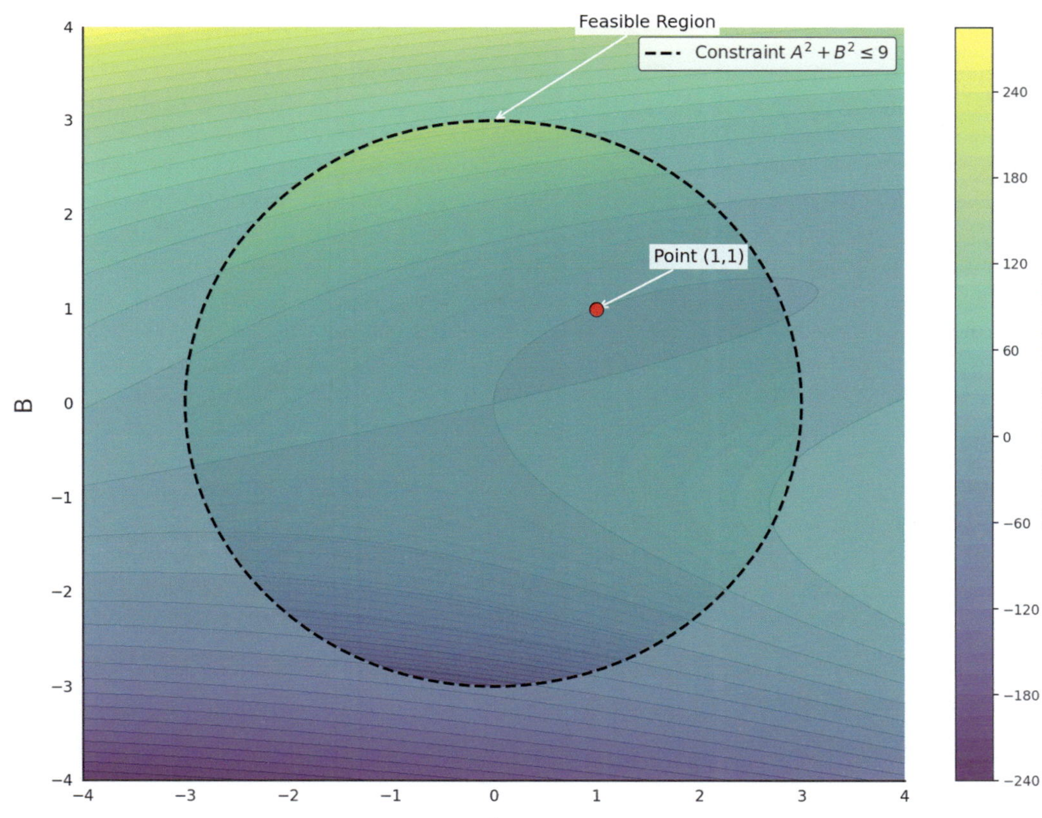

FIGURE 5.3 Non-linear optimization with constraints.

(a) *Pure Integer Programming (PIP):* All decision variables are integers.
(b) *Mixed Integer Programming (MIP):* Combines integer and continuous variables, addressing a more complete range of practical scenarios.
(c) *Binary Integer Programming (BIP):* Decision variables are binary (0 or 1), ideal for problems involving binary decisions like turning a process on or off.

In some optimization problems, the variables are required to be whole numbers (integers). The objective function, which we aim to maximize or minimize, can be linear or non-linear, leading to situations known as linear IP or non-linear IP. Similarly, the constraints in these problems can be linear or non-linear and may also demand that the variables be integers. Several techniques are employed to solve these IP problems. One method is branch and bound (B&B), which involves branching on the fractional parts of variables. This technique systematically explores different possible solutions and uses upper and lower bounds to reduce the search space, making it easier to find the optimal integer solution. Another approach is cutting plane method, which add extra constraints called cutting planes to the problem. These methods derive valid inequalities to eliminate fractional solutions, guiding the search toward integer solutions. Additionally, heuristic methods like rounding or diving provide quick ways to find feasible integer solutions. While these methods are faster, they may not always find the best possible solution but offer an acceptable one. A general IO problem can be formulated as:

$$\text{Maximize or minimize } Z = c_1 x_1 + c_2 x_2 + \ldots + c_n x_n$$

Subject to:

$$a_{11} x_1 + a_{12} x_2 + \ldots + a_{1n} x_n \le b_1, \quad x_1, x_2, \ldots, x_n \in \mathbb{Z}$$

This formulation captures the requirement that decision variables must be integers, and methods like B&B or cutting plane methods are used to solve these problems. Suppose a company needs to produce tables and chairs. The decision variables are the number of tables **T** and chairs **C** to produce. The profit is given by the linear objective function $Z = 20T + 15C$, and the constraint is that the total production cannot exceed 100 units $T + C \le 100$. As fractional tables or chairs cannot be produced, **T** and **C** must be integers. The IO problems are classified as NP-hard, meaning the computational difficulty can scale exponentially with problem size. The discrete nature of the solution space restricts the use of gradient-based optimization methods that are effective in continuous scenarios. Due to their complexity, IO problems often necessitate specialized computational tools.

Figure 5.4 illustrates the feasible region defined by the constraint $A + B \le 12$ and highlights the optimal solution that maximizes the objective function $P = 3A + 2B$. The plot features a grid of discrete points where both **A** and **B** range from 0 to 12, representing all possible combinations of these variables. The feasible solutions are depicted using a scatter plot with square markers, colored according to their corresponding objective function values **P**. The semitransparent light gray shading delineates the feasible region where the constraint **A** + **B** ≤ 12 is satisfied, providing a boundary that confines the optimization problem's solution space. At the heart of the feasible region lies the optimal solution, marked prominently with a large golden-yellow scatter point. This point represents the values of **A** and **B** that maximize the objective function **P** while adhering to the constraint.

5.3.4 Convex Optimization

Convex optimization is a specialized area within optimization where both the objective function and the constraints are convex. This property greatly simplifies the optimization process because any local minimum is also a global minimum, eliminating the uncertainty often found in non-convex optimization. A function **f** is considered convex over an interval **I** if, for any two points $\mathbf{x}_1$ and $\mathbf{x}_2$ in **I** and any $\boldsymbol{\lambda}$ in the interval [0, 1], the following inequality holds:

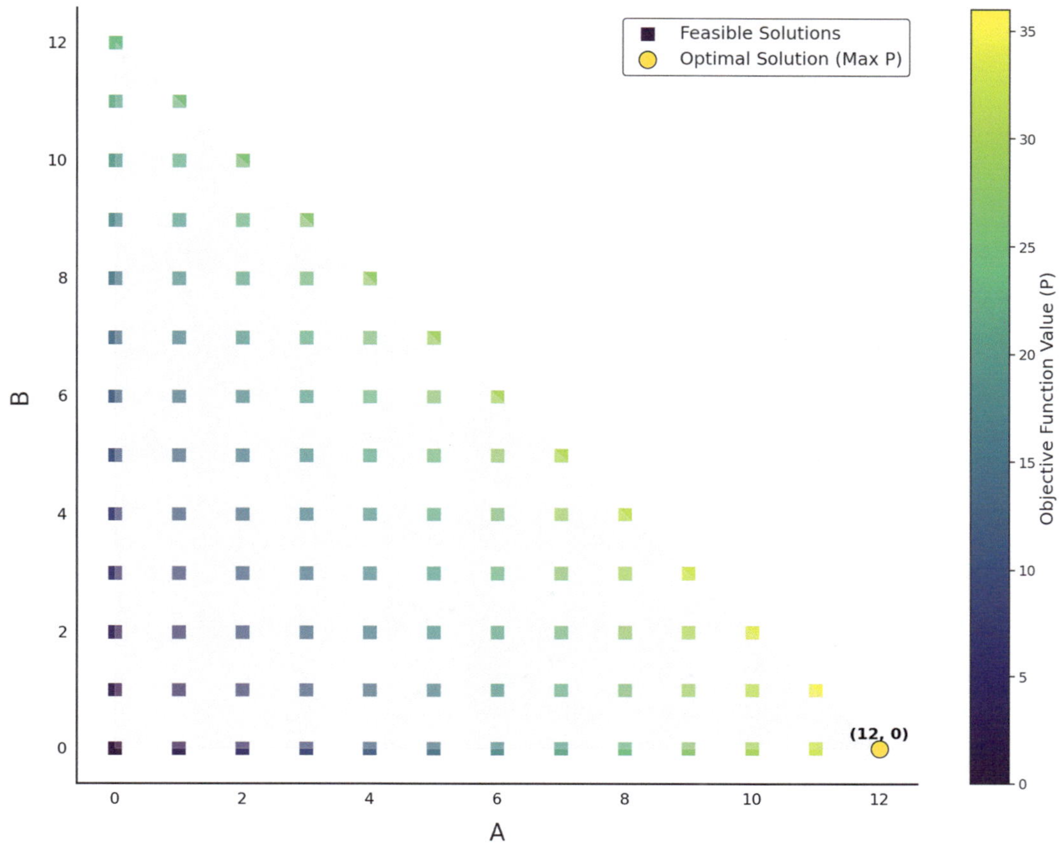

FIGURE 5.4 Integer optimization.

$$f(\lambda x_1 + (1-\lambda)x_2) \leq \lambda f(x_1) + (1-\lambda) f(x_2)$$

Graphically, this means that the line segment between any two points on the graph of the function lies above or on the graph itself. A set **S** is convex if, for every pair of points $\mathbf{x}_1$ and $\mathbf{x}_2$ in **S** and any $\boldsymbol{\lambda}$ in the interval [0, 1], the point $\lambda x_1 + (1-\lambda)x_2$ also belongs to **S**. This ensures that the entire line segment between any two points in **S** is contained within **S**. Convex optimization problems have several key features. First, if a solution exists, it is unique and globally optimal. Unlike non-convex problems, convex optimization does not have local optima that are not global. Additionally, the intersection of all constraints forms a convex feasible set, which simplifies the search for the optimal solution. Various techniques are used to solve convex optimization problems. Gradient descent is an iterative method that moves in the direction of the steepest decrease of the function. Interior point methods explore the inside of the feasible set to find the optimal solution. Subgradient and proximal methods are particularly useful for convex optimization problems that are not smooth. Despite their advantages, convex optimization problems can present challenges. Finding the global optimum might be computationally demanding, especially in high-dimensional spaces. Moreover, not all real-world problems are convex, and transforming non-convex problems into convex ones can be limiting and impractical. A general convex optimization problem can be formulated as:

$$\text{Minimize } f(x)$$

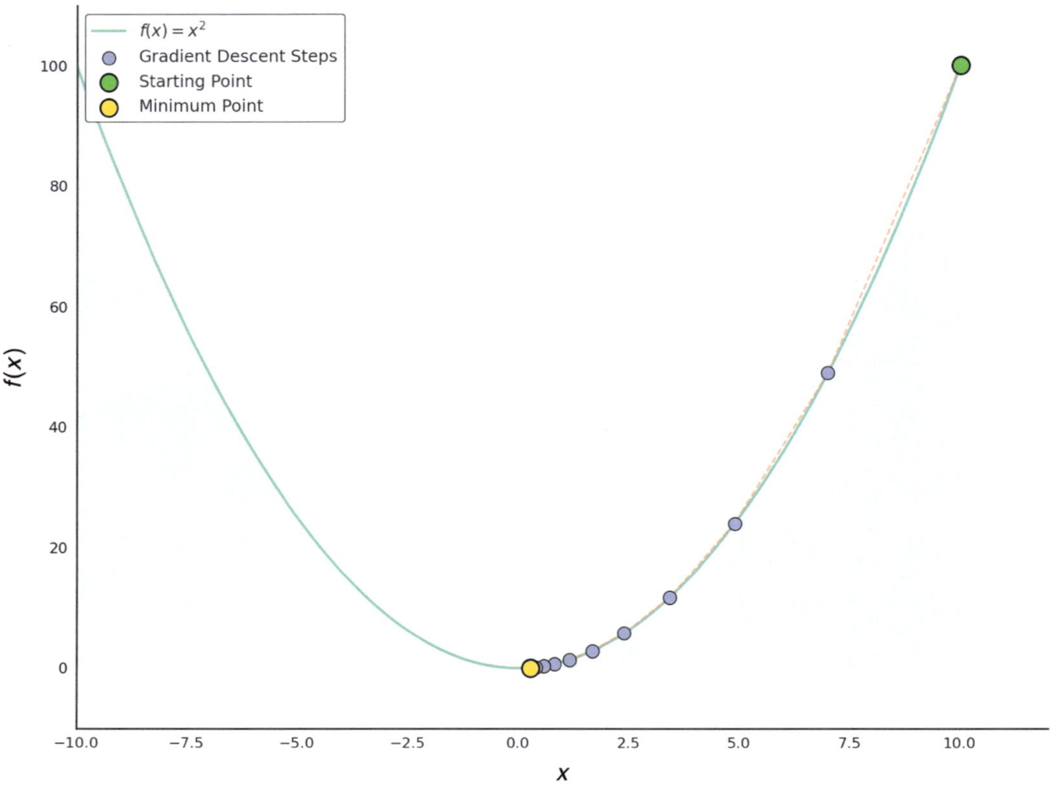

FIGURE 5.5 Convex optimization using gradient descent on $f(x) = x^2$.

Subject to:

$$g_i(x) \leq 0 \quad \forall i, \quad x \in C$$

where:

- **f(x)** is a convex function,
- $\mathbf{g_i(x)}$ represents convex constraints, and
- **C** is a convex set.

This ensures that any local minimum is also a global minimum, and techniques like gradient descent or interior point can efficiently solve such problems. Consider the convex function $\mathbf{f(x)} = \mathbf{x}^2 + 4\mathbf{x} + 4$, which has a global minimum at $\mathbf{x} = -2$. The objective is to minimize this function, subject to the constraint $\mathbf{x} \geq -3$. As **f(x)** is convex and the constraint forms a convex set, the optimal solution is $\mathbf{x} = -2$, where the function achieves its minimum value of 0.

In Figure 5.5, the function $f(x) = x^2$ is plotted as a smooth curve. This quadratic function represents a simple convex shape, making it suitable for demonstrating the gradient descent method. The gradient descent process begins at the starting point, marked by the green dot, located on the right side of the function. From here, gradient descent iteratively updates the value of **x** to minimize **f(x)**. At each step, the algorithm moves in the direction of the negative gradient (the slope of the function at the current point), adjusting **x** to decrease the function value. These steps are represented by the blue dots along the curve. Each dot shows a position where the algorithm evaluates and updates the function value, gradually moving toward the minimum. As the algorithm approaches the minimum point, highlighted in yellow, the steps become smaller. This behavior illustrates that as the gradient (slope) decreases near the minimum, the updates become less significant, allowing the algorithm to converge precisely at the minimum. The minimum of $f(x) = x^2$ occurs at $\mathbf{x} = 0$, where the function value is also zero. The shaded area under the curve provides visual emphasis on the function's domain, while the red dashed line shows the trajectory of the algorithm's updates, highlighting how it progressively converges toward the minimum.

5.3.5 Combinatorial Optimization

Combinatorial optimization is an important field within optimization that deals with problems in discrete and combinatorial settings, such as finding the shortest path in a network or graph. It addresses optimization challenges where the solution space consists of distinct, countable options, and the number of possible solutions can be immense, often increasing exponentially as the problem grows larger. A key feature of combinatorial optimization is the discrete nature of its solution space, meaning solutions are separate and countable. As the size of the problem increases, the complexity and the number of potential solutions expand dramatically. These problems often involve selecting the best combination or sequence of elements from a set, which is a defining characteristic of this type of optimization. Common examples of combinatorial optimization problems include the traveling salesman problem (TSP), where the objective is to find the shortest possible route that visits each city exactly once and returns to the starting point. Another example is graph coloring, where colors are assigned to the vertices of a graph so that no two adjacent vertices share the same color, aiming to use the fewest colors possible. Bin packing involves efficiently packing objects of different sizes into the smallest number of bins. Various techniques are used to solve combinatorial optimization problems. Exact algorithms, such as B&B or the Hungarian algorithm, guarantee an optimal solution but are often very computationally intensive. Heuristic algorithms, like the greedy method or hill climbing, aim to find a good solution quickly, though they do not guarantee the best possible solution. Metaheuristic algorithms, such as genetic algorithms or ant colony optimization, are higher-level strategies that guide simpler heuristics toward better solutions. The main challenges in combinatorial optimization include the NP-hard nature of many of these problems, meaning there is no known efficient (polynomial-time) solution, and it is uncertain if one exists. Additionally, the vast size of the solution space makes exhaustive searches impractical except for the smallest problems.

Figure 5.6 showcases an example of the TSP, which is a classic optimization problem where the objective is to determine the shortest possible route that visits each city exactly once and returns to the starting point. The figure illustrates five cities, labeled **A**, **B**, **C**, **D**, and **E**, positioned on a 2D coordinate plane. Each city is marked with a distinct shape and color to enhance clarity. The optimal path connecting these cities is shown by the dark line traversing from one city to the next. The total length of this optimal route is indicated in the legend as 20.06 units, representing the shortest path that satisfies the TSP criteria for these specific cities. The path begins at city **A** (blue circle), moves through city **B** (orange star), then continues to city **E** (purple plus), city **C** (green triangle), and finally reaches city **D** (red diamond) before looping back to city **A**.
The objective function for the TSP is:

$$\text{Minimize} \sum_{i=1}^{n} d\left(x_i, x_{i+1}\right)$$

where:

- $\mathbf{x}_i$ is the $\mathbf{i}^{\text{th}}$ city in the tour,
- $\mathbf{d}(\mathbf{x}_i, \mathbf{x}_{i+1})$ represents the distance between cities $\mathbf{x}_i$ and $\mathbf{x}_{i+1}$, and
- the solution space consists of all possible permutations of city visits, making the problem combinatorial.

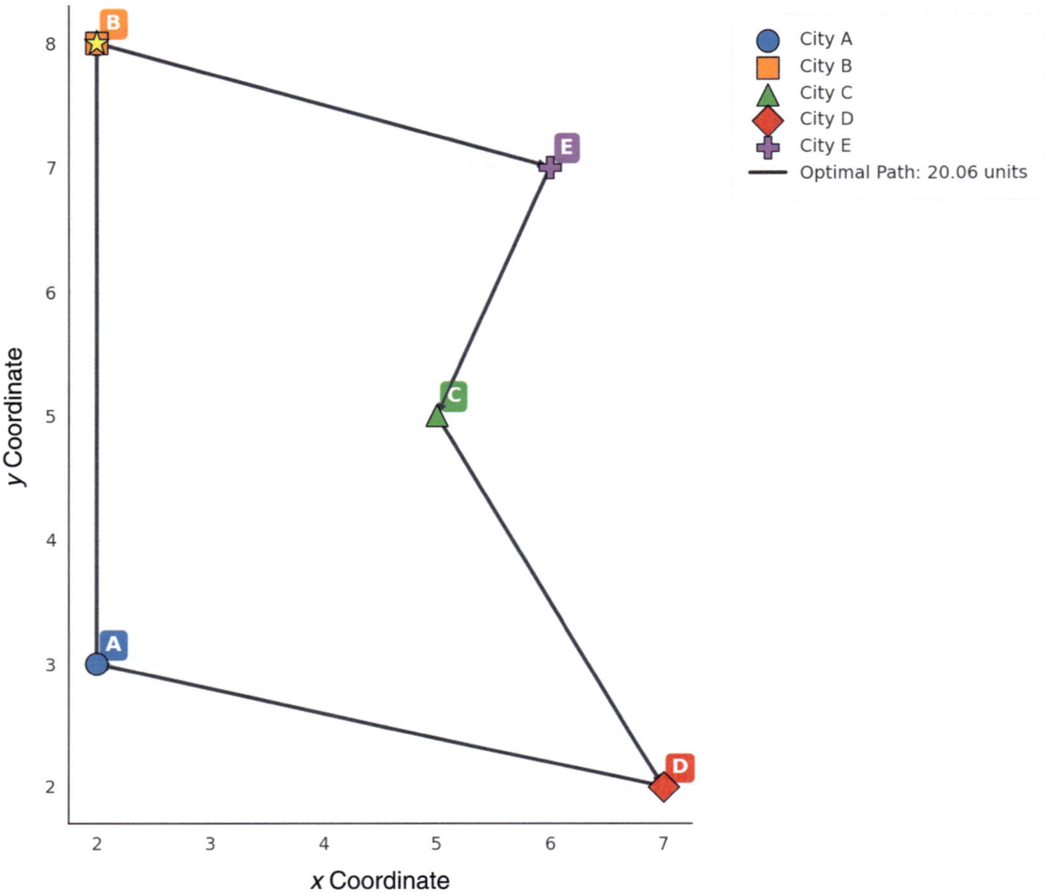

FIGURE 5.6 Traveling salesman problem.

This formulation represents the challenge of finding the optimal sequence of cities with an enormous solution space, often tackled using exact or heuristic methods. Consider the TSP with five cities. The goal is to find the shortest route that visits each city exactly once and returns to the starting city. If there are five cities, there are 120 possible routes (since 5! = 120). Using an exact algorithm like B&B, the shortest route can be found by systematically evaluating and pruning possible routes.

5.3.6 Gradient Descent

Gradient descent is a foundational optimization technique in machine learning, essential for solving complex problems where analytical solutions are either unavailable or computationally prohibitive. This iterative method is adept at navigating multi-dimensional data landscapes to find the local minimum of a function, typically a loss function in machine learning contexts. The primary objective of gradient descent is to minimize the loss function, which measures the discrepancy between the predicted outputs of a model and the actual outcomes. The principle behind gradient descent is analogous to descending a hill. Mathematically, this is achieved by moving in the direction opposite to the function's gradient at the current point, which points toward the steepest ascent. Let us go to the detailed mechanics of it:

- *Gradient:* For a function $f(x)$, where $\mathbf{x}$ is a vector of parameters, the gradient $\nabla f(x)$ is a vector of partial derivatives. It indicates the direction and rate of the steepest increase.

- *Update Rule:* The parameter vector **x** is updated iteratively using the rule: $x_{\text{new}} = x_{\text{old}} - \alpha\nabla_x f\left(x_{\text{old}}\right)$ where α is the learning rate.
- *Convergence:* Properly tuning the learning rate, and under certain conditions (e.g., convexity of the function), gradient descent can converge to the global minimum. In non-convex scenarios, it may settle at a local minimum or a saddle point.

Gradient descent comes in several variations, each designed to enhance efficiency, stability, or convergence when solving optimization problems. Stochastic gradient descent (SGD) uses a single randomly selected data point to estimate the gradient, which speeds up each iteration and can help avoid getting stuck in local minima. Mini-batch gradient descent uses a small subset of data to compute the gradient, offering a middle ground between the speed of SGD and the stability of using the full dataset. The momentum method adds a portion of the previous update to the current step. This approach aims to accelerate convergence and reduce oscillations, especially in areas where the loss function changes sharply. Adaptive learning rate methods, like Adagrad, RMSprop, and Adam, adjust the learning rate based on previous gradients. This allows for more precise convergence and improved stability during training. Setting the learning rate too high can cause the algorithm to diverge, while setting it too low might lead to slow convergence or getting stuck in suboptimal points. With non-convex functions, there's a risk of converging to local minima or saddle points instead of the global minimum. As gradient descent is sensitive to the scale of the features, normalizing or standardizing them is often necessary for effective optimization. Despite these challenges, gradient descent remains a fundamental tool in optimization, especially in machine learning. It supports a wide range of applications, from simple regression tasks to complex deep learning algorithms. Its flexibility and efficiency make it indispensable for researchers and practitioners.

The updated rule for gradient descent is given by:

$$\theta_{\text{new}} = \theta_{\text{old}} - \alpha\nabla J\left(\theta\right)$$

where:

- $\boldsymbol{\theta}$ is the parameter vector,
- $\boldsymbol{\alpha}$ is the learning rate, and
- $\boldsymbol{\nabla J(\theta)}$ is the gradient of the cost function with respect to $\boldsymbol{\theta}$.

This iterative process continues until the gradient converges, minimizing the loss function and finding the optimal parameters for the model. Suppose you are training a linear regression model to predict house prices. The cost function $J\left(\theta\right) = \frac{1}{2m}\Sigma(h_\theta\left(x_i\right) - y_i)^2$ needs to be minimized, where $\boldsymbol{\theta}$ represents the model parameters. Let the initial value of $\boldsymbol{\theta} = 0.5$, and the gradient at this point be calculated as $\boldsymbol{\nabla J(\theta)} = 1.2$. Using a learning rate $\boldsymbol{\alpha} = 0.01$, the update rule would adjust $\boldsymbol{\theta}$ as follows:

$$\theta_{\text{new}} = \theta - \alpha\nabla J\left(\theta\right) = 0.5 - 0.01 \times 1.2 = 0.488$$

Figure 5.7 shows the trajectory of gradient descent on the quadratic function $\mathbf{f(x) = x^2}$. The plot starts with a high initial value (the "Starting Point") and iteratively updates based on the gradient, moving toward the function's minimum. The objective is to reach the lowest point of the function, which is at $\mathbf{x} = 0$, also marked as the "Final Point." Each red dashed line and crimson scatter point represents a step in the gradient descent process, with every fifth step labeled for clarity. The goal is to illustrate how the gradient descent algorithm takes steps proportional to the negative of the gradient to minimize the function. The plot helps visualize the optimization process, showing how the gradient descent path narrows down towards the optimal solution.

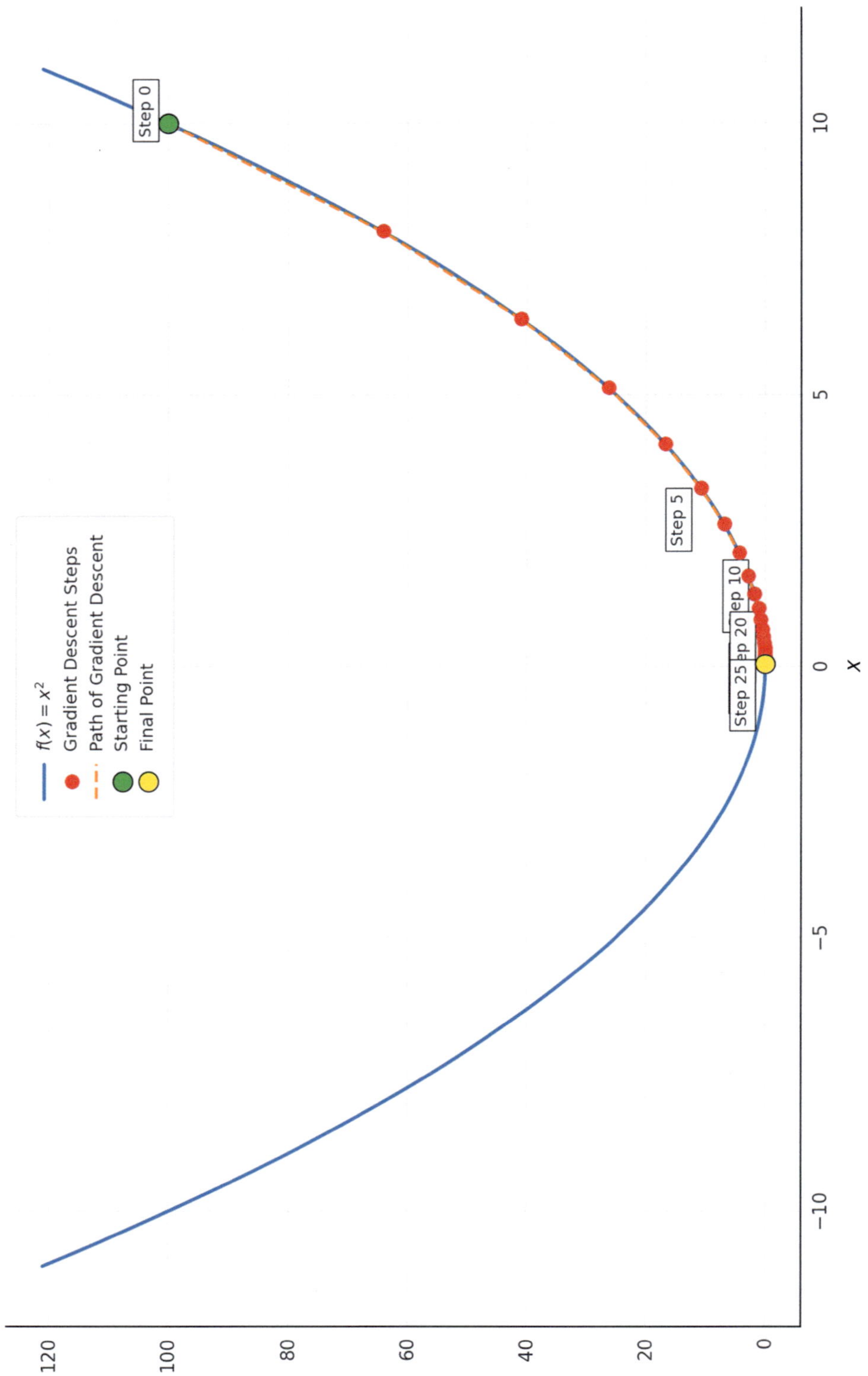

FIGURE 5.7 Gradient descent on a quadratic function.

5.3.7 Stochastic Optimization

Stochastic optimization is a field that deals with problems under uncertainty. In these cases, some parts of the model, such as parameters, constraints, or the objective function, include random variables. This uncertainty makes the optimization process more complex but also more reflective of real-world situations. The goal in stochastic optimization is often to find a feasible and optimal solution on average or with a high probability rather than one that is deterministically optimal. This approach is crucial for handling scenarios where exact information about the model is unavailable or when the system is subject to random fluctuations and noise. Several techniques are used in stochastic optimization. One method is sample average approximation (SAA), which involves solving the problem multiple times using different samples from the probability distributions of the uncertain parameters and then averaging the results. Another technique is SGD, an extension of the traditional gradient descent method that uses a randomly selected subset of data, called a mini-batch, to perform each update. This method is commonly used in machine learning for training models on large datasets. Monte Carlo simulation is also employed to assess the impact of risk and uncertainty in prediction and forecasting models by simulating a wide range of possible outcomes. Stochastic optimization faces several challenges. The inherent uncertainty in these problems makes solutions computationally complex and difficult to obtain. The quality of the solutions heavily depends on the quality and quantity of the data available about the uncertainties involved, leading to data dependency issues. Scalability becomes a concern as the size of the data and the number of uncertain parameters increase, requiring significantly more computational resources to find a solution efficiently. Despite these challenges, stochastic optimization is increasingly important in industries and sectors where uncertainty is a significant factor. By incorporating randomness directly into the decision-making process, organizations can develop strategies that are not only theoretically optimal but also practical and robust against real-world variability and unpredictability. This optimization approach improves decision-making by providing frameworks that anticipate and effectively manage the inherent uncertainties of various operational environments. The general form of a stochastic optimization problem is:

$$\text{Minimize } \mathbb{E}\left[f(x,\xi)\right]$$

where:

- **f(x, ξ)** is the objective function,
- ξ represents random variables (uncertainties), and
- $\mathbb{E}$ is the expectation over the random variables.

Techniques such as SAA estimate the objective by averaging over several samples of ξ, helping optimize decisions under uncertainty. Suppose a company wants to optimize its inventory management under uncertain demand. Using SGD, they aim to minimize the cost function $\mathbf{J(\theta)}$, where $\boldsymbol{\theta}$ represents the reorder point. With demand data subject to randomness, they update the reorder point iteratively using small mini-batches of demand data. If the gradient estimate at iteration 1 is $\boldsymbol{\nabla}\mathbf{J(\theta)} = 0.6$ and the learning rate is $\boldsymbol{\alpha} = 0.05$, the updated reorder point would be:

$$\theta_{\text{new}} = \theta_{\text{old}} - \alpha\nabla J(\theta) = \theta - 0.05 \times 0.6$$

Figure 5.8, illustrates the process of SGD for optimizing the objective function $f(x) = x^2 + \text{noise}$. The blue line represents the noisy objective function, $f(x) = x^2 + \text{noise}$, which includes randomness, making the curve less smooth. The path of SGD is traced with red dashed lines, highlighting the steps taken from the starting point to the final point. The green dot marks the starting point of

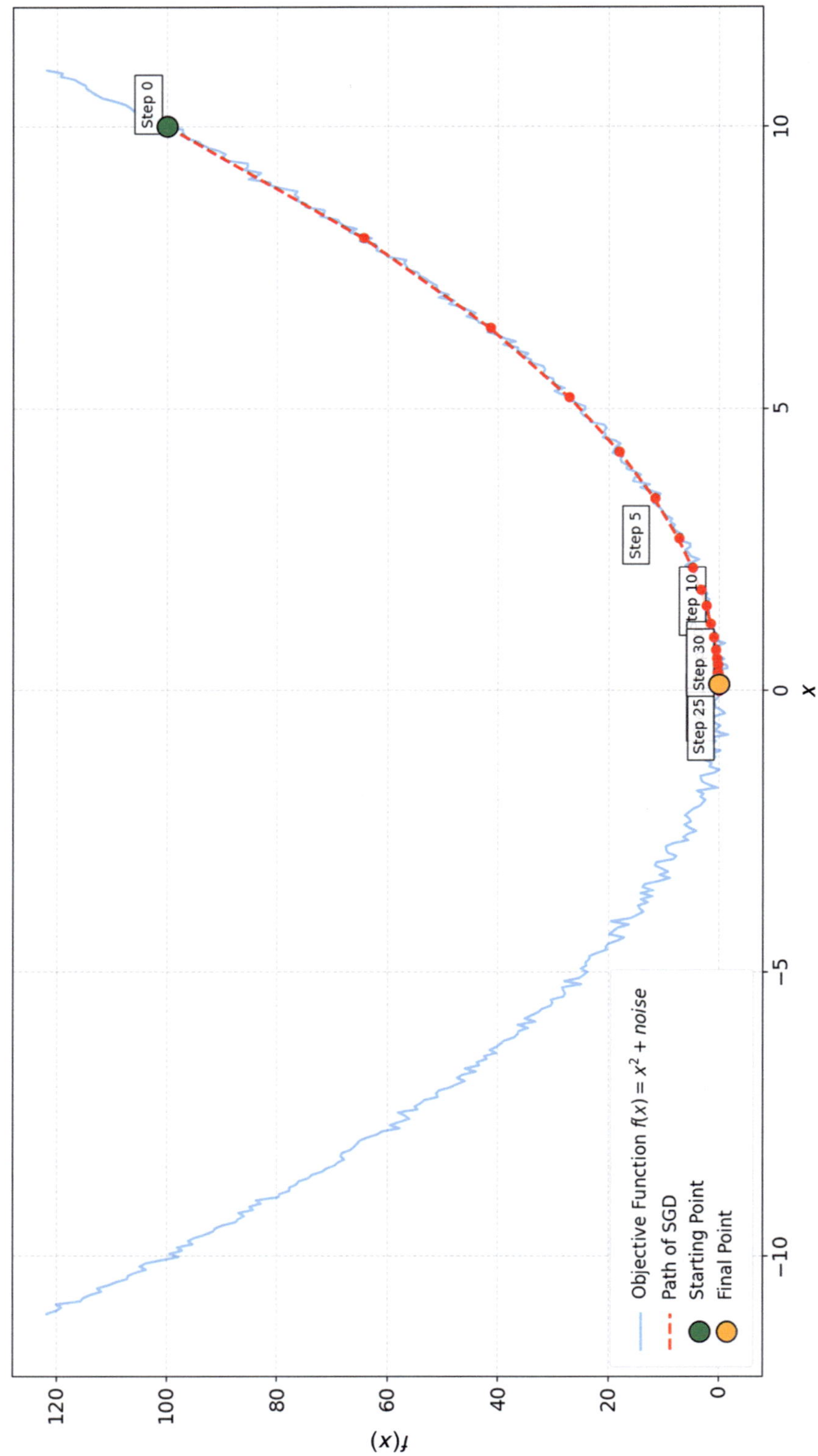

FIGURE 5.8 Stochastic gradient descent.

x = 10, where the optimization process begins. The orange dot indicates the final point after 30 iterations, showing where the algorithm has converged. Red dots along the dashed line indicate the positions visited by SGD at each iteration. To avoid clutter, every fifth step is annotated with "Step **X**", where **X** is the iteration number, showing the algorithm's progression.

5.3.8 Simplex Method

The Simplex method is a fundamental algorithm in linear optimization, widely used to solve linear programming problems. In linear programming, the primary objective is to maximize or minimize a linear function while satisfying a set of linear equality or inequality constraints. These constraints define the feasible region, which is typically a polyhedron or polytope where the solution must lie. If an optimal solution exists, it will be found at one of the vertices (corner points) of this feasible region. The algorithm starts at a vertex of the feasible region and moves from one vertex to another, checking at each step whether the move improves the value of the objective function. This process continues until it reaches a vertex where no adjacent vertex offers a better value, indicating that the optimal solution has been found. The Simplex method is generally very efficient in practice, even though its worst-case time complexity can be exponential. This means that while it could theoretically take a very long time for some problems, it usually performs exceptionally well in real-world scenarios. Sometimes, the method might cycle between the same vertices without making progress. To prevent this, strategies like Bland's Rule are used to ensure that the algorithm moves toward a solution without getting stuck in a loop. Every linear programming problem has a corresponding dual problem. Understanding the relationship between the original (primal) problem and its dual can provide deeper insights and can sometimes simplify finding the solution. The Simplex method has been adapted into various forms to handle more complex situations, such as the Two-Phase Simplex method and the Revised Simplex method. While the Simplex method remains a dominant technique in linear programming, it faces competition from interior-point methods, which can be more suitable for solving very large-scale linear programming problems. The linear programming problem in standard form is:

$$\text{Maximize } Z = c_1x_1 + c_2x_2 + \ldots + c_nx_n$$

Subject to:

$$Ax \leq b, \quad x \geq 0$$

where:

- $c_1, c_2, \ldots, c_n$ are the coefficients of the objective function,
- **A** is the matrix of constraint coefficients,
- **x** is the vector of decision variables, and
- **b** is the vector of constraint bounds.

This formulation represents the optimization problem, where the Simplex method navigates the feasible region defined by the constraints to find the optimal solution. Consider a linear programming problem to maximize the objective function $\mathbf{Z} = 3\mathbf{x}_1 + 2\mathbf{x}_2$, subject to the constraints:

$$x_1 + x_2 \leq 4, 2x_1 + x_2 \leq 6, x_1, x_2 \geq 0$$

The Simplex method starts at one vertex of the feasible region, say (0, 0), and iteratively moves to adjacent vertices, checking for improvement in the objective function. The method eventually reaches the optimal solution at $\mathbf{x}_1 = 2$, where $\mathbf{Z} = 10$.

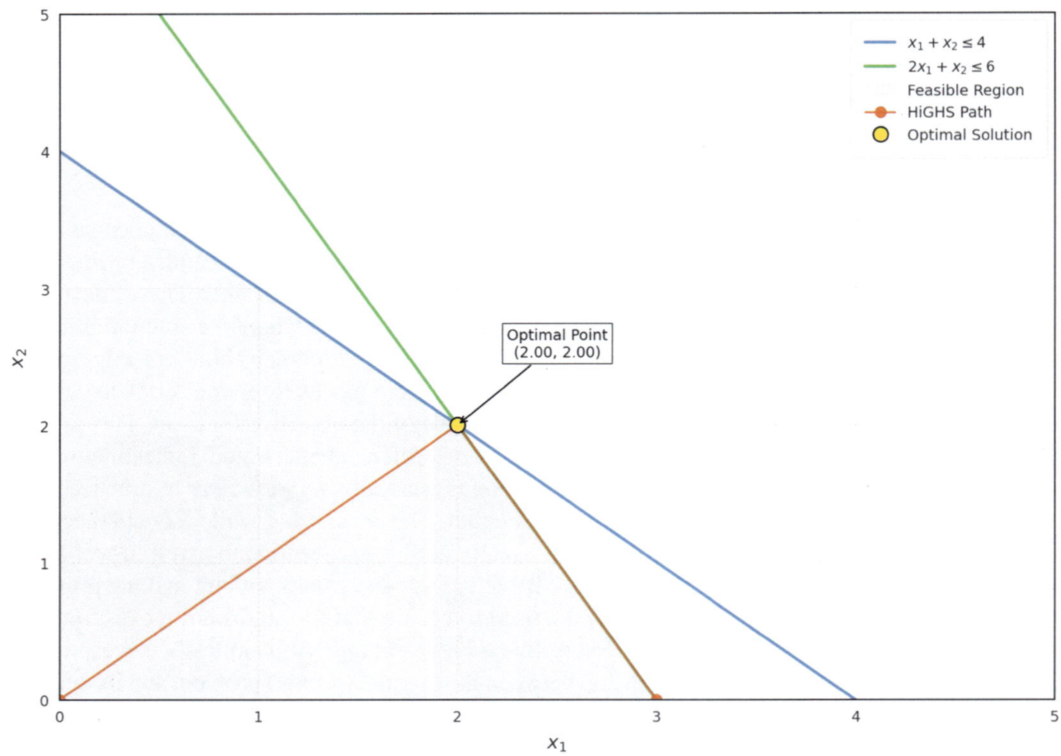

FIGURE 5.9 Simplex method.

Figure 5.9 demonstrates the application of the Simplex method in solving a linear programming problem. The objective is to find the optimal solution that maximizes or minimizes a linear objective function subject to constraints. In this illustration, the constraints are represented by the lines: $x_1 + x_2 \leq 4$ (blue line) and $2x_1 + x_2 \leq 6$ (green line). These constraints define the feasible region, shaded in light gray. The feasible region is the set of all points that satisfy both inequalities simultaneously, and it represents all possible solutions to the linear programming problem. The Simplex path is marked in brown and shows the sequence of steps taken by the Simplex algorithm as it moves from one vertex of the feasible region to another, improving the objective function value at each step. The path culminates at the optimal point (2, 2), highlighted in yellow, which represents the optimal solution that maximizes (or minimizes) the objective function under the given constraints.

5.3.9 Lagrangian Multipliers

Lagrange multipliers are a robust optimization method used to find a function's local maxima and minima subject to equality constraints. Named after Joseph-Louis Lagrange, this technique transforms constrained optimization problems into unconstrained ones, facilitating their solution. Consider the problem of finding the extremum (maximum or minimum) of a function/ subject to a constraint $g\left(x_1, x_2, \ldots, x_n\right) = 0$. The method introduces an auxiliary variable, $\boldsymbol{\lambda}$ (the Lagrange multiplier), to incorporate the constraint into the objective function. The Lagrangian, **L**, is formulated as:

$$\mathcal{L}\left(x_1, x_2, \ldots, x_n, \lambda\right) = f\left(x_1, x_2, \ldots, x_n\right) + \lambda g\left(x_1, x_2, \ldots, x_n\right)$$

To find the extremum of **f** subject to **g**, we solve for the points where the gradient of the Lagrangian is zero:

$$\nabla L = \nabla f + \lambda \nabla g = 0$$

This results in a system of equations that, when solved, provide the values of the variables and the Lagrange multiplier. The Lagrange multiplier, has a meaningful interpretation. Generally, it measures the rate of change in the maximum or minimum value of the objective function as the constraint varies. Lagrange multipliers are applied across various fields. In economics, they are used for utility maximization and cost minimization problems. In machine learning, algorithms like support vector machines use Lagrange multipliers for optimization with constraints. While robust, Lagrange multipliers are primarily suited for equality constraints. For problems involving inequality constraints, methods like the Karush–Kuhn–Tucker (KKT) conditions are more appropriate. Consider maximizing the function **f(x, y) = xy**, subject to the constraint **x + y** = 10. Using the Lagrange multiplier **λ**, we form the Lagrangian:

$$\mathcal{L}(x, y, \lambda) = xy + \lambda(10 - x - y)$$

To find the extremum, we take the partial derivatives with respect to **x**, **y**, and **λ** and set them equal to zero:

$$\frac{\partial \mathcal{L}}{\partial x} = y - \lambda = 0, \quad \frac{\partial \mathcal{L}}{\partial y} = x - \lambda = 0, \quad \frac{\partial \mathcal{L}}{\partial \lambda} = 10 - x - y = 0$$

Solving this system yields **x** = **y** = 5, giving the maximum value of **f(x, y)** = 25. The Lagrange multiplier method transforms a constrained optimization problem into an unconstrained one by introducing a new variable **λ**. The Lagrangian is defined as:

$$\mathcal{L}(x, \lambda) = f(x) + \lambda\left[g(x) - c\right]$$

where:

- **f(x)** is the objective function,
- **g(x)** = **c** is the constraint,
- **λ** is the Lagrange multiplier.

The solution is found by solving the system of equations obtained from setting the gradients of the Lagrangian to zero:

$$\nabla \mathcal{L}(x, \lambda) = 0$$

Figure 5.10 displays a contour plot of the function **f(x, y)= xy**, with contour lines representing levels of constant function values. The color gradient indicates the objective function's values, transitioning from blue (lower values) to red (higher values), giving a visual sense of how **f(x, y)** changes across the plane. The dashed blue circle represents the constraint $x^2 + y^2 = 1$, which defines the feasible region for this optimization problem. This constraint forms a circle of radius 1 centered at the origin, encompassing all points that satisfy the equation. The red dots marked as optimal points (0.71, 0.71 and −0.71, −0.71) indicate where the function reaches its maximum and minimum values within the constrained region. These points are located on the circle where the product **xy** is maximized and minimized, showing how the constraint influences the objective function's optimization.

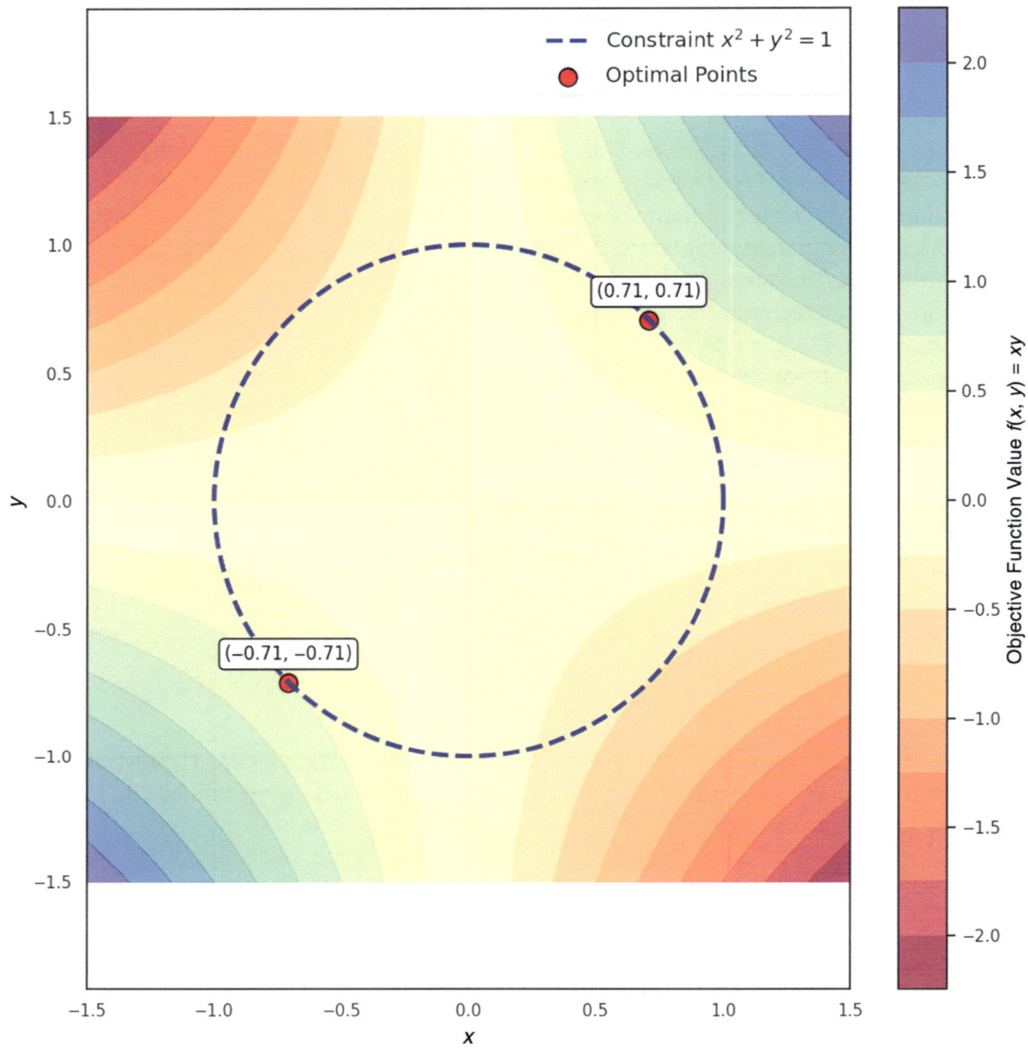

FIGURE 5.10 Optimization with constraints.

5.3.10 Branch and Bound and Cutting Plane Methods

B&B and cutting plane methods are crucial algorithms for solving optimization problems, particularly combinatorial and IP challenges. B&B is a systematic method for exploring possible solutions within a tree structure. The process begins with branching, where the main problem is divided into smaller subproblems, creating branches in a solution tree. In bounding, each subproblem is evaluated by calculating a bound, and if this bound is better than the current best solution, the subproblem is explored further; otherwise, it is discarded. Pruning occurs when subproblems that cannot improve upon the best-known solution are eliminated, enhancing efficiency by reducing unnecessary calculations. The cutting plane methods iteratively solve IP problems by refining the feasible region. Here are its general steps:

(a) *Starting Solution:* Start by solving the problem's linear programming relaxation, ignoring integer constraints.

(b) *Check & Cut:* Verify if the solution meets integer constraints. If not, a cut (linear inequality) is added to exclude the current fractional solution while retaining feasible integer points.
(c) *Iterate:* Repeat until an acceptable integer solution is found or proven that none exists.

The B&B method explores a solution tree, where each subproblem is a branch, and solutions are bounded. The cutting plane method iterates by adding constraints to refine the feasible region. For the cutting plane method, starting from a linear relaxation:

$$\text{Maximize } Z = c_1x_1 + c_2x_2 \text{ Subject to:} Ax \leq b, \quad x \geq 0$$

If the solution is not integer, add a cut (a linear inequality) to exclude fractional solutions while retaining feasible integer points. These methods iteratively refine and solve the problem, improving computational efficiency by avoiding exhaustive searches. Consider an IP problem where we aim to maximize $\mathbf{Z} = 3\mathbf{x}_1 + 4\mathbf{x}_2$, subject to the constraints $\mathbf{x}_1 + 2\mathbf{x}_2 \leq 8$ and $\mathbf{x}_1, \mathbf{x}_2 \geq 0$, with $\mathbf{x}_1$ and $\mathbf{x}_2$ required to be integers. Using B&B, we solve the linear relaxation (ignoring integer constraints), yielding a solution of $\mathbf{x}_1 = 4$, $\mathbf{x}_2 = 2$. We then branch by creating subproblems where $\mathbf{x}_1 \leq 3$ and $\mathbf{x}_1 \geq 4$, continuing to solve and prune subproblems until the optimal integer solution is found.

Figure 5.11 illustrates the B&B tree used to solve a knapsack problem, a combinatorial optimization challenge where the goal is to maximize the value of items placed in a knapsack without exceeding its weight capacity. The tree starts with the initial node labeled "Start" (value = 0, weight = 0), where no items have been considered yet. From this node, the algorithm branches out, evaluating two choices: Include item 1 (value = 40, weight = 2) or exclude item 1 (value = 0, weight = 0). Each choice generates a new branch, exploring the impact of either including or excluding the item. As the tree progresses, it evaluates further branches for item 2, following the same process of including or excluding the item. Nodes that exceed the weight capacity are marked as pruned/infeasible, indicating that these paths do not satisfy the problem's constraints and are thus discarded from further consideration. The algorithm eventually reaches the optimal node (value = 190, weight = 10), representing the best possible solution that maximizes the value while staying within the weight limit. This node is highlighted to signify that the optimal solution has been found.

Figure 5.12 illustrates the cutting plane method for solving an integer linear programming (ILP) problem. This approach refines the feasible region of a linear programming problem to efficiently identify integer solutions. The plot presents a system of constraints, each represented as a line: $-x - 2y \leq -3$ (blue line), $x + 2y \leq 8$ (green line), and $2x + y \leq 10$ (purple line). These constraints collectively form the feasible region, shaded in light gray, where all points satisfy the inequalities simultaneously. The objective is to find an integer solution within this region that optimizes the objective function. To achieve this, the figure shows two cutting planes represented by dashed lines. Cutting plane 1, marked as a red dashed line, adds an additional constraint to exclude non-integer solutions from the feasible region. Cutting plane 2, shown as an orange dashed line, further refines the solution space, ensuring that only integer solutions remain. The optimal solution, indicated by a yellow dot within the feasible region, represents the best integer solution that meets all constraints while optimizing the objective function.

5.3.11 Evolutionary Algorithms

Evolutionary algorithms (EAs) are optimization techniques inspired by natural selection. They simulate the biological principles of evolution, including survival of the fittest, to solve complex problems in various fields such as computer science and engineering. These are the basic concepts:

- *Population:* A group of potential solutions.
- *Chromosomes:* Each potential solution.
- *Genes:* Elements of a solution.
- *Fitness:* The quality or suitability of a solution.

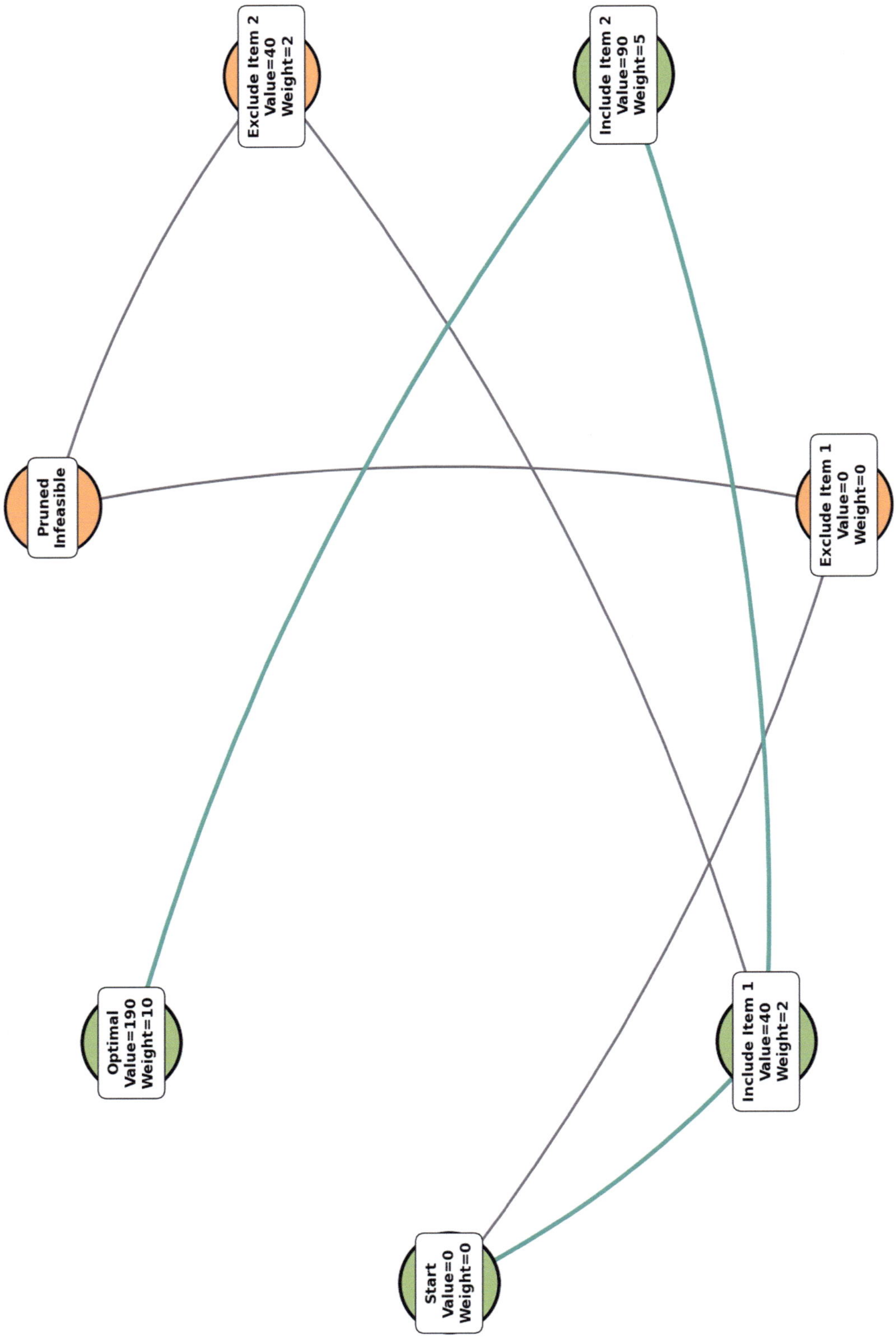

FIGURE 5.11 Branch and Bound tree for knapsack problem.

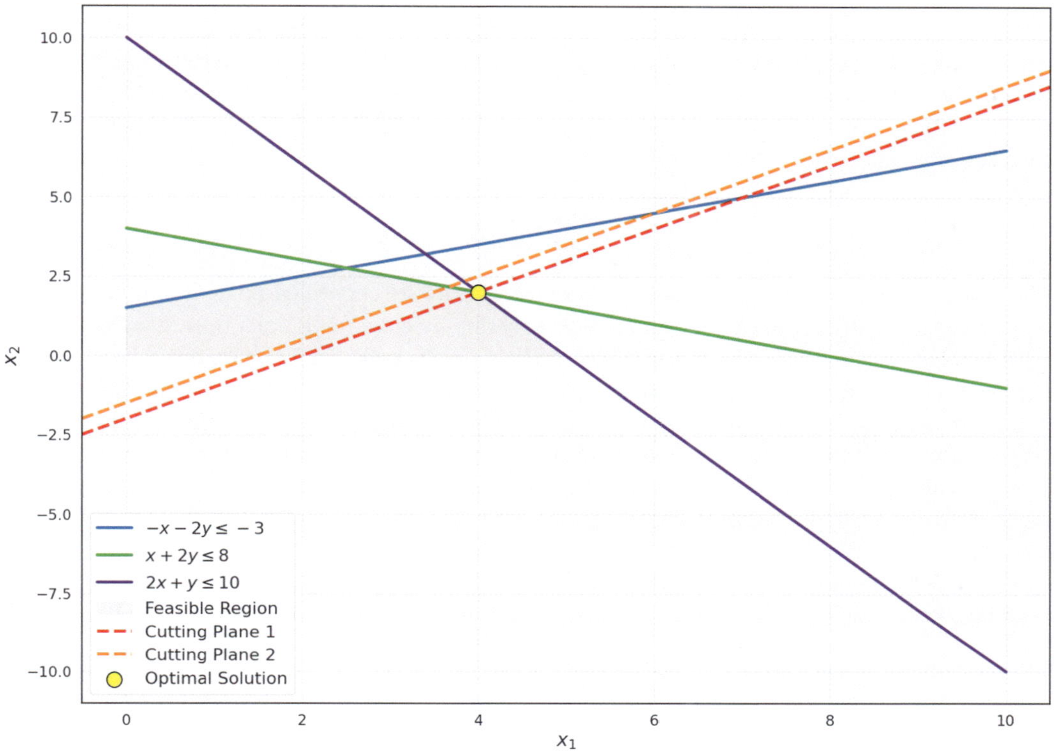

FIGURE 5.12 Cutting plane method.

The major steps in EAs are:

- *Initialization:* Generate an initial population of possible solutions randomly.
- *Selection:* Choose solutions based on fitness to contribute to the next generation. Higher fitness increases the likelihood of selection.
- *Crossover (Recombination):* Combine selected solutions to create new ones, mimicking reproduction.
- *Mutation:* Introduce small changes to new solutions with a certain probability to ensure genetic diversity.
- *Replacement:* Replace the old population with the new one.
- *Termination:* Repeat the process until a stopping criterion is met, such as a maximum number of generations or a satisfactory fitness level.

The types of EAs include genetic algorithms (GAs), which are the most popular type and use mutation, crossover, and selection to evolve potential solutions. Genetic programming evolves computer programs as solutions, while differential evolution is used for real-valued function optimization. Evolution strategies focus on strategy parameters such as mutation strength, and evolutionary programming focuses on the mutation of finite-state machines. The main applications of EAs include function optimization, where they are used to find the maximum or minimum of functions. In machine learning, they are used for feature selection, hyperparameter tuning, and neural network training. The advantages of EAs include flexibility, as they can be applied to a wide range of problems, and global search capabilities, as they have a higher likelihood of finding global optima than many traditional methods. They also support parallelism, allowing for the simultaneous evaluation of many solutions, which can be done in parallel with modern hardware. However, the limitations include being computationally intensive, as the required computational resources can grow significantly with problem size and complexity. Additionally, they do not provide certainty, often offering reasonable solutions without guaranteeing the global optimum. In GAs, the fitness function evaluates each solution **x**:

$$\text{Fitness}(x) = f(x)$$

The process involves applying selection, where individuals are chosen based on fitness scores. This is followed by crossover, which generates new offspring by combining traits from selected individuals. Finally, mutation is applied by randomly altering some genes to maintain diversity within the population and prevent premature convergence. This iterative process continues until the stopping criteria are met, typically optimizing the objective function. Suppose you are optimizing a function $\mathbf{f(x) = -x^2 + 4x}$. You initialize a population of 10 potential solutions (chromosomes) for the variable **x**, each with random values between 0 and 5. The fitness of each solution is calculated using the objective function **f(x)**, and after selection, crossover, and mutation steps, the best solution **x** = 2, with a fitness value of **f**(2) = 4, is found after 100 generations.

Figure 5.13, demonstrates the application of a GA to optimize the Rastrigin function, a common benchmark problem in evolutionary computation. The Rastrigin function, known for its large search space and many local minima, is defined as $f(\boldsymbol{x}) = 10 \cdot d + \sum_{i=1}^{d} \left(x_i^2 - 10\cos(2\pi x_i)\right)$, where **d** is the

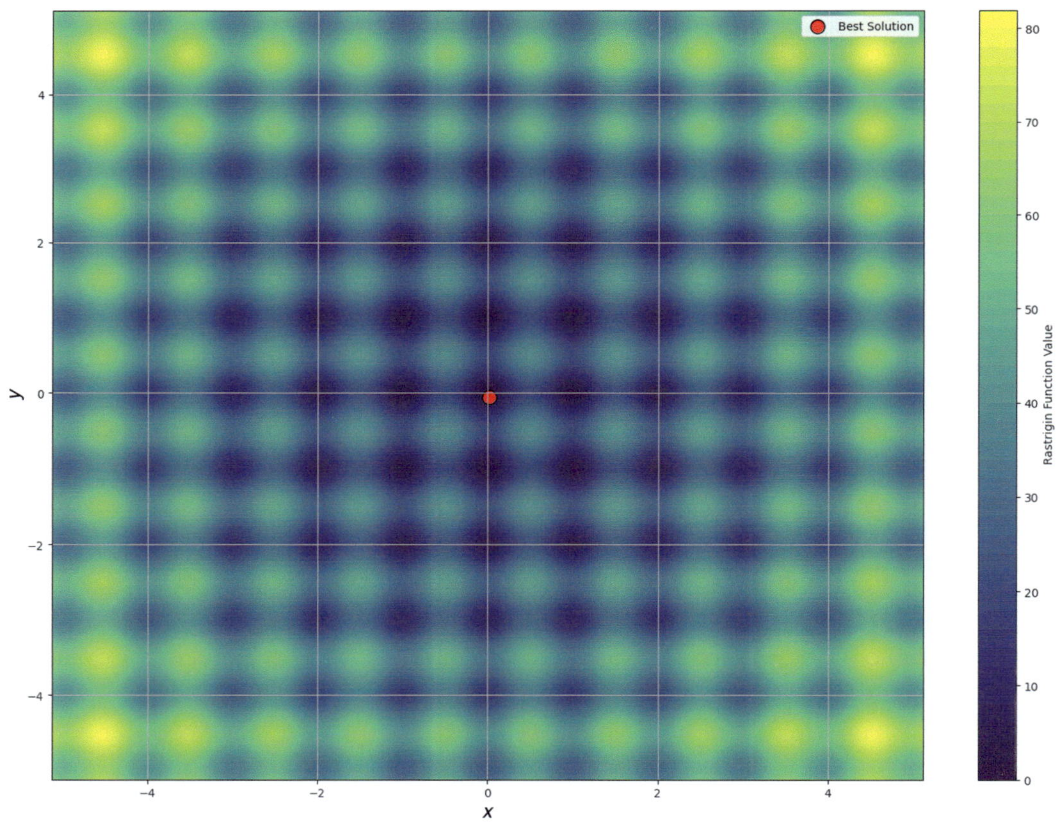

FIGURE 5.13 Genetic algorithm optimization of Rastrigin function.

number of dimensions. This example plots the function in two dimensions over the range [−5.12, 5.12] for **x** and **y**. The contours are filled using the colormap, visually representing the function's values across the grid. A color bar on the side indicates the corresponding function values. The GA is executed to find the minimum value of the Rastrigin function over 50 generations with a population size of 100, selecting 20 parents in each generation and applying a mutation rate of 0.1. The best solution the GA finds is highlighted with a red dot on the contour map, indicating the optimal point.

5.4 GLOBAL VERSUS LOCAL OPTIMA

Understanding the difference between global and local best solutions in optimization is very important because it affects how we find the best answer to a problem. This challenge is especially significant in complex problems where the solution landscape has many ups and downs. A global optimum is the absolute best solution among all possible options. For problems where we want to minimize something (like cost), it's the lowest point. For problems where we want to maximize something (like profit), it's the highest point. A local optimum, on the other hand, is a solution that's better than nearby options but might not be the best overall, like finding a small hill that isn't the tallest mountain. You can think of the function we're trying to optimize as a landscape with hills and valleys. The local optima are the smaller hills and valleys, while the global optimum is the highest hill or the deepest valley. Many optimization methods might get stuck at a local optimum, depending on where they start. When there are multiple local optima, the landscape becomes rugged, making it harder to find the global best solution. In simpler problems, called convex problems, any local optimum is also a global optimum, which makes finding the best solution easier. However, in more complicated, non-convex problems, it's hard to tell if the solution we found is the absolute best without knowing the whole landscape. To avoid getting stuck at local optima, we can use several strategies. One method is random restarts, where we run the optimization multiple times from different starting points to increase the chances of finding the global optimum. Another technique is simulated annealing, which is inspired by the process of slowly cooling metal to make its structure more uniform—a practice in metallurgy. In optimization, this means occasionally accepting worse solutions to escape local optima and explore more possibilities. We can also use GAs, which mimic natural selection by keeping a group of possible solutions and combining them in new ways, hoping to find better solutions through processes similar to mutation and crossover in biology. Swarm intelligence techniques, like particle swarm optimization, use multiple agents that move through the solution space, sharing information to find the best solution together. In real-world applications such as machine learning, finance, and engineering, optimization problems often have complex landscapes with many local optima. Whether we end up at a local or global optimum can greatly affect performance, costs, and outcomes. While we can't always guarantee we'll find the global best solution, using different techniques, monitoring how the solution progresses, and understanding the problem better can help us get closer to it.

A general optimization problem can be represented as:

$$\text{Minimize } f(x), \quad x \in \mathcal{S}$$

where **f(x)** is the objective function, and **S** is the solution space. A global minimum is defined as:

$$f(x^*) \leq f(x), \quad \forall x \in \mathcal{S}$$

A local minimum occurs when:

$$f(x^*) \leq f(x), \quad \forall x \text{ in the neighborhood of } x^*$$

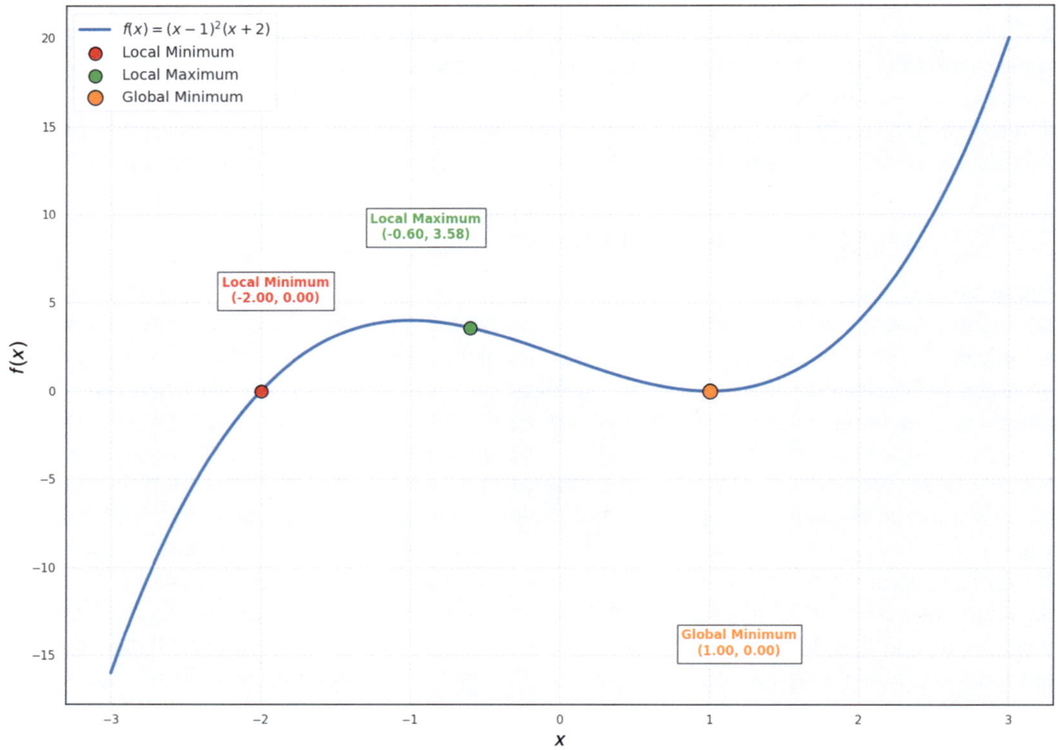

FIGURE 5.14 Global versus local optima.

Strategies like simulated annealing or GAs are used to escape local optima and find the global minimum in complex landscapes. Consider a non-convex function $f(x) = x^4 - 3x^3 + 2$. The function has two local minima and one global minimum. A local minimum occurs at $\mathbf{x} = 2$, where $\mathbf{f}(2) = 2$, but the global minimum occurs at $\mathbf{x} = 0.5$, where $\mathbf{f}(0.5) = -0.375$. If an optimization algorithm starts near $\mathbf{x} = 2$, it may converge to the local minimum instead of the global minimum.

Figure 5.14 presents the plot of the function $f(x) = (x-1)^2 \cdot (x+2)$ showing its behavior across the domain with key points. The graph illustrates the different types of extrema, including a local minimum, a local maximum, and a global minimum. The function initially rises to a local maximum at the point (–0.60, 3.58), marked in green. This point is where the function reaches a temporary peak before declining. As the graph continues, it reaches a local minimum at (–2, 0), highlighted in red. This local minimum indicates a point where the function temporarily decreases before increasing again. However, this is not the lowest value the function can attain within the entire domain. Further along, the function drops to the global minimum at the point (1, 0), shown in orange. This global minimum represents the lowest point of the function across its entire domain, indicating the true minimum value of $f(x)$. Beyond this point, the function rises again.

5.5 RECENT DEVELOPMENTS IN OPTIMIZATION

The rise of big data and more powerful computers has greatly changed the field of optimization. New developments such as distributed and real-time optimization, quantum computing, and advances in machine learning are key areas of current research. These advancements are driven by the need to solve large problems efficiently and make quick decisions. As data grows rapidly, optimization algorithms need to process huge amounts of information effectively. Big data often involves complex, high-dimensional spaces where traditional methods may not work well, so new strategies are needed to handle these challenges. Optimization tasks are now often spread out and run across multiple computers or nodes, which helps in dealing with large-scale problems. However, this brings challenges like ensuring data security, keeping everything synchronized, and managing communication between nodes. Optimization is also very important in situations that need immediate solutions or decisions in a short time, often as data is being created. Examples include self-driving cars and high-speed trading systems. These applications must balance speed and accuracy to compute quickly without losing solution quality. Quantum computers, which use principles from quantum mechanics, offer incredible computing power. Algorithms like the quantum approximate optimization algorithm (QAOA) are being developed to tap into this potential, possibly bringing big changes to fields like cryptography, materials science, and machine learning. Machine learning, especially deep learning, relies a lot on optimization to train models. At the same time, machine learning techniques are improving optimization algorithms. Automated systems help in choosing the best machine learning models and fine-tuning their settings. In real-world situations that often have uncertain or noisy data, robust optimization finds solutions that work well under different scenarios, while stochastic optimization deals with uncertainties in constraints and objectives. Moreover, real-world problems often have conflicting goals, leading to new methods that aim to find the best possible solutions where improving one goal doesn't make another worse. Optimization is going through a major change, driven by new technologies and the complexities of modern problems. As we move further into the digital age, optimization remains essential for finding efficient and effective solutions in a rapidly changing world. In distributed optimization, we often solve a problem by spreading it across multiple computers or nodes. This can be represented as:

$$\text{Minimize } f(x) = \sum_{i=1}^{n} f_i(x) \quad \mathrm{x} \in \mathrm{X}$$

where:

- $\mathbf{f_i(x)}$ is the local objective function on the $\mathbf{i^{th}}$ node,
- $\mathbf{X}$ is the shared feasible set across nodes.

Techniques such as distributed gradient descent are used to optimize functions across multiple machines, significantly speeding up the process for large-scale data. Consider an optimization problem in high-frequency trading that requires decision-making in less than a millisecond. Using real-time optimization, an algorithm balances speed and accuracy by processing 1 million transactions per second across 100 nodes, reducing latency from 5 milliseconds to 1 millisecond.

Figure 5.15 demonstrates the process of distributed optimization applied to the objective function $f(x) = x^2 + 10\sin(x)$. The function's behavior, represented by the blue curve, displays the combined effects of a quadratic term and a sinusoidal component, resulting in multiple local minima and maxima across the domain. The function's domain is divided into different regions, each assigned to a separate node, indicated by distinct shaded areas. Node 1, represented in dark gray, Node 2 in green, Node 3 in tan, Node 4 in pink, and Node 5 in light blue, each handle a specific segment of the function. The results from each node are marked with dots, showing where each node has found significant points such as local minima or maxima. For example, Node 1 identifies a point at (−6.00, 38.79), while Node 3 finds a local minimum at (−2.00, −5.09). The figure illustrates how distributed optimization techniques divide the problem into smaller, manageable parts, allowing each node to optimize the function locally within its region. This approach not only enhances computational efficiency but also allows for parallel processing, which is especially advantageous for large-scale or complex optimization problems.

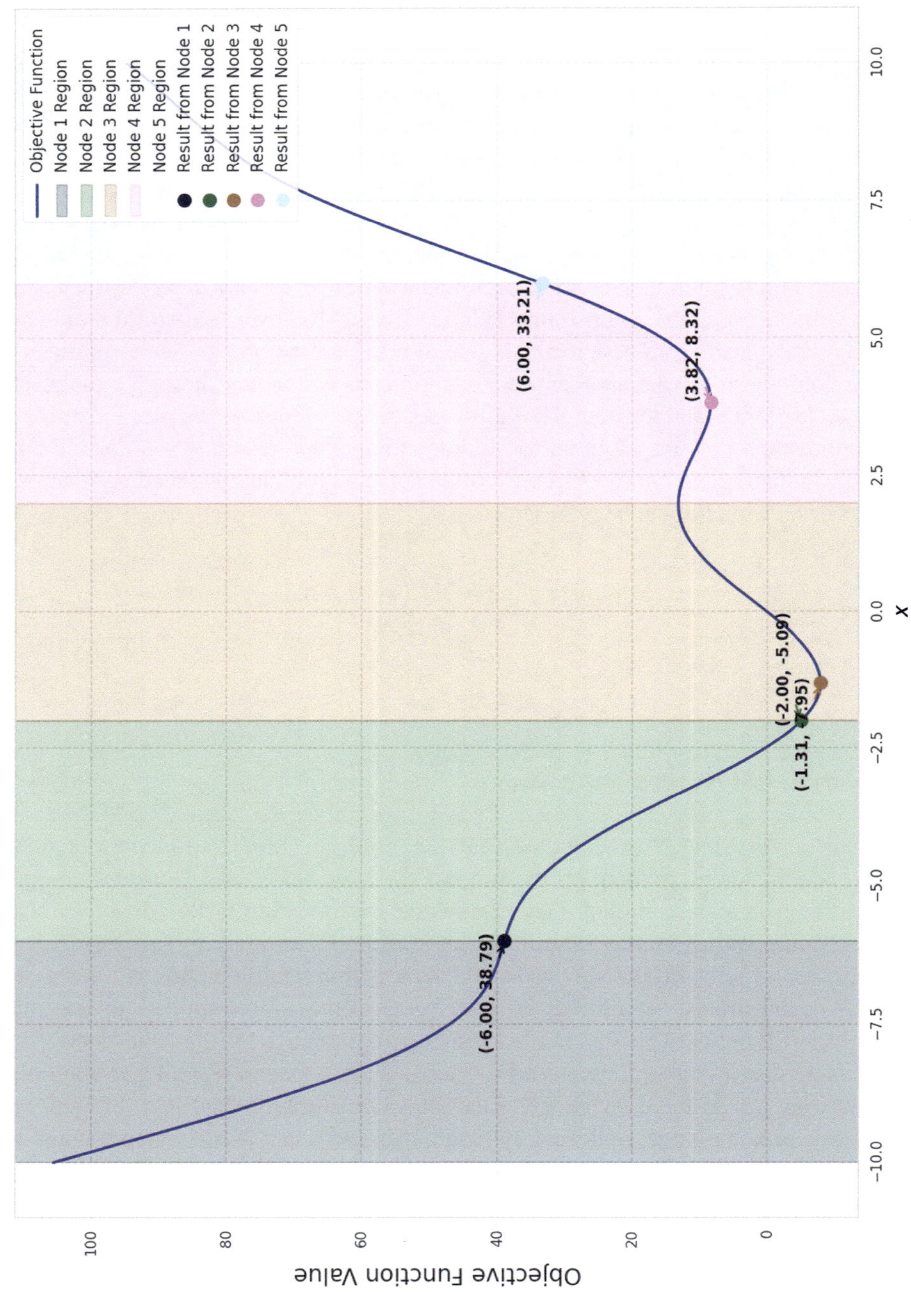

FIGURE 5.15 Distributed optimization.

5.6 OPTIMIZATION METHODS IN DEEP LEARNING

Optimization in deep learning is crucial because the goal is to minimize (or sometimes maximize) a loss function. This loss function measures how far our predictions are from the actual values. By optimizing this function, we ensure that our model generalizes well to new, unseen data.

5.6.1 Batch Gradient Descent (BGD)

BGD, specifically, calculates the gradient of the loss function with respect to each parameter for the entire training dataset at once. BGD guarantees convergence to the global optimum for convex functions and is straightforward to implement. It systematically reduces the loss function by taking steps proportional to the negative gradient of the loss function. The steps are controlled by a learning rate, which determines the size of each step. The algorithm works by first initializing the parameters randomly. Then, for each iteration, the gradient of the loss function with respect to each parameter is calculated using the entire dataset. The parameters are updated by subtracting the product of the learning rate and the gradient. This process is repeated until the parameters converge to values that minimize the loss function. However, BGD is not feasible for large datasets that do not fit in memory. It can also be slow on extensive datasets due to the need to process the entire dataset for each update. This means that every iteration can be time-consuming, especially for high-dimensional data. The update rule for BGD is:

$$\theta_{\text{new}} = \theta_{\text{old}} - \alpha \nabla J(\theta)$$

where:

- $\boldsymbol{\theta}$ is the parameter vector,
- $\boldsymbol{\alpha}$ is the learning rate, and
- $\nabla \mathbf{J}(\boldsymbol{\theta})$ is the gradient of the loss function $\mathbf{J}(\boldsymbol{\theta})$, calculated over the entire dataset.

This process is repeated until convergence, minimizing the loss function for the entire dataset. Suppose you are training a linear regression model to predict house prices. The loss function is the mean squared error (MSE), and you have a dataset with 1,000 houses. Using BGD, you calculate the gradient of the loss function for all 1,000 houses in each iteration. With a learning rate of $\boldsymbol{\alpha} = 0.01$, the parameters $\boldsymbol{\theta}$ are updated by subtracting $\boldsymbol{\alpha} \times \nabla \mathbf{J}(\boldsymbol{\theta})$ from the current values. For example, if the gradient of the loss function at a specific iteration is $\nabla J(\theta) = [1.5, -2.0]$, and the current parameter values are $\theta = [2.0, 3.0]$, the updated parameters will be calculated as:

$$\theta_{\text{new}} = \theta_{\text{current}} - \alpha \times \nabla J(\theta)$$

Substituting the values:

$$\theta_{\text{new}} = [2.0, 3.0] - 0.01 \times [1.5, -2.0] = [2.0 - 0.015, 3.0 + 0.02] = [1.985, 3.02]$$

This process continues until the parameters $\boldsymbol{\theta}$ converge to values that minimize the loss function, ensuring the model provides the best predictions for house prices based on the training data. BGD ensures smooth updates, but its computational cost can become a bottleneck for very large datasets.

5.6.2 Stochastic Gradient Descent (SGD)

SGD is an optimization algorithm used to minimize the loss function by iteratively adjusting the model parameters. Unlike BGD, which computes the gradient using the entire dataset, SGD computes the gradient using only a single sample (or a small batch) at each iteration. SGD is known for its faster convergence compared to BGD, especially when dealing with large datasets. As it updates the parameters more frequently, it often reaches a good solution faster. Additionally, the noisy updates from using individual samples can help SGD escape local optima, providing a better chance of finding a global optimum in non-convex optimization problems. The algorithm works by first initializing the parameters randomly. For each iteration, it selects a random sample from the dataset and computes the gradient of the loss function with respect to each parameter. The parameters are then updated by subtracting the product of the learning rate and the gradient. This process is repeated, and a new random sample is selected at each iteration until convergence. However, the high variance in updates can cause SGD to oscillate around the minimum, making convergence harder to control. This variability means that while SGD can quickly reach the vicinity of the optimal solution, it might struggle to settle precisely at the minimum. Techniques such as learning rate decay, momentum, and mini-batching are often employed to mitigate these issues and stabilize convergence. The update rule for SGD is:

$$\theta_{\text{new}} = \theta_{\text{old}} - \alpha \nabla J\left(\theta, x^{(i)}, y^{(i)}\right)$$

where:

- $\boldsymbol{\theta}$ is the parameter vector,
- $\boldsymbol{\alpha}$ is the learning rate, and
- $\nabla J\left(\theta, x^{(i)}, y^{(i)}\right)$ is the gradient of the loss function computed for the single sample $\left(x^{(i)}, y^{(i)}\right)$.

This process repeats, updating the model for each random sample, allowing faster convergence in large datasets. Suppose you are training a model to predict stock prices with a dataset of 10,000 samples. In each iteration, SGD randomly selects one sample and updates the parameters based on that sample. If the learning rate $\boldsymbol{\alpha}$=0.001 and the gradient for a specific iteration is 0.5, the parameter $\boldsymbol{\theta}$ is updated as:

$$\theta_{\text{new}} = \theta_{\text{old}} - 0.001 \times 0.5 = \theta_{\text{old}} - 0.0005$$

5.6.3 Mini-Batch Gradient Descent

Mini-batch gradient descent is an optimization algorithm that offers a compromise between BGD and SGD. It updates the model parameters based on smaller subsets, or "mini-batches," of the dataset. Mini-batch gradient descent combines the advantages of both BGD and SGD. By using mini-batches, it reduces the variance in parameter updates compared to SGD, leading to more stable convergence. At the same time, it does not require processing the entire dataset in each iteration, making it more efficient than BGD. This approach can also benefit from parallelized hardware, such as Graphics Processing Units (GPUs), which can process mini-batches concurrently. The algorithm works by first initializing the parameters randomly. Each iteration randomly selects a mini-batch of samples from the dataset and computes the gradient of the loss function with respect to the parameters using only this mini-batch. The parameters are then updated by subtracting the product of the learning rate and the gradient. This process is repeated for each mini-batch until the entire dataset has been processed, completing one epoch. The procedure is iterated over multiple epochs until convergence. One of the key considerations in mini-BGD is the choice of batch size. The batch size can significantly impact performance and convergence speed. A smaller batch size offers noisier updates and may benefit from the regularization effects similar to those of SGD. In contrast, a larger batch size provides a more accurate estimate of the gradient, leading to more stable updates, but at the cost of computational efficiency. The update rule for mini-BGD is:

$$\theta_{\text{new}} = \theta_{\text{old}} - \alpha \frac{1}{m} \sum_{i=1}^{m} \nabla J\left(\theta, x^{(i)}, y^{(i)}\right)$$

where:

- **θ** is the parameter vector,
- **α** is the learning rate,
- **m** is the mini-batch size, and
- $\nabla J\left(\theta, x^{(i)}, y^{(i)}\right)$ is the gradient of the loss function computed for the mini-batch samples $\left(x^{(i)}, y^{(i)}\right)$.

This method strikes a balance between the computational efficiency of BGD and the frequent updates of SGD. Suppose you are training a neural network with 10,000 samples, and you decide to use a mini-batch size of 100. In each iteration, you randomly select 100 samples from the dataset and compute the gradient. If the learning rate **α** = 0.01 and the average gradient for the mini-batch is 0.4, the parameter **θ** is updated as:

$$\theta_{\text{new}} = \theta_{\text{old}} - 0.01 \times 0.4 = \theta_{\text{old}} - 0.004$$

5.6.4 Momentum

Momentum is an optimization technique used to accelerate gradient descent by considering past gradients in the update process. This approach helps gradient descent algorithms converge faster and reduce oscillations, especially in cases where the objective function's surface has features of different scales or saddle points. The key idea behind momentum is to maintain a velocity vector that accumulates the gradient of the loss function over time. This accumulated gradient is used to update the model parameters, resulting in faster convergence and smoother trajectories. The velocity vector is updated using a combination of the current gradient and the previous velocity. The updated equations for gradient descent with momentum are as follows:

$$v_t = \beta v_{t-1} + \alpha \nabla f\left(\theta_t\right)$$

$$\theta_{t+1} = \theta_t - v_t$$

where:

- v_t is the velocity vector at iteration **t**,
- β is the momentum term (a hyperparameter typically set between 0 and 1),
- $\nabla f\left(\theta_t\right)$ is the gradient of the loss function with respect to the parameters at iteration **t**,
- θ_t are the model parameters at iteration **t**, and
- α is the learning rate.

The momentum term β controls the contribution of the previous gradients to the current update. A higher momentum value emphasizes the past gradients more, leading to faster convergence but potentially overshooting the minimum if set too high. Momentum offers several advantages. It accelerates convergence compared to vanilla SGD, particularly in the presence of saddle points or when the features have different scales. Reducing fluctuations helps the optimization process navigate the parameter space more efficiently and reach the optimum faster. However, momentum also introduces an additional hyperparameter, the momentum term β, which requires tuning. The choice of β can also significantly impact the performance of the optimization algorithm, and finding the optimal value often involves empirical experimentation. Assume we're optimizing a simple quadratic function using gradient descent with momentum. The function we want to minimize is:

$$f(x) = (x-3)^2$$

For example, if the parameters are:

- Learning rate (α): 0.1,
- Momentum (β): 0.9,
- Initial value of x: 0 (starting point), and
- Gradient at iteration t is $g_t = 2(x_t-3)$

At iteration 1, starting from $x_0 = 0$:

- Gradient is: $g_1 = 2(x_0-3) = 2(0-3) = -6$
- Velocity update is: $v_1 = \beta v_0 + \alpha g_1 = 0.9 \times 0 + 0.1 \times (-6) = -0.6$
- Parameter update is: $x_1 = x_0 - v_1 = 0 - (-0.6) = 0.6$

At iteration 2:

- Gradient is: $g_2 = 2(x_1-3) = 2(-0.6 - 3) = -7.2$
- Velocity update is: $v_2 = \beta v_1 + \alpha g_2 = 0.9 \times (-0.6) + 0.1 \times (-7.2) = -1.26$
- Parameter update is: $x_2 = x_1 - v_2 = -0.6 - (-1.26) = 1.86$

Here, you can see how the momentum term accelerates the descent by considering the past velocity $\mathbf{v}_1$ along with the new gradient $\mathbf{g}_2$, leading to a faster movement toward the minimum at $\mathbf{x} = 3$. As the iterations progress, the oscillations are reduced, allowing the algorithm to converge more efficiently.

5.6.5 Adagrad (Adaptive Gradient Algorithm)

Adagrad is an optimization algorithm that adapts the learning rates of all model parameters by scaling them inversely proportional to the square root of the sum of all historical squared gradients. This method is particularly useful for dealing with sparse data and features that have varying degrees of informativeness. The key idea behind Adagrad is to adjust the learning rate for each parameter individually, allowing for larger updates for infrequent parameters and smaller updates for frequent ones. This is achieved by accumulating the squared gradients over time and using this accumulated value to adjust the learning rates. The update rule for Adagrad is as follows:

$$g_{t,i} = g_{t-1,i} + (\nabla f(\theta_{t,i}))^2$$

$$\theta_{t,i} = \theta_{t-1,i} - \frac{\alpha}{\sqrt{g_{t,i}} + \epsilon} \nabla f(\theta_{t-1,i})$$

where:

- $\nabla f\left(\theta_{t,i}\right)$ is the gradient of the loss function with respect to parameter **i** at iteration **t**,
- $g_{t,i}$ is the sum of the squares of the gradients with respect to parameter **i** up to iteration **t**,
- α is the global learning rate, and
- is a small constant added to prevent division by zero.

Adagrad is particularly advantageous for dealing with sparse data, where some features are much more informative than others. By scaling the learning rate based on historical gradients, Adagrad ensures that infrequent but informative features receive larger updates, improving the model's ability to learn from these features. However, one of the main drawbacks of Adagrad is that the learning rate can decrease significantly over time, causing the algorithm to stop learning prematurely. As the sum of squared gradients grows, the effective learning rate diminishes, which can hinder further progress, especially in non-convex optimization problems or over long training periods. Assume we are optimizing a simple function:

$$f\left(x,y\right)=x^2+2y^2$$

For example, if the parameters are:

- Global learning rate (α): 0.1
- Small constant (ϵ): 10^{-8}
- Initial values of $x_0 = 2$, $y_0 = 2$

At iteration 1, we calculate the gradients for **x** and **y**:

- Gradient of **x**: $g_{x1} = 2x_0 = 2 \times 2 = 4$
- Gradient of **y**: $g_{y1} = 4y_0 = 4 \times 2 = 8$
- Accumulated squared gradients for **x** and **y**: $G_{x1} = g_{x1}^2 = 4^2 = 16, \quad G_{y1} = g_{y1}^2 = 8^2 = 64$
- Update rule for x_1: $x_1 = x_0 - \dfrac{\alpha}{\sqrt{G_{x1}+\epsilon}} \times g_{x1} = 2 - \dfrac{0.1}{\sqrt{16+10^{-8}}} \times 4 = 2 - 0.1 \times 1 = 1.9$
- Update rule for y_1: $y_1 = y_0 - \dfrac{\alpha}{\sqrt{G_{y1}+\epsilon}} \times g_{y1} = 2 - \dfrac{0.1}{\sqrt{64+10^{-8}}} \times 8 = 2 - 0.1 \times 1 = 1.9$

At iteration 2:

- Gradients at new values $x_1 = 1.9$, $y_1 = 1.9$: $g_{x2} = 2 \times 1.9 = 3.8, \quad g_{y2} = 4 \times 1.9 = 7.6$
- Accumulated squared gradients:

$$G_{x2} = G_{x1} + g_{x2}^2 = 16 + 3.8^2 = 30.44, \quad G_{y2} = G_{y1} + g_{y2}^2 = 64 + 7.6^2 = 121.76$$

- Update rule for x_2: $x_2 = 1.9 - \dfrac{0.1}{\sqrt{30.44+10^{-8}}} \times 3.8 \approx 1.9 - 0.069 = 1.831$
- Update rule for y_2: $y_2 = 1.9 - \dfrac{0.1}{\sqrt{121.76+10^{-8}}} \times 7.6 \approx 1.9 - 0.069 = 1.831$

As you can see, the learning rate for both **x** and **y** decreased slightly at the second iteration due to the accumulation of squared gradients. This helps balance the learning, allowing more updates to infrequent parameters while slowing down updates to frequent ones. However, over many iterations, the learning rate will continue to diminish.

5.6.6 Root Mean Square Propagation (RMSprop)

RMSprop is an optimization algorithm designed to address the issue of diminishing learning rates in Adagrad. By introducing a decay factor, RMSprop ensures that the learning rate does not decrease too aggressively, making it more suitable for training deep networks. RMSprop modifies Adagrad by maintaining a moving average of the squared gradients instead of accumulating all the past squared gradients. This moving average helps to prevent the learning rates from becoming excessively small, allowing the optimization process to continue learning effectively over time. The update rule for RMSprop is as follows:

$$g_{t,i} = \nabla f\left(\theta_{t,i}\right)$$

$$E\left[g_{t,i}^2\right] = \beta E\left[g_{t-1,i}^2\right] + (1-\beta)(g_{t,i})^2$$

$$\theta_{t+1,i} = \theta_{t,i} - \frac{\alpha}{\sqrt{E\left[g_{t,i}^2\right] + \epsilon}} g_{t,i}$$

where:

- $\mathbf{g_{t,i}}$ is the gradient of the function f with respect to the parameter $\boldsymbol{\theta_i}$ at iteration **t**,
- $\boldsymbol{\beta}$ is the decay rate (typically set to 0.9),
- $\boldsymbol{E\left[g_{t,i}^2\right]}$ is the moving average of squared gradients for parameter **i** up to iteration **t**,
- $\boldsymbol{\theta_{t+1,i}}$ is the updated parameter,
- $\boldsymbol{\alpha}$ is the global learning rate,
- $\boldsymbol{\epsilon}$ is a small constant to prevent division by zero.

RMSprop's decay rate γ controls how much the algorithm considers the recent gradients compared to the past gradients. By adjusting this parameter, RMSprop can balance the influence of the current gradient and the historical gradients, preventing the learning rate from diminishing too quickly. The main advantage of RMSprop is that it resolves Adagrad's problem of diminishing learning rates, making it more suitable for training deep neural networks. The controlled adjustment of learning rates allows RMSprop to maintain a more consistent and effective learning process, even over extended training periods. Assume we are optimizing a simple function:

$$f(x) = (x-2)^2$$

For example, if the parameters are:

- Global learning rate (α): 0.01,
- Decay rate (β): 0.9,
- Small constant (ϵ): 10^{-8}, and
- Initial value of $x_0 = 4$.

At iteration 1, we calculate the gradient for **x**:

- Gradient at x_0: $g_{x1} = 2(x_0 - 2) = 2(4-2) = 4$
- Exponential moving average of squared gradients (starting from 0):

$$E\left[g_{x1}^2\right] = \beta E\left[g_{x0}^2\right] + (1-\beta)g_{x1}^2 = 0.9 \times 0 + 0.1 \times 4^2 = 1.6$$

- Update rule for $\mathbf{x}_1$: $x_1 = x_0 - \dfrac{\alpha}{\sqrt{E\left[g_{x1}^2\right] + \epsilon}} \times g_{x1} = 4 - \dfrac{0.01}{\sqrt{1.6 + 10^{-8}}} \times 4 \approx 4 - 0.1 = 3.9$

At iteration 2:

- Gradient at $x_1 = 3.9$: $g_{x2} = 2(3.9-2) = 3.8$
- Update the exponential moving average of squared gradients:

$$E\left[g_{x2}^2\right] = 0.9 \times 1.6 + 0.1 \times 3.8^2 = 2.928$$

- Update rule for x_2: $x_2 = 3.9 - \dfrac{0.01}{\sqrt{2.928 + 10^{-8}}} \times 3.8 \approx 3.9 - 0.07 = 3.83$

In this example, RMSprop adjusts the learning rate dynamically by calculating the exponentially decaying average of past squared gradients. This prevents the learning rate from diminishing too rapidly, as seen with Adagrad, allowing for more controlled and consistent updates to **x**.

5.6.7 Adam (Adaptive Moment Estimation)

Adam is an optimization algorithm that combines the benefits of both Adagrad and RMSprop. It leverages the advantages of adaptive learning rates and incorporates momentum to accelerate convergence and stabilize updates. Adam maintains individual learning rates for each parameter by computing the first and second moments of the gradients. The first moment is the mean of the gradients, and the second is the gradients' uncentered variance. These moments are used to adapt the learning rates for each parameter dynamically. The updated rules for Adam are as follows:

$$g_t = \nabla f(\theta_t)$$

$$m_t = \beta_1 m_{t-1} + (1-\beta_1) g_t$$

$$v_t = \beta_2 v_{t-1} + (1-\beta_2) g_t^2$$

$$\hat{m}_t = \frac{m_t}{1-\beta_1^t}$$

$$\hat{v}_t = \frac{v_t}{1-\beta_2^t}$$

$$\theta_{t+1} = \theta_t - \frac{\alpha}{\sqrt{\hat{v}_t} + \epsilon} \hat{m}_t$$

where:

- g_t is the gradient of the loss function with respect to the parameters at iteration **t**
- m_t and v_t are the first-moment (mean) and second-moment (uncentered variance) estimates, respectively
- β_1 and β_2 are the decay rates for the moment estimates (commonly set to 0.9 and 0.999, respectively)
- $\hat{m}_t$ and $\hat{v}_t$ are the bias-corrected moment estimates
- α is the learning rate
- ϵ is a small constant added to prevent division by zero

Adam combines the adaptive learning rate benefits of Adagrad and the exponentially decaying average of squared gradients from RMSprop while also incorporating momentum to smooth out the updates. This results in a robust and efficient optimization algorithm that performs well across various problems. The main advantages of Adam are its suitability for problems with extensive data and parameter spaces and its general effectiveness in practice. Adam typically converges faster and more reliably than other optimization algorithms, making it a popular choice for training deep neural networks and other complex models. However, Adam also has some drawbacks. The choices of hyperparameters, such as learning and decay rates, can be critical and require fine-tuning. The default values for these hyperparameters often work well, but for specific problems, manual tuning may be necessary to achieve optimal performance. Assume we are optimizing a simple quadratic function:

$$f(x) = (x-3)^2$$

For example, if the parameters are:

- Global learning rate (α): 0.1,
- Decay rates for the first moment (β_1): 0.9,
- Decay rates for the second moment (β_2): 0.999,
- Small constant (ϵ): 10^{-8},
- Initial value of $x_0 = 0$,
- Gradient at iteration t: $g_t = 2(x_t - 3)$.

At iteration 1:

- Gradient is: $g_1 = 2(x_0 - 3) = 2(0-3) = -6$
- First-moment estimate (m_1): $m_1 = \beta_1 m_0 + (1-\beta_1) g_1 = 0.9 \times 0 + 0.1 \times (-6) = -0.6$
- Second-moment estimate (v_1): $v_1 = \beta_2 v_0 + (1-\beta_2) g_1^2 = 0.999 \times 0 + 0.001 \times (-6)^2 = 0.036$
- Bias-corrected first moment ($\hat{m}_1$): $\hat{m}_1 = \dfrac{m_1}{1-\beta_1^1} = \dfrac{-0.6}{1-0.9} = -6$
- Bias-corrected second moment ($\hat{v}_1$): $\hat{v}_1 = \dfrac{v_1}{1-\beta_2^1} = \dfrac{0.036}{1-0.999} = 36$
- Parameter update: $x_1 = x_0 - \dfrac{\alpha}{\sqrt{\hat{v}_1} + \epsilon} \times \hat{m}_1 = 0 - \dfrac{0.1}{\sqrt{36} + 10^{-8}} \times (-6) = 0 + 0.1 = 0.1$

At iteration 2:

- Gradient at $x_1 = 0.1$: $g_2 = 2(0.1-3) = -5.8$
- First-moment estimate (m_2): $m_2 = 0.9\times(-0.6)+0.1\times(-5.8) = -1.12$
- Second-moment estimate (v_2): $v_2 = 0.999\times 0.036 + 0.001\times(-5.8)^2 = 0.068$
- Bias-corrected first moment ($\hat{m}_2$): $\hat{m}_2 = \dfrac{m_2}{1-0.9^2} = -5.89$
- Bias-corrected second moment ($\hat{v}_2$): $\hat{v}_2 = \dfrac{v_2}{1-0.999^2} = 34.06$
- Parameter update: $x_2 = x_1 - \dfrac{0.1}{\sqrt{34.06}+\epsilon}\times(-5.89) \approx 0.1+0.1 = 0.2$

In this example, you can see how Adam combines the first and second moments (mean and variance) to adjust the learning rate dynamically. The algorithm benefits from both momentum (accelerating convergence) and adaptive learning rates (preventing small updates), making it highly effective for optimizing complex functions.

5.6.8 AdaMax

AdaMax is an Adam optimization algorithm variant based on the infinity norm (maximum norm) rather than the $\mathbf{L_2}$ norm used in Adam. By using the infinity norm, AdaMax provides a different way of adapting the learning rates, often leading to more stable and robust performance. The update rules for AdaMax are like Adam but with modifications to incorporate the infinity norm:

$$g_t = \nabla f(\theta_t)$$

$$m_t = \beta_1 m_{t-1} + (1-\beta_1) g_t$$

$$u_t = \max(\beta_2 u_{t-1}, |g_t|)$$

$$\hat{m}_t = \frac{m_t}{1-\beta_1^t}$$

$$\theta_{t+1} = \theta_t - \frac{\alpha}{u_t}\hat{m}_t$$

where:

- g_t is the gradient of the loss function with respect to the parameters at iteration **t**,
- m_t is the first-moment (mean) estimate,
- u_t is the exponentially weighted infinity norm,
- β_1 and β_2 are the decay rates for the moment estimates (commonly set to 0.9 and 0.999, respectively),
- $\hat{m}_t$ is the bias-corrected first-moment estimate, and
- α is the learning rate.

The key difference between AdaMax and Adam lies in the estimation of the second moment. Instead of using the uncentered variance, AdaMax uses the gradients' infinity norm. This change simplifies the denominator of the parameter update rule, leading to potentially more stable updates.

The main advantage of AdaMax is its stability and robustness compared to Adam. Using the infinity norm helps control the learning rate more effectively, especially in scenarios where the gradients can vary significantly. This can lead to more consistent performance and faster convergence in some cases. Assume we are optimizing a simple quadratic function:

$$f(x) = (x-3)^2$$

For example, if the parameters are:

- Global learning rate (α): 0.1,
- Decay rates for the first moment (β_1): 0.9,
- Decay rates for the second moment (β_2): 0.999,
- Initial value of $x_0 = 0$, and
- Gradient at iteration t: $g_t = 2(x_t - 3)$.

At iteration 1:

- Gradient: $g_1 = 2(x_0 - 3) = 2(0-3) = -6$
- First-moment estimate (m_1): $m_1 = \beta_1 m_0 + (1-\beta_1) g_1 = 0.9 \times 0 + 0.1 \times (-6) = -0.6$
- Infinity norm (maximum gradient value) (u_1): $u_1 = \max(\beta_2 u_0, |g_1|) = \max(0.999 \times 0, |-6|) = 6$
- Bias-corrected first moment ($\hat{m}_1$): $\hat{m}_1 = \dfrac{m_1}{1-\beta_1^1} = \dfrac{-0.6}{1-0.9} = -6$
- Parameter update: $x_1 = x_0 - \dfrac{\alpha}{u_1} \times \hat{m}_1 = 0 - \dfrac{0.1}{6} \times (-6) = 0 + 0.1 = 0.1$

At iteration 2:

- Gradient at $x_1 = 0.1$: $g_2 = 2(0.1-3) = -5.8$
- First-moment estimate (m_2): $m_2 = 0.9 \times (-0.6) + 0.1 \times (-5.8) = -1.12$
- Infinity norm update (u_2): $u_2 = \max(0.999 \times 6, |-5.8|) = 6$
- Bias-corrected first moment ($\hat{m}_2$): $\hat{m}_2 = \dfrac{m_2}{1-0.9^2} = -5.89$
- Parameter update: $x_2 = x_1 - \dfrac{0.1}{6} \times (-5.89) \approx 0.1 + 0.098 = 0.198$

In this example, AdaMax uses the infinity norm (maximum value) of the gradient, which makes the denominator stable over iterations. This ensures smoother and more stable parameter updates, even when gradients fluctuate. AdaMax's stability can lead to more consistent performance, particularly in problems where the gradients vary significantly.

5.6.9 Nadam (Nesterov-accelerated Adaptive Moment Estimation)

Nadam is an optimization algorithm combining RMSprop's and Nesterov momentum's benefits. By integrating the adaptive learning rate mechanism of RMSprop with the momentum-accelerated gradient updates of Nesterov momentum, Nadam aims to achieve faster and more reliable convergence in various optimization tasks. Nadam modifies the standard Adam update rules by incorporating Nesterov momentum. The key difference is in how the gradients are calculated, with Nadam computing the gradients at the predicted future position of the parameters. The updated rules for Nadam are as follows:

$$g_t = \nabla f(\theta_t)$$

$$m_t = \beta_1 m_{t-1} + (1-\beta_1) g_t$$

$$v_t = \beta_2 v_{t-1} + (1-\beta_2) g_t^2$$

$$\hat{m}_t = \frac{m_t}{1-\beta_1^t}$$

$$\hat{v}_t = \frac{v_t}{1-\beta_2^t}$$

$$\tilde{g}_t = g_t + \beta_1 \hat{m}_t$$

$$\theta_{t+1} = \theta_t - \frac{\alpha}{\sqrt{\hat{v}_t} + \epsilon} \tilde{g}_t$$

where:

- g_t is the gradient of the loss function with respect to the parameters at iteration **t**,
- m_t and v_t are the first-moment (mean) and second-moment (uncentered variance) estimates, respectively,
- β_1 and β_2 are the decay rates for the moment estimates (commonly set to 0.9 and 0.999, respectively),
- $\hat{m}_t$ and $\hat{v}_t$ are the bias-corrected moment estimates,
- α is the learning rate, and
- ϵ is a small constant added to prevent division by zero.

Nadam enhances the standard Adam algorithm by incorporating the look-ahead mechanism of Nesterov momentum. This mechanism anticipates the future position of the parameters based on their current velocity, leading to more informed and potentially more effective updates. The main advantage of Nadam is that it can converge faster than Adam in some scenarios due to the added momentum. By leveraging the predictive capability of Nesterov momentum, Nadam can accelerate the optimization process, especially in cases where the objective function has complex or noisy gradients. Assume we are optimizing a simple quadratic function:

$$f(x) = (x-3)^2$$

For example, if the parameters, be:

- Global learning rate (α): 0.1,
- Decay rates for the first moment (β_1): 0.9,

- Decay rates for the second moment (β_2): 0.999,
- Small constant (ϵ): 10^{-8},
- Initial value of $x_0 = 0$, and
- Gradient at iteration t: $g_t = 2(x_t - 3)$.

At iteration 1:

- Gradient: $g1 = 2(x_0 - 3) = 2(0-3) = -6$
- First-moment estimate (m_1): $m_1 = \beta_1 m_0 + (1-\beta_1) g_1 = 0.9 \times 0 + 0.1 \times (-6) = -0.6$
- Second-moment estimate (v_1): $v_1 = \beta_2 v_0 + (1-\beta_2) g_1^2 = 0.999 \times 0 + 0.001 \times (-6)^2 = 0.036$
- Bias-corrected first moment ($\hat{m}_1$): $\hat{m}_1 = \dfrac{m_1}{1-\beta_1^1} = \dfrac{-0.6}{1-0.9} = -6$
- Bias-corrected second moment ($\hat{v}_1$): $\hat{v}_1 = \dfrac{v_1}{1-\beta_2^1} = 36$
- Look-ahead gradient adjustment using Nesterov momentum: $\tilde{g}_1 = g_1 + \beta_1 \hat{m}_1 = -6 + 0.9 \times (-6) = -11.4$
- Parameter update: $x_1 = x_0 - \dfrac{\alpha}{\sqrt{\hat{v}_1} + \epsilon} \times \tilde{g}_1 = 0 - \dfrac{0.1}{\sqrt{36 + 10^{-8}}} \times (-11.4) = 0 + 0.19 = 0.19$

At iteration 2:

- Gradient at $x_1 = 0.19$: $g_2 = 2(0.19 - 3) = -5.62$
- First-moment estimate (m_2): $m_2 = 0.9 \times (-0.6) + 0.1 \times (-5.62) = -1.1$
- Second-moment estimate (v_2): $v_2 = 0.999 \times 0.036 + 0.001 \times (-5.62)^2 = 0.067$
- Bias-corrected first moment ($\hat{m}_2$): $\hat{m}_2 = \dfrac{m_2}{1-0.9^2} = -5.89$
- Bias-corrected second moment ($\hat{v}_2$): $\hat{v}_2 = \dfrac{v_2}{1-0.999^2} = 33.5$
- Look-ahead gradient adjustment: $\tilde{g}_2 = g_2 + 0.9 \times (-5.89) = -5.62 + (-5.3) = -10.92$
- Parameter update: $x_2 = x_1 - \dfrac{0.1}{\sqrt{33.5 + 10^{-8}}} \times (-10.92) = 0.19 + 0.19 = 0.38$

In this example, Nadam combines the adaptive learning rate of Adam and the look-ahead mechanism of Nesterov momentum. By anticipating the future position of the parameters, Nadam can accelerate convergence, especially when dealing with noisy gradients. The look-ahead adjustment helps make more informed updates, potentially improving performance over Adam in some optimization tasks.

5.6.10 Learning Rate Annealing or Decay

Limited-memory Broyden–Fletcher–Goldfarb–Shanno (L-BFGS) is an optimization algorithm approximating the BFGS process using limited computer memory. It is a quasi-Newton method that is particularly efficient for smaller datasets and certain types of optimization problems where memory usage is a concern. Unlike the standard BFGS algorithm, which requires storing and updating a full approximation of the Hessian matrix, L-BFGS maintains only a limited number of vectors representing the approximation. This makes it more memory-efficient while still benefiting from the quasi-Newton approach, which leverages curvature information to accelerate convergence. The key steps of L-BFGS include initialization, iteration, and updating. Initially, the parameters are set, and initial values for the limited-memory vectors are established. For each iteration, the parameters are updated using a direction derived from the limited-memory approximation of the inverse Hessian. The limited-memory vectors are then updated based on the latest parameter values and gradients. The algorithm repeats these steps until convergence, using a small set of historical gradients and parameter updates to approximate the curvature information. L-BFGS is efficient for smaller datasets and certain types of optimization problems where memory usage is a concern. It leverages curvature information to potentially accelerate convergence compared to first-order methods like gradient descent. It is well-suited for optimization problems with a relatively small number of parameters or when the full Hessian is impractical to compute and store. However, L-BFGS is less commonly used for very large-scale deep learning tasks due to memory constraints and the complexity of maintaining the limited-memory approximation in high-dimensional spaces. It may not perform as well as some specialized optimization algorithms designed for deep learning tasks, which can handle the unique challenges of training large neural networks. Suppose you are optimizing a function with 50 parameters and storing a full Hessian matrix (which would require 50 × 50 = 2500 elements) is too memory-intensive. Using L-BFGS, you limit the storage to only the five most recent updates, significantly reducing memory usage while still approximating the curvature information. After initializing the parameters, each iteration updates the parameters using the stored five vectors, allowing for faster convergence with minimal memory overhead. The L-BFGS update rule is derived from the BFGS update but uses limited memory. The direction $\mathbf{p_k}$ at iteration **k** is calculated as:

$$p_k = -H_k \nabla f_k$$

where:

- $\mathbf{H_k}$ is the inverse Hessian approximation,
- $\nabla \mathbf{f_k}$ is the gradient of the function at iteration **k**.

L-BFGS stores a small number of past gradients and updates to approximate $\mathbf{H_k}$ efficiently, updating the inverse Hessian approximation iteratively. This allows for memory-efficient optimization while still leveraging second-order information. Suppose you are optimizing the function:

$$f(x, y) = (x-2)^2 + (y+3)^2$$

For example, if the parameters are:

- The function has two parameters, **x** and **y**, to minimize.
- Let's assume you are using L-BFGS with a memory size of 2, meaning you store the 2 most recent gradient updates.
- Initial values: $x_0 = 0$, $y_0 = 0$.

Iteration 1 (Initial Setup):

1. Gradient is: $\nabla f(x_0, y_0) = (2(x_0 - 2), 2(y_0 + 3)) = (2(0-2), 2(0+3)) = (-4, 6)$. This gradient will be stored as the first historical gradient.
2. Initialize Hessian approximation $H_0 = I$ (the identity matrix, as L-BFGS starts with no curvature information).

3. Direction calculation: $p_0 = -H_0 \nabla f(x_0, y_0) = -I \times (-4, 6) = (4, -6)$
4. Parameter update: $(x_1, y_1) = (x_0, y_0) + \alpha p_0 = (0, 0) + 0.1 \times (4, -6) = (0.4, -0.6)$

Iteration 2:

1. Gradient is: $\nabla f(x_1, y_1) = (2(0.4 - 2), 2(-0.6 + 3)) = (-3.2, 4.8)$. This gradient is stored as the second historical gradient.
2. Approximate Hessian update: Using the difference between gradients and parameter updates from iteration 1 to iteration 2, the Hessian inverse approximation H_1 is updated using limited memory (only the last two updates are stored).
3. Direction calculation: $p_1 = -H_1 \nabla f(x_1, y_1)$
   The direction now incorporates curvature information from the stored gradients, allowing for a more efficient update.
4. Parameter update: $(x_2, y_2) = (x_1, y_1) + \alpha p_1$

The updated parameter values after this step will move toward the minimum more efficiently than gradient descent. Instead of storing the full Hessian matrix (which for two parameters would be 2 × 2), L-BFGS only stores the most recent two gradients and updates, making it much more memory-efficient. In a higher-dimensional example with 50 parameters, L-BFGS would store only a few historical gradients and parameter updates (e.g., 5 or 10), rather than the full Hessian, which would require 50 × 50 = 2500 elements. This reduction in memory allows for optimization in cases where the full Hessian is impractical to compute and store. This memory efficiency makes L-BFGS suitable for smaller-scale problems or problems with limited memory resources while still benefiting from second-order information for faster convergence.

5.7 REAL-WORLD APPLICATIONS AND EXAMPLES

5.7.1 Supply Chain Optimization

Optimization techniques are extensively used in supply chain management to streamline operations and reduce costs. For instance, linear programming is often employed to optimize the distribution of goods from warehouses to various retail outlets. The goal is to minimize transportation costs while meeting demand at each location. This involves solving a linear optimization problem where constraints include transportation capacities, demand requirements, and available inventory. The Simplex method is particularly effective here, providing an optimal solution that balances these factors and ensures efficient resource allocation across the supply chain.

5.7.2 Portfolio Optimization in Finance

In finance, optimization theory plays a crucial role in portfolio management, where the objective is to maximize returns while minimizing risk. Portfolio optimization typically involves solving a non-linear optimization problem where the return on investment is maximized under constraints such as budget limitations and risk tolerance. Techniques such as quadratic programming are used to allocate assets in a way that optimizes the expected return for a given level of risk, taking into account the covariance between different assets. This approach helps investors make informed decisions, balancing potential gains with associated risks.

5.7.3 Telecommunications Network Design

Designing efficient telecommunications networks requires solving complex optimization problems to ensure reliable and cost-effective service delivery. Integer programming is frequently used to determine the optimal placement of network nodes and the routing of data across the network. This involves minimizing the overall cost of the network infrastructure while ensuring sufficient capacity and coverage to meet user demand. The B&B method is particularly useful for these types of problems, where decision variables are often binary (e.g., whether to install a particular piece of equipment or not).

5.7.4 Energy Management and Power Grid Optimization

The management of power grids is a complex optimization problem that involves balancing supply and demand while minimizing operational costs and ensuring stability. Convex optimization is often applied to optimize the generation and distribution of electricity across the grid. The objective is to minimize the cost of power generation while meeting the demand and adhering to operational constraints, such as transmission limits and generation capacities. This is crucial for maintaining a reliable power supply and reducing energy costs.

5.7.5 Transportation and Logistics

Combinatorial optimization is widely used in transportation and logistics to solve problems like vehicle routing and scheduling. The TSP is a classic example where the goal is to find the shortest possible route that visits a set of locations and returns to the starting point. Solutions to this problem are crucial for delivery companies that need to minimize fuel costs and delivery times. Advanced algorithms like genetic algorithms or ant colony optimization are employed to find near-optimal solutions to these NP-hard problems, enabling companies to optimize their logistics operations.

5.8 HANDS-ON EXAMPLE

In this hands-on example, we'll explore advanced optimization techniques for deep learning, focusing on gradient descent and its variants.

Step 1: Import necessary libraries

In this step, we are importing the necessary libraries to build and train a neural network using TensorFlow's Keras API. First, we import the TensorFlow library and essential components from TensorFlow.Keras, which is an easy-to-use API for building deep learning models. The sequential class is used to create a linear stack of layers for the model, while dense layers are fully connected layers, commonly used in neural networks. We also import three popular optimizers: SGD, Adam, and RMSprop, each of which offers different strategies for adjusting the model's weights during training to minimize the loss function. Lastly, matplotlib.pyplot is imported as plt, which allows us to create visualizations and plot the model's performance.

```
import tensorflow as tf
from tensorflow.keras.models import Sequential
from tensorflow.keras.layers import Dense
from tensorflow.keras.optimizers import SGD, Adam, RMSprop
import matplotlib.pyplot as plt
```

5.7.2 Step 2: Prepare the dataset

In this step, we are loading and preprocessing the **MNIST** dataset, which is a widely used dataset of handwritten digits (0–9) for training and evaluating machine learning models. The tf.keras.datasets.

mnist is a built-in function in TensorFlow that provides easy access to this dataset. The dataset is split into training (x_train, y_train) and testing (x_test, y_test) sets. Each image is represented as a 28 × 28-pixel grayscale image, where the pixel values range from 0 to 255. To simplify the model's learning process, the images are normalized by dividing each pixel value by 255, resulting in pixel values within the range [0, 1]. This normalization step ensures that all input values are on the same scale, which improves convergence during training. Finally, the images are flattened from 2D arrays (28 × 28) into 1D arrays with 784 elements (28 × 28) to prepare them for input into a fully connected neural network. This preprocessing step transforms the dataset into a suitable format for training neural networks.

```
# Load the MNIST dataset
mnist = tf.keras.datasets.mnist
(x_train, y_train), (x_test, y_test) = mnist.load_data()
# Normalize the images to the range [0, 1]
x_train, x_test = x_train / 255.0, x_test / 255.0
# Flatten the images
x_train = x_train.reshape(-1, 28 * 28)
x_test = x_test.reshape(-1, 28 * 28)
```

5.7.3 Step 3: Build a simple neural network model

In this step, we define the function build_model that creates a simple neural network model using the Keras Sequential API. The model consists of two layers: the first is a dense layer with 128 neurons and the Rectified Linear Unit (ReLU) activation function, which processes the flattened 784-dimensional input (representing each 28 × 28 pixel **MNIST** image), and the second layer is a dense output layer with 10 neurons (one for each digit class, 0–9) and a softmax activation function. The softmax function converts the outputs into probability distributions, making it suitable for multi-class classification tasks. The model is compiled with the specified optimizer, the sparse categorical crossentropy loss function, which is used for multi-class classification, and the accuracy metric to monitor performance during training. The function returns the compiled model, ready to be trained on the MNIST dataset. This model is a simple feed-forward neural network designed to classify the handwritten digits.

```
 def build_model(optimizer):
model = Sequential([
Dense(128, activation='relu', input_shape=(784,)),
Dense(10, activation='softmax')
 ])
 model.compile(optimizer=optimizer,
   loss='sparse_categorical_crossentropy',
   metrics=['accuracy'])
 return model
```

5.7.4 Step 4: Train the model with different optimizers

In this step, we are defining a dictionary of three different optimizers (SGD, RMSprop, and Adam), each with specified learning rates. These optimizers control how the model updates its weights during training. The goal is to compare how different optimization algorithms perform on the same task. The for loop goes through each optimizer in the dictionary and builds a new model using the build_model function with that optimizer. For each model, training is carried out for 10 epochs on the **MNIST** dataset with an 80–20 split between training and validation data. The model.fit() function is responsible for training the model, and the training history, which includes details about the loss and accuracy over each epoch, is stored in the histories dictionary. This setup allows us to compare the performance of different optimization algorithms on the same model and dataset. The progress of each optimizer is printed out, although the training output is set to be quiet with verbose=0 to avoid cluttering the console.

```
optimizers = {
  'SGD': SGD(learning_rate=0.01),
  'RMSprop': RMSprop(learning_rate=0.001),
  'Adam': Adam(learning_rate=0.001)
}
histories = {}
for name, optimizer in optimizers.items():
  print(f"Training with {name}...")
  model = build_model(optimizer)
  history = model.fit(x_train, y_train, epochs=10, validation_
  split=0.2, verbose=0)
  histories[name] = history
```

5.7.5 Step 5: Evaluate the models

In this step, we are evaluating the performance of models trained with different optimizers on the test set. For each optimizer in the optimizers dictionary, a new model is created using the build_model function, and the model is trained for 10 epochs on the **MNIST** dataset. The validation_split=0.2 ensures that 20% of the training data is used for validation during training. After training, the model is evaluated on the test set (x_test, y_test) using the model.evaluate() function, which computes the test loss and test accuracy. The accuracy on the test set is printed for each optimizer, allowing us to compare how well each optimization algorithm generalizes to unseen data. The verbose=0 option ensures that the output is minimal, focusing on the final test accuracy results for each optimizer, formatted to four decimal places for clarity.

```
for name, optimizer in optimizers.items():
  model = build_model(optimizer)
  model.fit(x_train, y_train, epochs=10, validation_split=0.2,
  verbose=0)
  test_loss, test_acc = model.evaluate(x_test, y_test, verbose=0)
  print(f'Test accuracy with {name}: {test_acc:.4f}')
```

5.9 COMMON MISTAKES AND TROUBLESHOOTING TIPS IN OPTIMIZATION

5.9.1 Linear Optimization

- *Mistake:* Misinterpreting the feasible region in linear programming problems.
- *Tip:* Always clearly define and plot the constraints. Ensure the feasible region is correctly identified as the intersection of all constraint regions.

5.9.2 Non-linear Optimization

- *Mistake:* Ignoring the potential for multiple local optima.
- *Tip:* Use multiple starting points or global optimization techniques to increase the likelihood of finding the global optimum.
- *Mistake:* Assuming that gradient-based methods will always converge to the global optimum.
- *Tip:* Recognize that gradient-based methods can get stuck in local optima. Consider using hybrid methods that combine gradient-based approaches with heuristic or metaheuristic methods.

5.9.3 Integer Optimization

- *Mistake:* Treating integer variables as continuous.
- *Tip:* Ensure that the optimization algorithm respects the integer constraints. Use specific integer programming solvers like B&B or cutting plane methods.
- *Mistake:* Using standard linear programming methods for integer problems.
- *Tip:* Employ algorithms designed for integer programming, such as the Simplex method combined with B&B, to handle integer constraints effectively.

5.9.4 Convex Optimization

- *Mistake:* Misidentifying non-convex problems as convex.
- *Tip:* Verify the convexity of the objective function and the feasible region. Only convex problems guarantee that any local minimum is a global minimum.
- *Mistake:* Neglecting the role of constraints in defining convexity.
- *Tip:* Ensure that both the objective function and all constraints are convex so that convex optimization techniques can be applied correctly.

5.9.5 Combinatorial Optimization

- *Mistake:* Expecting exact solutions from heuristic methods.
- *Tip:* Understand that heuristic and metaheuristic methods provide good approximations but not necessarily exact solutions. Use them when exact methods are computationally infeasible.

5.9.6 Stochastic Optimization

- *Mistake:* Ignoring the impact of randomness on convergence.
- *Tip:* Use multiple runs with different random seeds to ensure robustness. Average the results to get a more reliable solution.
- *Mistake:* Failing to model uncertainty accurately.
- *Tip:* Incorporate accurate probabilistic models for uncertainty. Use methods such as SGD and SAA appropriately.

5.9.7 Gradient Descent

- *Mistake:* Using an inappropriate learning rate.
- *Tip:* Tune the learning rate carefully. A too-high learning rate can cause divergence, while a too-low rate can lead to slow convergence.

- *Mistake:* Not normalizing input features.
- *Tip:* Normalize or standardize input features to ensure that gradient descent converges more efficiently.

5.9.8 Simplex Method

- *Mistake:* Misinterpreting degeneracy and cycling.
- *Tip:* Use anti-cycling rules like Bland's rule to handle degeneracy and ensure convergence.
- *Mistake:* Ignoring the potential for large numbers of iterations in complex problems.
- *Tip:* For significant problems, consider alternative methods like Interior Point Methods that may be more efficient.

5.9.9 Lagrangian Multipliers

- *Mistake:* Misapplying Lagrangian multipliers to inequality constraints.
- *Tip:* Use Karush–Kuhn–Tucker (KKT) conditions for problems involving inequality constraints.

5.9.10 B&B and Cutting Plane Methods

- *Mistake:* Underestimating the computational effort required.
- *Tip:* Use hybrid and heuristic methods to reduce computational effort in large-scale problems.
- *Mistake:* Ignoring the importance of initial bounds.
- *Tip:* Start with reasonable initial bounds to enhance the efficiency of B&B methods.

5.9.11 Evolutionary Algorithms

- *Mistake:* Overfitting to the training data in GAs.
- *Tip:* Use techniques like cross-validation and regularization to ensure the generalizability of the solutions.
- *Mistake:* Neglecting parameter tuning.
- *Tip:* Tune parameters such as population size, mutation rate, and crossover rate to balance exploration and exploitation effectively.

5.10 REVIEW QUESTIONS

1. What are the critical differences between linear and non-linear optimization?
2. How do objective functions and constraints define an optimization problem?
3. What is the Simplex method, and how is it used to solve linear optimization problems?
4. How are feasible regions, constraints, and objective functions represented in linear programming?
5. What challenges are associated with non-linear optimization compared to linear optimization?
6. What standard methods are used to solve non-linear optimization problems, such as gradient-based and direct search methods?
7. What is integer programming, and in what scenarios is it advantageous?
8. How do B&B and cutting plane methods work, and what problems do they solve?
9. How do techniques like SAA and SGD handle uncertainty in optimization problems?
10. In what types of optimization problems are GAs particularly effective?

5.11 PROGRAMMING QUESTIONS

5.11.1 Easy

Implement a simple linear regression model using gradient descent to fit a line to a set of data points.

1. Generate or load a dataset with a single feature and target variable.
2. Set initial values for the slope (m) and intercept (b) of the line.
3. Calculate the predictions using the current values of m and b.
4. Calculate the gradients of the loss function with respect to m and b.
5. Repeat the gradient descent steps for the specified number of iterations.

5.11.2 Medium

Build a neural network to classify handwritten digits from the **MNIST** dataset.

1. Load the **MNIST** dataset. Normalize the images to the range [0, 1].
2. Split the dataset into training and testing sets.
3. Define a sequential model with input, hidden, and output layers.
4. Select an optimizer, loss function, and evaluation metric.
5. Include a validation split to monitor validation accuracy and loss.

5.11.3 Hard Problem

Create a **CNN** to classify images from the **CIFAR-10** dataset.

1. Load the **CIFAR-10** dataset. Normalize the images to the range [0, 1].
2. Define a sequential model with convolutional, pooling, and dense layers.
3. Select an optimizer, loss function, and evaluation metrics.
4. Train the model on the training data with validation.
5. Evaluate the model's performance on the test set.
6. Experiment with different architectures, hyperparameters, and regularization techniques to improve performance. Use techniques such as grid search or random search for hyperparameter optimization.

6 Information Theory

6.1 INTRODUCTION

Information theory is foundational in various areas, including communication, data compression, cryptography, etc. Information is paramount in our rapidly evolving technological landscape, bridging the gap between abstract thought and tangible computation. It defines how we communicate, store, and process data. The theory's ability to measure uncertainty, optimize data encoding, and ensure reliable communication even in noisy environments underpins today's digital infrastructure. As we delve into this subject, you will be introduced to foundational concepts such as entropy, which quantifies the unpredictability of information content, and mutual information, which reveals the shared knowledge between variables.

6.2 ENTROPY

Entropy, in the context of information theory, can be considered a measure of unpredictability or uncertainty. Regarding data or messages, entropy scales the average level of "surprise" contained in the potential outcomes. This might seem abstract but consider the flip of a fair coin. As it has an equal probability of landing heads or tails, the result is very uncertain, leading to higher entropy. In contrast, if you had a biased coin that almost always landed heads, its outcome's entropy (or uncertainty) would be much lower. For a discrete random variable **x** with a given probability distribution **P(x)**, where **x** is an outcome and **P(x)** is the probability of that outcome, the entropy **H(x)** is calculated as:

$$H(X) = -\sum_{x \in X} P(x) \log P(x)$$

where:

- $H(X)$ is the entropy of the discrete random variable X,
- $P(x)$ is the probability of outcome x,
- The summation runs over all possible outcomes x in the set X.

Here, the sum spans over all possible outcomes of **x**. The logarithm base often used in information theory is base 2, which means the entropy is measured in bits. However, natural or logarithms to the base 10 can also be used, resulting in entropy being measured in nats (when the logarithm base is **e** (natural logarithm)) or dits (or Hartleys (when the logarithm base is 10)), respectively. The main question is: Why does entropy matter? When considering transmitting messages or data, we want

DOI: 10.1201/9781032690742-6

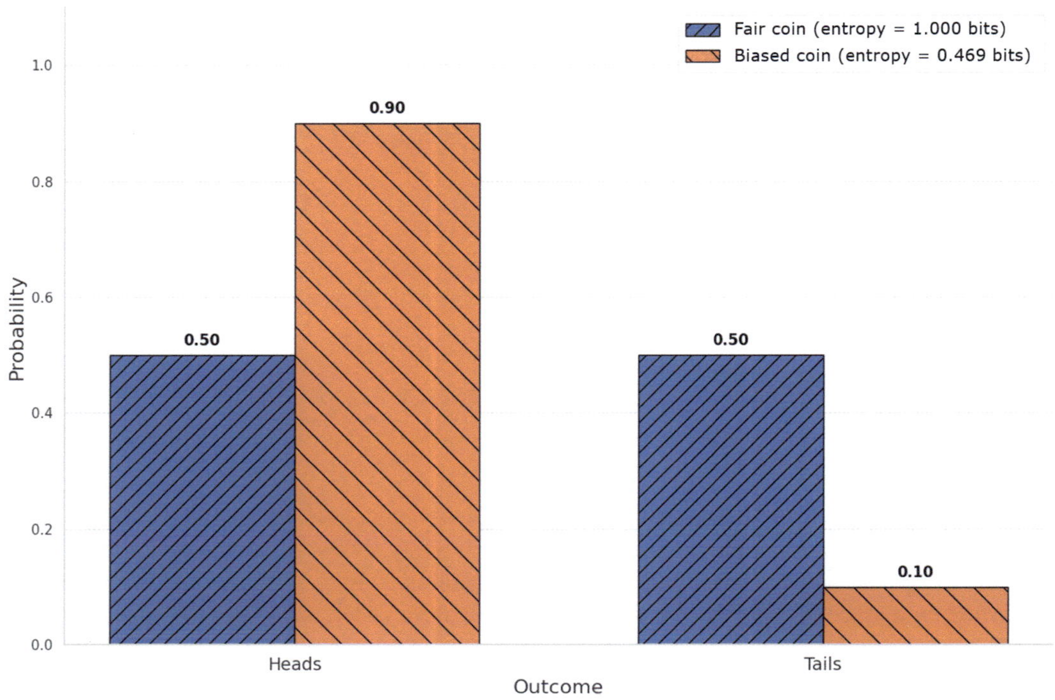

FIGURE 6.1 Probability distributions of a fair coin and a biased coin.

to use as few bits as possible. Entropy gives us a lower bound on the average number of bits needed to represent symbols from **x**. If symbols are encoded optimally, a more frequent symbol should be assigned a shorter code, while a rare symbol might get a more extended code. This principle is leveraged in data compression techniques like Huffman coding. Entropy **H(x)** tells us the minimum average number of bits we would need per symbol if we could design our encoding most efficiently. Entropy quantifies the uncertainty or unpredictability in a set of outcomes, guiding efficient data encoding and offering insights into data's inherent structure and randomness.

Figure 6.1 illustrates the probability distributions of a fair coin and a biased coin. The bar plot shows the probability of obtaining "Heads" and "Tails" for each type of coin. The fair coin, represented in blue with a striped pattern, has equal probabilities of 0.5 for both outcomes, resulting in higher entropy, indicating maximum uncertainty. In contrast, the biased coin, shown in orange with a slanted hatch, has a much higher probability for "Heads" (0.9) and a lower probability for "Tails" (0.1), leading to lower entropy and reduced uncertainty.

6.3 JOINT AND CONDITIONAL ENTROPY

Joint and conditional entropies offer richer insights into systems with multiple interacting variables. They provide a deeper understanding beyond individual uncertainties, offering a clearer picture of the system's dynamics and the relationships between its components.

6.3.1 Joint Entropy

When we talk about two random variables, let us say **X** and **Y**, their combined uncertainty can be represented by what we term "Joint Entropy." The joint entropy of **X** and **Y** measures the uncertainty (or unpredictability) associated with the pair (**X**, **Y**) when they are considered together.

Mathematically, for discrete random variables **X** and **Y** with joint probability distribution **P(x, y)**, the joint entropy **H(X, Y)** is defined as:

$$H(X,Y) = -\sum_{x\in X}\sum_{y\in Y} P(x,y)\log P(x,y)$$

where:

- *H(X, Y)* is the joint entropy of the random variables **X** and **Y**,
- *P(x, y)* is the joint probability of the outcomes **x** and **y**,

Here, the sum runs over all possible combinations of outcomes **x** from **X** and **y** from **Y**. Joint entropy provides a more general view of the system's unpredictability when both random variables are considered together. For instance, if we have two correlated variables, knowing the outcome of one might reduce the unpredictability of the other, leading to a joint entropy that is less than the sum of their entropies. Suppose we have two random variables, **X** and **Y**, where **X** represents the outcome of a coin toss (Heads or Tails), and **Y** represents the outcome of rolling a six-sided die (numbers 1–6). Let's assume that **X** has two possible outcomes: Heads (**H**) and Tails (**T**), each with a probability of **P(X = H)** = 0.5 and **P(X = T)** = 0.5. **Y** has six possible outcomes: **Y** = {1, 2, 3, 4, 5, 6}, with each outcome having a probability $P(Y = y) = \frac{1}{6}$ (fair die). The joint probability distribution **P(X, Y)** would give the probability of each combination of **X** and **Y**. As **X** and **Y** are independent, the joint probability **P(X = x, Y = y) = P(X = x) · P(Y = y)**. For example:

$$P(X = H, Y = 1) = 0.5\times\frac{1}{6} = \frac{1}{12}, \quad \text{and} \quad P(X = T, Y = 2) = 0.5\times\frac{1}{6} = \frac{1}{12}$$

and so on for all other combinations. Let's compute joint entropy **H(X, Y)** for this example:

$$H(X,Y) = -\sum_{x\in H}\sum_{y=1}^{6} P(x,y)\log_2 P(x,y)$$

As $P(x,y) = \frac{1}{12}$ for each combination, the joint entropy will be:

$$H(X,Y) = -12\times\left(\frac{1}{12}\log_2\frac{1}{12}\right) = -12\times\left(\frac{1}{12}\times(-3.585)\right) = 3.585 \text{ bits}$$

This value, 3.585 bits, represents the total uncertainty in the system when considering both the coin toss and die roll together. The joint entropy is smaller than the sum of the individual entropies of **X** and **Y**, reflecting the combined unpredictability of both variables.

6.3.2 Conditional Entropy

Conditional entropy, denoted as $H(Y|X)$, represents the average uncertainty remaining in **Y** once the value of **X** is known. In other words, it quantifies how much we still do not know about **Y** even after observing **X**. The conditional entropy $H(Y|X)$ can be defined as:

$$H(Y|X) = \sum_{x\in X} P(x)H(Y|X = x)$$

where $H\left(Y \middle| X=x\right)$ is the entropy of **Y** given a particular value **x** for **X**. Alternatively, it can be expressed in terms of joint and marginal probabilities:

$$H(Y|X) = -\sum_{x \in X} \sum_{y \in Y} P\left(x, y\right) \log P(y|x)$$

It is a powerful metric to determine how much one variable informs about another, providing insights into the interdependencies and relationships between variables. This measure is central to information theory and has significant implications in various domains, including data analysis and machine learning. Imagine we have two random variables: **X** represents the weather, with possible outcomes: Sunny **(S)** or Rainy **(R)**. **Y** represents whether a person will carry an umbrella, with possible outcomes: Yes **(Y)** or No **(N)**. Let's say the following probabilities are given based on past data:

- **P(X = S)** = 0.7, **P(X = R)** = 0.3 (the weather is sunny 70% of the time and rainy 30% of the time).
- **P(Y = Y | X = S)** = 0.2, **P(Y = N | X = S)** = 0.8 (if it's sunny, the person carries an umbrella 20% of the time).
- **P(Y = Y | X = R)** = 0.9, **P(Y = N | X = R)** = 0.1 (if it's rainy, the person carries an umbrella 90% of the time).

The conditional entropy **H(Y | X)** measures the uncertainty in **Y** (whether the person carries an umbrella) given that we already know **X** (the weather). We can calculate it using the formula:

$$H\left(Y|X\right) = -\sum_{x} P\left(X=x\right) \sum_{y} P(Y=y\,|X=x) \log_2 P(Y=y\,|X=x)$$

Now, let's compute it:

1. For **X = S** (Sunny): $P(Y=Y\,|\,X=\text{S}) = 0.2$, $P(Y=N\,|\,X=S) = 0.8$

   $$H(Y\,|\,X=S) = -\left(0.2\log_2(0.2) + 0.8\log_2(0.8)\right) = -\left(0.2\times(-2.322) + 0.8\times(-0.322)\right)$$
   $$= 0.2\times 2.322 + 0.8\times 0.322 = 0.4644 + 0.2576 = 0.722 \text{ bits}$$

2. For **X = R** (Rainy): $P(Y=Y\,|\,X=R) = 0.9$, $P(Y=N\,|\,X=R) = 0.1$

   $$H(Y\,|\,X=R) = -\left(0.9\log_2(0.9) + 0.1\log_2(0.1)\right) = -\left(0.9\times(-0.152) + 0.1\times(-3.322)\right)$$
   $$= 0.9\times 0.152 + 0.1\times 3.322 = 0.1368 + 0.3322 = 0.469 \text{ bits}$$

Now, using the total probability **P(X)**, we can compute the overall conditional entropy:

$$H\left(Y|X\right) = P\left(X=S\right)\cdot H\left(Y|X=S\right) + P\left(X=R\right)\cdot H(Y|X=R)$$

$$H(Y|X) = 0.7\cdot 0.722 + 0.3\cdot 0.469 = 0.5054 + 0.1407 = 0.646 \text{ bits}$$

Thus, the conditional entropy **H(Y | X)** = 0.646 bits tells us the remaining uncertainty about whether the person will carry an umbrella after knowing the weather. As this value is less than the entropy of **Y** alone, it indicates that knowing the weather reduces our uncertainty about **Y**.

Figure 6.2 illustrates two key concepts in information theory: joint probability distribution and conditional entropy. The left part of Figure 6.2a displays the joint probability distribution *P(X, Y)*,

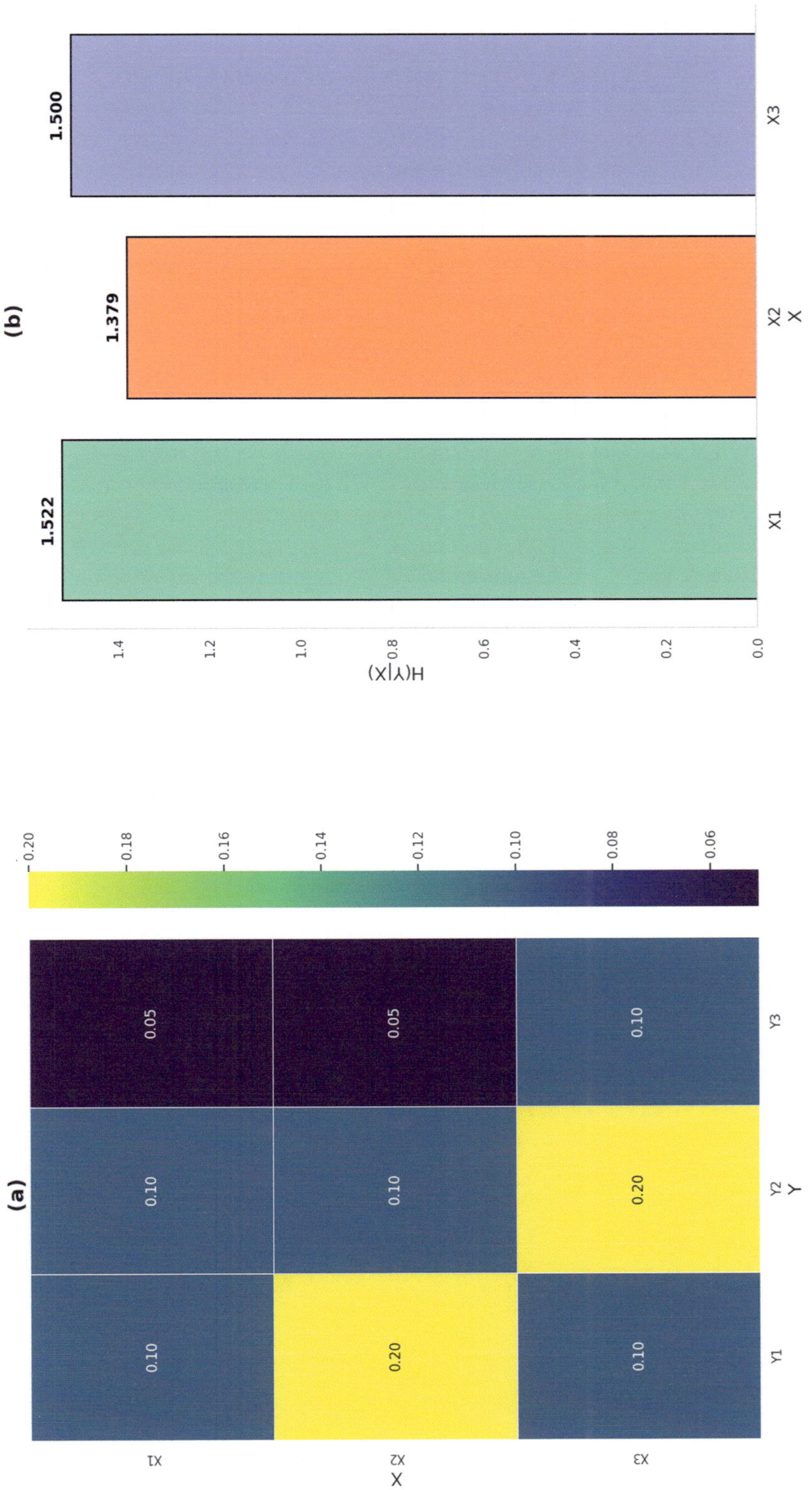

FIGURE 6.2 (a) Joint probability distribution $P(X, Y)$, and (b) conditional entropy $H(Y \mid X)$.

showing how two random variables, **X** and **Y**, are related. Each cell in the matrix represents the probability of a particular combination of values of X (X_1, X_2, X_3) and Y (Y_1, Y_2, Y_3). For instance, the probability of $\mathbf{X}_2$ occurring with $\mathbf{Y}_1$ is 0.20, indicating that this combination is more likely compared to others, such as $\mathbf{X}_1$ and $\mathbf{Y}_3$, which have a lower probability of 0.05. The color scale on the right provides a visual guide, with higher probabilities represented by yellow and lower probabilities by darker shades. The right part of Figure 6.2b shows the conditional entropy $H(Y \mid X)$ for each value of **X**. This measure quantifies the uncertainty or unpredictability of **Y** given that **X** has already occurred. For example, when $X = X_1$, the conditional entropy is approximately 1.522, indicating a moderate level of uncertainty about **Y**. In contrast, when $X = X_2$, the entropy value is lower at around 1.379, suggesting that knowing X_2 provides more information about Y compared to other cases.

6.4 INFORMATION GAIN

Information gain quantifies the reduction in uncertainty about one variable given knowledge of another variable. When the value of **X** significantly reduces the uncertainty of **Y**, the conditional entropy $H(Y \mid X)$ becomes much smaller than the entropy $H(Y)$. This reduction in uncertainty indicates that **X** provides information about **Y**, suggesting a strong dependency between them. Conversely, if $H(Y \mid X)$ is approximately equal to $H(Y)$, knowing **X** does not enhance the prediction of **Y** significantly, indicating that **X** and **Y** are largely independent regarding the information they convey. If $H(Y \mid X) = 0$, knowing **X** allows us to predict **Y** perfectly, indicating a deterministic relationship. If $H(Y \mid X)$ is high but not equal to $H(Y)$, **X** provides some information about **Y**, but significant uncertainty remains. In machine learning, information gain is crucial for feature selection and decision tree algorithms. During feature selection, conditional entropy helps identify features that reduce the uncertainty of the target variable. A feature **X** that significantly reduces $H(Y \mid X)$ compared to $H(Y)$ is valuable for the model. Algorithms like decision trees use information gain to select features that provide the highest reduction in entropy. At each step, the feature that offers the most significant information gain about the target variable is chosen, improving the tree's predictive accuracy. Consider a simple weather prediction scenario where we want to predict whether it will rain (**Y**) based on whether there are clouds in the sky (**X**).

- Outcomes for **Y** (Rain): Yes (1), No (0)
- Outcomes for **X** (Clouds): Yes (1), No (0)

Probabilities are:

- $P(Y = 1) = 0.4$ $P(\mathrm{Y} = 1) = 0.4$ $P(Y = 1) = 0.4$ (probability of rain)
- $P(Y = 0) = 0.6$ $P(Y = 0) = 0.6$ $P(Y = 0) = 0.6$ (probability of no rain)
- $P(X = 1) = 0.5$ $P(X = 1) = 0.5$ $P(X = 1) = 0.5$ (probability of clouds)
- $P(X = 0) = 0.5$ $P(X = 0) = 0.5$ $P(X = 0) = 0.5$ (probability of no clouds)
- $P(Y = 1 \mid X = 1) = 0.8$ $P(Y = 1 \mid X = 1) = 0.8$ $P(Y = 1 \mid X = 1) = 0.8$ (probability of rain given clouds)
- $P(Y = 1 \mid X = 0) = 0.2$ $P(Y = 1 \mid X = 0) = 0.2$ $P(Y = 1 \mid X = 0) = 0.2$ (probability of rain given no clouds)

Let us calculate the entropy:

1. Entropy of **Y** (Rain):

$$H(Y) = -[0.4\log_2 0.4 + 0.6\log_2 0.6] \approx -[0.4 \times (-1.322) + 0.6 \times (-0.737)] \approx 0.970 \text{ bits}$$

2. Conditional Entropy of **Y** given **X**:

$$H(Y|X) = P(X=1)\cdot H(Y|X=1) + P(X=0)\cdot H(Y|X=0)$$

$$H(Y \mid X=1) = -[0.8\log_2 0.8 + 0.2\log_2 0.2] \approx 0.722 \text{ bits}$$

$$H(Y \mid X=0) = -[0.2\log_2 0.2 + 0.8\log_2 0.8] \approx 0.722 \text{ bits}$$

$$H(Y|X) = 0.5\cdot 0.722 + 0.5\cdot 0.722 \approx 0.722 \text{ bits}$$

3. Information gain:

$$IG(Y;X) = \mathrm{H}(Y) - \mathrm{H}(Y|X) = 0.970 - 0.722 \approx 0.249 \text{ bits}$$

The information gain of 0.249 bits indicates that knowing whether there are clouds in the sky reduces the uncertainty about whether it will rain by 0.249 bits.

6.5 MUTUAL INFORMATION

Mutual information is a fundamental quantity in information theory that quantifies the information obtained about one random variable by observing another. It measures the dependence between two variables and has wide applications in feature selection, machine learning, and data analysis. Mutual information provides a quantitative measure of the relationship between two random variables. In contexts like feature selection, it can help determine which features (variables) carry the most information about the target variable, thereby being most relevant for tasks like classification or regression. There are three main basic properties of mutual information:

1. *Symmetry:* Mutual information is symmetric. This means that the amount of information gained about **X** by knowing **Y** is the same as the amount of information gained about **Y** by knowing **X**. It can be shown as: $I(X; Y) = I(Y; X)$
2. *Non-Negative:* Mutual information is always non-negative. If $I(X; Y) = 0$, it implies that the two random variables are independent, and knowing the value of one does not provide any information about the other.
3. *Range:* Mutual information can take a value between 0 (when the variables are independent) and the entropy of one of the variables (when one variable is a deterministic function of the other).

Here are the brief explanations of the mutual information formula:

$$I(X; Y) = H(X) - H(X \mid Y)$$

This formula means that the mutual information between **X** and **Y** is the difference between the entropy of **X** (the uncertainty in **X**) and the conditional entropy of **X** given **Y** (the remaining uncertainty in **X** when we know **Y**):

$$I(X; Y) = H(Y) - H(Y \mid X)$$

Similarly, this formula shows that the mutual information between **X** and **Y** is the difference between the entropy of **Y** (the total uncertainty in **Y**) and the conditional entropy $H(Y \mid X)$ (the remaining uncertainty in **Y** given that **X** is known).

Mutual information is a powerful tool because it captures non-linear dependencies between variables, unlike other measures, such as correlation, which capture only linear dependencies. If the mutual information between two variables is high, the variables are strongly related. Suppose we have two random variables: **X** represents the outcome of flipping a fair coin, with possible outcomes: Heads **(H)** or Tails **(T)**, each with a probability **P(X = H)** = 0.5 and **P(X = T)** = 0.5. **Y** represents whether or not a person decides to walk based on the weather, with possible outcomes: Walk **(W)** or Stay **(S)**, each with a probability dependent on the coin flip (to simulate a correlation).

Let's assume the probabilities based on the coin flip are:

If **X = H** (Heads), the person decides to walk with probability **P(Y = W | X = H)** = 0.8 and stays with **P(Y = S | X = H)** = 0.2.

If **X = T** (Tails), the person decides to stay with probability **P(Y = S | X = T)** = 0.9, and walks with **P(Y = W | X = T)** = 0.1.

Step 1: Calculate the individual entropies **H(X)** and **H(Y)**:

H(X): Since **X** is a fair coin flip, the entropy of **X** is:

$$H(X) = -\sum_{x} P(X = x)\log_2 P(X = x) = -(0.5\log_2 0.5 + 0.5\log_2 0.5) = 1 \text{ bit}$$

H(Y): To calculate the entropy of **Y**, we first need the marginal probabilities for **Y**. These are:

$$P(Y = W) = P(Y = W \mid X = H)\cdot P(X = H) + P(Y = W \mid X = T)\cdot P(X = T)P$$

$$P(Y = W) = (0.8\cdot 0.5) + (0.1\cdot 0.5) = 0.4 + 0.05 = 0.45$$

$$P(Y{=}S){=}1{-}P(Y{=}W){=}0.55$$

Now, we can calculate **H(Y)**:

$$H(Y) = -(0.45\log_2 0.45 + 0.55\log_2 0.55) = -(0.45\times -1.152 + 0.55\times -0.863) = 1.027 \text{ bits}$$

Step 2: Calculate the conditional entropy **H(X | Y)**:

We need the conditional probabilities **P(X | Y)**, but it's easier to calculate **H(X | Y)** using the joint probabilities of **X** and **Y**:

$$P(X = H, Y = W) = P(Y = W \mid X = H)\cdot P(X = H) = 0.8\times 0.5 = 0.4$$

$$P(X = H, Y = S) = P(Y = S \mid X = H)\cdot P(X = H) = 0.2\times 0.5 = 0.1$$

$$P(X = T, Y = W) = P(Y = W \mid X = T)\cdot P(X = T) = 0.1\times 0.5 = 0.05$$

$$P(X = T, Y = S) = P(Y = S \mid X = T)\cdot P(X = T) = 0.9\times 0.5 = 0.45$$

Using the joint probabilities, we calculate the conditional entropy **H(X | Y)**:

$$H(X \mid Y) = -\sum_{x,y} P(X = x, Y = y)\log_2 \frac{P(X = x, Y = y)}{P(Y = y)}$$

$$H(X \mid Y) = -\left(0.4\log_2\frac{0.4}{0.45} + 0.05\log_2\frac{0.05}{0.45} + 0.1\log_2\frac{0.1}{0.55} + 0.45\log_2\frac{0.45}{0.55}\right) \approx 0.56 \text{ bits}$$

Step 3: Calculate mutual information **I(X; Y)**:

Finally, mutual information **I(X; Y)** is the difference between the entropy of **X** and the conditional entropy **H(X | Y)**:

$$I(X;\, Y) = H(X) - H(X \mid Y) = 1 - 0.56 = 0.44 \text{ bits}$$

This result tells us that knowing **Y** provides 0.44 bits of information about **X**, meaning there is some dependency between the two variables (but not complete dependence). The mutual information quantifies this relationship and can be used to assess the strength of the dependency between **X** and **Y**.

Figure 6.3 provides a view of the relationship between variables **X** and **Y** using information-theoretic concepts. In subplot Figure 6.3a, the joint probability distribution $P(X, Y)$ is visualized as a heatmap. The distribution shows the probability values for combinations of **X** (with values X_1, X_2, X_3) and **Y** (with values Y_1, Y_2, Y_3). Each cell in the matrix displays the probability associated with a particular combination of **X** and **Y**. For example, $P(X_2, Y_2)$ has a higher probability value (0.20), highlighted in yellow, compared to other combinations. This heatmap provides insight into how the two variables interact, showing the likelihood of different outcomes. Subplot Figure 6.3b illustrates the conditional entropy $H(Y \mid X)$, which measures the uncertainty in **Y** given that **X** is known. The bar plot presents the conditional entropy values for each category of **X**. The values are 1.522 for $\mathbf{X}_1$, 1.379 for $\mathbf{X}_2$, and 1.500 for $\mathbf{X}_3$, indicating how much uncertainty remains in **Y** when each corresponding value of **X** is observed. Subplot Figure 6.3c shows the mutual information $I(X; Y)$, which quantifies the amount of information shared between **X** and **Y**. It is represented as a single bar with a value of 0.059 bits, indicating that there is some degree of dependency between **X** and **Y**, although it is relatively small.

6.6 DATA COMPRESSION

Data compression is a method to reduce the quantity of data used to represent information. Understanding the statistical properties of data makes it possible to represent it more compactly without losing essential information. With concepts like entropy, information theory offers a mathematical foundation to understand the limits and potential of data compression. The tools and principles derived from this theory have driven the development of many efficient algorithms and standards in digital communication and storage. In simple terms, entropy quantifies the amount of uncertainty or randomness in a source of information. It measures the "surprisal," how unexpected or uncertain a message is. When considering data compression, a high-entropy source means that the data is complex to predict and, hence, more challenging to compress, while a low-entropy source indicates the opposite. For a given source of data or a message, the entropy provides a theoretical lower bound on the average number of bits needed to encode symbols from that source. In other words, it gives the minimum bits required to represent the information without any loss. This understanding is pivotal in designing efficient encoding schemes. If an encoding scheme achieves this bound, it is considered optimal. There are several applications for it in encoding. For instance, if you are trying to encode a text written in English, not all letters or combinations of letters occur with equal frequency. Spaces, vowels like "**e**" and "**a**," are more frequent than letters like "**z**" or "**q**." One can achieve compression by using fewer bits to represent frequently occurring letters and more bits for less frequent ones. This is the basic idea behind Huffman coding, a popular lossless data

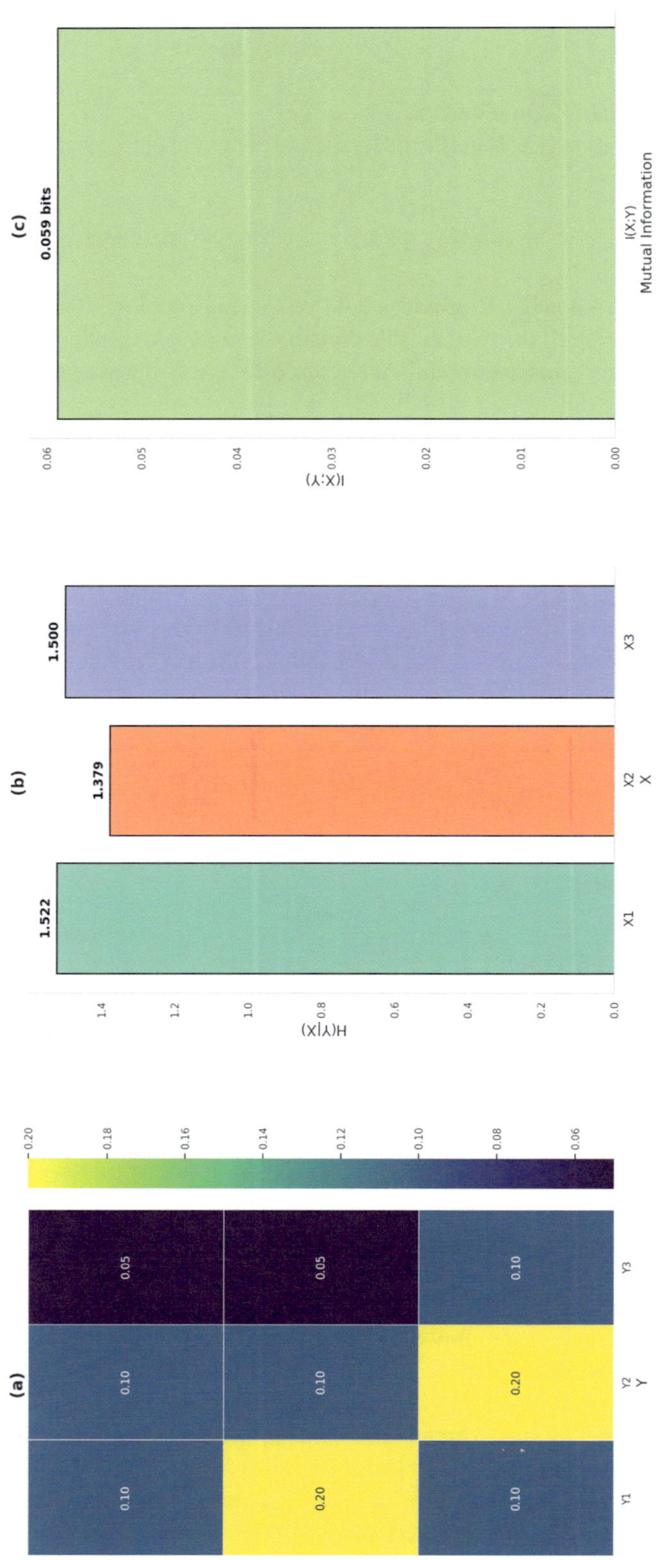

FIGURE 6.3 (a) Joint probability distribution $P(X, Y)$, (b) conditional entropy $H(Y \mid X)$, and (c) mutual information $I(X; Y)$.

compression algorithm. Let us have a review on lossless vs. lossy compression. It is essential to differentiate between two primary types of data compression: lossless and lossy.

1. *Lossless Compression:* This method ensures that the original data can be perfectly reconstructed from the compressed data. It is used in applications where the preservation of original data is crucial, like text files or certain types of image and audio files (e.g., PNG, FLAC).
2. *Lossy Compression:* This method sacrifices some data for higher compression rates, meaning the original data cannot be perfectly reconstructed. It is often acceptable for multimedia applications like audio, images, and videos where some loss of detail might not be perceptible to human senses (e.g., MP3, JPEG).

Consider a text written in English where each letter has a different frequency of occurrence. The letter "**e**" appears frequently, while the letter "**z**" appears much less often. Using Huffman coding, a lossless data compression technique, we can assign shorter codes to more frequent letters and longer codes to less frequent ones. Suppose the probabilities of four letters are as follows:

"**e**": 0.3; "**a**": 0.25; "**t**": 0.2; and "**z**": 0.05

Using Huffman coding, we can assign binary codes like this:

"**e**" → 1 (1 bit), "**a**" → 01 (2 bits), "**t**" → 001 (3 bits) and "**z**" → 000 (3 bits)

The most frequent letter, "**e**," gets the shortest code (1 bit), while the least frequent letter, "**z**," gets a longer code (3 bits), making the overall encoding more efficient. The entropy of a source, denoted as **H(X)**, quantifies the uncertainty or "randomness" in the source and provides a lower bound on the average number of bits required to encode the data. For a source **X** with probabilities $\mathbf{p(x_i)}$ for each symbol $\mathbf{x_i}$, the entropy is defined as:

$$H(X) = -\sum_{i=1}^{n} p(x_i)\log_2 p(x_i)$$

For the example above with letters "**e**," "**a**," "**t**," and "**z**," the entropy calculation would be:

$$H(X) = -(0.3\log_2 0.3 + 0.25\log_2 0.25 + 0.2\log_2 0.2 + 0.05\log_2 0.05) \approx 1.9 \text{ bits}$$

This means, on average, 1.9 bits are required to represent a letter from this source, and Huffman coding brings us close to this limit.

Figure 6.4 explores the frequency distribution of letters in English text and their corresponding encoding lengths using Huffman Encoding. In subplot Figure 6.4a, the letter probabilities are shown as a bar chart, representing the relative frequencies of each letter in English text. The chart reveals that some letters, like "*e*" (with a probability of 0.112) and "*t*" (with a probability of 0.080), occur more frequently than others, such as "*q*" and "*z*," which have much lower probabilities (0.001). This variation in frequencies reflects the nature of English, where certain letters are used far more often. Understanding these probabilities is fundamental for efficient data encoding, as frequent letters can be assigned shorter codes to reduce overall message length. Subplot Figure 6.4b illustrates the Huffman Encoding Lengths assigned to each letter based on the probabilities shown in subplot Figure 6.4a. In Huffman encoding, letters that appear more frequently are assigned shorter binary codes, while those that occur less frequently are given longer codes. For instance, the letter "*e*" with

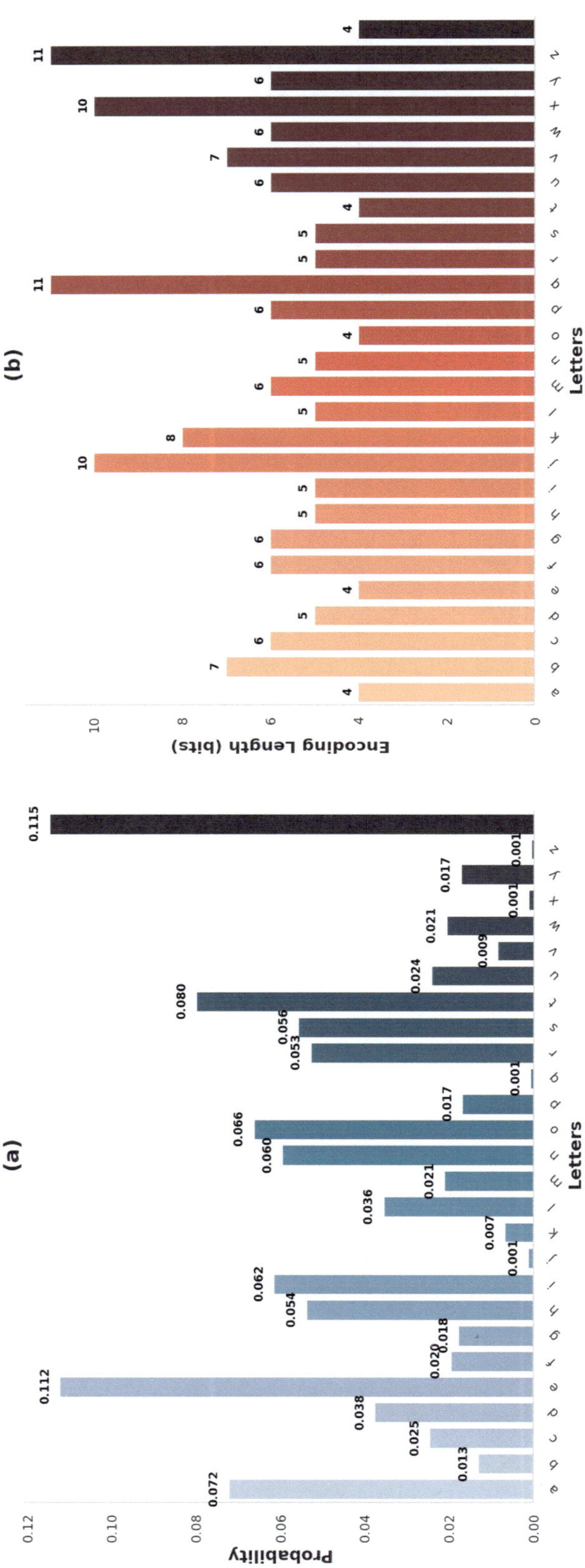

FIGURE 6.4 (a) Letter probabilities in English text, and (b) Huffman encoding lengths for letters.

the highest probability, has an encoding length of 4 bits, while infrequent letters like "*q*" and "*z*" require 11 bits.

6.7 CHANNEL CAPACITY AND SHANNON'S THEOREM

In practical terms, Shannon's theorem laid the theoretical foundation for modern digital communication systems. Many communication systems today, like those used in cellular networks or satellite communication, employ error-correcting codes to approach the channel capacity as closely as possible and ensure reliable data transfer. Channel capacity is a fundamental concept in information theory. It represents the highest rate at which information can be reliably transmitted over a given communication channel. Expressed in bits per second (bps), the maximum number of bits that can be sent over the channel per unit of time with an arbitrarily low error rate. Importantly, capacity is not about the speed of data transfer but how much information can be reliably conveyed. This means that even if a channel can transfer large amounts of data quickly, it might still have a low capacity if much of that data is redundant or corrupted. Factors that can impact the channel capacity include the channel's bandwidth, the signal's power, the environment's noise, and the channel's inherent characteristics. The Shannon theorem is a profound insight into the nature of communication over noisy channels. In essence, it states that reliable communication over a noisy channel is possible up to a specific maximum rate, known as the channel capacity. It is possible to encode messages below this rate so that the probability of error in decoding the message can be made arbitrarily small. However, above this rate, no encoding method will avoid a high probability of error. Given a noisy channel with a known capacity, it suggests that one can design encoding and decoding schemes (or error-correcting codes) to ensure almost error-free communication as long as the communication rate stays below the channel capacity. However, errors will become unavoidable once you try to transmit information at a rate higher than the channel capacity. Consider a cellular network where the channel has a bandwidth of 1 MHz (1,000,000 Hz), and the signal-to-noise ratio (**SNR**) is 10 (in linear scale, equivalent to 10 dB). Using Shannon's theorem, we can calculate the maximum channel capacity, which gives the upper limit on how much information can be transmitted reliably over this channel. Shannon's channel capacity theorem defines the maximum rate **C** (in bps) at which information can be transmitted over a noisy channel without error as long as the transmission rate is below this capacity. The formula for the channel capacity **C** is given by:

$$C = B\log_2\left(1 + \text{SNR}\right)$$

where:

- **C** is the channel capacity in bps,
- **B** is the bandwidth of the channel in hertz (Hz),
- **SNR** is the signal-to-noise ratio (in linear form, not dB).

For the cellular network example, bandwidth **B** = 1,000,000 Hz, and **SNR** = 10 (in linear scale).

Using the formula:

$$C = 1{,}000{,}000 \times \log_2\left(1 + 10\right) \approx 1{,}000{,}000 \times 3.459 \approx 3.459 \text{ Mbps}$$

This means the maximum channel capacity for this network is approximately 3.459 Mbps. As long as the transmission rate remains below this limit, reliable communication with very low error rates is possible.

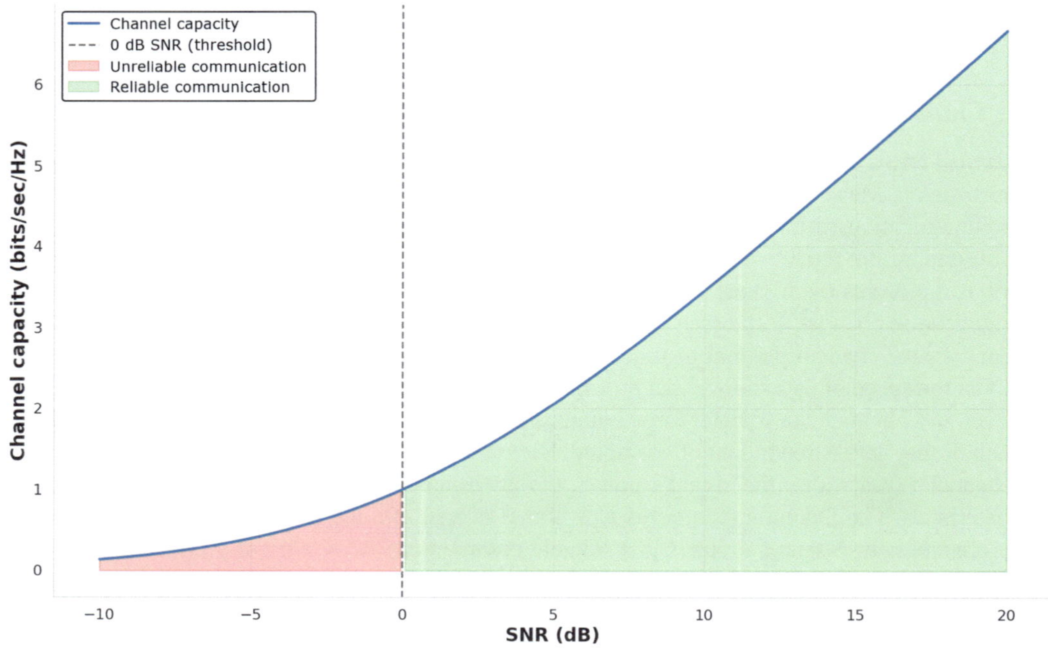

FIGURE 6.5 Channel capacity as a function of SNR.

Figure 6.5 represents the relationship between SNR and the maximum achievable data rate (in bps/Hz) of a communication channel, following Shannon's capacity formula. The blue curve shows how the channel capacity increases as the SNR improves. At lower SNR values, specifically when the SNR is below 0 dB (marked by the vertical dashed line), the capacity remains low, indicating the region of unreliable communication, highlighted in red. This area signifies that when the noise level is too high compared to the signal strength, achieving a reliable communication rate is challenging or impossible. As the SNR increases beyond the 0 dB threshold, the channel enters the reliable communication region, shaded in green. In this region, the capacity of the channel improves significantly with higher SNR values, reflecting that a stronger signal relative to noise allows for higher data transmission rates while maintaining reliability.

6.8 KULLBACK–LEIBLER DIVERGENCE

The Kullback–Leibler (**KL**) divergence is used to quantify the difference between two probability distributions. It is essential in statistics, information theory, and machine learning. The **KL** divergence is a powerful tool for understanding and quantifying differences between probability distributions, playing a central role in many machine learning algorithms and statistical measures. The **KL** divergence is defined for two discrete probability distributions, **P** and **Q**. Usually, **P** represents the "true" distribution of data, observations, or a precisely calculated theoretical distribution. In contrast, **Q** typically represents the approximation, hypothesis, or a model's predictions. The **KL** divergence of **Q** from **P** is defined as:

$$D_{\mathrm{KL}}(P \parallel Q) = \sum_{x} P(x) \log \frac{P(x)}{Q(x)}$$

The logarithm is typically taken in base two if we measure the divergence in bits or base **e** (natural logarithm) for nats. **KL** divergence is asymmetric, meaning $D_{\mathrm{KL}}(P \parallel Q)$ is not necessarily equal to $D_{KL}(Q \parallel P)$. This is a fundamental property, implying that the "cost" of approximating **P** with **Q** can differ from the "cost" of approximating **Q** with **P**. In machine learning, **KL** divergence is commonly used in algorithms that involve probability distributions. For instance, it is used in variational inference to measure the divergence between the actual posterior distribution and its approximation. We are training specific generative models like variational autoencoders (VAEs). Information retrieval is used to compare the distribution of terms in different documents. A **KL** divergence of 0 indicates that the two distributions are identical. As the divergence increases, the difference between the two distributions grows. However, be cautious: the **KL** divergence can be infinite if any value for which **P(x)** is non-zero and **Q(x)** is zero. This is because the logarithm of zero becomes undefined, highlighting a fundamental mismatch between the distributions. Consider two probability distributions **P** and **Q** for a simple coin flip scenario, where the coin can land on heads **(H)** or tails **(T)**. Let:

- **P(H)** = 0.6, **P(T)** = 0.4 (true distribution),
- **Q(H)** = 0.5, **Q(T)** = 0.5 (model's predicted distribution).

The **KL** divergence between **P** and **Q** quantifies how much information is lost when using **Q** to approximate **P**. The **KL** divergence between two discrete probability distributions, **P** and **Q**, is defined as:

$$D_{KL}(P \parallel Q) = \sum_{x} P(x) \log \frac{P(x)}{Q(x)}$$

where:

- $\mathbf{D_{KL}(P \parallel Q)}$ is the **KL** divergence of **Q** from **P**,
- **P(x)** is the probability of event **x** in the true distribution **P**,
- **Q(x)** is the probability of event **x** in the model's approximation **Q**.

For the coin flip example, the **KL** divergence is calculated as:

$$D_{KL}(P \parallel Q) = 0.6 \log \frac{0.6}{0.5} + 0.4 \log \frac{0.4}{0.5}$$

Calculating each term:

$$0.6 \log \frac{0.6}{0.5} \approx 0.6 \times \log(1.2) = 0.04751 \quad \text{and} \quad 0.4 \log \frac{0.4}{0.5} \approx 0.4 \times \log(0.8) = 0.03876$$

Thus:

$$D_{KL}(P \parallel Q) \approx 0.04751 - 0.03876 = 0.00875$$

In this case, the KL divergence ($D_{KL}(P \parallel Q)$) is approximately **0.00875**, quantifying the extent to which the distribution Q diverges from P. KL divergence of **0** would indicate that the two distributions are identical, while any positive value reflects the difference in information content between them.

Figure 6.6 represents these differences and quantifies them using KL divergence. Figure 6.6 subplot a illustrates the true distribution (*P*), where probabilities are assigned to each event as follows: Event 1 has a probability of 0.40, Event 2 is assigned 0.35, Event 3 is 0.20, and Event 4 is 0.05. This distribution serves as the baseline or observed probabilities for each event. In contrast,

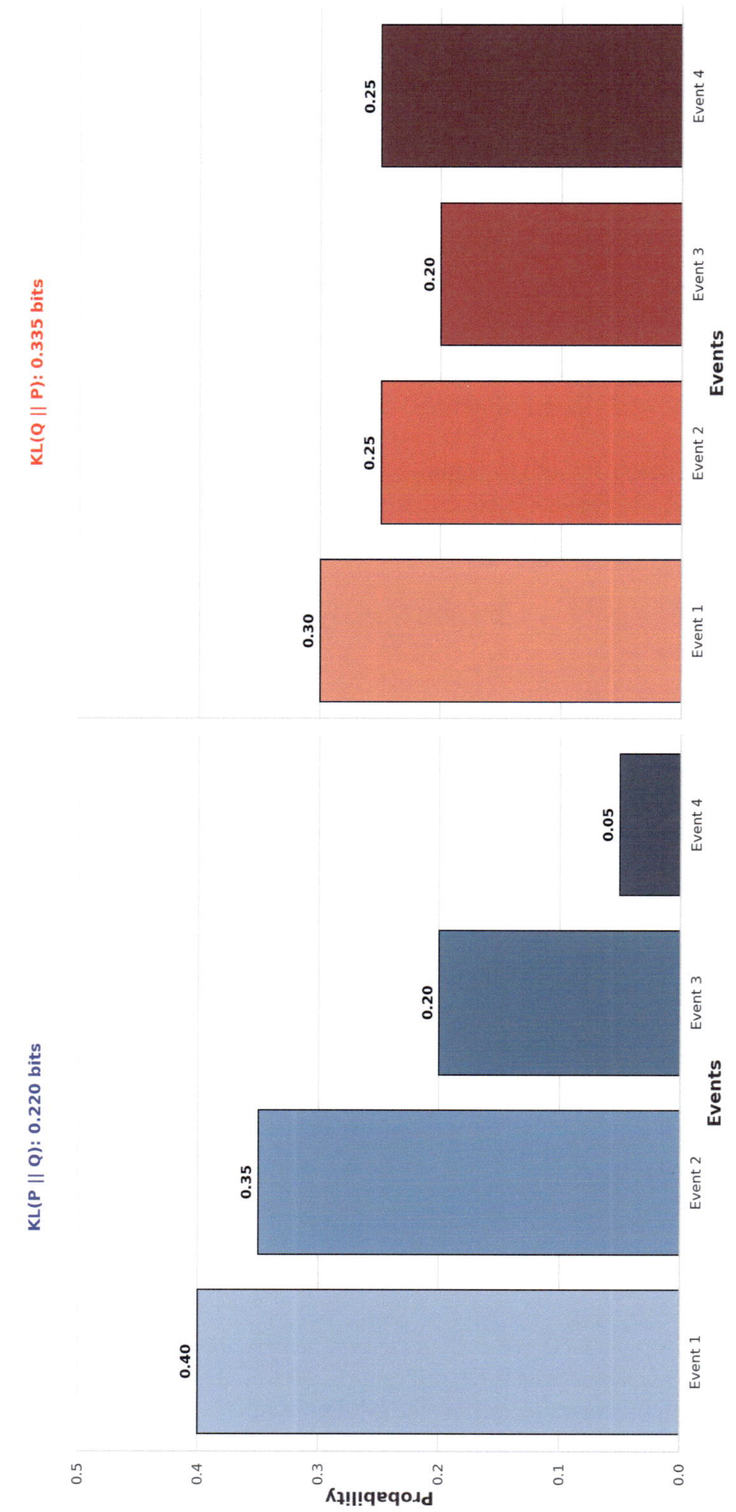

FIGURE 6.6 (a) True distribution **P** and (b) approximate distribution **Q**.

subplot Figure 6.6b displays the approximate distribution (Q), where the probabilities differ: Event 1 is assigned a probability of 0.30, Event 2 is 0.25, Event 3 remains at 0.20, and Event 4 increases significantly to 0.25. These changes indicate deviations between P and Q, suggesting that Q only partially approximates the true distribution. Above the bar plots, the figure presents the calculated KL divergence values. The KL($P \parallel Q$) is 0.220 bits (shown in blue), quantifying the information lost when Q is used to approximate P. The KL($Q \parallel P$) is 0.335 bits (shown in red), measuring the divergence in the reverse direction. The asymmetry in these values highlights that KL divergence is not symmetric. The difference in magnitude reflects the distinct perspectives of approximating P using Q vs. Q using P. The smaller value for KL($P \parallel Q$) suggests that Q provides a relatively reasonable approximation of P, but some discrepancies remain.

6.9 INFORMATION THEORY IN MACHINE LEARNING AND DEEP LEARNING

In machine learning, information gain, a concept based on entropy, is used in decision trees to decide which feature to split on at each node. Clustering algorithms utilize mutual information to measure cluster similarity, enhancing the grouping of similar data points. In neural networks, especially in areas like VAEs, concepts such as **KL** divergence are employed to measure the difference between the learned representation and the actual data distribution, thereby optimizing model performance and representation accuracy. As deep learning continues to evolve, the foundational principles of information theory will likely remain integral in providing insights and tools for advancing the field. Information theory provides a framework to quantify the maximum amount of information that a network can store or process, which is intricately linked to its architecture and size. The architecture of a neural network, including the number of layers and the number of neurons per layer, determines its capacity to represent complex functions. Information theory helps quantify this capacity, offering a precise measure of how much information the network can capture from the input data. This understanding is crucial for designing networks that are neither underfitting nor overfitting the data. Activation functions play a pivotal role in how information flows through a network. From an information-theoretic perspective, activation functions can be analyzed based on their ability to maintain or transform information as it propagates through the layers. For instance, functions like ReLU (Rectified Linear Unit) can help preserve the gradient during backpropagation, preventing issues like vanishing gradients and ensuring efficient information flow. Optimizing the selection and design of activation functions can significantly enhance the network's performance and efficiency. The information bottleneck principle provides another layer of understanding by examining how networks compress input information into a more compact representation while retaining essential features. This principle aids in developing models that balance complexity and generalization, ensuring that the network captures the most relevant information for the task at hand. Techniques such as mutual information can measure the amount of information shared between different network layers. This measurement can guide the adjustment of network parameters to maximize the flow of relevant information, thereby improving the learning process and the model's overall effectiveness. Consider a neural network with three layers: an input layer with 10 neurons, a hidden layer with 50 neurons, and an output layer with 5 neurons. The architecture defines the network's capacity to store and process information. Using information theory, we can analyze how much information this network can theoretically capture from the input data and how efficiently it can transform that information across layers. The capacity of a neural network can be related to the number of weights (parameters) it has. For a simple, fully connected neural network with **L** layers, where each layer **i** has $\mathbf{n}_\mathbf{i}$ neurons, the total number of weights is:

$$\text{Total weights} = \sum_{i=1}^{L-1} n_i \times n_{i+1}$$

In our example:

- Input layer: 10 neurons,
- Hidden layer: 50 neurons,
- Output layer: 5 neurons.

The number of weights between layers is:

- From input to hidden: $10 \times 50 = 500$ weights,
- From hidden to output: $50 \times 5 = 250$ weights.

Thus, the total number of weights is:

$$500 + 250 = 750 \text{ weights}$$

This means the network can store a significant amount of information across these 750 connections, which directly influences its capacity to represent complex functions.

6.9.1 Regularization and Optimization

Regularization and optimization in neural networks often utilize concepts from information theory to improve model performance and generalization. For example, **VAEs** are a generative model that incorporates the **KL** divergence as a regularization term. This information-theoretic measure ensures that the learned latent variable distribution remains close to a prior distribution, typically Gaussian, thereby enhancing the generative capabilities of the model. Another method, the information bottleneck, aims to retain as much relevant information about the input while compressing or removing irrelevant information. This method quantifies compression and retention using mutual information, balancing the trade-off between preserving essential information and reducing redundancy. By focusing on these principles, neural networks can achieve more efficient and effective learning outcomes. Regularization techniques, such as dropout or weight decay, are also informed by information theory, aiming to control the amount of information stored in the network weights and promoting better generalization to unseen data. Consider training a **VAE** on the MNIST dataset, where the latent space is represented by two dimensions, and the goal is to generate digit images. A **KL** divergence regularization term is used to ensure that the latent space $\mathbf{z}$ follows a Gaussian distribution with a mean of 0 and variance of 1. Let's assume the learned latent variable distribution has a mean $\boldsymbol{\mu} = 0.5$ and variance $\boldsymbol{\sigma}^2 = 0.25$. The **KL** divergence between this learned distribution and the prior $\mathbf{N}(0, 1)$ (a standard normal distribution) is calculated as:

$$D_{\text{KL}}\left(q(z|x) \parallel p(z)\right) = \log\frac{1}{0.25} + \frac{0.25 + (0.5)^2}{2} - 1 = 1.386$$

This **KL** divergence value of 1.386 indicates that the learned distribution deviates from the Gaussian prior, and the model will be penalized accordingly.

6.9.2 Generalization and Overfitting

The capacity of a model is intrinsically related to its ability to fit noise in the data. A model with a higher capacity, such as a deep neural network, can fit the training data more closely but is also more prone to overfitting. Overfitting occurs when the model captures not only the underlying patterns but also the random noise in the data, leading to poor generalization on new, unseen data.

Information-theoretic measures provide a quantitative framework to assess this capacity and the trade-off between fitting the data and overfitting. By using concepts such as entropy and mutual information, researchers can evaluate how much information the model retains and whether this information is relevant or merely noise. This approach helps in designing models that balance complexity and generalization, ultimately enhancing their performance on real-world tasks. Consider a neural network trained on a dataset of 1000 points. The model has 100,000 parameters (weights), which gives it a high capacity to fit the data. During training, the model achieves near-perfect accuracy, with a training loss of 0.02. However, when tested on unseen data, the test loss is much higher, at 1.5, indicating overfitting. The model has captured not only the true patterns in the training data but also the noise, leading to poor generalization.

6.9.3 Model Interpretability

Model interpretability is crucial for understanding how a deep learning model makes its predictions. Using mutual information, we can quantify the importance of different features or inputs by measuring how much information each feature shares with the output. This information-theoretic approach provides a clear and quantitative method to identify which features contribute most to the model's predictions, enhancing our ability to interpret and trust the model. By analyzing mutual information, researchers can gain insights into the relationships between inputs and outputs, leading to more transparent and explainable models. This interpretability is essential for applications where understanding the decision-making process is as important as the predictions themselves. Suppose a neural network is trained to predict housing prices using features like square footage, number of rooms, and distance to the city center. Using mutual information, we can calculate that square footage has a mutual information score of 0.8, while the distance to the city center has a mutual information score of 0.4. This tells us that square footage contributes more to the model's predictions than the distance to the city center, helping us interpret the model.

6.9.4 Transfer Learning and Domain Adaptation

Information theory offers valuable tools for measuring and managing the differences between source and target domains in transfer learning and domain adaptation. A crucial aspect of this process is measuring domain shift, which involves quantifying the difference between source and target distributions using measures like **KL** divergence. These information-theoretic metrics help in understanding how much the domains differ, allowing models to be adapted accordingly. Optimal transport techniques, such as the Wasserstein distance, which is rooted in information geometry, provide a more nuanced approach to measuring the difference between distributions. These techniques are particularly effective in guiding the alignment of source and target domains, ensuring that the model can generalize well to the target domain by minimizing the discrepancies between the distributions. By leveraging these information-theoretic approaches, transfer learning, and domain adaptation can be more precisely tuned, leading to better model performance in new and diverse environments. Imagine a model trained on a source domain of animal images (e.g., cats and dogs) and then applied to a target domain of wild animals (e.g., lions and tigers). The **KL** divergence between the source and target distributions is calculated as 1.2, indicating a significant shift in data. By minimizing this divergence, the model can be adapted to perform better on the target domain.

6.9.5 Adversarial Attacks and Robustness

Information theory can be a powerful tool for measuring the robustness of neural networks to adversarial attacks. By quantifying how much an adversarial input perturbs the information flow within the network, we can assess the network's vulnerability. Specifically, information-theoretic measures can help determine the extent to which an adversarial input disrupts the network's internal representations and decision-making processes. This quantification allows for a deeper understanding of the network's resilience and provides insights into developing more robust models that can withstand adversarial perturbations, ultimately leading to more secure and reliable neural network applications. Suppose a neural network is trained to classify images of digits (0–9), and an adversarial attack perturbs an input image, changing the model's classification from 3 to 7. Using mutual information, we calculate that the perturbed input reduces the information shared between the network's internal layers and the true label by 0.6 bits, indicating a significant disruption in the network's decision-making process.

6.9.6 Neural Architecture Search

Information-theoretic concepts can significantly enhance the process of Neural Architecture Search (NAS) by guiding the exploration of efficient neural architectures with an optimal trade-off between accuracy and complexity. By applying measures such as entropy and mutual information, researchers can evaluate and compare different architectures to identify those that capture the most relevant information while maintaining a manageable level of complexity. This approach helps in designing neural networks that achieve high performance without unnecessary computational overhead, leading to models that are both accurate and efficient. Utilizing information theory in NAS allows for a more systematic and theoretically grounded search for the best possible neural architectures. Consider two neural architectures: Architecture **A** with 5 million parameters and Architecture **B** with 2 million parameters. Using mutual information, we find that Architecture **A** captures 1.2 bits of relevant information from the input, while Architecture **B** captures 1.1 bits. Despite the slight difference in information captured, Architecture **B** is more efficient due to its lower parameter count, making it a better choice in terms of balancing accuracy and complexity.

6.9.7 Layer-Wise Relevance Propagation

Layer-wise relevance propagation (**LRP**) is an interpretability technique that uses concepts rooted in information theory to explain the decisions of deep networks. By propagating relevance scores from the output back to the input, **LRP** provides a measure of the contribution of each input feature to the final decision. This backward propagation of relevance scores helps in understanding which features are most influential in the model's predictions, offering valuable insights into the inner workings of the network. This technique enhances model transparency and trustworthiness, making it easier to diagnose model behavior and improve its design based on the relevance of different features. Consider a neural network that classifies images of cats and dogs. For an image of a dog, **LRP** assigns relevance scores to input pixels. The pixels corresponding to the dog's ears and eyes receive high relevance scores of 0.9 and 0.8, respectively, indicating that these features are the most influential in the model's decision. **LRP** computes relevance scores for each neuron in the network. The relevance score $\mathbf{R_j}$ for neuron j in layer l is propagated from the relevance scores $\mathbf{R_k}$ of neurons in the next layer **l + 1**:

$$R_j = \sum_k \frac{a_j w_{jk}}{\sum_j a_j w_{jk}} R_k$$

where:

- $\mathbf{a_j}$ is the activation of neuron **j**,
- $\mathbf{w_{jk}}$ is the weight between neuron **j** and neuron **k** in the next layer.

6.10 REAL-WORLD APPLICATIONS

6.10.1 Data Compression

One of the most direct applications of information theory is in data compression. The concept of entropy, which measures the unpredictability or information content of a source, is foundational in developing efficient compression algorithms. For instance, Huffman coding, a widely used method for lossless data compression, leverages entropy to assign shorter codes to more frequent symbols and longer codes to less frequent ones. This approach minimizes the average length of the data representation, reducing storage requirements without losing any information. This principle is vital in formats such as **PNG** for images and **FLAC** for audio, where exact reproduction of the original data is crucial.

6.10.2 Cryptography

Information theory plays a crucial role in cryptography, ensuring secure communication by measuring the uncertainty and unpredictability of data. The entropy of a cryptographic key, for instance, quantifies its strength: the higher the entropy, the more unpredictable and secure the key is. Information theory also guides the design of encryption algorithms, helping to balance the trade-offs between security and performance. Mutual information, which measures the dependency between variables, is used to assess the strength of encryption schemes by ensuring that encrypted data reveals minimal information about the original data.

6.10.3 Telecommunications and Error Correction

Shannon's noisy channel coding theorem, a cornerstone of information theory, underpins the design of reliable communication systems. This theorem quantifies the maximum rate at which information can be transmitted over a noisy channel while still being reliably decoded. In practice, this principle is applied in developing error-correcting codes, such as Reed-Solomon or Turbo codes, which are used in **CDs**, **DVDs**, and mobile communications. These codes add redundancy to the transmitted data, allowing the receiver to correct errors caused by noise, ensuring the integrity of the information despite transmission errors.

6.10.4 Network Security and Anomaly Detection

Information theory aids in detecting anomalies in network traffic, a crucial aspect of cybersecurity. By analyzing the entropy of network packets, security systems can identify unusual patterns that may indicate a security breach or attack. For example, a sudden decrease in entropy might suggest that an attacker is using a predictable, repetitive pattern to infiltrate the network. Similarly, mutual information can be used to detect relationships between seemingly unrelated data streams, uncovering hidden channels or covert communication attempts within the network.

6.10.5 Biological Data Analysis

In bioinformatics, information theory is employed to analyze complex biological data, such as gene expression patterns. Entropy measures the diversity and variability within gene expression profiles, helping researchers identify genes that are most variable across conditions or diseases. Mutual information is used to assess the dependency between different genes or between genes and phenotypic traits, providing insights into the underlying regulatory mechanisms. This application is crucial for understanding complex biological systems and for identifying potential targets for drug development.

6.10.6 Finance and Risk Management

In finance, information theory assists in portfolio optimization and risk management by analyzing the dependency and information flow between different financial assets. Mutual information can be used to measure the strength of the relationship between different stocks or asset classes, helping investors diversify their portfolios effectively. By understanding the information shared between assets, financial models can predict market movements more accurately and manage risk more effectively, leading to more robust investment strategies.

6.11 HANDS-ON EXAMPLE

In this section, we will explore the concept of entropy and mutual information within the context of neural networks.

6.11.1 Step 1: Import Necessary Libraries

First, we are importing essential libraries for building and evaluating machine learning models. First, NumPy is imported as np for numerical operations such as array manipulation. Then, matplotlib.pyplot is imported as plt for plotting and visualizing data. TensorFlow (tensorflow) is imported, along with Keras modules (Sequential and Dense) to define and build neural network models. Additionally, we import mutual_info_score from the sklearn.metrics library, which is used to compute the mutual information between two variables, a measure of how much information one variable provides about another. Finally, entropy from scipy.stats is imported to compute the entropy of distributions, which quantifies the uncertainty or randomness in the data.

```
import numpy as np
import matplotlib.pyplot as plt
import tensorflow as tf
from tensorflow.keras.models import Sequential
from tensorflow.keras.layers import Dense
from sklearn.metrics import mutual_info_score
from scipy.stats import entropy
```

6.11.2 Step 2: Generate Sample Data

We define a function generate_data to create a simple synthetic dataset for binary classification. The function takes n_samples as an input, which specifies the number of data points to generate. Inside the function, **X** is a 2D array of shape (n_samples, 2) containing random values between 0 and 1, generated using np.random.rand(). Each row of **X** represents a sample with two features. The target labels **y** are then computed based on a simple rule: if the sum of the two features in a sample is greater than 1, the corresponding label is set to 1; otherwise, it's set to 0. This creates a classification problem where the decision boundary is the line $x_0 + x_1 = 1$. The dataset consists of **X** (features) and **y** (labels), and after generating the data for 1000 samples, it is stored in the variables **X** and **y** for further analysis or model training.

```
# Generate a simple dataset
def generate_data(n_samples):
  X = np.random.rand(n_samples, 2)
```

```
    y = (X[:, 0] + X[:, 1] > 1).astype(int)
    return X, y
X, y = generate_data(1000)
```

6.11.3 Step 3: Calculate Entropy

Here, we define a function calculate_entropy that calculates the entropy of a given set of labels. Entropy is a measure of the uncertainty or randomness in the distribution of labels. The function uses np.unique() to find the unique values in the labels array and count their occurrences. The entropy is then computed using the entropy() function from the scipy.stats library, which takes the counts of each unique value as input. Higher entropy means the labels are more evenly distributed, while lower entropy indicates that the labels are more skewed toward a particular value. After defining the function, we calculate and print the entropy of the labels **y**, which were generated in the previous dataset, with the result formatted to four decimal places. This allows us to quantify the uncertainty or distribution balance of the labels in the dataset.

```
def calculate_entropy(labels):
    value, counts = np.unique(labels, return_counts=True)
    return entropy(counts)
print(f'Entropy of y: {calculate_entropy(y):.4f}')
```

6.11.4 Step 4: Mutual Information Calculation

Now, we define a function calculate_mutual_information to compute the mutual information between the features **X** and the labels **y**. Mutual information is a measure of how much information one variable provides about another, quantifying the dependency between them. In this case, we are using mutual_info_score from the sklearn.metrics library to calculate the mutual information between **X** (the features) and **y** (the labels). The X.ravel() function flattens the 2D array **X** into a 1D array to align with the expected input format for the mutual information calculation. After defining the function, we compute and print the mutual information between **X** and **y**, formatted to four decimal places. This measure helps us understand how much information the features **X** contain about the target labels **y**, thereby indicating the strength of the relationship between them.

```
def calculate_mutual_information(X, y):
    return mutual_info_score(X.ravel(), y)
print(f'Mutual Information between X and y: {calculate_mutual_
information(X, y):.4f}')
```

6.11.5 Step 5: Build and Train a Simple Neural Network

Finally, we are building and training a simple neural network using TensorFlow's Keras API. The network is defined using the Sequential model, which stacks layers sequentially. The first layer is a Dense layer with 10 neurons, a ReLU activation function, and an input dimension of 2 (as the input data X has two features). The second layer is a Dense output layer with 1 neuron and a sigmoid activation function, which is commonly used for binary classification problems. The model is compiled with the Adam optimizer, a popular optimization algorithm for training neural networks, and the binary cross-entropy loss function, which is appropriate for binary classification tasks. The performance is tracked using accuracy as the evaluation metric. Finally, the model is trained on the dataset *X* and *y* for 100 epochs using model.fit(), with the training progress being silent (verbose = 0). This neural network is designed to learn the relationship between the two input features and the binary target labels in the dataset generated earlier.

```
# Build a simple neural network
model = Sequential([
  Dense(10, input_dim=2, activation='relu'),
  Dense(1, activation='sigmoid')
])
model.compile(optimizer='adam',      loss='binary_crossentropy',
metrics=['accuracy'])
history = model.fit(X, y, epochs=100, verbose=0)
```

6.12 COMMON MISTAKES AND TROUBLESHOOTING TIPS

6.12.1 Misinterpreting Entropy

- *Mistake:* Confusing entropy with variance or other measures of spread.
- *Tip:* Remember that entropy measures the unpredictability or uncertainty of a random variable, not the variability of its outcomes. Entropy quantifies the average amount of information produced by a stochastic data source.

6.12.2 Incorrect Calculation of Entropy

- *Mistake:* Using incorrect probabilities or forgetting to sum over all possible outcomes.
- *Tip:* Ensure that the probability distribution sums to 1 and carefully calculate entropy by summing the product of each probability and its logarithm.

6.12.3 Misunderstanding Joint and Conditional Entropies

- *Mistake:* Treating joint and conditional entropies as independent of each other.
- *Tip:* Joint entropy accounts for the combined uncertainty of two variables, while conditional entropy measures the remaining uncertainty of one variable given the other. Always consider their relationship and dependencies.

6.12.4 Overlooking the Asymmetry of KL Divergence

- *Mistake:* Assuming KL divergence is symmetric and misinterprets the results.
- *Tip:* Remember that $D_{KL}(P \| Q) \neq D_{KL}(Q \| P)$. The divergence from *P* to *Q* is not the same as from *Q* to *P*. Always interpret the direction correctly.

6.12.5 Confusing Mutual Information with Correlation

- *Mistake:* Equating mutual information with linear correlation.
- *Tip:* Mutual information measures any dependency between variables, not just linear relationships. It captures both linear and non-linear dependencies, unlike Pearson correlation.

6.12.6 Ignoring the Basis of Logarithms in Entropy Calculations

- *Mistake:* Using inconsistent logarithm bases when calculating entropy.
- *Tip:* Use base 2 logarithms for entropy measured in bits. Ensure consistency in the logarithm base throughout your calculations.

6.12.7 Misapplying Shannon's Noisy Channel Coding Theorem

- *Mistake:* Misinterpreting the channel capacity or ignoring noise effects.
- *Tip:* Understand that channel capacity is the maximum reliable transmission rate over a noisy channel. Ensure that your encoding and decoding schemes are designed to operate below this capacity to minimize errors.

6.13 REVIEW QUESTIONS

1. Define entropy. How does it measure the unpredictability or randomness of information?
2. Explain the significance of entropy in the context of data transmission and compression.
3. How is entropy mathematically defined for a discrete random variable? Provide an example using a binary source.
4. What are joint entropy and conditional entropy? How do they differ from and relate to standard entropy?
5. Provide an example where knowing one variable significantly reduces the uncertainty of another variable.
6. Define mutual information and explain its importance in feature selection and machine learning models.
7. How is mutual information used to assess the relationship between two variables?
8. How does Shannon's theorem guide the design of error-correcting codes?
9. Explain the concept of the information bottleneck method. In what ways is this method utilized in deep learning?
10. Describe the KL divergence and its significance in comparing probability distributions.

6.14 PROGRAMMING QUESTIONS

6.14.1 Easy

Calculate the entropy of a given discrete random variable and plot its probability distribution.

1. Define a discrete random variable with given probabilities for each outcome.
2. Calculate the entropy of the random variable using the entropy formula.
3. Create a bar plot to visualize the probability distribution of the random variable.

6.14.2 Medium

Calculate the mutual information between two discrete random variables and interpret the results.

1. Define two discrete random variables with given joint probabilities.
2. Calculate the marginal probabilities of each variable.
3. Compute the joint entropy, marginal entropies, and conditional entropies.
4. Calculate the mutual information using the formula: $I(X; Y) = H(X) + H(Y) - H(X, Y)$.
5. Interpret the mutual information value and explain what it signifies about the relationship between the two variables.

6.14.3 Hard

Implement and train a neural network on a synthetic dataset, then analyze the change in entropy and mutual information of the hidden layers' activations during training.

1. Generate a synthetic dataset for binary classification.
2. Build a neural network with multiple hidden layers using a deep learning framework (e.g., TensorFlow or PyTorch).
3. Train the neural network on the synthetic dataset, recording the activations of each hidden layer at each epoch.
4. Calculate the entropy of the activations for each hidden layer at different epochs.
5. Calculate the mutual information between the input and the activations and between the activations and the output at different epochs.
6. Visualize the change in entropy and mutual information over the training epochs using line plots.

7 Graph Theory

7.1 INTRODUCTION

In the world of deep learning, mathematical structures like graphs have emerged as an effective tool to model complex relationships and patterns. Graph theory, with its ability to depict pairwise relations between entities, has proven to be required, bridging gaps that traditional linear methods in machine learning could not address. As our digital universe grows increasingly connected, the webs of relationships, from social networks to molecular structures, demand a more subtle approach to data representation and analysis. Neural networks, the backbone of many modern artificial intelligence (AI) systems, have started integrating principles from graph theory, giving birth to innovative models tailored to handle this newfound complexity. The union of graph theory and neural networks is not just a theoretical attraction; it carries significant practical implications. By treating data as nodes and their relationships as edges, a crowd of real-world applications, from social media analysis to drug discovery, has been unlocked. This chapter explores the role of graph theory in deep learning.

7.2 GRAPH THEORY FOR DEEP LEARNING

Understanding the basic types of graphs and their properties is fundamental for modeling and solving problems in various domains, including computer science, biology, social sciences, and more. Graphs provide a useful and powerful way to represent and analyze the relationships and structures within data.

7.2.1 Graph

A graph is a fundamental mathematical structure consisting of a set of nodes (vertices) and edges. The key components of the graph are nodes and edges.

7.2.1.1 Nodes (Vertices)

Nodes represent entities or objects within the graph. Each node can contain attributes or features that provide additional information about the entity it represents.

7.2.1.2 Edges (Connections)

Edges represent the relationships or connections between nodes. Edges can be directed or undirected, weighted or unweighted, depending on the nature of the relationship they model.

DOI: 10.1201/9781032690742-7

7.2.2 Directed Graph

A directed graph is a type of graph in which the edges have a specific direction. This means each edge points from one node to another, indicating a one-way relationship. Directed graphs are represented visually by arrows that show the direction of the connection between nodes. Directed graphs provide a clear and effective way to model and analyze systems where directionality is a critical component of the relationships between entities. They enable the study of influence, flow, and hierarchy within a network. Its key characteristics are directional edges and asymmetry.

7.2.2.1 Directional Edges

Each edge has a direction, typically represented by an arrow pointing from the source node (start) to the target node (end).

7.2.2.2 Asymmetry

The relationship between nodes is not necessarily equal. For instance, if there is a directed edge from Node **A** to Node **B**, it does not imply there is an edge from **B** to **A**.

7.2.2.3 Directed Graph Example

Consider a citation network in academia. Each research paper is represented as a node, and a citation from one paper to another forms a directed edge. For instance, if Paper **A** cites Paper **B**, we have a directed edge from Node **A** to Node **B**. This means that the influence flows in one direction—Paper **A** is influenced by Paper **B**. However, this does not imply the reverse relationship (i.e., Paper **B** may not cite Paper **A**). A directed graph is defined as $\mathbf{G} = (\mathbf{V}, \mathbf{E})$, where $\mathbf{V}$ is a set of vertices (or nodes), and $\mathbf{E} \subseteq \mathbf{V} \times \mathbf{V}$ is a set of directed edges. Each edge $(\mathbf{u}, \mathbf{v}) \in \mathbf{E}$ represents a directed connection from Node $\mathbf{u}$ to Node $\mathbf{v}$. This directionality distinguishes directed graphs from undirected graphs. For instance, let's define a small citation network with three papers: Paper **A** cites Paper **B** and Paper **C**, and Paper **B** cites Paper **C**. This directed graph can be represented as: $\mathbf{V} = \{\mathbf{A}, \mathbf{B}, \mathbf{C}\}$ and $\mathbf{E} = \{(\mathbf{A}, \mathbf{B}), (\mathbf{A}, \mathbf{C}), (\mathbf{B}, \mathbf{C})\}$.

7.2.3 Undirected Graph

An undirected graph is a type of graph in which the edges have no direction. This means that each edge simply connects two nodes without implying any order or hierarchy between them. In undirected graphs, the relationships represented by the edges are mutual and bidirectional. Its key characteristics are bidirectional edges and symmetry.

7.2.3.1 Bidirectional Edges

Each edge connects two nodes without any direction, indicating that the relationship goes both ways.

7.2.3.2 Symmetry

If there is an edge between Node **A** and Node **B**, it implies a two-way relationship, meaning **A** is connected to **B**, and **B** is connected to **A**.

7.2.3.3 Undirected Graph Example

Undirected graphs are ideal for modeling scenarios where the relationships are inherently mutual. Consider a social network. In this network, each person is represented by a node, and each friendship is represented by an undirected edge between two nodes. For example, if Alice and Bob are friends, there is an undirected edge between the nodes representing Alice and Bob. This edge implies a mutual relationship, meaning Alice is connected to Bob, and Bob is equally connected to Alice. An undirected graph is defined as **G = (V, E)**, where **V** is a set of vertices (or nodes), and **E ⊆ {{u, v} | u, v ∈ V}** is a set of edges, where each edge **{u, v}** connects Node **u** and Node **v**, with no direction implied. For example, consider a simple social network with three people: Alice **(A)**, Bob **(B)**, and Carol **(C)**. Suppose Alice is friends with both Bob and Carol, and Bob is friends with Carol. The undirected graph representing this network can be expressed as: **V = {A, B, C}** and **E = {{A, B}, {A, C}, {B, C}}.** In this graph, the edge **{A, B}** means that Alice and Bob are friends, and the relationship is bidirectional.

7.2.4 Weighted Graph

A weighted graph is a type of graph in which each edge has an associated weight. These weights typically represent some quantitative attribute of the connection between nodes, such as cost, length, capacity, or strength. Weighted graphs provide a more detailed representation of relationships by incorporating the magnitude of the connections. One of the key characteristics of weighted graphs is the presence of edge weights. Each edge in the graph has a numerical value, or weight, that quantifies the strength, cost, or capacity of the connection between the nodes. These weights play a crucial role in determining the significance of relationships within the graph. Weighted graphs are particularly useful in scenarios where the strength or cost of connections needs to be considered. Consider a transportation network where cities are nodes and roads between them are edges. A weighted graph is defined as **G = (V, E, w)**, where **V** is the set of vertices (or nodes), **E ⊆ V × V** is the set of edges, and **w: E → R** is a function that assigns a weight to each edge. In this case, the weight **w(e)** represents the quantitative value associated with edge **e**. For example, consider a transportation network with three cities: **A**, **B**, and **C**. Let the distances between the cities be: **A** to **B**: 200 km, **A** to **C**: 150 km, and **B** to **C**: 300 km. This weighted graph can be represented as follows: **V = {A, B, C}**, **E = {(A, B), (A, C), (B, C)}**, **w(A, B)** = 200, **w(A, C)** = 150, and **w(B, C)** = 300.

Figure 7.1 illustrates the structure and transformation of graphs under different configurations. In Figure 7.1a, an undirected graph is displayed, where nodes **A** through **E** are connected by edges that do not have any directional information. This layout is useful for modeling scenarios where the relationship between entities is bidirectional or symmetrical, such as social connections or mutual collaborations. Figure 7.1b shows the same set of nodes in a directed graph configuration. In this graph, the edges are directed, indicated by arrows pointing from one node to another, demonstrating that the relationship now has directionality. This change is essential in cases where the flow of information or influence is one way, such as in citation networks or communication pathways. Nodes **A** through **E** are connected in a manner that shows the direction of influence or dependency among them, providing insights into the asymmetry of these relationships. The third Figure 7.1c introduces a weighted directed graph, where both the direction and the magnitude of influence are represented. The edges not only show the direction (as indicated by arrows) but also have associated weights, which are indicated by the thickness of the edges and labeled numbers along the arrows. These weights quantify the strength or significance of the connections, which is crucial for understanding the intensity of relationships or the capacity of flows between nodes, such as in transportation networks or neural connections in brain models.

7.3 GRAPH NEURAL NETWORKS

Graph neural networks (GNNs) seamlessly blend classical graph algorithms with deep learning techniques, showcasing neural networks' incredible flexibility and adaptability. They represent a promising future where diverse data types are naturally integrated into AI models. Unlike images or sequences with structured formats, graph data is inherently irregular. GNNs rise to this challenge, offering a class of neural networks designed explicitly for graph-structured data. At the heart of GNNs lies the principle of neighborhood aggregation, also known as message passing. Each node

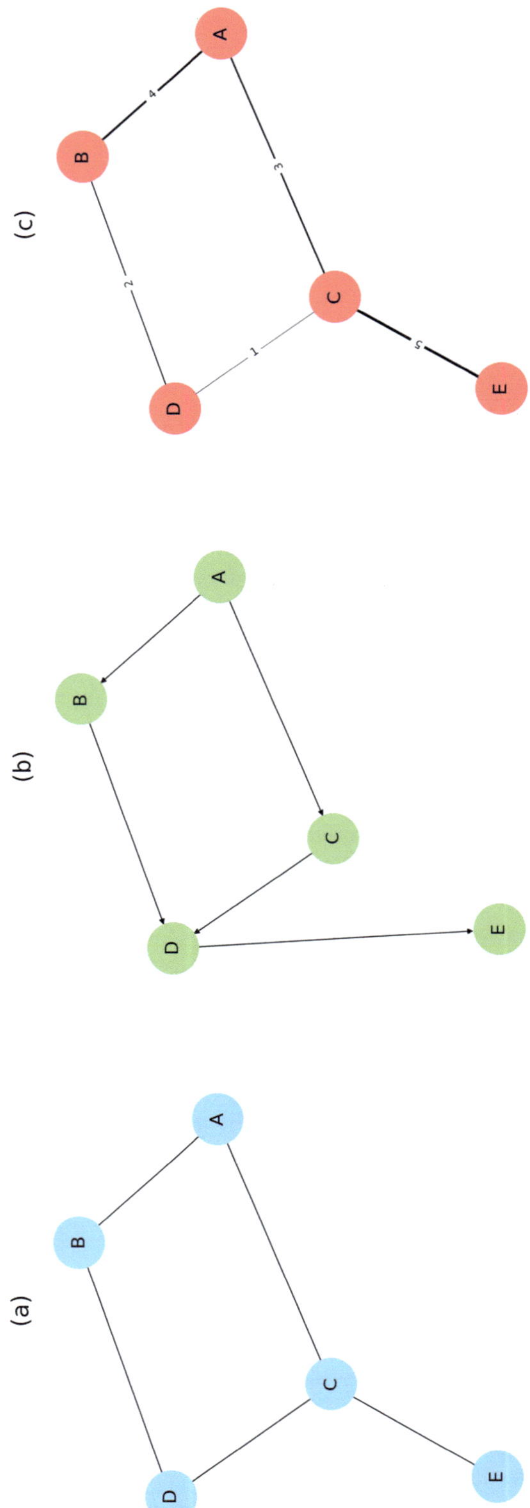

FIGURE 7.1 (a) Undirected graph, (b) directed graph, and (c) ss weighted graph.

aggregates information from its immediate neighbors to update its representation in this process. This iterative approach allows information to flow through the graph, ensuring each node gains contextual knowledge about its broader neighborhood. The update rule is mathematically expressed as:

$$h_v^{(k)} = \sigma\left(W^{(k)} \cdot \text{AGGREGATE}\left(\left\{h_u^{(k-1)} : u \in N(v)\right\}\right)\right)$$

where:

- $h_v^{(k)}$ is the representation of Node **v** at the **k**[th] iteration,
- σ is a nonlinear activation function,
- **W** is a learnable weight matrix,
- **N(v)** denotes the neighbors of Node **v**, and
- **AGGREGATE** is an aggregation function, such as sum, mean, or max.

Consider a simple graph with four nodes: **A**, **B**, **C**, and **D**. The connections are as follows:

- Node **A** is connected to Node **B** and Node **C**,
- Node **B** is connected to Node **A** and Node **D**,
- Node **C** is connected to Node **A**, and
- Node **D** is connected to Node **B**.

The adjacency list representation is **N(A) = {B, C}**, **N(B) ={A, D}**, **N(C) ={A}**, and **N(D) = {B}**, and each node has an initial feature vector (representation): $h_A^{(0)} = [1]$, $h_B^{(0)} = [2]$, $h_C^{(0)} = [3]$, and $h_D^{(0)} = [4]$; for example, the parameters are as follows:

- Aggregation function: Sum
- Activation function (**σ**): ReLU (Rectified Linear Unit), defined as $\sigma(x) = \max(0, x)$
- Weight matrix (**W**): For simplicity, use a scalar **W** = [0.5].

Here, we'll perform one iteration (**k** = 0 to **k** = 1).

1. *Aggregate Neighbor Features:* For each node, sum the features of its neighbors.
   - *Node* **A***:*
     a. Neighbors: **N(A) = {B, C}**
     b. Neighbor features: $h_B^{(0)} = [2], h_C^{(0)} = [3]$
     c. Aggregated features: $\text{AGG}_A = h_B^{(0)} + h_C^{(0)} = [2]+[3]=[5]$.
   - *Node* **B***:*
     a. Neighbors: **N(B) = {A, D}**
     b. Neighbor features: $h_A^{(0)} = [1], h_D^{(0)} = [4]$
     c. Aggregated features: $\text{AGG}_B = h_A^{(0)} + h_D^{(0)} = [1]+[4]=[5]$.
   - *Node* **C***:*
     a. Neighbors: **N(C) = {A}**
     b. Neighbor features: $h_A^{(0)} = [1]$
     c. Aggregated features: $\text{AGG}_C = h_A^{(0)} = [1]$.
   - *Node* **D***:*
     a. Neighbors: **N(D) = {B}**
     b. Neighbor features: $h_B^{(0)} = [2]$
     c. Aggregated features: $\text{AGG}_D = h_B^{(0)} = [2]$.

2. *Apply Weight Matrix and Activation Function:* Compute the new representation for each node.
   - *Node* **A**:
     a. Before activation: $W \cdot \text{AGG}_A = [0.5] \cdot [5] = [2.5]$
     b. After activation: $h_A^{(1)} = \sigma([2.5]) = \max(0, [2.5]) = [2.5]$.
   - *Node* **B**:
     a. Before activation: $W \cdot \text{AGG}_B = [0.5] \cdot [5] = [2.5]$
     b. After activation: $h_B^{(1)} = \sigma([2.5]) = [2.5]$.
   - *Node* **C**:
     a. Before activation: $W \cdot \text{AGG}_C = [0.5] \cdot [1] = [0.5]$
     b. After activation: $h_C^{(1)} = \sigma([0.5]) = [0.5]$.
   - *Node* **D**:
     a. Before activation: $W \cdot \text{AGG}_D = [0.5] \cdot [2] = [1.0]$
     b. After activation: $h_D^{(1)} = \sigma([1.0]) = [1.0]$.

Updated node features after one iteration: $h_A^{(1)} = [2.5]$, $h_B^{(1)} = [2.5]$, $h_C^{(1)} = [0.5]$, and $h_D^{(1)} = [1.0]$. Node **A** and Node **B** now have the same updated feature, reflecting the influence of their neighbors. Node **C** has a lower updated feature because it is only connected to Node **A**, which had an initial lower feature. Finally, Node **D**'s updated feature reflects its connection to Node **B**.

Figure 7.2 depicts a graph with 20 nodes and 40 edges. Each node is represented as a circle, with its color indicating the predicted class determined by the GNN. The layout uses a spring-based algorithm, ensuring an even distribution of nodes for better visualization. The dashed edges

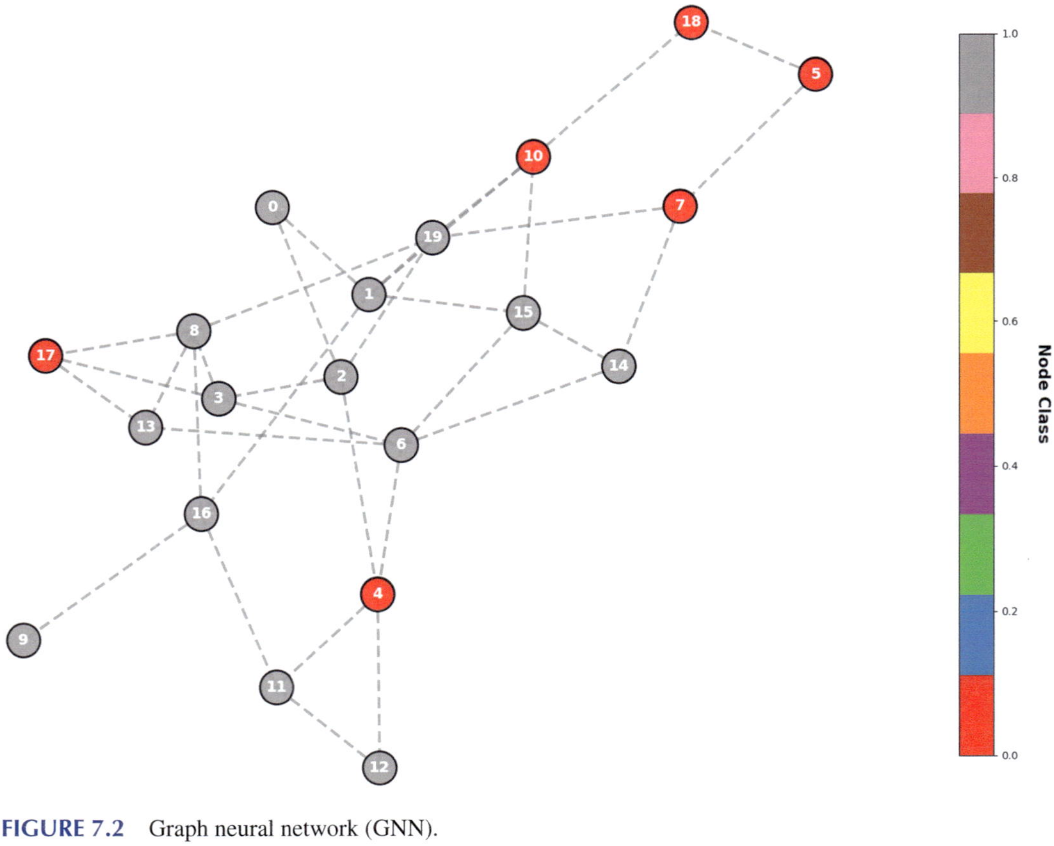

FIGURE 7.2 Graph neural network (GNN).

connect related nodes and emphasize the graph's topology. A color bar on the side maps the colors to corresponding node classes, making each node's class after classification clear. The node classification relies on node features and numerical attributes specific to each node. These features are processed through the GNN layers, which learn patterns based on both the node features and their connections in the graph. After the second layer of the GNN, each node is assigned to one of two classes (class 0 or class 1), visualized by distinct colors in the figure. The process demonstrates how GNNs capture structural information and use it for classification tasks. Here are the node features after GNN propagation:

[[–1.2524378, –0.33660233], [–1.3791587, –0.29007196], [–1.0469463, –0.43233487]

[–0.92821044, –0.5029573], [–0.73885995, –0.6494331], [–0.9682473, –0.47762945]

[–2.123791, –0.12735331], [–1.6736698, –0.2077101], [–0.6209714, –0.7709404]

[–0.5346627, –0.88156784], [–1.7315028, –0.19482112], [–0.51204944, –0.91445786]

[–0.5225032, –0.89902604], [–0.90361524, –0.5193661], [–2.2530715, –0.11101644]

[–2.3229516, –0.10312293], [–0.5895047, –0.80878687], [–0.6153719, –0.77748555]

[–1.3487849, –0.30050445], [–0.92802536, –0.50307834]]

The output numbers represent the transformed node features after passing through the GNN. Each pair corresponds to a specific node, with two values indicating the log probabilities of the node belonging to two different classes (class 0 and class 1). Below is an example of how nodes are classified:

- *Example 1 (Node 0):* The output is [–1.2524378, –0.33660233]. The first value (–1.2524378) corresponds to class 0, and the second value (–0.33660233) corresponds to class 1. As –0.33660233 is greater (less negative) than –1.2524378, Node 0 is assigned to class 1.
- *Example 2 (Node 4):* The output is [–0.73885995, –0.6494331]. The first value (–0.73885995) corresponds to class 0, and the second value (–0.6494331) corresponds to class 1. As –0.6494331 is greater (less negative) than –0.73885995, Node 4 is also assigned to class 1.

The classification process determines the class of each node by selecting the class with the higher log probability value. These assignments are visualized in Figure 7.2, where nodes are color coded according to their respective class.

7.4 WHY GRAPHS FOR DEEP LEARNING?

Data in many real-world applications, such as social networks, transportation systems, and biological networks, is inherently graph-structured. Traditional neural networks aren't designed to handle this type of data effectively. Graph-based deep learning methods, like GNNs, address this challenge by modeling relationships and interactions between entities, capturing underlying structures that traditional methods cannot. Here, we discuss why graphs are crucial for deep learning in these contexts.

7.4.1 Intuitive Representation of Complex Relationships

Graphs effectively represent complex relationships and dependencies between entities. In social networks, for example, users are nodes connected by edges representing friendships or follows. This structure is fundamental for analyzing user behavior, predicting community formation, and studying information propagation. By examining these connections, we gain valuable insights into individual and collective behaviors, which is essential for tailoring services and targeting audiences.

7.4.2 Encoding Relational Information

Graphs inherently encode relational information by directly representing connections between entities. This allows models to leverage these relationships for analytical tasks. A key application is link prediction, which infers missing or future connections by analyzing existing patterns. This is useful for suggesting new friends in social networks, recommending products in e-commerce, or hypothesizing protein interactions in biology. Another application is node classification, where nodes are labeled based on their attributes and connections, improving accuracy in tasks like classifying research papers or categorizing users into interest groups.

7.4.3 Flexibility and Versatility

Graphs can represent diverse data types and relationships, making them suitable for various applications. They can include weighted edges with numerical values like strength or cost, useful in transportation networks (distances and travel times) and communication networks (bandwidth and latency). Directed edges indicate one-way relationships, essential for modeling web links, citation networks, and social media interactions. Graphs can also represent heterogeneous relationships, modeling different types of connections in complex systems like biological networks or recommendation systems. For instance, in transportation networks, cities are nodes, and roads are edges with weights representing distances to calculate efficient routes; in airline networks, directed edges represent one-way flights.

7.4.4 Efficiency and Scalability with GNNs

GNNs excel at efficiently propagating information through graphs, capturing both local and global structures. They allow nodes to aggregate information from neighbors, learning features from immediate connections and distant nodes, resulting in a complete representation of the graph. Unlike traditional methods that struggle with scalability due to complex connections, GNNs reduce computational complexity by focusing on neighborhood aggregation. This makes them well-suited for processing large-scale datasets like social networks or biological networks without compromising speed or accuracy. Their ability to capture intricate relationships enhances performance in tasks like node classification, link prediction, and recommendation systems.

7.4.5 Uncovering Advanced Insights

Graph-based methods enable deep learning models to uncover advanced insights that are valuable for decision-making and predictive modeling. They help identify influential nodes, those with many connections or those bridging different communities. In social networks, these influential users effectively spread information and are central to multiple groups; targeting them is invaluable for marketing campaigns or strategic dissemination. Graph algorithms also detect communities within the network by revealing clusters where nodes are densely connected. Understanding these communities provides insights into the network's structure and behavior, aiding in discovering user groups with similar interests or identifying functional modules in biological networks. Additionally, analyzing how graphs evolve over time enhances understanding of network dynamics, including information spread and relationship changes, which is critical for predicting future trends and behaviors.

Figure 7.3 offers an exploration of various graph configurations and processes, demonstrating how different graph structures and analyses reveal distinct aspects of networked systems. In Figure 7.3a, an undirected graph is shown, where each connection between nodes is mutual, representing a scenario where relationships are bidirectional, such as friendships in a social network. Figure 7.3b displays a directed graph, introducing arrows that indicate the direction of relationships between nodes. This structure is essential for modeling systems where interactions are not reciprocal, such as citations in academic papers or one-way communication channels. Nodes like **D** and **C** show directional influence over other nodes, highlighting how information or influence flows through the network. Figure 7.3c progresses to a weighted directed graph, where each edge not only has a direction but also a weight that quantifies the strength of the relationship. The weights, labeled and represented by the thickness of the edges, provide additional depth by illustrating the intensity or capacity of each connection, which is crucial in contexts like transportation networks or influence spread models. Figure 7.3d illustrates information propagation within the graph, showing how information can travel from one node to others based on their connections. Figure 7.3e depicts community detection within the graph, where nodes are grouped based on their connectivity patterns. Nodes that are closely interconnected are grouped together, representing clusters or communities. Figure 7.3f focuses on identifying influential nodes in the graph. The nodes are colored based on their influence, with red indicating the most influential Node **A**.

7.5 NODE AND GRAPH CLASSIFICATION

7.5.1 Node Classification

Graph convolutional networks (GCNs) effectively predict labels for nodes within a graph by leveraging both the features of each node and the attributes of its neighbors, a task known as node classification. This application is crucial across many domains, offering valuable insights and enhancing predictive capabilities. For example, in social networks, GCNs can categorize users based on behavior and connections, distinguishing between influencers, regular users, or potential spammers and predicting user interests for personalized recommendations and targeted advertising. The classification process involves aggregating features from a node's neighbors and combining them with its own features to capture the local structure and context within the graph. Through multiple layers of convolution, GCNs learn meaningful representations of nodes that encode both their attributes and relational context. These representations are then used by a classifier to predict the label or category of each node. Imagine a small social network where we want to classify users into two categories: "Regular Users" or "Influencers" based on their connections and attributes. Assume, we have a graph with five nodes representing users: Node **A**, Node **B**, Node **C**, Node **D**, and Node **E**. The connections (edges) are as follows:

- Node **A** is connected to Node **B** and Node **C**.
- Node **B** is connected to Node **A**, Node **C**, and Node **D**.
- Node **C** is connected to Node **A**, Node **B**, and Node **E**.
- Node **D** is connected to Node **B**.
- Node **E** is connected to Node **C**.

The adjacency list is as follows: **N(A) = {B, C}**, **N(B) = {A, C, D}**, **N(C) = {A, B, E}**, **N(D) = {B}**, and **N(E) = {C}**. Each node has a two-dimensional feature vector representing user attributes (e.g., activity level and content quality):

- $h_A^{(0)} = [0.9, 0.1]$ (high activity and low content quality)
- $h_B^{(0)} = [0.8, 0.2]$

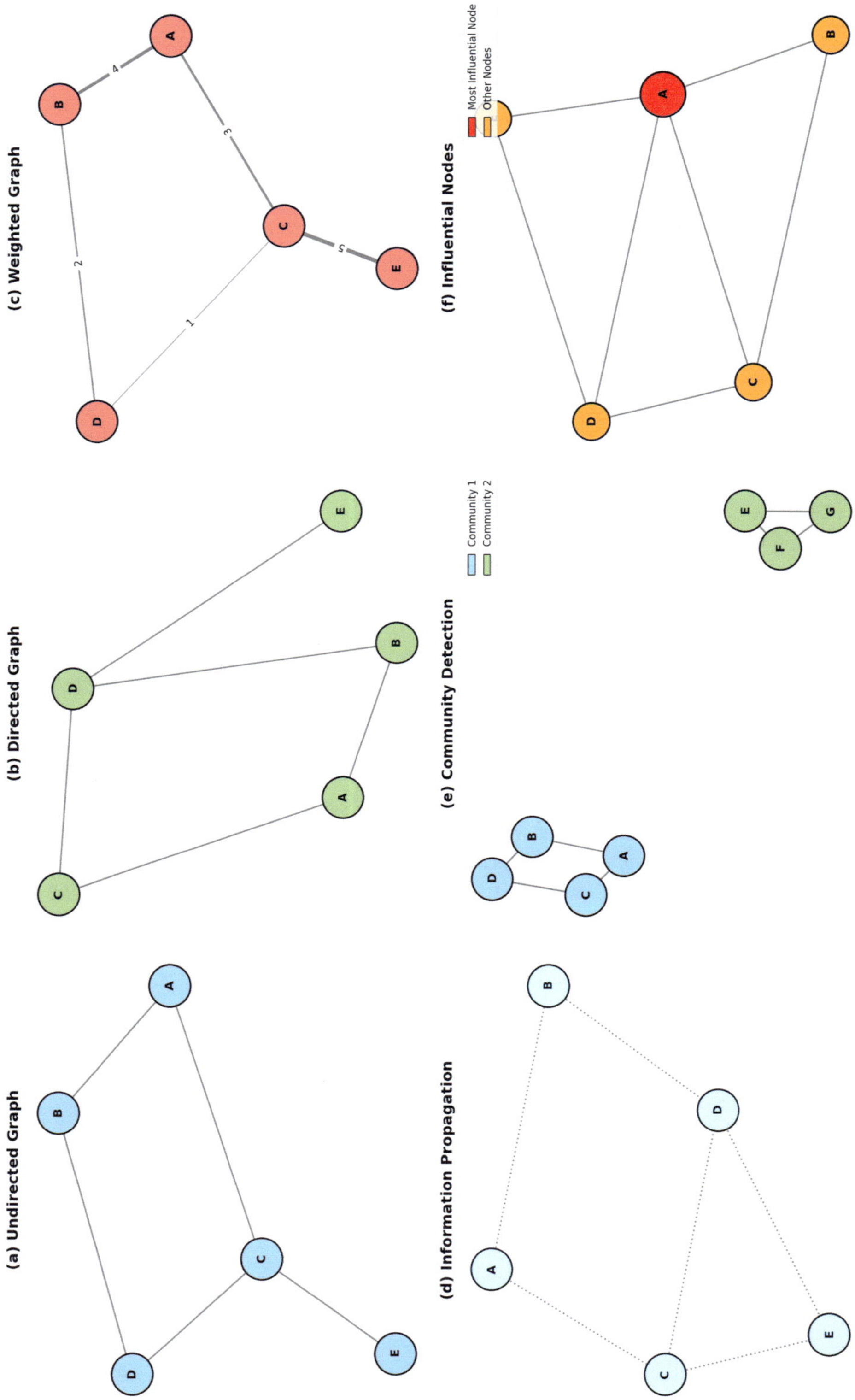

FIGURE 7.3 (a) Undirected graph, (b) directed graph, (c) weighted graph, (d) information propagation, (e) community detection, and (f) influential nodes.

- $h_C^{(0)} = [0.3, 0.7]$
- $h_D^{(0)} = [0.2, 0.8]$
- $h_E^{(0)} = [0.1, 0.9]$ (low activity and high content quality).

Node labels are as follows:

- *Labeled Nodes:* Node **A**: Regular user (Label 0) and Node **D**: Influencer (Label 1)
- *Unlabeled Nodes:* Node **B**:?, Node **C**:?, Node **E**:?.

Our goal is to predict the labels for Nodes **B**, **C**, and **E**. GCN layer equation is as follows:

$$H^{(k+1)} = \sigma\left(\tilde{D}^{-1/2}\tilde{A}\tilde{D}^{-1/2}H^{(k)}W^{(k)}\right)$$

where:

- $H^{(k)}$: Node features at layer **k,**
- $\tilde{A} = A + I$: Adjacency matrix with self-loops (**I** is the identity matrix),
- $\tilde{D}$: Diagonal node degree matrix of $\tilde{A}$,
- $\mathbf{W^{(k)}}$: Weight matrix at layer **k,**
- **σ**: Activation function (e.g., ReLU).

We'll perform one GCN layer (**k** = 0 to **k** = 1) and use ReLU as the activation function: **σ(x)** = max(0,x). For simplicity, we will use the mean aggregation instead of normalized adjacency and the same weight matrix **W** for all nodes. Here are the step-by-step computations:

1. *Build the Adjacency Matrix with Self-Loops:* The adjacency matrix (**A**) is:

$$A = \begin{bmatrix} 0 & 1 & 1 & 0 & 0 \\ 1 & 0 & 1 & 1 & 0 \\ 1 & 1 & 0 & 0 & 1 \\ 0 & 1 & 0 & 0 & 0 \\ 0 & 0 & 1 & 0 & 0 \end{bmatrix}$$

A is Adjacency matrix (without self-loops) and With self-loops added ($\tilde{A}$ = A + I):

$$\tilde{A} = A + I = \begin{bmatrix} 1 & 1 & 1 & 0 & 0 \\ 1 & 1 & 1 & 1 & 0 \\ 1 & 1 & 1 & 0 & 1 \\ 0 & 1 & 0 & 1 & 0 \\ 0 & 0 & 1 & 0 & 1 \end{bmatrix}$$

2. *Compute Degree Matrix:* Calculate the degree (number of connections) for each node, including self-loops.
   - Node **A**: Degree = 1 (self-loop) + 2 (**B** and **C**) = 3,
   - Node **B**: Degree = 1 (self-loop) + 3 (**A, C, D**) = 4,
   - Node **C**: Degree = 1 (self-loop) + 3 (**A, B, E**) = 4,
   - Node **D**: Degree = 1 (self-loop) + 1 (**B**) = 2,
   - Node **E**: Degree = 1 (self-loop) + 1(**C**) = 2.

Degree matrix ($\tilde{D}$) is as follows:

$$\tilde{D} = \begin{bmatrix} 3 & 0 & 0 & 0 & 0 \\ 0 & 4 & 0 & 0 & 0 \\ 0 & 0 & 4 & 0 & 0 \\ 0 & 0 & 0 & 2 & 0 \\ 0 & 0 & 0 & 0 & 2 \end{bmatrix}$$

3. *Compute Normalized Adjacency Matrix:* Using mean aggregation (simplified here), we'll divide each row of $\tilde{A}$ by the degree of the node. The normalized adjacency matrix $\hat{A}$ is:

$$\hat{A}_{ij} = \frac{\tilde{A}_{ij}}{\tilde{D}_{ii}}$$

Compute each element:

- For Node **A** (Row 1): $\hat{A}_{A,A} = \frac{1}{3}$, $\hat{A}_{A,B} = \frac{1}{3}$, and $\hat{A}_{A,C} = \frac{1}{3}$.
- For Node **B** (Row 2): $\hat{A}_{B,A} = \frac{1}{4}$, $\hat{A}_{B,B} = \frac{1}{4}$, $\hat{A}_{B,C} = \frac{1}{4}$, and $\hat{A}_{B,D} = \frac{1}{4}$.

Continue similarly for other nodes. For shortness, we'll represent $\hat{A}$as a matrix:

$$\hat{A} = \begin{bmatrix} \frac{1}{3} & \frac{1}{3} & \frac{1}{3} & 0 & 0 \\ \frac{1}{4} & \frac{1}{4} & \frac{1}{4} & \frac{1}{4} & 0 \\ \frac{1}{4} & \frac{1}{4} & \frac{1}{4} & 0 & \frac{1}{4} \\ 0 & \frac{1}{2} & 0 & \frac{1}{2} & 0 \\ 0 & 0 & \frac{1}{2} & 0 & \frac{1}{2} \end{bmatrix}$$

4. *Define Weight Matrix:* Assume a weight matrix **W** for transforming features:

$$W = \begin{bmatrix} 0.5 & 0.1 \\ 0.3 & 0.7 \end{bmatrix}$$

As our initial features are two-dimensional and we aim to obtain two-dimensional outputs, **W** should be a 2×2 matrix.

5. *Compute Updated Node Features:* Compute $H^{(1)} = \sigma\left(\hat{A}H^{(0)}W\right)$. Let's compute this step by step for each node. Compute the product $H^{(0)}W$ for each node:

- For Node **A**: $h_A^{(0)}W = [0.9,\ 0.1]\begin{bmatrix} 0.5 & 0.1 \\ 0.3 & 0.7 \end{bmatrix}$
$= [(0.9)(0.5)+(0.1)(0.3), (0.9)(0.1)+(0.1)(0.7)] = [0.48,\ 0.16]$

- For Node **B**: $h_B^{(0)}W = [0.8, 0.2]W = [0.44, 0.22]$.
- For Node **C**: $h_C^{(0)}W = [0.3, 0.7]W = [0.36, 0.52]$.
- For Node **D**: $h_D^{(0)}W = [0.2, 0.8]W = [0.34, 0.58]$.
- For Node **E**: $h_E^{(0)}W = [0.1, 0.9]W = [0.32, 0.64]$.

Assemble $H^{(0)}W$ matrix:

$$H^{(0)}W = \begin{bmatrix} 0.48 & 0.16 \\ 0.44 & 0.22 \\ 0.36 & 0.52 \\ 0.34 & 0.58 \\ 0.32 & 0.64 \\ & \end{bmatrix}$$

Compute $\hat{A}H^{(0)}W$: Compute the weighted sum of neighbor features for each node. For Node **A**:

$$h_A^{(1)} = \sigma\left(\sum_{j \in N(A) \cup \{A\}} \hat{A}_{A,j} \cdot h_j^{(0)}W\right)$$

Compute the contributions:

- Self-loop contribution: $\hat{A}_{A,A} \cdot h_A^{(0)}W = \frac{1}{3} \cdot [0.48, 0.16] = [0.16, 0.053]$.
- From Node **B**: $\hat{A}_{A,B} \cdot h_B^{(0)}W = \frac{1}{3} \cdot [0.44, 0.22] = [0.147, 0.073]$.
- From Node **C**: $\hat{A}_{A,C} \cdot h_C^{(0)}W = \frac{1}{3} \cdot [0.36, 0.52] = [0.12, 0.173]$.

Sum them up:

$$\begin{aligned} h_A^{(1)} &= \sigma([0.16, 0.053] + [0.147, 0.073] + [0.12, 0.173]) \\ &= \sigma([0.427, 0.299]) = [0.427, 0.299] \end{aligned}$$

For Node **B,** the contributions are:

- Self-loop: $\hat{A}_{B,B} \cdot h_B^{(0)}W = \frac{1}{4} \cdot [0.44, 0.22] = [0.11, 0.055]$.
- From Node **A**: $\hat{A}_{B,A} \cdot h_A^{(0)}W = \frac{1}{4} \cdot [0.48, 0.16] = [0.12, 0.04]$.
- From Node **C**: $\hat{A}_{B,C} \cdot h_C^{(0)}W = \frac{1}{4} \cdot [0.36, 0.52] = [0.09, 0.13]$.
- From Node **D**: $\hat{A}_{B,D} \cdot h_D^{(0)}W = \frac{1}{4} \cdot [0.34, 0.58] = [0.085, 0.145]$.

Sum them up:

$$h_B^{(1)} = \sigma\left(\left[0.11, 0.055\right] + \left[0.12, 0.04\right] + \left[0.09, 0.13\right] + \left[0.085, 0.145\right]\right)$$
$$= \sigma\left(\left[0.405, 0.37\right]\right) = \left[0.405, 0.37\right]$$

For Node **C,** the contributions are:

- Self-loop: $\hat{A}_{C,C} \cdot h_C^{(0)} W = \frac{1}{4} \cdot \left[0.36, 0.52\right] = \left[0.09, 0.13\right]$.
- From Node **A**: $\hat{A}_{C,A} \cdot h_A^{(0)} W = \frac{1}{4} \cdot \left[0.48, 0.16\right] = \left[0.12, 0.04\right]$.
- From Node **B**: $\hat{A}_{C,B} \cdot h_B^{(0)} W = \left[0.11, 0.055\right]$.
- From Node **E**: $\hat{A}_{C,E} \cdot h_E^{(0)} W = \frac{1}{4} \cdot \left[0.32, 0.64\right] = \left[0.08, 0.16\right]$.

Sum them up:

$$h_C^{(1)} = \sigma\left(\left[0.09, 0.13\right] + \left[0.12, 0.04\right] + \left[0.11, 0.055\right] + \left[0.08, 0.16\right]\right)$$
$$= \sigma\left(\left[0.4, 0.385\right]\right) = \left[0.4, 0.385\right]$$

For Node **D,** the contributions are:

- Self-loop: $\widehat{A}_{D,D} \cdot h_D^{(0)} W = \frac{1}{2} \cdot \left[0.34, 0.58\right] = \left[0.17, 0.29\right]$.
- From Node **B**: $\hat{A}_{D,B} \cdot h_B^{(0)} W = \frac{1}{2} \cdot \left[0.44, 0.22\right] = \left[0.22, 0.11\right]$.

Sum them up:

$$h_D^{(1)} = \sigma\left(\left[0.17, 0.29\right] + \left[0.22, 0.11\right]\right) = \sigma\left(\left[0.39, 0.4\right]\right) = \left[0.39, 0.4\right]$$

For Node **E**, the contributions are:

- Self-loop: $\hat{A}_{E,E} \cdot h_E^{(0)} W = \frac{1}{2} \cdot \left[0.32, 0.64\right] = \left[0.16, 0.32\right]$.
- From Node **C**: $\hat{A}_{E,C} \cdot h_C^{(0)} W = \frac{1}{2} \cdot \left[0.36, 0.52\right] = \left[0.18, 0.26\right]$.

Sum them up:

$$h_E^{(1)} = \sigma\left(\left[0.16, 0.32\right] + \left[0.18, 0.26\right]\right) = \sigma\left(\left[0.34, 0.58\right]\right) = \left[0.34, 0.58\right]$$

6. *Classification:* Now, we'll use the updated node features $\mathbf{H}^{(1)}$ to classify the unlabeled nodes. We'll use a simple linear classifier:

Logits $= H^{(1)} W_c$, where $\mathbf{W}_c$ is a weight matrix mapping node features to class scores.

Assume:

$$W_c = \begin{bmatrix} 1.0 & -1.0 \\ -1.0 & 1.0 \end{bmatrix}$$

This is a simple weight matrix for illustrative purposes. Compute logits for each node.

- *Node* **B**: $\text{Logits}_B = h_B^{(1)} W_c = [0.405, 0.37]\begin{bmatrix} 1.0 & -1.0 \\ -1.0 & 1.0 \end{bmatrix} = [(0.405)(1.0)+(0.37)(-1.0), (0.405)(-1.0)+(0.37)(1.0)] = [0.035, -0.035]$.
- *Node* **C**: $\text{Logits}_C = h_C^{(1)} W_c = [0.4, 0.385] W_c = [0.015, -0.015]$.
- *Node* **E**: $\text{Logits}_E = h_E^{(1)} W_c = [0.34, 0.58] W_c = [-0.24, 0.24]$.

Apply SoftMax to get probabilities:

$$\text{Softmax}(z_i) = \frac{e^{z_i}}{\sum_j e^{z_j}}$$

Compute for Node **B**:

- $z = [0.035, -0.035]$.
- *Exponentials:* $e^{0.035} \approx 1.0357$, $e^{-0.035} \approx 0.9658$.
- *Sum:* $1.0357 + 0.9658 = 2.0015$.
- *Probabilities:* Class 0 (regular user): $\frac{1.0357}{2.0015} \approx 0.5178$, Class 1 (Influencer): $\frac{0.9658}{2.0015} \approx 0.4822$.

Similarly, compute for Nodes **C** and **E**. For Node **C**, the probabilities are similar to Node **B** due to close logits. These probabilities will be around 0.5075 for Class 0 and 0.4925 for class 1. The Node **E** probabilities are:

- $z = [-0.24, 0.24]$
- Exponentials: $e^{-0.24} \approx 0.7866$, $e^{0.24} \approx 1.2712$
- Sum: $0.7866 + 1.2712 = 2.0578$

The probability of class 0 is $\frac{0.7866}{2.0578} \approx 0.3823\ 0$ and that of class 1 is $\frac{1.2712}{2.0578} \approx 0.6177$.

Assign predicted labels:

- *Node* **B**: Class 0 (regular user) with probability ~51.78%
- *Node* **C**: Class 0 (regular user) with probability ~50.75%
- *Node* **E**: Class 1 (influencer) with probability ~61.77%

Here, Node **B** and Node **C** are predicted to be regular users, which aligns with their connections to Node **A**, a regular user. Node **E** is predicted to be an influencer due to its strong feature vector, which reflects high content quality, and its connection to Node **C**.

7.5.2 Graph Classification

GCNs can predict labels for entire graphs, a task known as graph classification. This is particularly valuable in fields where the overall graph structure determines its classification. In fraud detection, for instance, financial transactions can be represented as graphs, where the nodes are bank accounts or individuals, and the edges represent transactions between these accounts. GCNs classify these graphs as either "Fraudulent" or "Legitimate." The methodology involves representing each graph with nodes, edges, and features. GCNs aggregate features from all nodes and edges to capture the graph's global structure. After multiple convolutional layers, a global pooling operation generates a fixed-size representation of the entire graph. This pooled representation is then fed into a classifier to predict the graph's label. Let's consider an example of financial transaction data for fraud detection. Each transaction network is represented as a graph. The nodes represent bank accounts, and the edges represent the transactions between them. Our goal is to classify these networks as either "Fraudulent" or "Legitimate" based on the graph structure. Let us work with a small dataset consisting of three transaction networks: Network **A**, Network **B**, and Network **C**. Network **A** is a legitimate transaction network (Label 0), Network **B** is fraudulent (Label 1), and Network **C** is legitimate (Label 0). The graph structures for these networks are as follows: In Network **A**, there are three nodes representing three accounts. Account 1 has transactions with Account 2 and Account 3. In Network **B**, Account 1 is linked to a potentially fraudulent account (Account 4), which in turn transacts with Account 3. Lastly, Network **C** consists of four nodes, with Account 1 connected to Accounts 2 and 3, and Account 2 connected to Account 4. The graph structure allows us to represent the relationships between accounts. We begin by representing each account using a one-hot encoding based on its type: regular accounts are encoded as [1, 0], and potentially fraudulent accounts are encoded as [0, 1]. The initial node features for each network are as follows:

- *Network* **A**: Account 1, Account 2, and Account 3 are regular accounts, so their initial features are [1, 0].
- *Network* **B**: Account 1 is regular [1, 0], Account 4 is potentially fraudulent [0, 1], and Account 3 is regular [1, 0].
- *Network* **C**: Accounts 1, 2, 3, and 4 are all regular accounts, so their initial features are [1, 0].

Next, we define the adjacency matrices to represent the connectivity between nodes. For Network **A**, the adjacency matrix would look like this:

$$A_{\text{Network A}} = \begin{bmatrix} 0 & 1 & 1 \\ 1 & 0 & 0 \\ 1 & 0 & 0 \end{bmatrix}$$

This matrix shows that Account 1 is connected to both Account 2 and Account 3, while Account 2 and Account 3 are not directly connected to each other. Similarly, the adjacency matrices for Networks **B** and **C** would represent their respective connections. We then perform graph convolutions to update the node features. A GCN layer aggregates the feature vectors of each node's neighbors. The aggregation function we use here is mean aggregation, where each node's feature is updated by averaging its own features and those of its neighboring nodes. The GCN layer equation for Node **v** at layer **k + 1** is:

$$h_v^{(k+1)} = \sigma\left(\frac{1}{|N(v)|}\sum_{u \in N(v)} W^{(k)} h_u^{(k)}\right)$$

where:

- $h_v^{(k)}$ is the feature vector of Node **v** at layer **k**,
- $\mathbf{W}^{(\mathbf{k})}$ is the weight matrix for layer **k**,
- **N(v)** is the set of neighbors of Node **v**, and
- **σ** is the activation function (ReLU).

We start with Network **A**, where the initial node features for all nodes are [1, 0]. We use the following weight matrix for the first GCN layer:

$$W^{(0)} = \begin{bmatrix} 0.5 & 0.1 \\ 0.3 & 0.7 \end{bmatrix}$$

Step 1: Initial Node Features for Network A: Account 1: [1, 0], Account 2: [1, 0], and Account 3: [1, 0].

Step 2: Linear Transformation: Using the weight matrix $\mathbf{W}^{(\mathbf{0})}$, we perform the linear transformation for each node:

Account 1: $[1, 0] \cdot W^{(0)} = [0.5, 0.1]$,
Account 2: $[1, 0] \cdot W^{(0)} = [0.5, 0.1]$,
Account 3: $[1, 0] \cdot W^{(0)} = [0.5, 0.1]$.

Step 3: Mean Aggregation: Now, we aggregate the node features using the mean of the neighbors' features (including self-loops). For Account 1, its neighbors are Account 2 and Account 3. The aggregated feature for Account 1 is:

$$\frac{[0.5, 0.1] + [0.5, 0.1] + [0.5, 0.1]}{3} = [0.5, 0.1]$$

Similarly, for Accounts 2 and 3, the aggregation process results in the same feature: [0.5, 0.1].

Step 4: Apply Activation Function (ReLU): The ReLU activation function keeps the features unchanged as they are all positive. The updated node features after this GCN layer remain [0.5, 0.1] for all nodes in Network **A**.

We repeat this process for Network **B** and Network **C** by applying the same linear transformation and aggregation steps and adjusting for the different adjacency structures. After updating the node features, we use global pooling to generate graph-level representations. We apply sum pooling, which adds the feature vectors of all nodes in a graph. Here is the sum of the features of all three nodes, known as the graph-level representation:

- For Network **A**, the graph-level representation is: $\mathbf{g}_\mathbf{A} = [1.5, 0.3]$.
- For Network **B**, the graph-level representation is: $\mathbf{g}_\mathbf{B} = [1.3, 0.9]$.
- For Network **C**, the graph-level representation is: $\mathbf{g}_\mathbf{C} = [2.0, 0.4]$.

Finally, we classify the graphs using a simple classification layer. We define a weight vector $\mathbf{W}_\mathbf{c} = [-1.0, 1.0]$ and a bias $\mathbf{b} = 0$. We compute the logits for each network:

- Network **A**: $\text{logit}_A = W_c \cdot g_A + b = -1.2$
- Network **B**: $\text{logit}_B = -0.4$

Applying the sigmoid function to get probabilities:

- Probability of Fraud for Network **A**: $P(\text{Fraud}) = \frac{1}{1+e^{1.2}} \approx 0.23$ (predicted as legitimate).
- Probability of Fraud for Network **B**: $P(\text{Fraud}) = \frac{1}{1+e^{0.4}} \approx 0.40$ (predicted as legitimate).

If these predictions don't match the actual labels, the model needs further training to adjust the weights and improve accuracy.

Figure 7.4 illustrates different network structures, distinguishing between legitimate and fraudulent patterns through node interactions and configurations. In Figure 7.4a, the graph shows a variety of node colors, each indicating different account types or groups within a network of legitimate interactions. Nodes are connected in a way that suggests diverse, healthy exchanges with no apparent signs of suspicious clustering. Figure 7.4b depicts a linear, chain-like structure where each node connects to the next in a sequence. This pattern is often associated with fraudulent activity as it may signify a controlled sequence of interactions or transactions, a common tactic in creating synthetic behavior that mimics legitimate patterns. Figure 7.4c shows a fully connected triangle, where each node is interconnected with others in a balanced and symmetrical manner. Figure 7.4d presents a quadrilateral structure, suggesting a looped pattern where nodes engage in orchestrated interactions. The symmetry and closed shape indicate a coordinated fraud ring, which could be used to manipulate interactions, such as forming closed loops to hide the origin and destination of transactions. Figure 7.4e highlights a combination of legitimate and fraudulent nodes. The red nodes (A, B, C, and D) represent fraudulent entities embedded within a chain of blue nodes, which may be legitimate. This setup shows how fraudulent actors could leverage legitimate accounts to facilitate fraudulent transactions or disguise their activities, creating a mixed network where legitimate and fraudulent behaviors are intertwined. Figure 7.4f illustrates a scenario where a central node (B1) serves as a hub, connected to various other nodes that represent legitimate accounts (green nodes). This central hub potentially manages fraudulent operations by utilizing these connections to distribute activities or transactions, making it harder to trace. The presence of the central fraud node indicates a strategic attempt to leverage legitimate connections for malicious purposes, a hallmark of organized fraud operations.

7.6 CHALLENGES

Addressing the challenges associated with GCNs is crucial for their broader adoption and effectiveness in real-world applications. Researchers are actively developing techniques to improve scalability, handle dynamic graphs, and manage heterogeneous graphs more effectively. These advancements aim to make GCNs more robust, efficient, and capable of dealing with the complexities of real-world graph data. In this section, we discuss some of these challenges.

7.6.1 Scalability

One of the significant challenges with GCNs is scalability. Deep learning models, particularly those involving graphs, can be computationally intensive. As the size of the graph increases, the amount of data and the number of computations required also grow, leading to high memory and processing demands. Large graphs, such as social networks or biological networks with millions of nodes and edges, can be particularly challenging to process efficiently. To address this issue, researchers are developing methods that help manage the scalability of GCNs. One such approach is graph sampling, as exemplified by techniques like GraphSAGE. This method involves sampling a subset of nodes and edges to create mini-batches for training, which reduces the computational burden while preserving the graph's structural information. Another approach is mini-batch training,

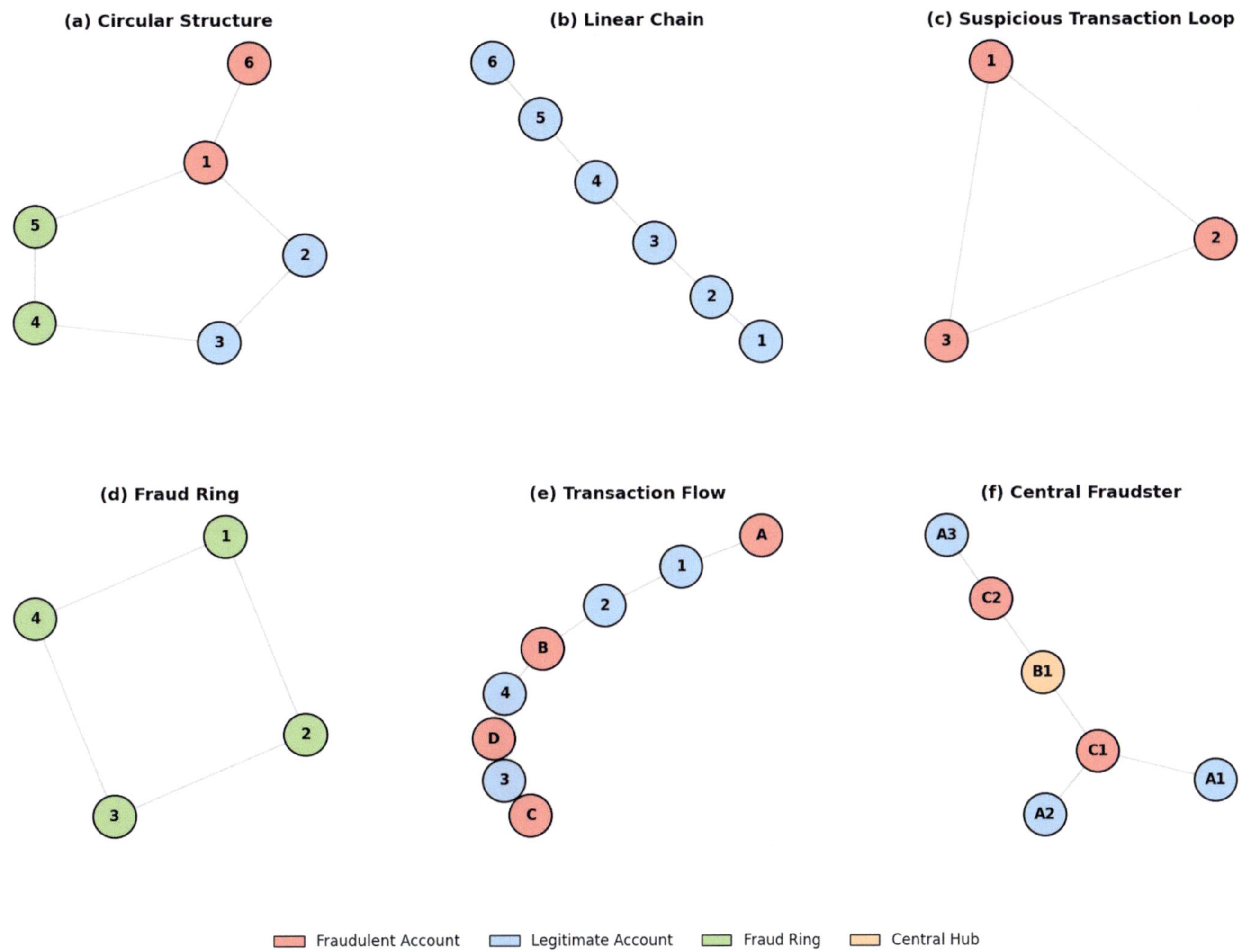

FIGURE 7.4 These graphs represent various account interactions, including (a, c) legitimate loops, (b, d) potential fraud rings, (e) mixed legitimate and fraudulent networks, (f) centralized fraud operations.

which processes the graph incrementally rather than all at once. This technique allows for incremental updates, making it possible to train on large graphs without exceeding memory limits. These methods are crucial for enabling the use of GCNs on large-scale data, ensuring that the models remain efficient and effective even as the size and complexity of the graphs increase. Consider a social network where millions of users (nodes) and billions of connections (edges) form a massive graph. Processing this entire graph at once using traditional GCNs would require enormous computational resources, making it difficult to scale effectively.

7.6.2 Dynamic Graphs

Many real-world graphs are not static but dynamic, with nodes and edges frequently being added or removed. Handling such dynamic graphs poses a significant challenge because the model must continuously update its structure and parameters to accommodate these changes. This requires algorithms that can adapt to changes in the graph in real time without necessitating a complete retraining of the model. One key approach to addressing this challenge is incremental learning. This method involves developing techniques that update the model incrementally as new data arrives, allowing the GCN to adapt to changes without starting from scratch. Another approach is processing graph data in a streaming fashion, known as streaming graphs. This allows for real-time updates and ensures that the model maintains its accuracy over time as the graph evolves. These approaches are essential for effectively managing dynamic graphs, enabling the models to remain responsive and accurate as the underlying data changes. Consider a social network where new users (nodes) are constantly joining, existing users are unfollowing or following others, and posts (edges) are continuously being posted or deleted. This creates a dynamic graph that evolves over time. To ensure that the GCN handling this data remains accurate and up-to-date, it must be capable of adjusting to these changes without needing to be retrained from scratch every time the graph changes.

7.6.3 Heterogeneous Graphs

Another challenge is dealing with heterogeneous graphs, where nodes and edges can be of different types. For example, in a knowledge graph, nodes might represent various entities such as people, places, and organizations, while edges might represent different types of relationships such as friendships, locations, and affiliations. Heterogeneous graphs require models capable of handling multiple types of nodes and edges, capturing the complex interactions between them. One solution to this challenge is the use of heterogeneous GNNs (HetGNN). These specialized architectures are designed to process and learn from heterogeneous graphs, effectively capturing the diverse interactions and relationships within the data. Another approach is the use of metapath-based methods, which leverage metapaths to capture the relationships between different types of nodes and edges, enhancing the model's ability to learn from heterogeneous data.

7.7 OTHER GRAPH-BASED DEEP LEARNING MODELS

These models represent significant advancements in graph-based deep learning, each addressing specific challenges associated with processing graph-structured data. By incorporating techniques such as neighborhood sampling, attention mechanisms, and polynomial approximations, these models enhance the scalability, expressiveness, and efficiency of GNNs, making them more suitable for various real-world applications.

7.7.1 GraphSAGE (Graph Sample and Aggregation)

GraphSAGE is a scalable technique designed to address the computational challenges of GCNs when applied to large graphs. By sampling a fixed-size subset of neighbors for each node, GraphSAGE reduces computational complexity while preserving essential structural information. This approach ensures that memory and processing requirements remain manageable, even as the graph size grows, making it ideal for large-scale graphs, such as social networks or biological networks. A key feature of GraphSAGE is its neighborhood sampling method. Rather than using all neighbors of a node, it samples a fixed number, allowing the model to scale without sacrificing important local structure. The model introduces several aggregation functions to combine the features of the sampled neighbors. The mean aggregation function calculates the mean of the neighbors' features, providing a smooth representation of the local structure. The long short-term memory network (LSTM) aggregation function uses a LSTM to capture sequential information and dependencies between neighbors. The pooling aggregation function, such as max pooling, highlights the most significant features in the neighborhood. The GraphSAGE methodology involves several steps. First, a fixed number of neighbors is sampled for each node, reducing neighborhood size while capturing representative structural information. Next, an aggregation function (mean, LSTM, or pooling) is applied to combine the sampled neighbors' features, which are then used to update the node's representation. Through multiple layers of sampling and aggregation, GraphSAGE learns meaningful node representations that incorporate both node features and the structural information from their neighborhood. These learned representations are then used to perform tasks such as node classification, link prediction, or graph classification. For a given Node **v**, GraphSAGE first samples a fixed-size set of neighbors **N(v)**. The feature vectors of these neighbors are then aggregated using one of the aggregation functions. For example, using mean aggregation, the new representation for Node **v** at layer **l + 1** is computed as:

$$h_v^{l+1} = \sigma\left(W^l \cdot \text{mean}\left(\left\{h_v^l\right\} \cup \left\{h_u^l, \forall u \in N(v)\right\}\right)\right)$$

where:

- h_v^l is the feature vector of Node **v** at layer **l**,
- W^l is a weight matrix,
- **N(v)** is the set of sampled neighbors of **v**, and
- $\boldsymbol{\sigma}$ is a nonlinear activation function, such as ReLU.

This process is repeated over multiple layers, enabling the model to incorporate multi-hop neighborhood information. Consider a simple undirected graph with six nodes: **A, B, C, D, E**, and **F**. Connections (edges) are as follows:

- Node **A** is connected to Nodes **B** and **C**.
- Node **B** is connected to Nodes **A**, **D**, and **E**.
- Node **C** is connected to Nodes **A** and **F**.
- Node **D** is connected to Node **B**.
- Node **E** is connected to Node **B**.
- Node **F** is connected to Node **C**.

The adjacency list is**: N(A) = {B,C}**, **N(B) = {A,D,E}**, **N(C) = {A,F}**, **N(D) = {B}**, **N(E) = {B}**, and **N(F) = {C}**. Each node has a two-dimensional feature vector representing certain attributes (e.g., properties or characteristics): $h_A^{(0)} = [1, 0]$, $h_B^{(0)} = [0, 1]$, $h_C^{(0)} = [1, 1]$, $h_D^{(0)} = [0,0]$, $h_E^{(0)} = [0, 1]$, and $h_F^{(0)} = [1, 0]$. The parameters, for example, are:

- *Sampling Size (K):* At each layer, sample up to two neighbors per node
- *Number of Layers:* Two layers (we will update node representations over two iterations)

- *Aggregation Function:* Mean aggregation
- *Activation Function (σ):* ReLU
- *Weight Matrices:* $W^{(1)} = \begin{bmatrix} 0.5 & 0.1 \\ 0.3 & 0.7 \end{bmatrix}$ and $W^{(2)} = \begin{bmatrix} 0.6 & 0.2 \\ 0.4 & 0.8 \end{bmatrix}$.

For each node and at each layer, these three steps should be done:

Step 1: Sampling Neighbors: Randomly sample up to **K** neighbors from the node's immediate neighbors.
Step 2: Aggregate Neighbor Features: Apply the aggregation function to the sampled neighbors' features.
Step 3: Update Node Representation: Combine the node's own feature with the aggregated neighbor features, apply the weight matrix, and pass it through the activation function to get the new node representation.

The step-by-step computation is as follows. At the first layer (from $\mathbf{h}^{(0)}$ to $\mathbf{h}^{(1)}$):

Step 1. Sampling Neighbors: For this example, we'll fix the sampled neighbors for consistency:

- *Node* **A**: Sampled Nodes **B** and **C**
- *Node* **B**: Sampled Nodes **A** and **D**
- *Node* **C**: Sampled Nodes **A** and **F**
- *Node* **D**: Sampled Node **B** (only neighbor)
- *Node* **E**: Sampled Node **B** (only neighbor)
- *Node* **F**: Sampled Node **C** (only neighbor).

Step 2. Aggregate Neighbor Features: Using mean aggregation, compute the aggregated neighbor features for each node.

- *Node* **A**:
  a. Sampled neighbors: **B** and **C**
  b. Neighbor features: $h_B^{(0)} = [0, 1]$ and $h_C^{(0)} = [1, 1]$
  c. Aggregated feature: $\text{agg}_A^{(1)} = \text{mean}\left(\left\{h_B^{(0)}, h_C^{(0)}\right\}\right) = \dfrac{[0, 1]+[1, 1]}{2} = [0.5, 1]$.

- *Node* **B**:
  a. Sampled neighbors: **A** and **D**
  b. Neighbor features: $h_A^{(0)} = [1, 0]$ and $h_D^{(0)} = [0, 0]$
  c. Aggregated feature: $\text{agg}_B^{(1)} = \dfrac{[1, 0]+[0, 0]}{2} = [0.5, 0]$.

- *Node* **C**:
  a. Sampled neighbors: **A** and **F**
  b. Neighbor features: $h_A^{(0)} = [1, 0]$ and $h_F^{(0)} = [1,0]$
  c. Aggregated feature: $\text{agg}_C^{(1)} = \dfrac{[1, 0]+[1, 0]}{2} = [1, 0]$.

- *Node* **D**:
  a. Sampled neighbor: **B**
  b. Aggregated feature: $\text{agg}_D^{(1)} = h_B^{(0)} = [0,1]$.

- *Node* **E**:
  a. Sampled neighbor: **B**
  b. Aggregated feature: $\text{agg}_E^{(1)} = h_B^{(0)} = [0, 1]$.
- *Node* **F**:
  a. Sampled neighbor: **C**
  b. Aggregated feature: $\text{agg}_F^{(1)} = h_C^{(0)} = [1, 1]$.

Step 3. Update Node Representations: Update each node's representation using $h_v^{(1)} = \sigma\left(W^{(1)} \cdot \left(h_v^{(0)} \| \text{agg}_v^{(1)}\right)\right)$. Here, || denotes the concatenation of the node's own features with its aggregated neighbor features, resulting in a four-dimensional vector. Accordingly, $\mathbf{W}^{(1)}$ should be a 2 × 4 matrix. For simplicity, we'll adjust $\mathbf{W}^{(1)}$ to match the dimensions. Let us redefine $\mathbf{W}^{(1)}$ as a 2 × 4 matrix:

$$W^{(1)} = \begin{bmatrix} 0.5 & 0.1 & 0.4 & 0.2 \\ 0.3 & 0.7 & 0.6 & 0.8 \end{bmatrix}$$

- *Node* **A**:
  a. Concatenated features: $h_A^{(0)} \| \text{agg}_A^{(1)} = [1, 0, 0.5, 1]$
  b. Multiply by $\mathbf{W}^{(1)}$:

$$h_A^{(1)} = \sigma\left(W^{(1)} \cdot [1, 0, 0.5, 1]^{\top}\right) = \sigma\left(\begin{bmatrix} (0.5)(1)+(0.1)(0)+(0.4)(0.5)+(0.2)(1) \\ (0.3)(1)+(0.7)(0)+(0.6)(0.5)+(0.8)(1) \end{bmatrix}\right)$$

$$= \sigma\left(\begin{bmatrix} 0.5+0+0.2+0.2 = 0.9 \\ 0.3+0+0.3+0.8 = 1.4 \end{bmatrix}\right) = [0.9, 1.4]$$

  c. Apply ReLU: $h_A^{(1)} = [0.9, 1.4]$
- *Node* **B**:
  a. Concatenated features: $h_B^{(0)} \| \text{agg}_B^{(1)} = [0, 1, 0.5, 0]$
  b. Multiply by $\mathbf{W}^{(1)}$:

$$h_B^{(1)} = \sigma\left(\begin{bmatrix} (0.5)(0)+(0.1)(1)+(0.4)(0.5)+(0.2)(0) \\ (0.3)(0)+(0.7)(1)+(0.6)(0.5)+(0.8)(0) \end{bmatrix}\right)$$

$$= \sigma\left(\begin{bmatrix} 0+0.1+0.2+0 = 0.3 \\ 0+0.7+0.3+0 = 1.0 \end{bmatrix}\right) = [0.3, 1.0]$$

Nodes **C**, **D**, **E**, and **F** can be computed similarly. Also, similar steps should be done for layer 2 (from $\mathbf{h}^{(1)}$ to $\mathbf{h}^{(2)}$).

7.7.2 Graph Attention Networks

Graph attention networks (GATs) enhance GCNs by introducing an attention mechanism that allows nodes to assign different levels of importance to their neighbors' features. Instead of treating all neighboring nodes equally, each node dynamically learns which neighbors are most relevant during the training process. This adaptive focus enables the model to capture intricate relationships within the graph more effectively, improving its expressiveness and performance, especially in heterogeneous and complex graphs where nodes and edges vary in type and significance.

By selectively concentrating on the most pertinent neighbors, GATs provide more accurate and insightful representations of the data. The methodology includes the following steps:

Step 1. Graph Representation: Each node is initially represented by its feature vector, capturing the node's attributes.

Step 2. Attention Calculation: For each node, calculate the attention coefficients with its neighbors. These coefficients represent the importance of each neighbor's features. The attention coefficient $\boldsymbol{a}_{ij}$ between nodes **i** and **j** is computed as:

$$\boldsymbol{a}_{ij} = \text{softmax}_j\left(\text{LeakyReLU}\left(\boldsymbol{a}^{\top}\left[\boldsymbol{Wh}_i \parallel \boldsymbol{Wh}_j\right]\right)\right)$$

where **a** is the attention vector, **W** is the weight matrix, $\mathbf{h}_i$ and $\mathbf{h}_j$ are the feature vectors of nodes **i** and **j**, ‖ denotes concatenation, softmax_j is the softmax function applied over all neighbors **j** of Node **i**, and LeakyReLU is a nonlinear activation function applied to the computed value before the softmax.

Step 3. Feature Aggregation: Aggregate the features of the neighbors using the attention coefficients. This results in a new feature representation for each node that emphasizes the more relevant neighbors:

$$h_{i'} = \sigma\left(\sum_{j \in N(i)} a_{ij} W h_j\right)$$

where $\boldsymbol{N}(\boldsymbol{i})$ denotes the neighbors of Node **i**, and $\boldsymbol{A}$ is an activation function such as ReLU.

Step 4. Multihead Attention: To stabilize the learning process and improve performance, GATs often use multihead attention, where multiple attention mechanisms run in parallel, and their outputs are concatenated or averaged.

7.7.3 ChebNet (Chebyshev Networks)

ChebNet utilizes Chebyshev polynomials to generalize convolution operations on graphs, enabling efficient and localized computations. By approximating the graph Laplacian's eigenfunctions with Chebyshev polynomials, ChebNet avoids the explicit computation of eigenvectors, addressing computational challenges associated with spectral methods. This approach enhances scalability and performance, particularly for large and sparse graphs. ChebNet employs Chebyshev polynomials, orthogonal polynomials that approximate spectral graph convolutions, to make the convolution process more efficient. This method allows the network to perform localized operations, effectively capturing the local structure of the graph by focusing on the immediate neighborhood of each node. This focus preserves important local details while maintaining computational efficiency. By leveraging polynomial approximations, ChebNet significantly reduces computational costs, streamlining spectral graph convolutions and making it suitable for large-scale graphs. Its scalability allows it to handle large and sparse graphs efficiently, making it well-suited for real-world datasets without overwhelming computational resources. Additionally, ChebNet's approach enables the model to capture multiscale features within the graph, providing a better understanding of the graph's structure and properties. This multiscale feature capture enhances the model's ability to perform various complex graph-based tasks effectively. The methodology includes the following steps:

Step 1. Graph Representation: Each node in the graph is represented by its feature vector. The graph's structure is captured by its adjacency matrix **A** and the degree matrix **D**.

Step 2. Chebyshev Polynomial Approximation: ChebNet approximates the spectral convolution operation using Chebyshev polynomials of the graph Laplacian **L**. The graph Laplacian is defined as **A** (adjacency matrix), **D** (degree matrix), and **L = D–A** (graph Laplacian). The scaled graph Laplacian is: $\tilde{\boldsymbol{L}} = \frac{2L}{\lambda_{\max}} - \boldsymbol{I}$, where $\lambda_{\max}$ is the largest eigenvalue of **L**, and **I** is the identity matrix. The convolution operation is defined as:

$$x' = g_\theta * x = \sum_{k=0}^{K} \theta_k T_k\left(\tilde{L}\right)x$$

- $\mathbf{x'}$ is the output feature vector after applying the convolution,
- $\mathbf{g}_\theta$ represents the filter parameterized by the Chebyshev coefficient $\boldsymbol{\theta}_k$,
- $\boldsymbol{\theta}_k$ is the Chebyshev coefficient to be learned,
- $\boldsymbol{T}_k\left(\tilde{\boldsymbol{L}}\right)$ is the Chebyshev polynomial evaluated at the scaled Laplacian $\tilde{L} = \frac{2L}{\lambda_{\max}} - I$, and

  x is the input feature vector.

Step 3. Localized Convolution: Using a finite number of Chebyshev polynomials, the convolution operation becomes localized, meaning that the filter only considers a local neighborhood of nodes.

Step 4. Model Training: Train the ChebNet model using the localized convolution operations to learn meaningful node representations for various tasks such as node classification or graph classification.

7.7.4 Improved Scalability and Efficiency

Traditional graph algorithms are often designed to be highly efficient and scalable, particularly for large graphs. By combining these algorithms with deep learning models, it is possible to preprocess or simplify the graph structure using classical methods, reducing the computational burden on the deep learning model. For example, a graph clustering algorithm can be used to partition the graph into smaller, more manageable subgraphs, which can then be processed by a GNN. Deep learning models can be used to predict important graph properties or optimize certain aspects of the graph, which can then guide the application of classical algorithms. For instance, a GNN can predict the critical nodes or edges in a network, which can be prioritized in a shortest path or maximum flow algorithm. This combination can lead to more targeted and efficient solutions for problems such as network optimization, routing, and resource allocation. Many real-world problems require both the predictive power of deep learning and the optimization capabilities of classical algorithms. By integrating these approaches, it is possible to tackle complex, multifaceted problems more effectively. For example, in transportation networks, a GNN can predict traffic patterns and congestion, while the shortest path algorithm can optimize routes based on these predictions, providing a solution for traffic management. Consider a transportation network where cities are represented as nodes and roads as edges, and the goal is to optimize traffic management. Traditional graph algorithms like shortest path or maximum flow can effectively solve specific optimization problems, such as finding the quickest route or maximizing traffic throughput. However, predicting future traffic conditions or identifying critical points in the network requires more advanced models. By combining GNNs with classical algorithms, we can achieve both predictive power and optimization. For instance, a GNN can be trained to predict traffic patterns based on historical data, identifying critical nodes where congestion is likely to occur. These predictions can then inform the shortest path algorithm, allowing it to prioritize specific routes that avoid congested areas. In this way, the integration of deep learning

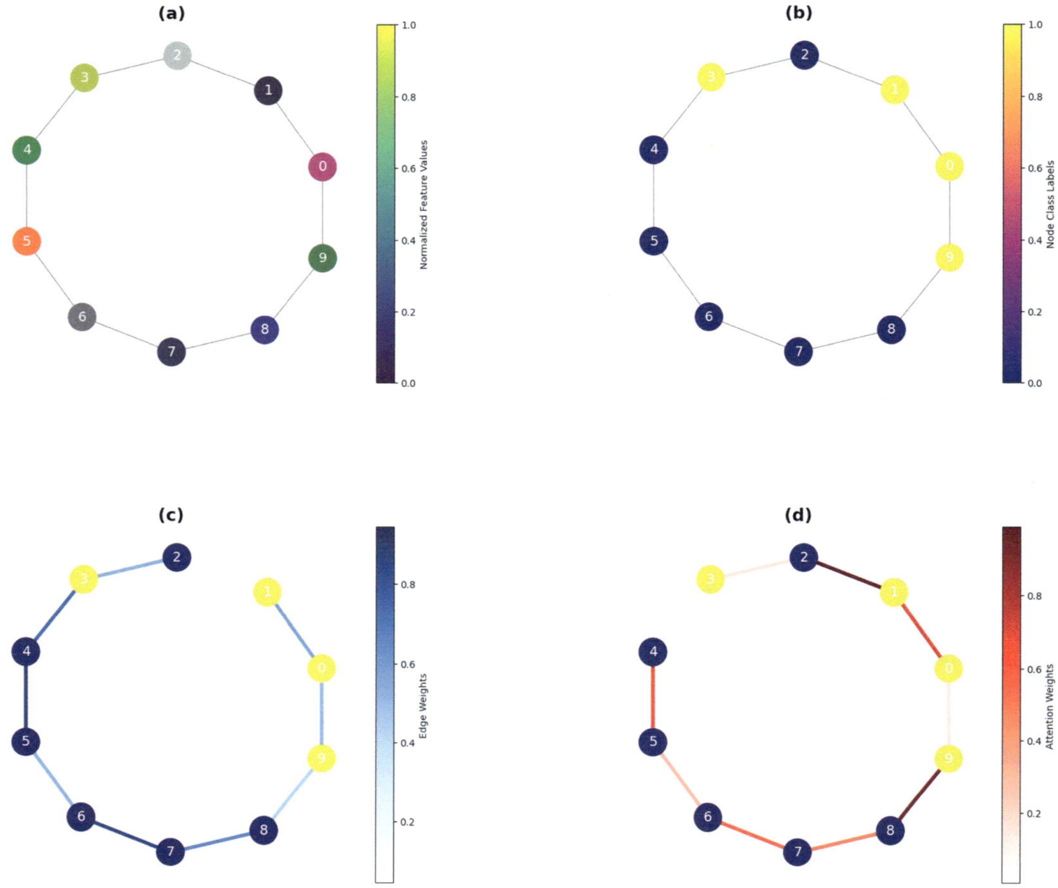

FIGURE 7.5 (a) Initial node features, (b) node classification using GCN, (c) graph with weighted edges, and (d) graph with attention mechanisms.

with classical methods leads to a more scalable and efficient solution for optimizing transportation networks.

Figure 7.5 offers a visualization of the different stages and components involved in the GNN process, specifically illustrating how it transforms graph data for node classification. In Figure 7.5a, initial node features, the graph is presented in its original state, with each node colored based on the average of its normalized feature values. Moving to Figure 7.5b, node classification using GCN, the graph displays the results of node classification after being processed by a GCN. In Figure 7.5c, a graph with weighted edges, the focus shifts to the relationships between nodes, with edges assigned random weights to simulate varying degrees of connection strength. The thickness and color intensity of the edges represent the magnitude of these weights, utilizing the "Blues" colormap. This gradient ranges from light blue for weaker connections to deep blue for stronger ones, visually conveying the influence of each connection between nodes. Thicker edges denote higher weights, emphasizing more influential relationships within the graph. Finally, Figure 7.5d, a graph with attention mechanisms, incorporates attention mechanisms into the graph, illustrating how the GNN selectively focuses on certain edges over others. This gradient highlights the edges that the GNN deems more influential or relevant for node classification tasks. Thicker and more intensely colored edges indicate higher attention weights, showcasing the GNN's ability to prioritize significant node relationships while downplaying less important ones.

Figure 7.6 provides a visualization of the GNN process, showcasing various stages and methodologies employed in transforming and analyzing graph-structured data for node classification. In Figure 7.6a, a directed graph, a foundational directed graph is depicted, consisting of five nodes interconnected by several directed edges. Transitioning to Figure 7.6b, an undirected graph, the same set of nodes and connections from Figure 7.6a is presented without the directional arrows. This undirected version underscores mutual relationships or bidirectional connections between nodes, suggesting that interactions are reciprocal and not confined to a single direction. Figure 7.6c, a weighted graph, introduces an additional layer of complexity by assigning random weights to the previously unweighted directed edges. These weights symbolize the strength or significance of the connections between nodes. In Figure 7.6d, initial node features, the focus shifts to the attributes of the nodes themselves. Figure 7.6e, node classification using GCN, showcases the impact of the GCN on node classification. After undergoing two convolutional layers, each node's features are updated by aggregating information from their immediate neighbors. In Figure 7.6f, node classification using GAT, the results of node classification after processing with a graph attention network (GAT) are illustrated. Figure 7.6g, node classification using GraphSAGE, presents the outcomes of node classification using the GraphSAGE algorithm. GraphSAGE aggregates features from a node's local neighborhood to generate its representation, which is then utilized for classification.

7.8 REAL-WORLD APPLICATIONS

7.8.1 Social Network Analysis

Social networks are a prime example where graph theory is extensively applied. Users are represented as nodes, and their interactions, such as friendships or follows, are depicted as edges. By analyzing the structure of these networks, we can gain insights into social behaviors, community formation, and information propagation. For example, social network platforms use graph-based deep learning models to recommend new friends or connections by predicting potential links between users. These models analyze the existing connections and suggest new links based on patterns observed in the network, enhancing user engagement and expanding the social graph.

7.8.2 Recommendation Systems

Graph-based recommendation systems leverage user–item interaction graphs to provide personalized suggestions. In these systems, nodes represent users and items (e.g., movies, products), while edges represent interactions such as ratings or purchases. GNNs process these interactions to predict which items a user might be interested in, leading to more accurate and relevant recommendations. This approach is widely used by platforms, where understanding the relationships between users and items is crucial for delivering a tailored experience.

7.8.3 Biological Network Analysis

In biology, graph theory is essential for understanding the interactions within complex biological systems, such as protein–protein interaction networks or gene regulatory networks. By representing these interactions as graphs, researchers can apply GNNs to predict unknown interactions or classify proteins based on their functions. This method is particularly valuable in drug discovery, where identifying potential interactions between proteins can lead to the development of new therapeutic targets.

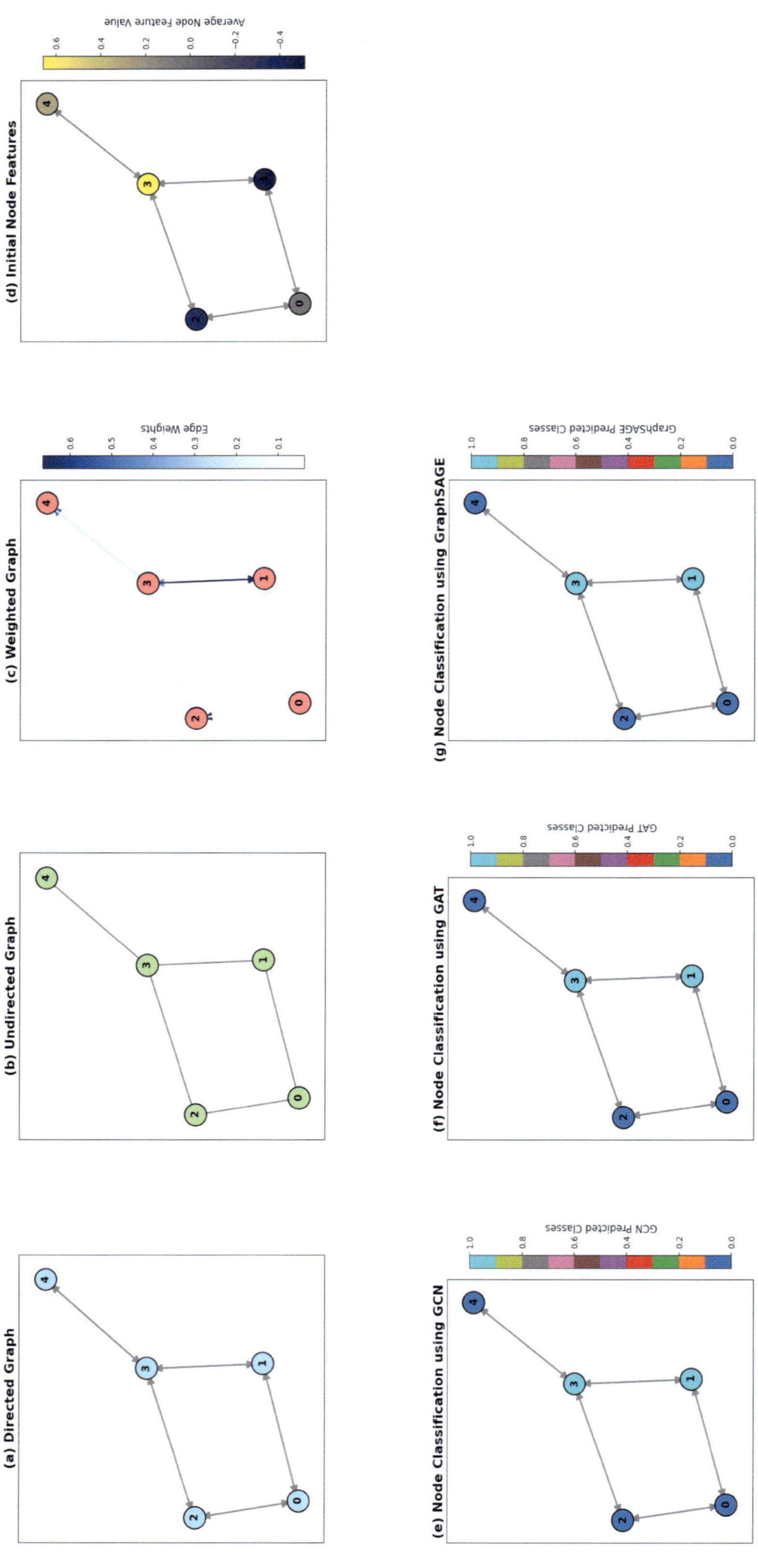

FIGURE 7.6 (a) Directed graph, (b) undirected graph, (c) weighted directed graph, (d) initial node features, (e) node classification using GNN, (f) node classification using GAT, and (g) node classification using GraphSAGE.

7.8.4 Transportation and Logistics

Graph theory also plays a vital role in optimizing transportation networks. In these networks, cities or transportation hubs are represented as nodes, and the routes between them as edges. The efficiency of these networks can be improved by applying graph-based algorithms to optimize routes, reduce travel times, and minimize costs. For example, delivery companies use these models to optimize their logistics operations, ensuring timely and cost-effective deliveries.

7.8.5 Fraud Detection

In finance, graph theory is employed to detect fraudulent activities by analyzing transaction networks. In these networks, accounts are represented as nodes and transactions as edges. By identifying unusual patterns or anomalies in the graph, such as clusters of accounts with suspicious interactions, financial institutions can detect and prevent fraudulent activities more effectively. This approach is particularly useful in combating complex financial crimes, where traditional methods may fall short.

7.8.6 Healthcare and Epidemic Modeling

Graph theory is increasingly used in healthcare to model and predict the spread of diseases. In epidemic modeling, individuals are represented as nodes, and their interactions as edges. By analyzing these graphs, public health officials can predict how diseases spread through populations and identify critical nodes (individuals or locations) where interventions can be most effective.

7.9 HANDS-ON EXAMPLE

The objective is to build a simple **GNN** model that can classify nodes in a graph based on their features and connections.

7.9.1 Step 1: Import Required Libraries

In this code snippet, we import various libraries necessary for building a GCN using TensorFlow and the Spektral library, along with utilities for data manipulation and visualization. NumPy is imported for numerical operations, while tensorflow.keras is used to build neural network models. The input and dense layers from Keras will help define the network structure. Spektral's **GCNConv** layer is a specialized layer for handling graph data, which operates by learning features for nodes in a graph based on their connections. The normalized_adjacency function helps in normalizing the adjacency matrix of the graph, a preprocessing step often required in graph-based learning tasks. The adjacency matrix will be handled as a sparse matrix using csr_matrix from scipy.sparse, which efficiently stores large, sparse matrices. For visualization, matplotlib.pyplot is used to plot graphs, and networkx is imported for graph generation and manipulation.

```
import numpy as np
import tensorflow as tf
from tensorflow.keras.layers import Input, Dense
from tensorflow. keras.models import Model
from spektral.layers import GCNConv
from spektral.utils import normalized_adjacency
from scipy.sparse import csr_matrix
import matplotlib. pyplot as plt
import networkx as nx
```

7.9.2 Step 2: Create the Graph

In this part of the code, we create an adjacency matrix representing a simple undirected graph using a sparse matrix format. The matrix adj_matrix is constructed using the csr_matrix function from scipy.sparse, which is efficient for storing large matrices with many zeros. In this example, the adjacency matrix represents a graph with three nodes. The value 1 in position (i, j) indicates that Node **i** is connected to Node **j**, while 0 indicates no connection. For this specific matrix, Node **0** is connected to Node **1**, Node **1** is connected to both Node **0** and Node **2**, and Node **2** is connected to Node **1**. This symmetric structure of the matrix reflects the undirected nature of the graph. The dtype= np.float32 ensures that the values are stored as 32-bit floating-point numbers, which is common in deep learning applications. This adjacency matrix will later be used in a GCN to define the structure of the graph on which the model operates.

```
adj_matrix = csr_matrix([[0, 1, 0],
                         [1, 0, 1],
                         [0, 1, 0]], dtype=np.float32)
```

7.9.3 Step 3: Normalize the Adjacency Matrix

In this section, we prepare the normalized adjacency matrix for use in a **GCN** by converting it into a format suitable for TensorFlow. First, the normalized_adjacency() function normalizes the adjacency matrix, a common preprocessing step in GNNs to ensure the graph's structure is properly scaled. Here, we use symmetric normalization (symmetric = True), which adjusts the adjacency matrix symmetrically to account for the degree of each node. Next, the adjacency matrix is converted to **COO** (coordinate list) format using tocoo(). **COO** is an efficient format for sparse matrices, which stores the nonzero elements by their row and column coordinates. The indices variable is created by stacking the row and column indices of the nonzero elements, which will be used to build a sparse tensor in TensorFlow. The adjacency matrix is then converted into a TensorFlow SparseTensor (A_tf), where indices hold the coordinates of nonzero elements, values hold the actual nonzero values from the adjacency matrix, and dense_shape represents the overall shape of the matrix. Finally, tf.sparse.reorder() ensures that the sparse tensor is correctly ordered internally for efficient computations in TensorFlow. This processed sparse adjacency matrix will be used as input for graph-based deep learning models like a **GCN**, allowing the model to learn from the graph structure.

```
adj_matrix = normalized_adjacency(adj_matrix, symmetric=True)
# Convert to COO format if not already
adj_matrix = adj_matrix.tocoo()
indices = np.column_stack((adj_matrix.row, adj_matrix.col))
A_tf = tf.sparse.SparseTensor(indices=indices, values=adj_
matrix.data, dense_shape=adj_matrix.shape)
# Ensure the sparse tensor is in the correct order
A_tf = tf.sparse.reorder(A_tf)
```

7.9.4 Step 4: Define Node Features

In this line of code, we create a feature matrix for the graph's nodes using np.eye(3, dtype= np.float32), which generates a 3×3 identity matrix. The identity matrix is often used as a simple feature representation in graph-based learning tasks, where each node has a unique feature vector. For a graph with three nodes, each row of the matrix represents the features of a node, and the identity matrix ensures that each node is represented by a unique one-hot encoded feature vector. For instance, Node **0** has the feature vector [1, 0, 0][1, 0, 0][1, 0, 0], Node **1** has [0, 1, 0][0, 1, 0][0, 1, 0], and Node **2** has [0, 0, 1][0, 0, 1][0, 0, 1]. The dtype=np.float32 ensures the matrix is stored as 32-bit floating-point numbers, which is common in machine learning tasks. This feature matrix will be used as input to the **GCN**, allowing the model to learn relationships between the graph's nodes based on their features.

```
features = np.eye(3, dtype=np.float32)
```

7.9.5 Step 5: Define Node Labels

In this line of code, we are creating a label matrix for a classification task associated with the nodes of the graph. The labels array is a 3 × 2 matrix, where each row corresponds to the one-hot encoded label for a node. The np.array() function is used to create the matrix, and dtype=np.float32 ensures the data is stored as 32-bit floating-point numbers. For this example, Node 0 and Node 2 have the label [1, 0][1, 0][1, 0], meaning they belong to class 0, and Node 1 has the label [0, 1][0, 1][0, 1], meaning it belongs to class 1. Each label is a one-hot encoded vector representing the class membership of the corresponding node. This label matrix will be used during training of the GCN, allowing the model to learn to classify each node into the appropriate class based on its features and the graph structure.

```
labels = np.array([[1, 0], [0, 1], [1, 0]], dtype=np.float32)
```

7.9.6 Step 6: Build the GCN Model

In this part of the code, we are building a **GCN** model using TensorFlow and the Spektral library. The model takes both node features and the adjacency matrix as inputs to learn from graph-structured data.

```
input_features = Input(shape=(3,), dtype='float32')
input_adj = Input(shape=(None,), sparse=True, dtype='float32')
gc = GCNConv(16, activation='relu')([input_features, input_adj])
output = Dense(2, activation='softmax')(gc)
model = Model(inputs=[input_features, input_adj], outputs=
output)
model.compile(optimizer='adam', loss='categorical_crossentropy',
metrics=['accuracy'])
```

7.9.7 Step 7: Train the Model

In this line, we are training the **GCN** model on the graph data using the model.fit() function. The inputs to the model are the features matrix and the adjacency matrix (in sparse tensor format), and the target labels are provided in the labels array. During training, the **GCN** learns to classify nodes by propagating and transforming information from node features and neighboring nodes' features based on the structure of the graph. The model will adjust its weights to minimize the categorical cross-entropy loss, and the accuracy will be evaluated after each epoch to track progress.

```
model.fit(x=[features, A_tf], y=labels, epochs=10, verbose=1)
```

7.9.8 Step 8: Predict and Visualize

Finally, the graph is displayed with plt.show(), providing a visual representation of the nodes and their predicted classes, demonstrating the classification results of the **GCN** model.

```
# Predict the node classes
predictions = model.predict([features, A_tf])
predicted_classes = np.argmax(predictions, axis=1)
# Create a simple graph for visualization
G = nx.Graph()
for i in range(len(predicted_classes)):
   G.add_node(i, label=predicted_classes[i])
# Add edges
edges = [(int(i), int(j)) for i, j in zip(adj_matrix.row, adj_
matrix.col) if i < j]
G.add_edges_from(edges)
# Color map for visualization
color_map = ['blue' if label == 0 else 'red' for label in
predicted_classes]
# Draw the graph
plt.figure(figsize=(8, 8))
pos = nx.spring_layout(G, seed=42)  # for consistent layout
nx.draw(G, pos, node_color=color_map, with_labels=True,
labels={i: f'Node {i}\nClass {predicted_classes[i]}' for i in
G.nodes()},
       node_size=700, font_color='white', font_weight='bold',
edge_color='gray', linewidths=2, alpha=0.9)
plt.title('Node Classification with GCN', fontsize=16)
plt.show()
```

7.10 COMMON MISTAKES AND TROUBLESHOOTING TIPS

7.10.1 Improper Node and Edge Representation

- *Mistake:* Incorrectly representing node features or edge connections, leading to errors in graph construction and analysis.
- *Tip:* Verify the structure of your graph data by visualizing it. Ensure that node features and edge connections are accurately represented and that any preprocessing steps (like normalization) are correctly applied.

7.10.2 Inadequate Data Preprocessing

- *Mistake:* Failing to preprocess data appropriately, which can lead to poor model performance or training issues.
- *Tip:* Normalize node features and ensure that the graph's adjacency matrix is correctly constructed. Handle missing data, and ensure consistency in the representation of nodes and edges.

7.10.3 Over-Smoothing in GNNs

- *Mistake:* Over-smoothing where node representations become too similar after multiple layers of graph convolutions.
- *Tip:* Limit the number of **GNN** layers, and consider using skip connections or residual connections to mitigate over-smoothing. Experiment with different aggregation functions and regularization techniques.

7.10.4 Ignoring Graph Size and Complexity

- *Mistake:* Applying standard **GNN** models directly to very large or complex graphs without considering scalability.
- *Tip:* Use models like **GraphSAGE** or sampling techniques to handle large graphs. Consider hierarchical approaches or graph coarsening to reduce complexity.

7.10.5 Overlooking Edge Weights and Attention Mechanisms

- *Mistake:* Not leveraging edge weights or attention mechanisms, leading to suboptimal performance in tasks requiring nuanced relational understanding.
- *Tip:* Incorporate edge weights into your model if your graph data includes them. Use graph attention networks (GATs) to apply attention mechanisms, which can weigh the importance of different edges.

7.10.6 Insufficient Model Evaluation

- *Mistake:* Failing to evaluate models thoroughly, leading to misleading conclusions about model performance.
- *Tip:* Use cross-validation, and ensure that evaluation metrics are appropriate for the specific graph-related task. Visualize model predictions and errors to gain insights into performance.

7.10.7 Misapplying Classical Graph Algorithms

- *Mistake:* Incorrectly integrating classical graph algorithms with neural network models, leading to errors in analysis or suboptimal solutions.
- *Tip:* Thoroughly understand the principles of classical graph algorithms before integrating them with neural networks. Ensure compatibility and correctness in implementation.

7.10.8 Neglecting Dynamic and Heterogeneous Graphs

- *Mistake:* Ignoring the dynamic nature of graphs or the presence of multiple types of nodes and edges, leading to incomplete or incorrect analysis.
- *Tip:* Use dynamic **GNN** models and account for the heterogeneity in your graph data. Develop custom models or preprocessing steps to handle dynamic and heterogeneous graphs effectively.

7.11 REVIEW QUESTIONS

1. What are the fundamental components of a graph, and how do they contribute to the graph's structure and functionality?
2. How do directed and undirected graphs differ, and in which scenarios is each type particularly useful?
3. What is a weighted graph, and how do edge weights influence the outcomes of graph analysis?
4. What are the limitations of traditional neural networks when processing graph-structured data, and why are specialized models like GNNs necessary?
5. How do GNNs process graph-structured data, and what is the significance of their architecture in handling complex networks?
6. How do GCNs generalize the convolution operation to effectively handle graph data?
7. What are the typical real-world applications of **GCNs**, and how do they provide value in those contexts?
8. What are the primary challenges encountered when implementing graph-based deep learning models, particularly in large-scale or complex datasets?
9. What is **GraphSAGE**, and how does it enhance the scalability and efficiency of **GNNs** in processing large graphs?
10. How does **ChebNet** utilize Chebyshev polynomials to generalize the convolution operation for graphs, and what advantages does this method offer?

7.12 PROGRAMMING QUESTIONS

7.12.1 Easy

Implement a single GCN.

1. Define the adjacency matrix and node feature matrix for a small graph.
2. Implement the **GCN** layer function, which includes normalizing the adjacency matrix and applying the **GCN** formula to the node features.
3. Apply the **GCN** layer to the graph data, and print the updated node features.

7.12.2 Medium

Use a **GCN** to perform node classification on the Cora citation network dataset.

1. Load and preprocess the Cora dataset, including the adjacency matrix and feature matrix.
2. Split the dataset into training and test sets.
3. Define a **GCN** model using TensorFlow/Keras or PyTorch with multiple **GCN** layers.
4. Train the **GCN** model on the training set and evaluate its performance on the test set.
5. Visualize the learned node embeddings and the classification results.

7.12.3 Hard

Implement and train a GAT for node classification on a large graph dataset.

1. Load and preprocess a large graph dataset, such as PubMed or Reddit, including adjacency matrix and feature matrix.

2. Implement the **GAT** layer, focusing on computing attention coefficients and applying attention to the node features.
3. Define a **GAT** model with multiple **GAT** layers.
4. Train the **GAT** model using appropriate techniques for large graphs.
5. Evaluate the model's performance on a test set and visualize the results, comparing them with a baseline **GCN** model.

8 Differential Geometry

8.1 INTRODUCTION

As we move into the digital age, with rapid advances in technology, differential geometry is finding new importance in deep learning. As deep learning becomes more complex, the need to understand its inner workings grows. This is where differential geometry becomes crucial. By interpreting neural networks using concepts like manifolds and curvature, we can gain valuable insights into how these networks train, optimize, and represent information. This chapter reviews these concepts in more detail.

8.2 BASICS OF DIFFERENTIAL GEOMETRY

Differential focuses on properties of space that remain unchanged under smooth transformations, like bending or stretching, and its understanding of the geometry of data and model parameters is crucial in deep learning and neural networks.

8.2.1 MANIFOLDS

A manifold is a topological space that, when examined closely, looks like a flat piece of Euclidean space. To illustrate, think of an ant walking on the surface of a basketball. From the ant's perspective, the surface appears flat, as it can only see a small portion at any given time, even though the surface is curved. This idea of being "locally flat" is the essence of a manifold. The number of coordinates required to describe a point on the manifold determines its dimension. For example, while the surface of a sphere curves in three-dimensional (3D) space, it is a 2D manifold because any small region looks flat, like a plane. In deep learning, data often exists in high-dimensional spaces, and understanding the structure and dimensionality of this data is essential for tasks like visualization, dimensionality reduction, and feature extraction. The manifold hypothesis suggests that high-dimensional data often resides on or near a lower-dimensional manifold. For instance, images of faces might occupy a lower-dimensional space within the vast high-dimensional space of all possible images. Consider a set of handwritten digits. Each image may be high-dimensional, such as a 28×28 pixel grid, but the variations between different images of the same digit can often be captured by just a few parameters, like thickness or slant. This implies that the images of handwritten digits sit on a lower-dimensional manifold within the higher-dimensional space of all possible pixel combinations. Methods such as Principal Component Analysis (PCA) and t-Distributed Stochastic Neighbor Embedding (t-SNE) help identify these lower-dimensional manifolds, making data analysis and visualization more manageable. Understanding manifolds helps deep learning

DOI: 10.1201/9781032690742-8

practitioners grasp the underlying structure of data, leading to more effective models and deeper insights. Let $X \subset R^n$ be a high-dimensional dataset, and assume that X lies on a manifold $M \subset R^d$, where $d \ll n$. The goal of techniques like PCA or autoencoders is to find a mapping f: $R^n \to R^d$, such that:

$$x = f(z), \quad \text{where } z \in \mathbb{R}^d \quad \text{and} \quad x \in \mathbb{R}^n.$$

Here, z represents the coordinates on the lower-dimensional manifold, and f maps these coordinates back to the high-dimensional space. For example, in PCA, the data are projected onto the first few principal components, which form the lower-dimensional manifold.

8.2.2 Tangent Space

To understand a tangent space, imagine picking a point on a curved surface, like a sphere. Now picture a flat plane that touches the surface at that exact point but doesn't cut through it. This plane represents tangent space, and it captures all possible directions from which you could move. For a 2D surface, the tangent space is a flat plane, but for higher-dimensional manifolds, the concept extends into more complex spaces. In deep learning, the idea of a tangent space becomes important when we consider optimization methods like gradient descent. When we compute the gradient of a loss function at a given point, that gradient can be viewed as a vector lying in the tangent space of the loss landscape. This vector points in the steepest direction of increase, and by following its opposite direction, we perform gradient descent, moving toward a local or global minimum. The tangent space provides a useful linear approximation of the curved manifold at a specific point. This approximation allows gradient-based optimization methods to update parameters effectively, even in the complex, curved spaces that arise in deep learning. For example, imagine standing at the North Pole of a sphere; the tangent space at this point would be a flat plane, representing all possible directions you could step. Though the surface is curved, the tangent plane gives a simple, local approximation of the surface's behavior. In the context of deep learning, the loss landscape can be seen as a high-dimensional manifold. The tangent space at any point on this manifold provides a way to navigate the loss landscape using gradient information. The gradient vector within this space shows the direction of the steepest ascent, but in optimization, the algorithm moves in the opposite direction, down the slope, toward minimizing the loss. For a manifold $\mathbf{M} \subset \mathbf{R}^n$ and a point $\mathbf{p} \in \mathbf{M}$, the tangent space at $\mathbf{p}$, denoted as $\mathbf{T_pM}$, is a vector space that contains all the possible directions in which one can move from $\mathbf{p}$ on the manifold. Formally, the tangent space at $\mathbf{p}$ is spanned by the partial derivatives of the coordinate functions at $\mathbf{p}$. In the context of gradient descent, at each iteration, the parameter update is given by:

$$\theta_{t+1} = \theta_t - \eta \nabla L(\theta_t),$$

where

- θ_t represents the parameters at time step t,
- η is the learning rate,
- $\nabla L(\theta_t)$ is the gradient of the loss function L at θ_t.

The gradient $\nabla L(\theta_t)$ can be interpreted as a vector in the tangent space of the loss surface at θ_t. This vector points in the steepest direction of increase, but gradient descent moves in the opposite direction to minimize the loss.

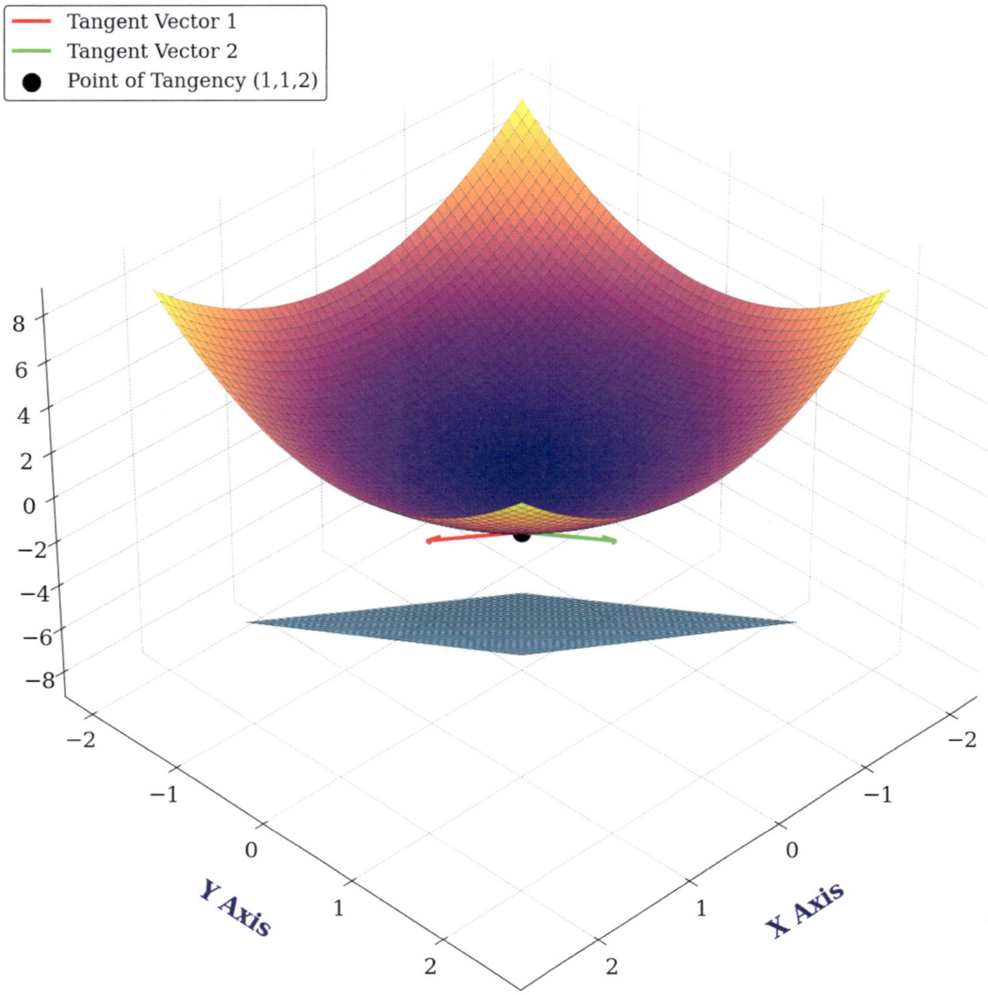

FIGURE 8.1 3D paraboloid surface with tangent plane and vectors at (1, 1).

Figure 8.1 illustrates a 3D visualization of a paraboloid surface defined by the function $z = x^2 + y^2$. The plot highlights the geometric and differential properties of the surface at a specific point of tangency located at coordinates (1, 1, 2). The paraboloid surface extends upward, demonstrating a smooth, continuous curvature characteristic of this quadratic function. At the point of tangency (1, 1, 2), a tangent plane is shown in light blue, providing a linear approximation of the surface at this point. This plane represents the best local approximation of the surface around this point, illustrating how the differential geometry concept of tangent spaces applies to this curved surface.

Additionally, two tangent vectors, labeled Tangent Vector 1 and Tangent Vector 2, are visualized on the plane. Tangent Vector 1 is depicted in red, while Tangent Vector 2 is in green. These vectors lie within the tangent plane, showing the directions along which the surface changes most rapidly at the point of tangency. The vectors highlight how the surface's behavior can be understood locally through these directional derivatives, capturing the rate and direction of change. The point of tangency itself is marked by a black dot, emphasizing the specific location where the tangent plane and vectors interact with the surface.

8.2.3 Metric Tensor

The metric tensor is a key concept that helps us understand how to measure distances on curved surfaces or manifolds. It extends the idea of the dot product, which we use in flat, Euclidean spaces, to more complex, curved spaces. This allows us to calculate distances and angles on those curved surfaces. Essentially, the metric tensor gives a manifold its unique geometric structure by defining how lengths and angles are measured at each point. Mathematically, the metric tensor works by taking two tangent vectors at any point on the manifold and returning a number (a scalar). This number represents the inner product of those two vectors, which helps determine their lengths and the angle between them. Through this process, the metric tensor defines the local geometric properties of the manifold. In the context of neural networks, the metric tensor is closely related to something called the Fisher Information Matrix. The Fisher Information Matrix helps us understand the shape (or curvature) of the parameter space, showing how changes in the model's parameters influence its predictions. It essentially measures how much information a random variable provides about an unknown parameter. Understanding the metric tensor and Fisher Information Matrix can be beneficial in several ways. In optimization, knowing the geometry of the parameter space can lead to better strategies; for example, natural gradient descent leverages the Fisher Information Matrix to adjust parameter updates, enabling faster and more efficient convergence. Regularization also benefits from this understanding, as the curvature of the parameter space defined by the metric tensor provides insights into designing regularization methods. By assessing the complexity of a model's geometry, techniques can be developed to prevent overfitting by penalizing overly complex models. Additionally, the metric tensor aids in interpreting how sensitive a model is to changes in its parameters. By analyzing the parameter space geometry, we can pinpoint areas where the model remains stable or becomes highly sensitive to small changes, which can inform efforts to refine model performance. Consider a 2D surface, like the surface of a sphere. To measure the distance between two points on this curved surface, we need more than the usual Euclidean distance formula because the surface is not flat. The metric tensor defines how distances and angles are measured on this curved surface, giving us the ability to compute the length of curves and the angle between two directions at a given point. In 2D Euclidean space, the metric tensor is simply the identity matrix, but on a curved surface like a sphere, the metric tensor is more complex, adapting to the curvature of the space. For a manifold **M** with coordinates $(\mathbf{x_1}, \mathbf{x_2}, \ldots, \mathbf{x_n})$, the metric tensor **g** is a function that assigns a matrix $\mathbf{g_{ij}}$ to each point on the manifold. This matrix is used to compute the inner product of two tangent vectors **u** and **v** at a point **p** on the manifold:

$$u, v = \sum_{i,j} g_{ij}(p) u^i v^j,$$

where u^i and v^j are the components of the tangent vectors, and g_{ij} are the components of the metric tensor at point p. For example, on the surface of a 2D sphere, the metric tensor **g** might look like:

$$g = \begin{pmatrix} 1 & 0 \\ 0 & \sin^2(\theta) \end{pmatrix},$$

where θ is the angle in spherical coordinates. This metric tensor adjusts how distances and angles are measured depending on where you are on the sphere.

Figure 8.2 provides a 3D visualization of a curved surface, illustrating the impact of the metric tensor on vector lengths across different regions of the surface. The surface itself is defined by a function that creates a wavelike structure, showing variations in height represented by the color gradient. The surface height ranges from lower elevations in deep purple to higher elevations in yellow, as indicated by the color bar on the right. Vectors are placed at various points on the surface,

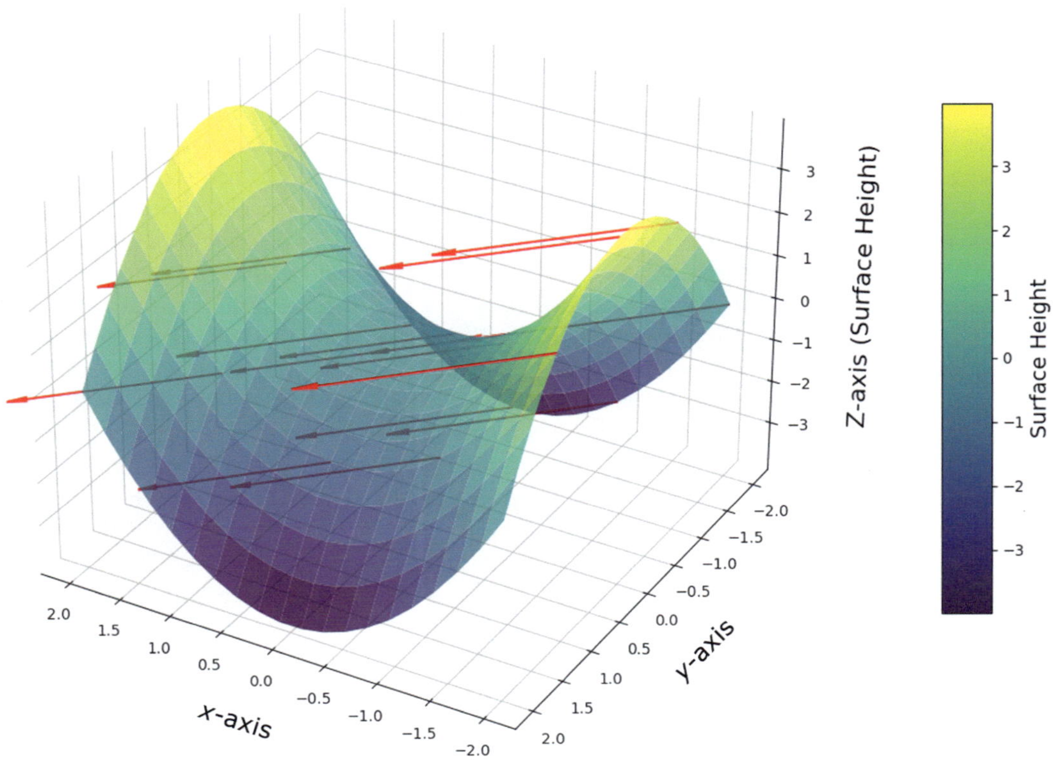

FIGURE 8.2 Influence of metric tensor on vector lengths on a curved surface.

and their lengths are determined by the metric tensor, which accounts for the local curvature of the surface. The vectors, shown in red, indicate the directions and magnitudes at these points. As the surface curves, the vectors change in length, reflecting the influence of the metric tensor in scaling them according to the surface's geometric properties.

8.2.4 Curvature

Curvature is a concept that describes how much a surface or manifold bends or deviates from being flat. For example, a sheet of paper has zero curvature because it's perfectly flat, while a sphere has positive curvature because it curves uniformly in all directions. Curvature gives us insight into the local shape and geometry of a manifold, and it can be measured in different ways. One common method is Gaussian curvature, which considers the bending in two principal directions at a point, while Ricci curvature generalizes this concept to higher dimensions, helping us understand the overall shape of the space. In the context of deep learning, curvature plays an important role when analyzing the loss landscape, the complex surface that represents how the model's performance changes with different parameter values. Understanding the curvature of this surface is crucial for several reasons. In optimization, areas of the loss landscape with high curvature represent steep slopes and sharp valleys, which can make it challenging for optimization algorithms like gradient descent to find the optimal solution. The steepness in these regions can cause unstable updates or slow progress. Conversely, flatter regions with lower curvature are easier to navigate, enabling smoother and faster convergence. Curvature also plays a role in parameter sensitivity; in high-curvature areas, even small changes to the model's parameters can lead to significant changes in the loss function, making the model more sensitive and potentially unstable. Recognizing these high-curvature regions allows for adjustments in optimization strategies, such as lowering the learning rate, to prevent instability. Furthermore, curvature affects generalization, as flat regions in the loss landscape are typically linked to better generalization. In these areas, the model is less sensitive to small variations in the data, resulting in more stable and robust performance. In contrast, high-curvature regions may suggest that the model is overly fine-tuned to the training data, which can be a sign of overfitting and lead to poor performance on unseen data.

Consider a simple loss landscape shaped like a bowl. If the bowl is steep and narrow (high curvature), it will have a sharp minimum, and gradient-based optimization methods like gradient descent may struggle to converge quickly or smoothly. Conversely, if the bowl is shallow and wide (low curvature), the optimization process is easier, and the parameter updates are more stable and less sensitive to small changes. One common way to measure curvature is through Gaussian curvature, which is the product of the principal curvatures k_1 and k_2 at a given point on a 2D surface:

$$K = k_1 \cdot k_2$$

- If $K > 0$, the surface has positive curvature, like a sphere,
- If $K = 0$, the surface is flat, like a plane,
- If $K < 0$, the surface has negative curvature, like a saddle.

For higher-dimensional manifolds, Ricci curvature is used to generalize this concept. The Ricci curvature at a point provides a way to quantify how much the manifold deviates from being flat in various directions.

Figure 8.3 visualizes a 2D manifold represented by a sphere, with a tangent plane positioned at the point (1, 0, 0). The sphere, shown in light blue, models the manifold, emphasizing its curved nature and continuous surface. At the point (1, 0, 0), marked in red, the tangent plane is illustrated as a flat, yellow surface intersecting the sphere. This plane serves as the best local approximation of the manifold at that specific point, demonstrating the concept of a tangent space in differential geometry. It is depicted as being tangent to the sphere's surface only at the red point, illustrating how the manifold's curvature is locally linearized at this location. Two tangent vectors, labeled Tangent Vector 1 (in blue) and Tangent Vector 2 (in green), lie within this tangent plane. These vectors represent the directions along which the surface changes most rapidly at the point of tangency. They provide insight into the local geometry of the sphere, showing how different directional derivatives describe the manifold's behavior at this point. The vectors highlight how, even in a higher-dimensional curved space, local analysis can be simplified using linear approximations.

8.3 DIFFERENTIAL GEOMETRY IN DEEP LEARNING

Differential geometry offers a robust mathematical framework that provides profound insights into the behavior, optimization, and utility of neural network models when applied to deep learning. As deep learning continues to evolve, integrating these fields is expected to deepen, leading to more robust and interpretable models.

8.3.1 Loss Landscapes

In deep learning, the loss function is critical as it measures how far a model's predictions deviate from the true values. Training a model means adjusting its parameters iteratively to minimize this loss. As models grow in complexity, often with millions or billions of parameters, the loss landscape becomes a highly intricate, multi-dimensional surface. You can imagine it as a vast, uneven terrain

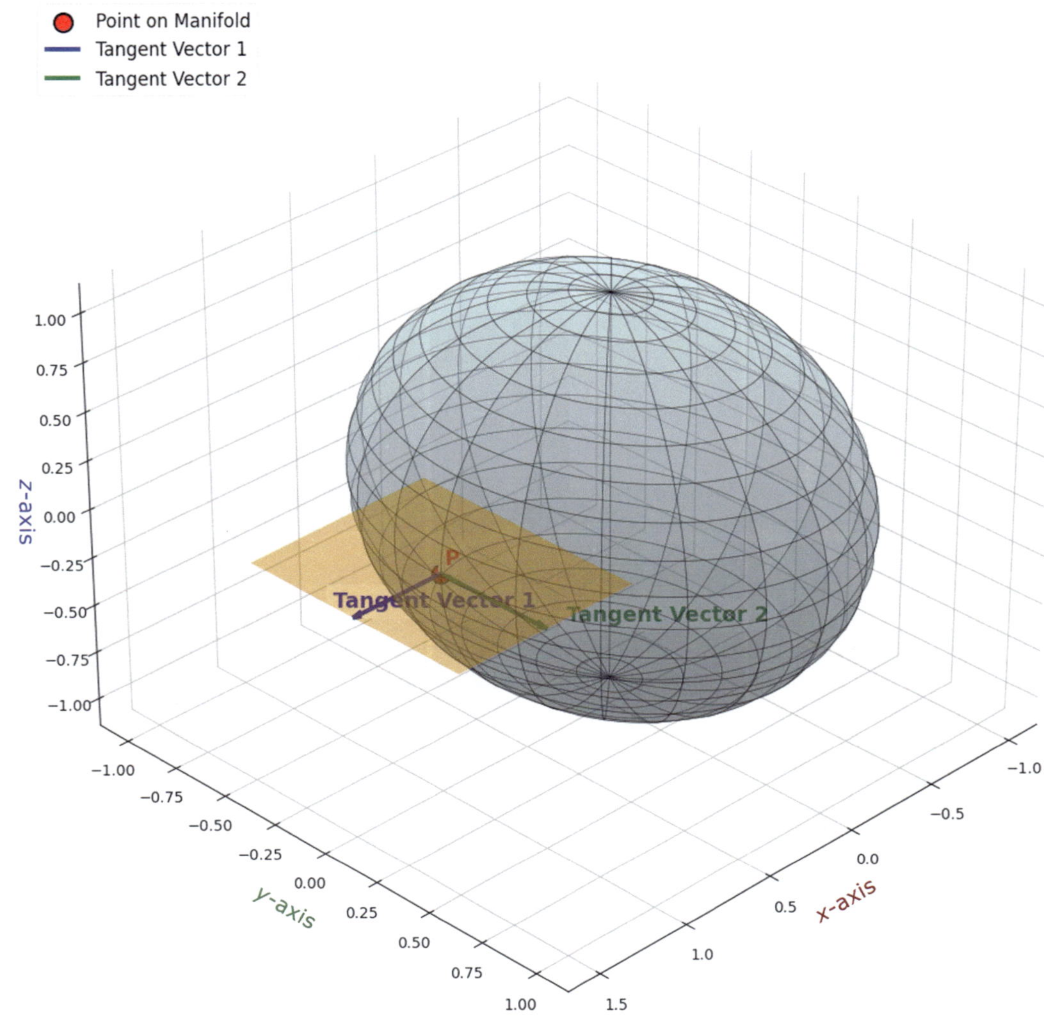

FIGURE 8.3 Visualization of a 2D manifold (sphere) with a tangent plane at the point (1, 0, 0).

filled with valleys, mountains, plateaus, and ridges. Navigating this terrain to find the lowest point (the minimum loss) is key to improving model performance, and this is where concepts from differential geometry, like curvature, become essential. Curvature provides insights into the shape of the loss landscape. For instance, in convex regions, imagine the inside of a bowl, the path to the minimum is smooth and straightforward. Gradient descent can reliably find the minimum in these areas, which leads to faster and more predictable convergence. These are the regions we want our optimization methods to find. However, deep neural networks rarely have simple, convex landscapes. Instead, they often feature saddle points, places on the surface that are neither a peak nor a valley but a combination of both, like a mountain pass. These points have a mix of positive and negative curvature along different directions. Because the gradient is close to zero in these regions, training can stall, with gradient descent making painfully slow progress. Understanding the curvature of the loss landscape is critical for developing more efficient optimization techniques. For example, methods like Newton's method use second-order information (such as the Hessian matrix) to capture the curvature more accurately, allowing for smarter updates to the parameters. However, these methods can be computationally expensive for large networks. Alternatively, adaptive learning rate algorithms like Adam or RMSprop adjust the learning rate based on the observed gradients, which allows them to implicitly respond to the local curvature, making them more efficient in navigating complex loss landscapes. Let's say you're training a neural network to classify images, and your loss function is cross-entropy. As the model updates its parameters, it effectively "travels" across this multi-dimensional loss landscape. Curvature tells you a lot about where you are on the landscape. In a convex region, like the inside of a bowl, gradient descent works smoothly because the slope always leads toward the minimum. But if you hit a concave region (the outside of a dome) or a saddle point, optimization becomes more difficult. Saddle points, where the gradient is nearly zero, but you're not at the lowest point, are particularly tricky in deep learning because they can slow training and confuse the optimizer. In higher-dimensional spaces, saddle points become even more common. This is because there are many directions in which the curvature can be either negative (concave) or zero. For instance, imagine a deep network's loss surface has a saddle point where the gradient is nearly zero, but the surrounding areas have varying curvatures. This can stall the training process, as the gradient descent algorithm may mistakenly interpret this point as being close to an optimum, even though it is not. Let $L(\theta)$ represent the loss function, where θ is the vector of model parameters. The gradient $\nabla L(\theta)$ gives the direction of steepest ascent, and the Hessian matrix $H(\theta)$ provides second-order information about the curvature of the loss landscape:

$$H(\theta) = \frac{\partial^2 L(\theta)}{\partial \theta^2}$$

At a convex point, all eigenvalues of the Hessian are positive, indicating that the surface curves upwards in every direction. At a saddle point, some eigenvalues are positive, and others are negative, indicating mixed curvature.

Figure 8.4 visualizes a hypothetical loss landscape, illustrating the presence of different critical points, a global minimum (convex), a local maximum (concave), and a saddle point. The global minimum, marked with a red sphere, represents the lowest point on the surface, indicating where the loss function reaches its minimum value. This convex region is where optimization algorithms ideally aim to converge, as it signifies the most optimal solution in the parameter space. The surrounding contour lines further emphasize this low point, showing concentric circles that decrease in height as they approach the minimum. The local maximum, indicated by a blue triangle, is a peak in the landscape where the function reaches a temporary high value within a specific region. This concave region demonstrates how optimization paths might be trapped if they encounter this point, thinking they have reached a maximum when, in reality, it is not the global extremum. The blue arrow points to this maximum, showing its prominence within its neighborhood. The saddle point, marked by a yellow square, is another critical feature of the landscape. This point illustrates where the surface curves upward in one direction and downward in another, forming a mix of convex and concave characteristics. Saddle points are significant in optimization because they can mislead algorithms, as gradients might not clearly indicate which direction moves toward the global minimum. The labeled yellow square highlights its position, and its location shows how it interrupts smooth descent or ascent paths on the surface.

Suppose we are training a simple neural network with one parameter to minimize a loss function. We'll explore different loss functions to see how their landscapes affect the optimization process.

Step 1. Quadratic Loss Function (Convex Region)

- *Loss Function:* For this $L(\theta) = (\theta - 2)^2$ as a simple convex function (a parabola) with a minimum at $\theta = 2$. The loss landscape is "bowl-shaped," and gradient descent should efficiently find the minimum.
- *Gradient and Hessian:* the gradient is $\nabla L(\theta) = 2(\theta - 2)$ and Hessian is $H(\theta) = 2$

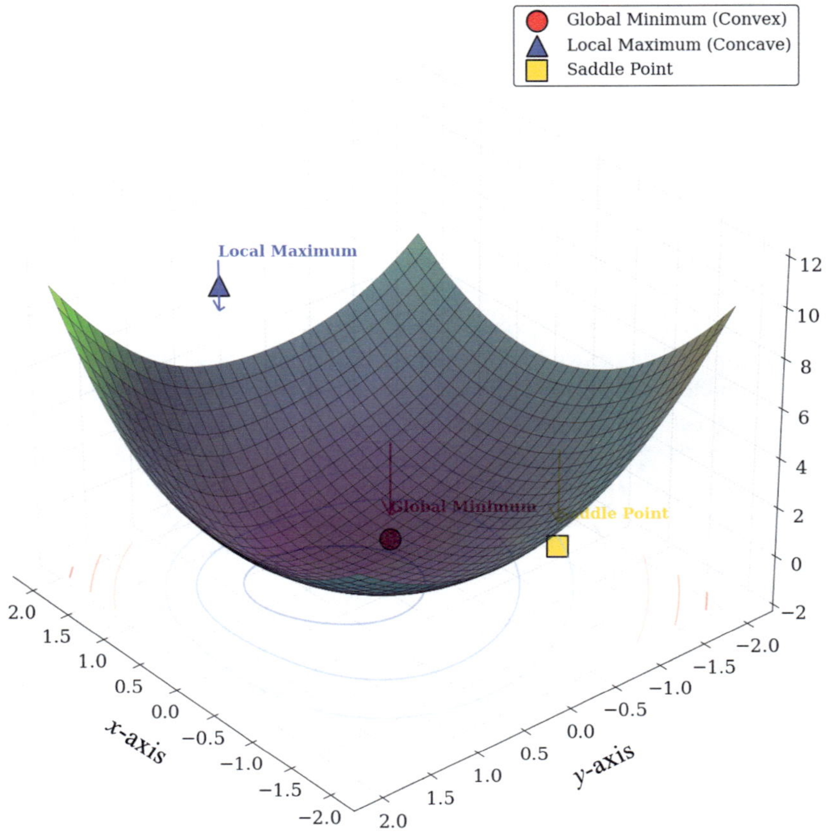

FIGURE 8.4 Hypothetical loss landscape with convex, concave, and saddle point regions.

Optimization Steps: Let's start with an initial parameter $\theta_0 = -4$.

1. Iteration 1:
   a. Compute gradient: $\nabla L(\theta_0) = 2(-4-2) = 2(-6) = -12$
   b. Update parameter (using a learning rate $\alpha = 0.1$): $\theta_1 = \theta_0 - \alpha\nabla L(\theta_0) = -4 - 0.1(-12) = -4 + 1.2 = -2.8$
   c. Compute loss: $L(\theta_1) = (-2.8-2)^2 = (-4.8)^2 = 23.04$
2. Iteration 2:
   a. Compute gradient: $\nabla L(\theta_1) = 2(-2.8-2) = 2(-4.8) = -9.6$
   b. Update parameter: $\theta_2 = -2.8 - 0.1(-9.6) = -2.8 + 0.96 = -1.84$
   c. Compute loss: $L(\theta_2) = (-1.84-2)^2 = (-3.84)^2 = 14.7456$

   The parameter θ is moving toward the minimum at $\theta = 2$. The loss decreases with each iteration, and the convex landscape allows gradient descent to converge efficiently.

Step 2. Saddle Point Example

- *Loss Function:* This function $L(\theta) = \theta^3$ has a saddle point at $\theta = 0$. The gradient is zero at $\theta = 0$, but it's neither a minimum nor a maximum. The loss landscape is not convex or concave throughout.
- *Gradient and Hessian:* The gradient is $\nabla L(\theta) = 3$ and the Hessian is $H(\theta) = 6\theta$
  At $\theta = 0$: $\nabla L(0) = 0$ and $H(0) = 0$

- *Optimization Steps:* Starting with $\theta_0 = 0.01$:
  1. Iteration 1:
     a. Gradient: $\nabla L(\theta_0) = 3(0.01)^2 = 3 \times 0.0001 = 0.0003$
     b. Update: $\theta_1 = 0.01 - 0.1(0.0003) = 0.01 - 0.00003 = 0.00997$
     c. Loss: $L(\theta_1) = (0.00997)^3 \approx 9.910 \times 10^{-7}$
  2. Iteration 2:
     a. Gradient: $\nabla L(\theta_1) = 3(0.00997)^2 \approx 0.0002982$
     b. Update: $\theta_2 = 0.00997 - 0.1(0.0002982) = 0.00997 - 0.00002982 \approx 0.00994018$
     c. Loss: $L(\theta_2) = (0.00994018)^3 \approx 9.821 \times 10^{-7}$

The gradient is very small near the saddle point. Parameter updates are tiny, causing slow progress. Even though the gradient is near zero, we're not at a minimum. Gradient descent struggles to escape the saddle point region.

Step 3. Non-Convex Function with Multiple Minima

- *Loss Function:* This function $L(\theta) = (\theta^2 - 1)^2$ has two minima at $\theta = -1$ and $\theta = 1$. There is a saddle point at $\theta = 0$.
- *Gradient and Hessian:* The gradient is $\nabla L(\theta) = 4\theta(\theta^2 - 1)$ and the Hessian is $H(\theta) = 4(3\theta^2 - 1)$
  At $\theta = 0$: $\nabla L(0) = 0$ and $H(0) = -4$ (negative curvature)
- *Optimization Steps:* Starting with $\theta_0 = 0.1$:
  1. Iteration 1:
     a. Gradient: $\nabla L(0.1) = 4(0.1)(0.01 - 1) = 4(0.1)(-0.99) = -0.396$
     b. Update: $\theta_1 = 0.1 - 0.1(-0.396) = 0.1 + 0.0396 = 0.1396$
     c. Loss: $L(\theta_1) = (0.1396^2 - 1)^2 \approx 0.738$
  2. Iteration 2:
     a. Gradient: $\nabla L(0.1396) = 4(0.1396)(0.0195 - 1) = 4(0.1396)(-0.9805) \approx -0.5467$
     b. Update: $\theta_2 = 0.1396 - 0.1(-0.5467) = 0.1396 + 0.05467 = 0.19427$
     c. Loss: $L(\theta_2) = (0.19427^2 - 1)^2 \approx 0.648$

     The gradient descent is moving away from the saddle point at $\theta = 0$ toward one of the minima. Depending on the starting point, it could converge to either $\theta = -1$ or $\theta = 1$. The saddle point at $\theta = 0$ can slow down the optimization if the starting point is near it.

Step 4. Hessian Matrix and Curvature

In higher dimensions, we consider the Hessian matrix $H(\theta)$, which contains second-order partial derivatives of the loss function with respect to the parameters.

- *Loss Function:* For example, it has two parameters $\theta = [\theta_1, \theta_2]$:$L(\theta_1, \theta_2) = \theta_1^2 - \theta_2^2$
- *Gradient:* $\nabla L(\theta) = \begin{bmatrix} \frac{\partial L}{\partial \theta_1} \\ \frac{\partial L}{\partial \theta_2} \end{bmatrix} = \begin{bmatrix} 2\theta_1 \\ -2\theta_2 \end{bmatrix}$
- *Hessian Matrix:* $H(\theta) = \begin{bmatrix} \frac{\partial^2 L}{\partial \theta_1^2} & \frac{\partial^2 L}{\partial \theta_1 \partial \theta_2} \\ \frac{\partial^2 L}{\partial \theta_2 \partial \theta_1} & \frac{\partial^2 L}{\partial \theta_2^2} \end{bmatrix} = \begin{bmatrix} 2 & 0 \\ 0 & -2 \end{bmatrix}$

The eigenvalues of $H(\theta)$ are $\lambda_1 = 2$ and $\lambda_2 = -2$. Positive eigenvalue indicates convexity in the θ_1 direction. A negative eigenvalue indicates concavity in the θ_2 direction. This confirms that $L(\theta)$ has a saddle point at $\theta = [0, 0]$. Gradient descent may oscillate or make slow progress near the saddle point because the gradient points in different directions. The optimizer might need many iterations to move away from the saddle point.

Step 5. Improving Optimization with Adaptive Methods

Improving optimization with adaptive methods helps navigate complex loss landscapes that include saddle points and varying curvature. Algorithms like Adam or RMSprop are particularly effective in these scenarios. The Adam algorithm uses adaptive learning rates, adjusting the rate for each parameter based on the first and second moments of the gradients. This adjustment accelerates training by enabling larger steps in directions with small gradients, which are common near saddle points. For instance, when using Adam with default parameters, during iterations near a saddle point, the algorithm increases the effective learning rate for parameters with small gradients. This helps the optimizer move past regions where gradient descent might otherwise stall. The result is faster convergence, as the model escapes saddle points more efficiently, and greater stability, as adaptive methods handle varying curvature better, leading to more stable training.

8.3.2 Feature Space Analysis

As data passes through a neural network, it changes form, moving from raw input to increasingly abstract representations. Early layers pick up on simple features, like edges in an image, while deeper layers capture more complex patterns, shapes, objects, or even higher-level concepts. This transformation shifts the data into what's called a "feature space," a high-dimensional space where each point represents the processed version of the original input. Understanding this feature space gives us important insights into what the network has learned. In this space, data points that are similar to one another are mapped closer together, while points that are different are pushed apart. This is crucial for tasks like classification, where the goal is to group similar items and separate those that belong to different categories. By studying how data is arranged in this feature space, we can get a sense of how well the model is performing its task. A key tool in understanding the feature space is the concept of a metric tensor, which essentially measures the distances between points in the space. By calculating these distances, we can tell how well the network can differentiate between different classes of data. For instance, if points from the same class are grouped tightly together, it means the network has learned to recognize important features of that class. If points from different classes are well-separated, it suggests the network is good at distinguishing between categories. Here are some practical uses for feature space analysis. Techniques like t-SNE and Uniform Manifold Approximation and Projection (UMAP) can simplify the high-dimensional feature space into a 2D or 3D map, making it easier to see how the network is organizing the data. This lets us check whether the network is correctly grouping similar data points (such as different images of the same object) and separating different classes. By analyzing how far apart the clusters of data from different classes are, we can evaluate the network's ability to separate categories. If the clusters overlap, it might indicate problems with the model, while large separations suggest strong performance. Looking at misclassified points in the feature space can help pinpoint where the model is going wrong and how to improve it. When a network has learned features that generalize well, they can be reused for other tasks. By analyzing the feature space, we can figure out if the network's learned features can be transferred to new problems or datasets, saving time and effort by avoiding retraining from scratch. Adversarial examples, intentionally crafted inputs designed to trick a model, can be studied in the feature space. A robust model should keep real and adversarial examples far apart in this space. By examining how these inputs are mapped, we can develop strategies to protect against such attacks. Let us have an example for better understanding. Consider a simple image classification task using the MNIST dataset of handwritten digits. Each input image is a 28 × 28 pixel grid, resulting in a high-dimensional input space (784 dimensions). As the data moves through the layers of a neural network, it is transformed into different representations, with each layer extracting increasingly complex features. In the earlier layers, the network might focus on low-level features such as edges, while in the deeper layers, it learns more abstract features like digit shapes. By the time the data reaches the final layer, each input image is represented as a point in a high-dimensional feature space, where similar digits (e.g., "1"s) are mapped closer together, and dissimilar digits (e.g., "1"s and "8"s) are pushed farther apart. Let's define the feature space mathematically. Suppose $x \in R^n$ is the input (an image), and $f_\theta(x)$ is the neural network that transforms this input into a high-dimensional feature vector. After several layers of transformation, the feature vector $z = f_\theta(x)$ lies in a high-dimensional space R^d, where $d \ll n$. The distance between two feature vectors z_1 and z_2 (representing two different images) can be computed using the Euclidean distance:

$$\text{Distance}\left(z_1, z_2\right) = \left\|z_1 - z_2\right\|_2 = \sqrt{\sum_{i=1}^{d}(z_{1,i} - z_{2,i})^2}$$

This distance quantifies how similar or dissimilar the representations of two images are in the feature space. If the network has learned good features, images of the same digit (e.g., "1") will have smaller distances between their feature vectors, while images of different digits (e.g., "1" and "8") will have larger distances. Suppose we have a simple neural network trained to classify two types of data points: Class **A**: Points clustered around [2, 2] and Class **B**: Points clustered around [−2, −2]. The Class **A** data points are [2, 2], [2.1, 1.9], [1.9, 2.2], and Class **B** data points are: [−2, −2], [−1.8, −2.1], and [−2.2, −1.9]. The neural network structure is as follows:

- Input layer: 2 neurons (features x_1 and x_2)
- Hidden layer: 2 neurons with ReLU activation
- Output layer: 1 neuron with sigmoid activation (for binary classification)

The weights and biases are as follows:

- Weights from input to hidden layer (W_1): $W_1 = \begin{bmatrix} 1 & 0 \\ 0 & 1 \end{bmatrix}$
- Biases for hidden layer (b_1): $b_1 = \begin{bmatrix} 0 \\ 0 \end{bmatrix}$
- Weights from hidden to output layer (W_2): $W_2 = \begin{bmatrix} 1 \\ 1 \end{bmatrix}$
- Bias for output layer (b_2): 0

Here is a forward pass example. Let's compute the network's output for a data point from Class **A**, say [2, 2]:

- Hidden layer activation (h):

$$h = \text{ReLU}\left(W_1 \cdot x + b_1\right) = \text{ReLU}\left(\begin{bmatrix} 1 & 0 \\ 0 & 1 \end{bmatrix} \cdot \begin{bmatrix} 2 \\ 2 \end{bmatrix} + \begin{bmatrix} 0 \\ 0 \end{bmatrix}\right) = \text{ReLU}\left(\begin{bmatrix} 2 \\ 2 \end{bmatrix}\right) = \begin{bmatrix} 2 \\ 2 \end{bmatrix}$$

- Output layer activation ($\hat{y}$):

$$\hat{y} = \sigma\left(W_2^{\mathrm{T}} \cdot h + b_2\right) = \sigma\left(\begin{bmatrix}1 & 1\end{bmatrix} \cdot \begin{bmatrix}2\\2\end{bmatrix} + 0\right) = \sigma(4) \approx 0.982,$$

where $\sigma(z) = \dfrac{1}{1+\mathrm{e}^{-z}}$ is the sigmoid function. The output $\hat{y} \approx 0.982$ indicates a high probability that the input belongs to Class **A**.

8.3.3 Neural Network Generalization

Generalization refers to how well a trained model performs on new, unseen data. It's a measure of the model's usefulness; if it can't generalize beyond the data it was trained on, it won't be effective in real-world applications. Improving generalization is a key goal when building machine learning models. Interestingly, the shape of the loss landscape, which reflects how the model's performance changes with different parameter settings, has a strong link to generalization. Research shows that models tend to generalize better when the loss landscape has wide, flat regions (minima). In contrast, sharp, narrow minima often lead to models that perform well on the training data but struggle with new data. Here's how the geometry of the loss landscape relates to generalization. Flat regions in the loss landscape indicate that small changes in the model's parameters don't drastically affect its predictions. This stability means the model is less sensitive to slight variations in the data, making it more robust for new inputs. Models that find flat minima are better at capturing general patterns in the data rather than overfitting to specific examples in the training set. In practice, this results in better performance when the model encounters unseen data. Sharp, narrow minima suggest that the model is highly sensitive to its parameters. In these regions, the model's parameters are finely tuned to the training data, which may lead to excellent performance during training but poor results on new data. In these cases, even small changes in the input or the parameters can cause the model's output to vary dramatically, making it less reliable when faced with unfamiliar data. This is often a sign that the model has overfitted, memorizing the training data rather than learning general patterns. Let's have an example for better understanding. Consider two neural networks trained on the same dataset for a classification task. Network **A** converges to a flat minimum, while network **B** converges to a sharp minimum on the loss landscape. Network **A** might not achieve the lowest possible training error but performs well on unseen test data, indicating strong generalization. In contrast, network **B** achieves near-perfect performance on the training set but struggles with test data due to overfitting. Flat minima correspond to regions in the loss landscape where the loss function remains relatively constant over a broad range of parameter values. This implies that the model's predictions are stable and not overly sensitive to small changes in the parameters. Mathematically, this flatness can be captured by the eigenvalues of the Hessian matrix H, which is the second derivative of the loss function with respect to the model parameters:

$$H(\theta) = \frac{\partial^2 L(\theta)}{\partial \theta^2}.$$

For flat minima, the eigenvalues of the Hessian are small, indicating low curvature. This suggests that the model has found a stable set of parameters, making it more robust to small perturbations in the input or noise in the data. A model that converges to a flat minimum is likely to have better generalization because it has learned general patterns in the data rather than memorizing the training examples. In contrast, sharp minima are regions where the loss function changes rapidly with small parameter adjustments. This is characterized by large eigenvalues of the Hessian matrix, indicating high curvature. A model that converges to a sharp minimum is typically more sensitive to small variations in the data, meaning that a slight change in the input can lead to significant changes in the model's predictions. This sensitivity is often a sign of overfitting, where the model has learned specific details of the training data that do not generalize well to new, unseen data. Let $L(\theta)$ represent the loss function, where θ is the vector of model parameters. Generalization can be influenced by the sharpness of the minimum found during training, which is often measured by the Hessian matrix $H(\theta)$. If the Hessian's eigenvalues are small, we are in a flat region, which is beneficial for generalization. Conversely, large eigenvalues indicate sharp regions:

$$\lambda_{\max}(H) \approx 0 \quad (\text{flat region}) \quad \text{vs.} \quad \lambda_{\max}(H) \gg 0 \quad (\text{sharp region}),$$

where $\lambda_{\max}(H)$ is the largest eigenvalue of the Hessian. Suppose we are training a simple neural network to perform binary classification on a small dataset. We want to classify whether a number is even or odd.

Training Data

Input (x)	Label (y)
1	1
2	0
3	1
4	0

Validation Data

Input (x)	Label (y)
5	1
6	0

We will compare two models with the same architecture but different training outcomes: Model **A** to find a flat minimum in the loss landscape and Model **B** to Find a sharp minimum in the loss landscape. The loss function is mean squared error (MSE) and the architecture is as follows: input layer; 1 neuron and output layer; 1 neuron with Sigmoid activation (outputs probability). Let do train the models:

1. *Model* **A** *(Flat Minimum):* Uses **L2** regularization to prevent overfitting. The training outcomes are weight (w): 0.5 and bias (b): 0.
2. *Model* **B** *(Sharp Minimum):* No regularization; the model may overfit the training data and the training outcomes are weight (w): 10 and bias (b): –25.

Now, let us do model predictions

1. *Model* **A** *Predictions:* Using $\hat{y} = \sigma(wx + b)$, where σ is the sigmoid function.
   a. Training Data:
      - For $x = 1$: $\hat{y} = \sigma(0.5 \times 1 + 0) = \sigma(0.5) \approx 0.622$
      - For $x = 2$: $\hat{y} = \sigma(0.5 \times 2 + 0) = \sigma(1.0) \approx 0.731$

      Predictions are around 0.6–0.7, not exactly matching labels but reasonable.
   b. Validation Data:
      - For $x = 5$: $\hat{y} = \sigma(0.5 \times 5 + 0) = \sigma(2.5) \approx 0.924$
      - For $x = 6$: $\hat{y} = \sigma(0.5 \times 6 + 0) = \sigma(3.0) \approx 0.953$

      Predictions are high, indicating even numbers, which may not perfectly match validation labels.

2. *Model* **B** *Predictions:*
   a. Training Data:
      - For $x = 1$: $\hat{y} = \sigma(10 \times 1 - 25) = \sigma(-15) \approx 3 \times 10^{-7}$
      - For $x = 2$: $\hat{y} = \sigma(10 \times 2 - 25) = \sigma(-5) \approx 0.0067$

      The model outputs are extremely low or high, closely matching labels.
   b. Validation Data:
      - For $x = 5$: $\hat{y} = \sigma(10 \times 5 - 25) = \sigma(25) \approx 1.0$
      - For $x = 6$: $\hat{y} = \sigma(10 \times 6 - 25) = \sigma(35) \approx 1.0$

      Predictions are very high, indicating odd numbers, which is incorrect for $x = 6$.

In flat minimum (Model **A**), the loss doesn't change drastically with small changes in weights, and the model is not overly sensitive to exact weight values. In sharp minimum (Model **B**), small changes in weights cause a significant loss increase, and the model memorizes training data but doesn't generalize well. But for generalization performance, Model **A**, the training loss is slightly higher due to less precise fitting, and validation accuracy is better, as the model makes reasonable predictions on unseen data on Model **B**, the training loss is very low, almost zero, the model fits training data perfectly, and validation accuracy is poor, as the model fails to predict correct labels on new data (e.g., misclassifies $x = 6$). The model's performance is stable under small data or parameter changes and gives better generalization as handling unseen data more effectively. On sharp minima, sensitivity, and minor variations can lead to large errors and overfitting, and it performs well on training data but poorly on new data.

8.3.4 Information Geometry

Information geometry is a fascinating intersection of probability theory and differential geometry. It studies random variables and probability distributions using geometric structures. One prominent concept from information geometry is the Fisher Information Metric. In deep learning, this metric helps us understand the sensitivity of a model's predictions to its parameters. In Bayesian deep learning, where we consider a distribution over neural network weights instead of fixed values, information geometry plays a crucial role in understanding the model's uncertainty and guiding the learning process. Imagine a neural network trained for image classification. Instead of treating the weights of the network as fixed values, we treat them as random variables, representing our uncertainty about the optimal weights. In this probabilistic framework, Information Geometry helps us understand the geometry of the parameter space in terms of probability distributions. The Fisher Information Metric is a key tool here, as it measures how sensitive the model's predictions are to small changes in the weights. The Fisher Information Metric defines a Riemannian metric on the space of probability distributions. It essentially measures how much information a random variable carries about the unknown parameters of a distribution. In the context of deep learning, it helps quantify how much influence a small change in the model's parameters θ has on the predicted probability distribution. Mathematically, if $p(x|\theta)$ is the probability distribution of the data x given the parameters θ, the Fisher Information Matrix $I(\theta)$ is defined as:

$$I(\theta) = \mathbb{E}\left[\left(\frac{\partial \log p(x|\theta)}{\partial \theta}\right)\left(\frac{\partial \log p(x|\theta)}{\partial \theta}\right)^T\right],$$

where:

- $p(x \mid \theta)$ is the likelihood function.
- The expectation **E** is taken with respect to the data distribution.

This matrix tells us how sensitive the likelihood is to changes in the parameters. If the Fisher Information Matrix has large values, it indicates that small changes in the parameters will significantly affect the likelihood, suggesting that the model is highly sensitive to those parameters.

In Bayesian deep learning, where we model a distribution over the neural network weights instead of using fixed weights, information geometry provides insights into uncertainty. The Fisher Information Metric can guide how the weight distribution is updated during training. Instead of taking fixed steps in parameter space, we adjust the steps based on the geometry of the space, leading to more informed updates. Consider the Fisher Information Matrix as a measure of the local geometry around a point in parameter space. In natural gradient descent, the Fisher Information Matrix is used to scale the parameter updates in a way that takes into account the underlying geometry of the probability distributions. The natural gradient update is given by:

$$\theta_{t+1} = \theta_t - \eta I(\theta_t)^{-1} \nabla_\theta L(\theta_t),$$

where:

- $I(\theta_t)^{-1}$ is the inverse of the Fisher Information Matrix,
- $\nabla_\theta L(\theta_t)$ is the gradient of the loss function,
- η is the learning rate.

This approach ensures that parameter updates are adapted to the sensitivity of the model to changes in the parameters, improving convergence, especially in high-dimensional or complex models.

Figure 8.5 provides a comparative visualization of a loss landscape Figure 8.5a and feature space clusters Figure 8.5b to demonstrate how different parameter configurations and feature distributions impact classification and optimization in machine learning. In Figure 8.5a, the 3D loss landscape illustrates how the value of the loss function changes across different parameter settings. The surface is colored using a gradient from blue (low loss) to red (high loss), indicating convex, concave, and saddle regions as the parameter values vary. The contour lines highlight the elevation levels, making it easier to visualize the gradient flow and pathways toward minima. The convex region (highlighted in red) represents areas where the function value is higher, typically seen in regions away from the global minimum. The concave region (blue) indicates a valley where the loss value is minimized. The green dot marks a saddle point, where the surface curves upward in one direction and downward in another, demonstrating the challenges faced during optimization when traversing such regions. Figure 8.5b presents a 2D scatter plot of feature space, showing two clusters corresponding to different classes: Class 1 (orange circles) and Class 2 (blue squares). The clusters are well-separated, indicating that the features provide enough distinction for classification. Arrows point to the centers of each class cluster, demonstrating the centroids where the mean feature values for each class lie.

8.4 PRACTICAL IMPLICATIONS

8.4.1 Regularization

Building on our earlier example of Model **A**, which converges to a flat minimum, and Model **B**, which settles into a sharp minimum, regularization techniques play a crucial role in guiding the optimization process toward flatter regions of the loss landscape. To explore these concepts, we begin by creating a simple graph with three nodes arranged in a chain structure. The adjacency matrix is normalized and converted into a sparse tensor format suitable for processing by a Graph Neural Network (GCN). Each node's features are represented by an identity matrix, and the labels are encoded using one-hot vectors for two distinct classes. After training the GCN, we predict the node

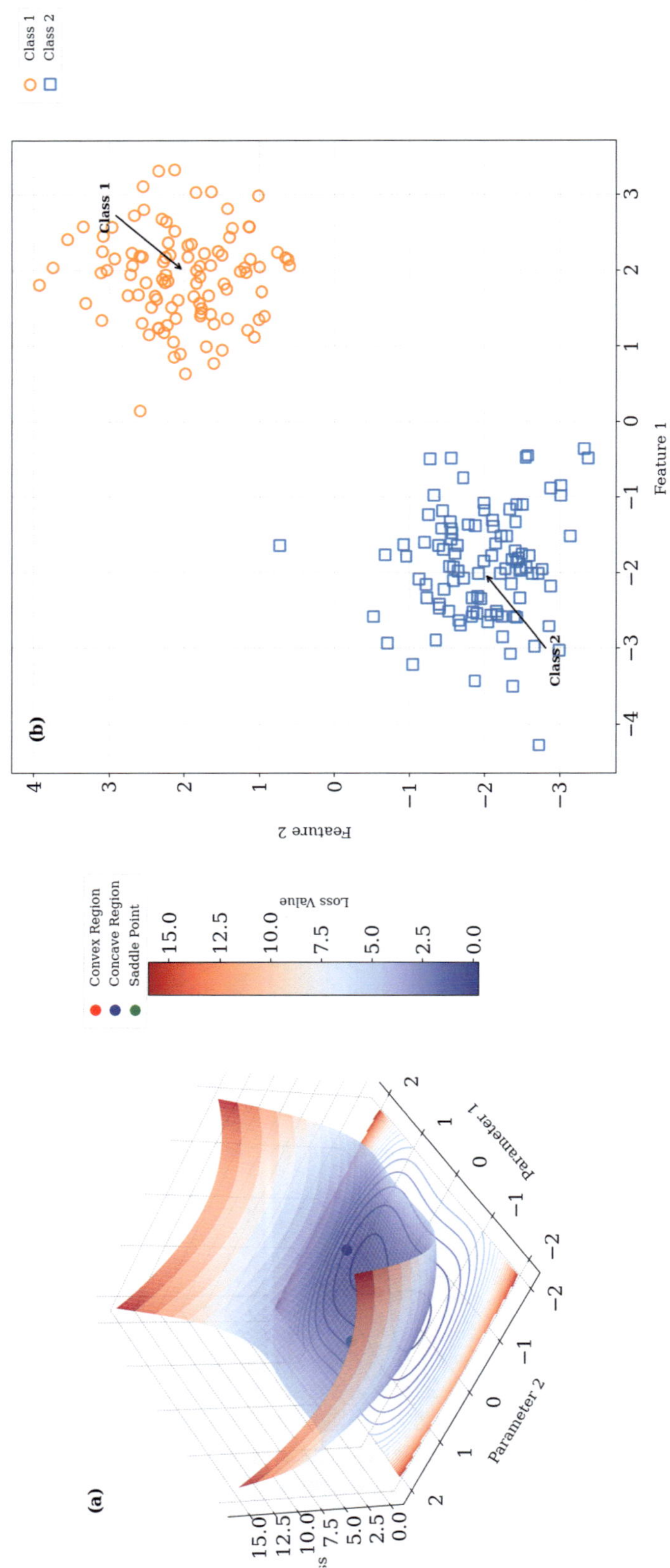

FIGURE 8.5 (a) Loss landscape, and (b) feature space clusters.

classes and visualize the graph with nodes colored based on their predicted classes. To further demonstrate manifold learning, we generate a Swiss roll dataset with 1,000 samples and a small amount of noise. We apply t-SNE and UMAP, two popular dimensionality reduction techniques, to project the high-dimensional Swiss roll data into two dimensions. Additionally, we compute the pairwise distance matrix, derive the Laplacian matrix, and perform singular value decomposition (SVD) to understand the data's geometric structure.

Figure 8.6 provides a visual exploration of different data transformation and dimensionality reduction techniques. Figure 8.6a shows node classification results from a graph convolutional network (GCN). The blue dots likely represent nodes in a graph that have been classified into distinct categories after training the model. This visualization demonstrates how the GCN has grouped nodes based on their learned features. In Figure 8.6b, we see the popular 3D Swiss roll dataset, which is often used to test algorithms that deal with nonlinear data structures. The Swiss roll is a twisted 3D manifold that provides a challenging structure for machine learning algorithms to untangle and understand. Figure 8.6c and d display the results of applying two different dimensionality reduction techniques to the Swiss roll dataset. In Figure 8.6c, the t-SNE method has been used to reduce the high-dimensional data into a 2D space. t-SNE focuses on preserving the local relationships between points, meaning points that were close in the original high-dimensional space are still close in this 2D projection. In Figure 8.6d, the UMAP method has been applied to the same dataset. UMAP, like t-SNE, reduces high-dimensional data into two dimensions but often does so more efficiently, preserving both local and global relationships between points.

8.4.2 Optimization

Optimizing deep neural networks involves finding effective minima within the complex, high-dimensional loss landscapes they inhabit. Differential geometry provides powerful tools to better understand and navigate these landscapes, leading to improved optimization strategies. Saddle points are locations in the loss landscape where the curvature changes direction; they are neither purely convex nor concave. In high-dimensional spaces, these points can significantly delay the training process, causing slow convergence. At saddle points, the curvature of the loss landscape is positive in some directions and negative in others. This variation in curvature means that conventional optimization algorithms may struggle, as they might incorrectly interpret the local geometry. By considering curvature information, optimization algorithms can be designed to rapidly escape saddle points. Techniques such as second-order optimization methods or algorithms incorporating curvature estimations can effectively navigate away from these problematic regions. Understanding the geometry of the loss landscape can inform the design of optimization techniques that achieve faster convergence. By adapting to the local curvature of the landscape, these methods can make more informed updates to the model parameters. Metrics derived from differential geometry, such as the curvature of the loss surface, can guide the adjustment of step sizes. This adaptive approach helps maintain an efficient trajectory through the landscape, avoiding areas of slow progress and making the optimization process more robust. Algorithms like Adaptive Moment Estimation (Adam) exemplify this approach. Adam adjusts the optimization trajectory based on the geometry of the loss landscape, effectively using curvature information to adapt learning rates. This results in more efficient and stable convergence, even in the presence of varying curvature. By leveraging the insights provided by differential geometry, deep learning practitioners can develop optimization algorithms that not only avoid saddle points but also achieve faster convergence. This dual focus on geometry-aware optimization techniques leads to more efficient training processes and improved performance of deep neural networks.

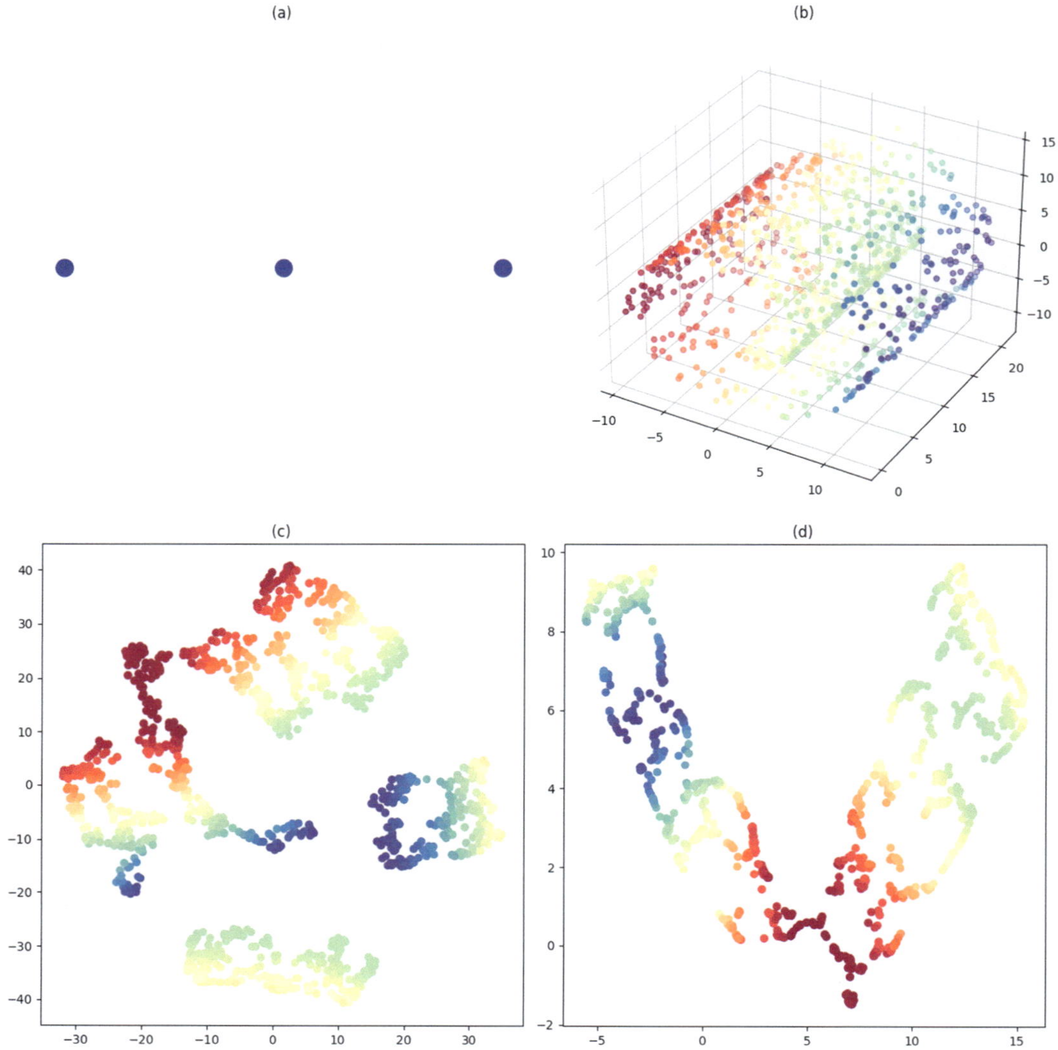

FIGURE 8.6 (a) Node classification results (GCN), (b) 3D Swiss roll dataset, (c) t-SNE embedding, (d) UMAP embedding.

8.4.3 MODEL INTERPRETABILITY

As neural networks become more complex, understanding their decision-making process becomes crucial. Deep neural networks transform input data into abstract representations in hidden layers known as feature spaces. The geometry of these feature spaces can provide insights into the relationships between different inputs and the learned features. For example, points close in the feature space might be similar in a meaningful way. Techniques like t-SNE, a dimensionality reduction method, can visualize these high-dimensional feature spaces in 2D or 3D. It tries to preserve the local structure, allowing us to see clusters of similar data points. By understanding the geometry of feature spaces, we can gain insight into what features the network considers essential. The curvature or shape of these spaces at different layers can tell us about the complexity and hierarchy of the learned features. Activation maximization or feature inversion techniques can be employed to study the feature spaces. By examining what excites the neurons the most, we can visualize the patterns or features the network has learned. This approach helps in understanding the network's learning process and the importance of various features in decision-making.

Figure 8.7 illustrates the practical implications of differential geometry in deep learning: The left plot shows a loss landscape with sharp (blue) and flat (red) minima. Penalizing regions of the loss landscape with high curvature (sharp minima) can guide the training process toward flatter areas, leading to better generalization. The middle plot shows a lost landscape with a saddle point (green). Saddle points can slow down the training process as the loss landscape is not purely convex or concave at these points. By understanding the curvature of the loss landscape, optimization algorithms can be designed to quickly escape from saddle points, improving the overall training efficiency. The right plot shows the feature space clusters for two classes using t-SNE. Data progresses through a neural network and is transformed into various representations in hidden layers. The clusters represent how the data for each class is grouped in the feature space, providing insights into what the network has learned.

8.5 CHALLENGES

8.5.1 High Dimensionality

Deep neural networks can have millions, if not billions, of parameters. This results in a highly high-dimensional parameter space, which introduces a significant level of complexity when trying to understand or visualize the behavior of the network. Differential geometric concepts like curvature, tangent spaces, and manifolds become increasingly complex in high dimensions, making their study and application non-trivial. The challenges in high-dimensional spaces are as follows.

8.5.1.1 Curvature

In high-dimensional spaces, curvature becomes a multi-faceted concept that is challenging to compute and interpret. Curvature can vary dramatically in different directions, making the loss landscape highly intricate. Understanding curvature in high-dimensional spaces is crucial for optimization. Techniques that use second-order information, such as Newton's method, rely on understanding the curvature to adjust the optimization path. However, computing and storing the Hessian matrix (which contains second-order derivatives) becomes impractical for large networks due to their size and computational cost. Tangent spaces provide a local linear approximation of the manifold at a point. In high dimensions, the concept of a tangent space helps in understanding local behavior but becomes harder to visualize and compute. The gradient, which lies in the tangent space, guides the optimization process. In high-dimensional settings, the gradient can point in complex directions, and small changes can have significant impacts, making the optimization path difficult to predict and control. Data and parameter spaces in neural networks often form high-dimensional manifolds. Understanding the structure and geometry of these manifolds is crucial for tasks like optimization, regularization, and generalization. Techniques like PCA, t-SNE, and UMAP attempt to reduce high-dimensional manifolds to lower dimensions for visualization and analysis. However, these techniques can lose important structural details, making it challenging to capture the true complexity of the manifold.

8.5.2 Visualization

One of the most discussed topics about differential geometry in deep learning is the loss landscape. However, visualizing a high-dimensional loss landscape is inherently challenging. Simple visualizations, like 2D or 3D plots, provide only a limited view, which might not capture the intricacies of the loss landscape of deep networks. While techniques like dimensionality reduction can help, they may not retain all the essential geometric features of the original space. The challenges

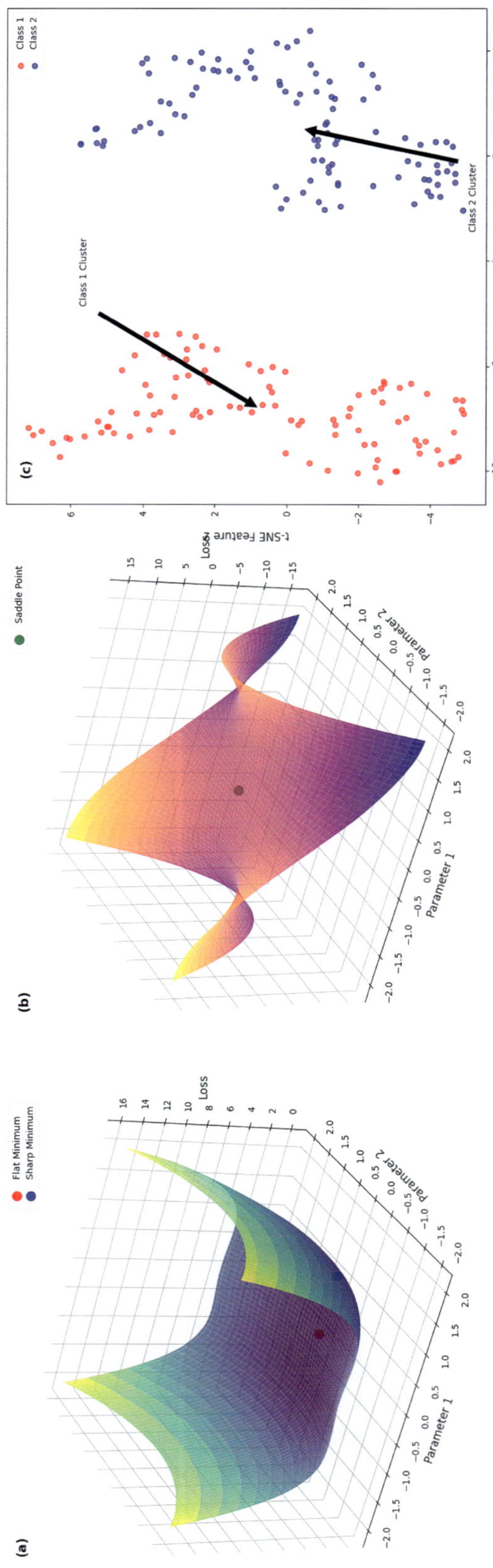

FIGURE 8.7 (a) Regularization: penalizing sharp minima, (b) optimization: avoiding saddle points, and (c) model interpretability: feature space clusters.

in visualizing high-dimensional loss landscapes. Traditional visualizations are limited to two or three dimensions, whereas the loss landscape in deep learning is typically situated in a parameter space with millions or billions of dimensions. These simple visualizations can only provide a very narrow perspective on the true complexity of the landscape. Projecting high-dimensional data into lower dimensions often leads to a loss of important structural information. Critical features such as the curvature and the arrangement of minima and saddle points might not be accurately represented. Methods like PCA, t-SNE, and UMAP attempt to reduce the dimensionality of the data while preserving its most significant features. While these techniques can highlight certain aspects of the loss landscape, they might fail to retain all the essential geometric properties. For example, they might distort distances or lose information about the local curvature and the shape of valleys and ridges.

8.5.3 Computational Cost

Computing specific differential geometric properties, such as curvature, for high-dimensional spaces can be computationally expensive. In a training loop where computations need to be efficient, introducing these calculations might significantly slow down the training process. This high computational cost can make the application of these concepts infeasible for extensive models or large datasets. Challenges of computational cost are as follows. Calculating curvature involves second-order derivatives, such as those found in the Hessian matrix. For high-dimensional spaces, the Hessian is extremely large, making its computation and storage impractical. The time complexity of computing the Hessian matrix scales quadratically with the number of parameters, resulting in significant computational overhead for large networks. In practice, training loops are designed to be as efficient as possible to handle large datasets and complex models. Introducing additional computations for differential geometric properties can hinder the overall training speed. Modern deep learning relies on batch processing to optimize resource utilization. Incorporating curvature computations within each batch can disrupt this balance and slow down the training process significantly.

8.5.4 Theoretical vs. Practical Gap

Differential geometry is a mathematical field that deals with abstract structures and concepts. In the context of deep learning, there is often a gap between theoretical insights and practical implementation. While differential geometric properties such as curvature can provide valuable insights into the behavior of neural networks, translating these insights into concrete steps for training and optimizing models can be challenging. The challenges of the theoretical vs. practical gap are as follows. Differential geometric concepts like curvature, manifolds, and tangent spaces are inherently complex and abstract. Understanding these concepts requires a solid foundation in advanced mathematics, which can be a barrier for many practitioners. While these concepts can offer deep intuition about the behavior of neural networks, applying them in a practical setting, such as during the training of a model, is not always straightforward. Theoretical insights suggest that regions with flat minima in the loss landscape correlate with better generalization. However, designing algorithms or modifying training procedures to explicitly find these flat regions is challenging. While understanding the curvature of the loss landscape can guide the development of optimization algorithms, implementing these ideas in a way that is computationally efficient and effective for large-scale neural networks is non-trivial.

8.5.5 Scalability

Modern deep learning often involves massive models and huge datasets. Even if specific differential geometric tools prove beneficial for smaller problems, scaling them to handle state-of-the-art models and datasets can be challenging. The ability to apply these tools efficiently at scale is critical for their practical adoption in contemporary deep learning applications. The challenges of scalability are as follows. State-of-the-art deep learning models can have millions or even billions of parameters. The sheer size of these models makes computing differential geometric properties, such as curvature or the Hessian matrix, computationally intensive and often impractical. Storing and manipulating large parameter sets and their associated geometric properties require substantial memory, which can exceed the capacity of even high-end hardware. Training on large datasets involves processing vast amounts of data in each training iteration. Integrating differential geometric computations into this process can significantly slow down the training pipeline. Efficient training relies on processing data in batches. Adding complex geometric calculations to each batch increases the computational overhead, making it difficult to maintain the required throughput for timely model training.

8.5.6 Emerging Insights

Deep learning research is vibrant and ever-evolving, with new architectures, techniques, and best practices emerging regularly. As the field progresses, the relevance and applicability of specific differential geometric insights might change. Keeping geometric tools and analyses updated and relevant in this dynamic environment is a constant challenge. The challenges of emerging insights are as follows. The introduction of new neural network architectures, such as transformers, graph neural networks, and neural ordinary differential equations (ODEs), can shift the focus of geometric analysis. Methods that were effective for earlier architectures may need adaptation or re-evaluation for these newer models. Advances in training techniques, such as self-supervised learning, transfer learning, and meta-learning, introduce new dynamics in the loss landscape and optimization process, impacting the application of geometric insights. The geometric tools and methods must evolve alongside advancements in deep learning. This requires ongoing research to refine and adapt these tools to new contexts and challenges. Collaboration between mathematicians, computer scientists, and domain experts is essential to ensure that geometric tools remain relevant and effective for emerging deep learning paradigms.

The plots in Figure 8.8 illustrate some of the challenges of applying differential geometry concepts to deep learning: The left plot shows a simplified high-dimensional loss landscape in 2D. This highlights the difficulty of understanding and visualizing complex, high-dimensional, deep-learning concepts. The middle plot shows the impact of additional computational cost on training time. The shaded area represents the additional time required for these computations. The right plot illustrates the gap between theoretical insights (green) and practical impact (orange) on training due to curvature.

8.6 REAL-WORLD APPLICATIONS

8.6.1 Autonomous Systems and Robotics

In the realm of robotics, differential geometry is instrumental in path planning and control. Autonomous systems, such as drones or self-driving cars, must navigate through complex environments while making real-time decisions. The concept of manifolds, which represent the possible states or configurations of a system, is crucial in this context. For instance, the configuration space of a robot arm, which includes all possible positions and orientations, can be modeled as a manifold. Understanding the curvature and topology of this space allows the robot to plan efficient and collision-free paths. This geometric insight is particularly valuable in dynamic environments, where the robot must adapt to changing conditions while maintaining optimal performance.

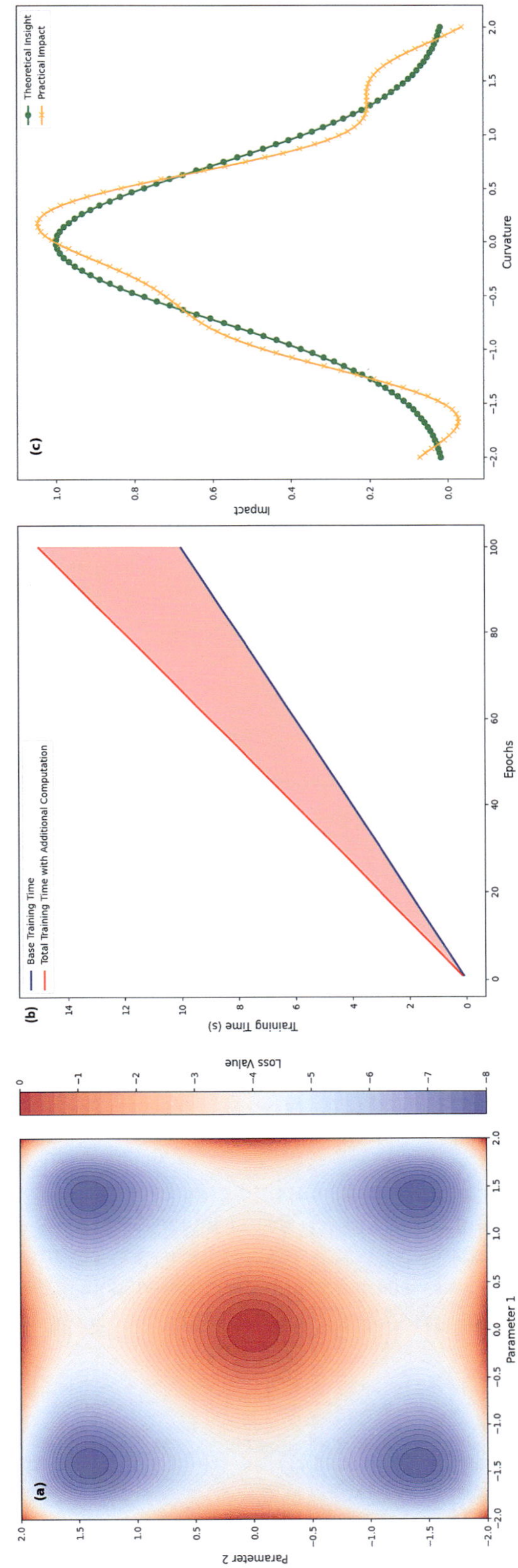

FIGURE 8.8 (a) High dimensionality: simplified loss landscape, (b) computational cost impact on training time, and (c) theoretical vss. practical gap.

8.6.2 Medical Image Analysis

Differential geometry is also pivotal in the analysis of medical images, where it aids in the segmentation and interpretation of complex anatomical structures. In medical imaging, the surfaces of organs or tissues can be modeled as manifolds, and the curvature of these surfaces provides critical information about their shape and structure. For instance, in brain imaging, the cortex can be represented as a 2D manifold embedded in 3D space. By analyzing the curvature of the cortical surface, researchers can detect abnormalities such as tumors or atrophy, which are indicative of neurological disorders. The ability to model and analyze these geometric properties enables more accurate diagnoses and better treatment planning.

8.6.3 Computer Vision and Image Recognition

In computer vision, differential geometry provides the mathematical foundation for understanding how images are processed and recognized by neural networks. The concept of manifolds is particularly important in image recognition tasks, where the high-dimensional space of all possible images is often constrained to a lower-dimensional manifold that captures the essential features of specific objects. For example, face recognition systems rely on the fact that images of the same person under different conditions (e.g., lighting, pose) lie on a manifold in the space of all possible images. By learning the geometry of this manifold, the system can accurately identify individuals across a wide range of variations. This approach has been successfully applied in various domains, from security systems to social media platforms.

8.6.4 Signal Processing and Communications

Differential geometry is also used in signal processing, where it helps in the analysis and compression of complex signals. For instance, in wireless communications, signals transmitted through the air can be affected by the curvature of the Earth's surface and other obstacles. By modeling the signal propagation using geometric principles, engineers can design more efficient communication systems that minimize interference and maximize data transmission rates.

8.7 HANDS-ON SECTION

In this hands-on section, we will explore the concepts of manifold learning and curvature in high-dimensional spaces using differential geometry.

8.7.1 Step 1: Import Libraries

In this part of the code, we are installing and importing the necessary packages for dimensionality reduction and visualization. The command!pip install umap-learn scikit-learn matplotlib ensures that the required libraries are installed. After installation, we import key functions and classes. These tools are essential for exploring and visualizing high-dimensional data in a lower-dimensional space, making it easier to understand patterns, clusters, and relationships in the data.

```
# Install necessary packages
!pip install umap-learn scikit-learn matplotlib
import numpy as np
import matplotlib.pyplot as plt
from sklearn.datasets import make_swiss_roll
```

```
from sklearn.decomposition import PCA
from sklearn.manifold import Isomap, TSNE
import umap
```

8.7.2 Step 2: Set Random Seed for Reproducibility

In this section, we are generating a synthetic dataset known as the Swiss roll using make_swiss_roll from scikit-learn. The Swiss roll is a common 3D dataset used to test and demonstrate dimensionality reduction techniques because it has a complex, curved structure that is challenging for linear methods to handle.

```
np.random.seed(42)
# Generate the Swiss Roll dataset
n_samples = 1000
noise = 0.05
X, color = make_swiss_roll(n_samples, noise=noise)
```

8.7.3 Step 3: Apply PCA for Dimensionality Reduction

In this section, we are applying four different dimensionality reduction techniques to project the high-dimensional Swiss roll dataset into 2D space. Dimensionality reduction is crucial for visualizing complex datasets in lower dimensions, making patterns or clusters easier to interpret.

```
pca = PCA(n_components=2)
X_pca = pca.fit_transform(X)
# Apply Isomap for Dimensionality Reduction
isomap = Isomap(n_components=2, n_neighbors=10)
X_isomap = isomap.fit_transform(X)
# Apply t-SNE for Dimensionality Reduction
tsne = TSNE(n_components=2, random_state=42)
X_tsne = tsne.fit_transform(X)
# Apply UMAP for Dimensionality Reduction
try:
  umap_reducer = umap.UMAP(random_state=42)
  X_umap = umap_reducer.fit_transform(X)
except Exception as e:
  print(f"Error with UMAP: {e}")
  X_umap = np.zeros((X.shape[0], 2))   # Fallback to prevent
  plot from being empty
```

8.7.4 Step 4: Plotting All Graphs in One Frame

In this section, we are visualizing the results of dimensionality reduction techniques applied to the Swiss roll dataset in a series of subplots.

```
fig, axs = plt.subplots(2, 3, figsize=(18, 12))
# Plot a: Swiss Roll Dataset in 3D
ax = fig.add_subplot(2, 3, 1, projection='3d')
ax.scatter(X[:, 0], X[:, 1], X[:, 2], c=color, cmap=plt.
cm.Spectral)
ax.set_title("a) 3D Swiss Roll Dataset")
# Plot b: PCA projection
axs[0, 1].scatter(X_pca[:, 0], X_pca[:, 1], c=color, cmap=plt.
cm.Spectral)
axs[0, 1].set_title("b) PCA Projection")
axs[0, 1].set_xlabel("PCA 1")
axs[0, 1].set_ylabel("PCA 2")
# Plot c: Isomap embedding
axs[0, 2].scatter(X_isomap[:, 0], X_isomap[:, 1], c=color,
cmap=plt.cm.Spectral)
axs[0, 2].set_title("c) Isomap Embedding")
axs[0, 2].set_xlabel("Isomap 1")
axs[0, 2].set_ylabel("Isomap 2")
# Plot d: t-SNE embedding
axs[1, 0].scatter(X_tsne[:, 0], X_tsne[:, 1], c=color, cmap=
plt.cm.Spectral)
axs[1, 0].set_title("d) t-SNE Embedding")
axs[1, 0].set_xlabel("t-SNE 1")
axs[1, 0].set_ylabel("t-SNE 2")
# Plot e: UMAP embedding
if X_umap.any():
   axs[1, 1].scatter(X_umap[:, 0], X_umap[:, 1], c=color,
cmap=plt.cm.Spectral)
else:
   axs[1, 1].text(0.5, 0.5, 'UMAP Failed', horizontalalignment=
'center', verticalalignment='center')
axs[1, 1].set_title("e) UMAP Embedding")
axs[1, 1].set_xlabel("UMAP 1")
axs[1, 1].set_ylabel("UMAP 2")
```

Figure 8.9 presents a comparative analysis of various dimensionality reduction techniques applied to the Swiss roll dataset. Figure 8.9a depicts the original 3D Swiss roll, illustrating its complex and nonlinear structure. Figure 8.9b shows the results of **PCA**, a linear technique, which flattens the data and loses much of the intrinsic geometric structure of the manifold, revealing only the global variance. Figure 8.9c displays the Isomap embedding, which successfully captures the underlying manifold by preserving geodesic distances, offering a more accurate low-dimensional representation. Figure 8.9d presents the **t-SNE** embedding, which excels in maintaining local relationships and clusters within the data but might distort global distances, offering a view of fine-grained structures. Figure 8.9e shows the **UMAP** embedding, which provides a balanced representation that preserves both local and global structures, giving a clear and meaningful visualization of the data's manifold structure.

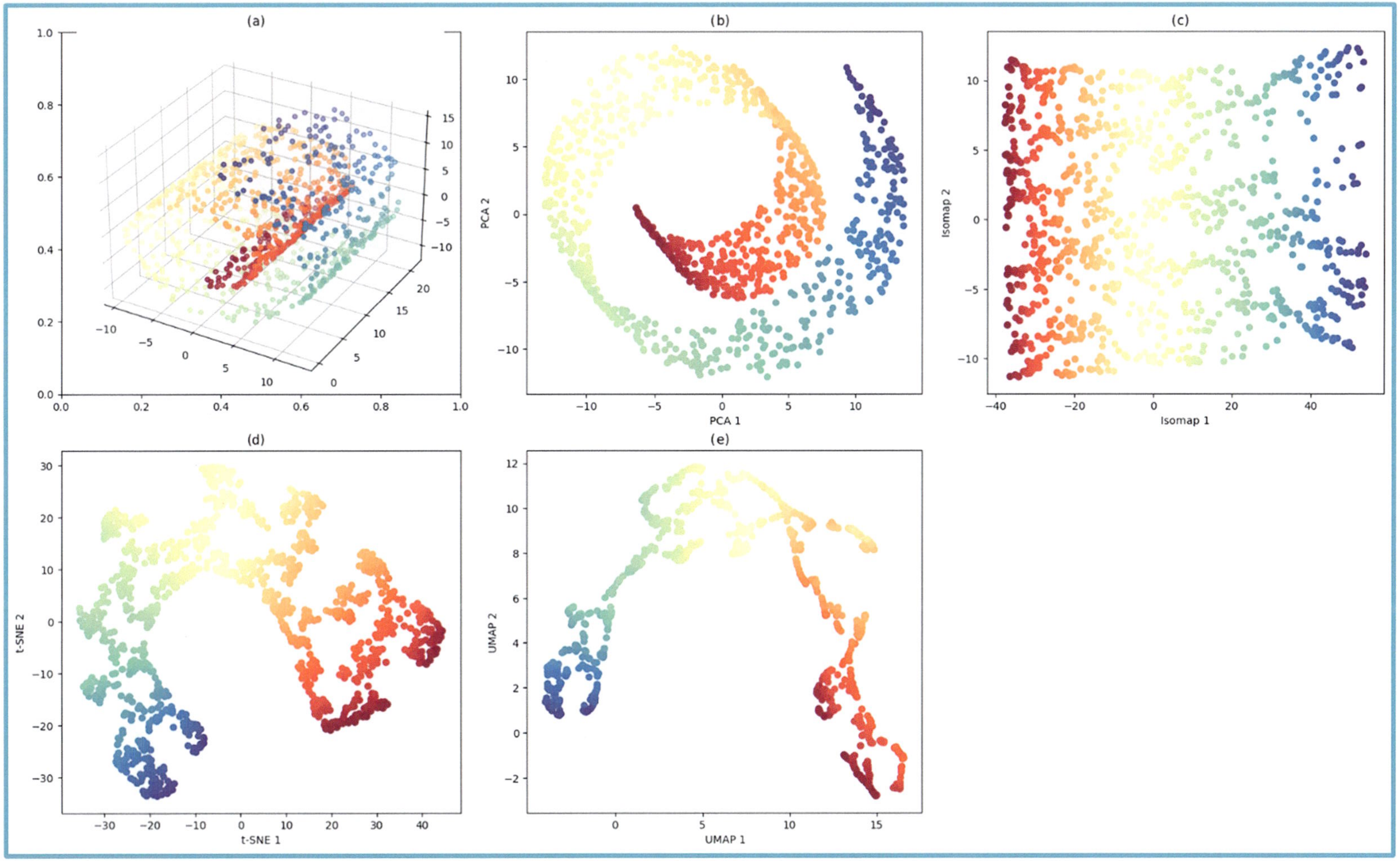

FIGURE 8.9 Comparing dimensionality reduction techniques on the Swiss roll dataset. (a) 3D Swiss roll dataset, (b) PCA projection, (c) Isomap embedding, (d) t-SNE embedding, and (e) UMAP embedding.

8.8 COMMON MISTAKES AND TROUBLESHOOTING TIPS

8.8.1 Misinterpreting Geometric Concepts

- *Mistake:* Misunderstanding fundamental geometric concepts like manifolds, tangent spaces, and curvature can lead to incorrect assumptions and implementations.
- *Tip:* Review basic differential geometry textbooks and online resources. Visual aids and interactive tools can help solidify understanding.

8.8.2 Visualizing High-Dimensional Spaces

- *Mistake:* Attempting to visualize high-dimensional loss landscapes or feature spaces without appropriate techniques can result in misleading interpretations.
- *Tip:* Use dimensionality reduction techniques like **t-SNE** or **PCA** to visualize high-dimensional data. Be aware of the limitations and what information might be lost during dimensionality reduction.

8.8.3 Ignoring Curvature in Optimization

- *Mistake:* Neglecting the curvature of the loss landscape during optimization can lead to poor convergence or getting stuck at saddle points.
- *Tip:* Implement optimization algorithms that consider curvature, such as those based on second-order derivatives or adaptive learning rates. Regularly evaluate and adjust these methods based on empirical performance.

8.8.4 Overfitting and Generalization

- *Mistake:* Focusing too much on achieving low training loss without considering the geometry of the loss landscape can lead to overfitting.
- *Tip:* Use regularization techniques that penalize sharp minima. Monitor validation performance closely and use early stopping to prevent overfitting.

8.8.5 Computational Overhead

- *Mistake:* Incorporating complex differential geometric calculations without considering computational cost can slow down training significantly.
- *Tip:* Balance the computational cost with the benefits. Use approximations or heuristics where possible to reduce the computational burden.

8.8.6 Bridging Theory and Practice

- *Mistake:* Failing to translate theoretical insights from differential geometry into practical applications can limit the usefulness of these concepts.
- *Tip:* Focus on practical implementations and empirical validation. Start with simple models and gradually incorporate more complex geometric insights as you gain confidence and understanding.

8.8.7 Handling Large-Scale Models

- *Mistake:* Applying techniques that work well on small models to large-scale models without considering scalability issues can lead to inefficiencies.

- *Tip:* Test on smaller subsets of your data and incrementally scale up. Use distributed computing and parallel processing techniques to handle large-scale models effectively.

8.9 REVIEW QUESTIONS

1. What are the fundamental concepts of differential geometry, and how do they apply to neural networks?
2. What is the difference between flat and sharp minima, and why are flat minima preferred for better generalization?
3. How do neural networks transform input data into high-dimensional feature spaces?
4. What insights can be gained from analyzing the geometry of these feature spaces?
5. Why is understanding the curvature of the loss landscape crucial for improving a model's generalization to unseen data?
6. What is the Fisher information metric, and how does it help understand the sensitivity of a model's predictions?
7. How does information geometry apply to Bayesian deep learning?
8. What is the impact of penalizing sharp minima on the training and generalization of neural networks?
9. What are saddle points, and why is it essential to design optimization techniques that avoid them?
10. How does the geometry of feature spaces enhance the interpretability of neural networks?

8.10 PROGRAMMING QUESTIONS

8.10.1 Easy: Implementing Basic Manifold Learning

1. Generate a synthetic high-dimensional dataset.
2. Apply PCA to reduce the dimensions to 2.
3. Visualize the reduced dataset using a scatter plot.

8.10.2 Medium: Comparing Manifold Learning Techniques

1. Use a complex dataset such as MNIST or CIFAR-10.
2. Apply t-SNE and UMAP separately to reduce the dimensions to 2.
3. Visualize the results side by side and compare their ability to preserve the structure.

8.10.3 Hard: Analyzing Curvature in High-Dimensional Dataset

1. Implement Laplacian eigenmaps to reduce the dimensions of a high-dimensional dataset.
2. Calculate the Laplacian matrix and perform eigenvalue decomposition.
3. Visualize the eigenvalues and analyze the curvature based on their distribution.

9 Topology in Deep Learning

9.1 INTRODUCTION

In the vast field of deep learning, the design and layout of neural networks are crucial in shaping how they work and how effective they are. This design, often referred to as "topology" in deep learning, is the base that allows algorithms to learn and improve. However, the word "topology" isn't used only for neural networks. It originally comes from a detailed area of mathematics that studies the properties of spaces that stay the same even when they are stretched or bent without breaking. This link between deep learning and mathematical topology helps us better understand how neural network algorithms come together and behave. This chapter reviews this concept in more detail.

9.2 BASIC TOPOLOGY

Topology is a part of mathematics that looks at the ways spaces can stay the same even when they are changed in certain ways, especially smooth and continuous changes. It began by trying to understand the basic nature of shapes and spaces, focusing on important features rather than minor details. In the end, although the word "topology" is used differently in pure mathematics and deep learning, both areas are all about structure and the properties that come from that structure.

9.2.1 Continuous Transformations and Invariance

A key idea in topology is the concept of homeomorphism: a continuous, one-to-one, onto map whose inverse is also continuous. Intuitively, a homeomorphism lets you bend or stretch one shape into another without cutting or attaching new pieces. If two shapes are connected by a homeomorphism, they are considered the same in topology, or "homeomorphic." This concept ignores measures like distances or angles, focusing instead on the fundamental properties that remain unchanged under continuous deformations. For example, topologically, a circle can be deformed into an ellipse without breaking or merging parts, so they are homeomorphic. A classic illustration is the coffee mug and the donut, each has exactly one hole, making them topologically equivalent despite looking quite different in everyday terms.

Figure 9.1a shows a circle (solid blue line) and an ellipse (dashed green line), which can be stretched or compressed into each other without cutting or gluing. This ability to transform smoothly shows that the circle and ellipse are topologically equivalent, they share the same fundamental structure in topology. Figure 9.1b illustrates a 3D view of a torus (donut shape) characterized by a single hole. This hole is a key topological feature that distinguishes it from shapes like the circle or ellipse, which have none. Together, these plots demonstrate how topology focuses on properties like connectedness and the number of holes rather than exact shapes.

DOI: 10.1201/9781032690742-9

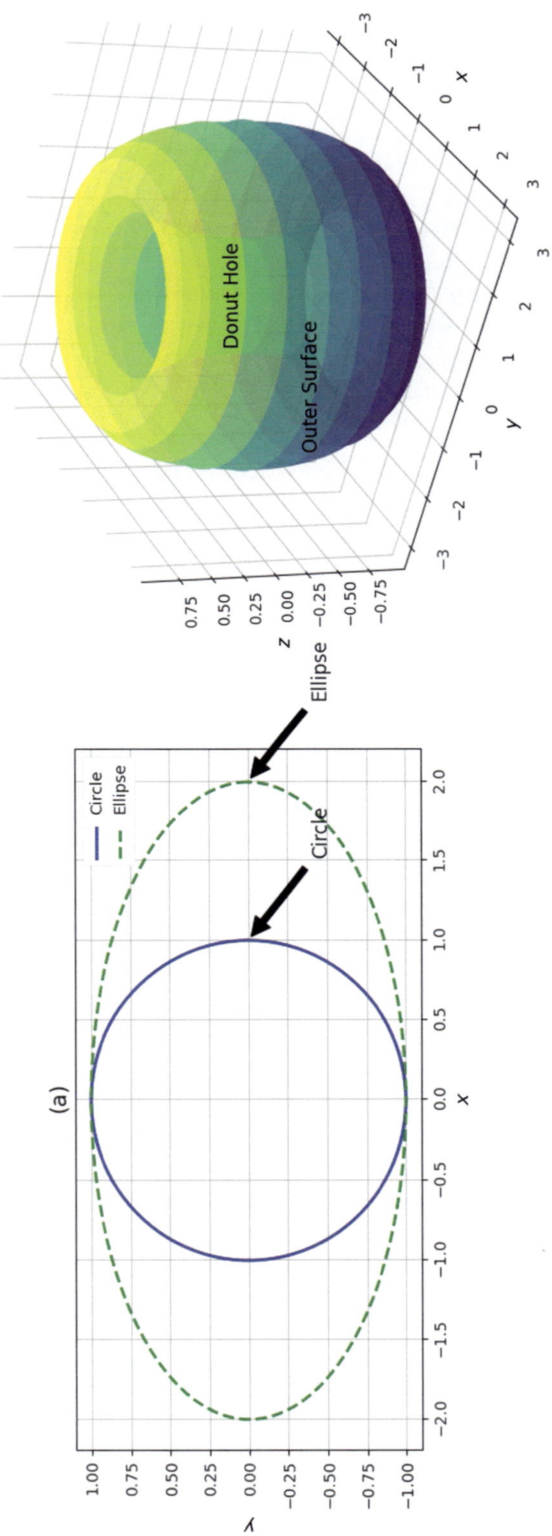

FIGURE 9.1 (a) Circle and ellipse, topologically equivalent due to smooth transformations. (b) Torus, highlighting its single hole as a distinct topological feature.

9.2.2 Neural Networks and Topological Structure

In deep learning, when we talk about the topology of a neural network, we mean its structure or design. This includes how many nodes (or neurons) there are, how they are arranged into layers, and how they are connected to each other. This "structure" acts like a plan that guides how the network works. However, comparing the topology of neural networks to mathematical topology is not straightforward. In mathematics, topology is about properties that stay the same even when shapes are stretched or bent. In deep learning, topology is more about the basic setup of the network, such as the number of layers and how the neurons are linked. The "shape" of the network affects how powerful it is, how it processes information, and the types of problems it can solve. There is also a deeper link between them. The way you design the topology of a neural network can affect how well it learns and how it can apply what it has learned to new situations. In some ways, the network's design and the way it learns work, like the rules in mathematical topology, deciding what parts can change and what parts stay the same as the network learns from data.

9.3 RELATION TO CONVERGENCE OF LEARNING ALGORITHMS

The topology of a neural network sets the foundation for how well it performs. However, factors like the number of layers (depth), the number of neurons in each layer (width), how the neurons are connected, and the activation functions used, all along with the type of data and the specific task, play a crucial role in how effectively the learning algorithm works and how quickly it converges. Good design, proper ways to start the network (initialization), and techniques to prevent overfitting (regularization) are essential, especially as the network becomes more complex. The way a neural network is structured can greatly affect how successfully the learning algorithm can find the best solution.

9.3.1 Depth and Width

9.3.1.1 Depth (Number of Layers)

The depth of a network, meaning the number of layers it has, is very important for its ability to understand and learn complex patterns and ideas. The layers in a deep network learn features in a hierarchical way:

1. *Initial Layers:* The first layers of a deep network usually learn simple and basic features, like edges, corners, and textures in images. These simple parts are the building blocks for more complicated patterns.
2. *Intermediate Layers:* As the data move through the network, the middle layers combine these simple features to create more complex shapes and outlines. This stage provides a more detailed and richer understanding of the input data.
3. *Deep Layers:* The deepest layers of the network capture very complex and abstract features. For example, in image recognition, these layers might identify specific objects, scenes, or intricate patterns that are important for high-level understanding and making decisions.

Adding more layers to neural networks has clear benefits for both how they represent data and how well they perform. First, having more layers helps the network learn and show complex patterns in data, which is especially useful for tasks like recognizing images and speech or processing language. With more layers, deep networks can understand features at different levels, allowing them to grasp data in a hierarchical way, which often works better than networks with fewer layers. Second, deeper networks have led to major improvements in performance. For example, very deep networks have achieved significant increases in accuracy for image recognition, setting new records in various tasks and demonstrating how depth can enhance a neural network's performance. However, deeper networks also bring some challenges, such as overfitting, problems with training, and the need for a lot of computing power. First, deeper networks with more parameters are more likely to overfit, especially when there is not enough data. Overfitting happens when the model learns the noise in the data instead of the useful patterns, which makes it worse at handling new data. Second, as networks become deeper, they can have trouble converging during training. This means that the gradients used to update the network's parameters can either become very small (vanish) or very large (explode), making it hard to optimize the model effectively. Finally, deeper networks need more computational power, which makes training them time-consuming and resource-heavy. These networks often require powerful GPUs or TPUs and distributed computing systems, which can be a challenge for those who do not have access to advanced hardware.

Figure 9.2a shows the decision boundary created by a shallow neural network with a single hidden layer containing five neurons. The network is limited in its ability to capture complex patterns, leading to a relatively simple decision boundary. This demonstrates how initial layers in a shallow network primarily capture basic features, resulting in a straightforward separation of the data. Figure 9.2b shows the decision boundary of a neural network with two hidden layers, each containing 10 neurons. This medium-depth network captures more complex patterns than the shallow network, leading to a smoother and more detailed decision boundary. The plot illustrates how intermediate layers allow the network to learn and combine more intricate features, improving its ability to model the underlying structure of the data. Figure 9.2c shows the decision boundary produced by a deep neural network with three hidden layers, each containing 50 neurons. The deep network is capable of capturing highly complex and abstract patterns, resulting in a very refined and intricate decision boundary.

9.3.1.2 Width (Number of Nodes per Layer)

The width of a layer in a neural network refers to how many nodes or neurons are in that layer. The width affects how well the network can learn from data and recognize complex patterns. Making the network wider can sometimes mean you don't need to make it deeper, but it also brings its own challenges. One of the main benefits of having wider neural networks is that they have a better ability to learn. Wider layers have more parameters, which allows the network to learn more features and handle more complicated tasks. This increased capacity can help the network fit the training data better. Sometimes, a wider network can perform just as well as a deeper one while being simpler in design. Another advantage of wider layers is that they enable the model to learn a more diverse set of features from the data. This variety helps the model work better with different types of data, especially complex datasets, and can improve performance without needing to add more layers. However, there are also challenges when increasing the width of neural networks. One major issue is overfitting. Wider layers can make the network too complex, causing it to memorize the training data instead of learning to work well with new, unseen data. This can lead to poor performance when the network is tested with real-world data. Making the network wider also means there are more parameters, which requires more computing power and longer training times. This can slow down the training process and may need stronger hardware like GPUs or TPUs. Additionally, more parameters use more memory, which can be a problem for very wide networks. Lastly, there comes a point where making the network wider won't significantly improve its performance. Adding more nodes might not make the network more accurate or effective. It's important to balance the width of the network with how efficiently it uses resources. Finding the right design is often better than simply adding more nodes. For example, imagine a neural network that classifies images into 10 categories, such as handwritten digits from 0 to 9. In this case, adjusting the width of the network can help it learn to recognize the different digits more effectively without necessarily making the network deeper. Let's look at two different situations: In the first situation, called a narrow network, the hidden layer has 50 nodes. If the input layer has 784 nodes, which represent a 28 by 28 image, then the number of connections between the input layer and the hidden layer is 784 multiplied by 50.

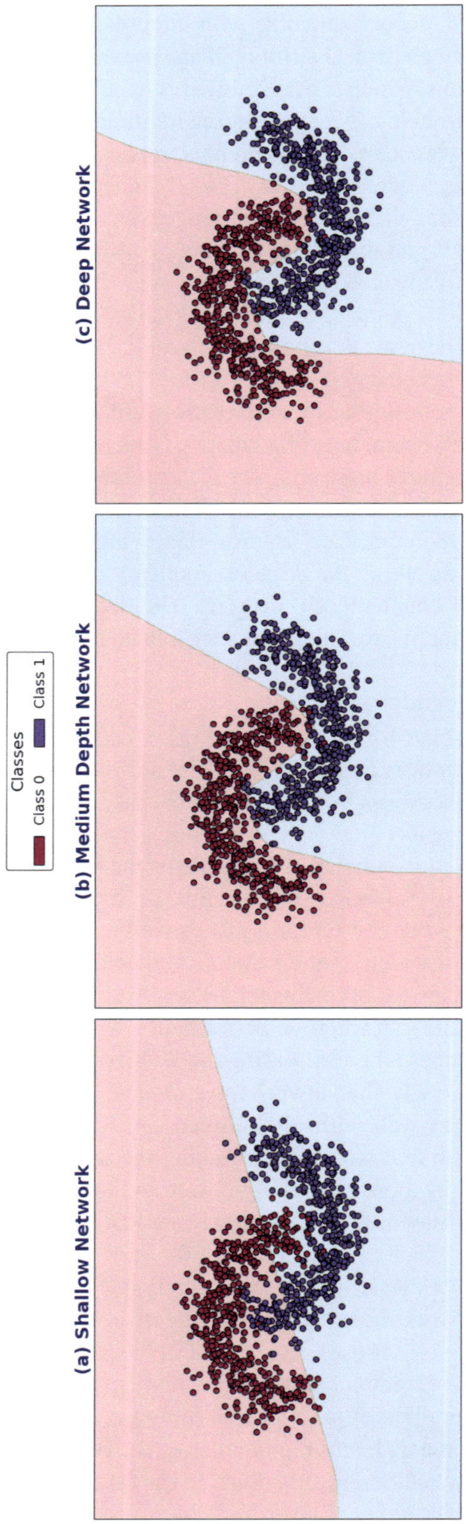

FIGURE 9.2 Hierarchical feature learning across network depths, (a) initial layers, (b) intermediate layers, (c) deep layers.

This equals 39,200 connections. In the second situation, called a wide network, the hidden layer has 500 nodes instead of 50. With the same input layer of 784 nodes, the number of connections between the input and hidden layers becomes 784 multiplied by 500, which is 392,000 connections. In this example, the wider network has ten times more connections than the narrow network. This larger number of connections allows the network to recognize more detailed and complex patterns in the training data. However, having so many connections also increases the chance that the network will overfit. Overfitting happens when the network learns the training data too well, including any noise or specific details, which makes it perform poorly when it encounters new, unseen data.

9.3.2 Skip Connections

Skip connections, also known as residual connections, allow information to move directly across one or more layers by creating a straight path around certain layers. This design is important for training very deep neural networks because it helps solve some of the main problems that come with having many layers. Skip connections have several benefits. They help prevent the vanishing gradient problem, which is when the signals used to train the network become too small as they move through many layers. By providing a direct path for these signals, skip connections keep them strong, making the training process more stable and efficient. This allows deeper networks to be trained effectively. Another advantage of skip connections is that they support a type of learning called ensemble learning. This means that each layer can focus on learning the difference between what it currently predicts and what it should predict. This approach allows each layer to build on what the previous layers have learned, leading to better performance and faster learning.

Skip connections also make the network more flexible. They let different layers adjust their outputs on their own, which helps the network handle various levels of detail and complexity in the data. This makes the network more adaptable to different tasks. However, skip connections also come with some challenges. They make the network more complex by creating multiple paths for information to flow, which requires careful design to keep everything stable. Additionally, they use a bit more memory and processing power because the network needs to combine the outputs from both the direct and the bypassed paths. Despite these challenges, the advantages of skip connections usually outweigh the drawbacks, especially when using efficient designs and modern hardware. Figure 9.3 illustrates the impact of skip connections on the gradient flow through a neural network's layers, comparing scenarios with and without skip connections. On the y-axis, the plot represents the gradient magnitude on a logarithmic scale, while the x-axis displays the layers of the network. Without skip connections, represented by the red dashed line, there is a significant reduction in gradient magnitude across layers, a phenomenon known as the vanishing gradient. This issue is prominent in deeper networks, where gradients become too small to effectively update weights, leading to poor learning. In contrast, the blue solid line shows the preserved gradient flow when skip connections (residuals) are introduced. These connections enable the gradient to remain relatively stable even as the network deepens, addressing the vanishing gradient problem. As seen in the plot, the gradient magnitude remains higher and more stable, preventing degradation of the learning process. The shaded gray region between the lines emphasizes the difference in gradient behavior, highlighting the improvement achieved through skip connections in deep neural networks.

9.3.3 Recurrent Connections

Recurrent neural networks (RNNs) are built to work with data that comes in a sequence by using loops that let information carry over from one step to the next. This setup allows RNNs to handle sequences of different lengths and understand how things change over time and depend on each other in order. The main benefits of these looping connections in neural networks are seen when processing sequential data and keeping information over time. First, the loops let RNNs handle data in

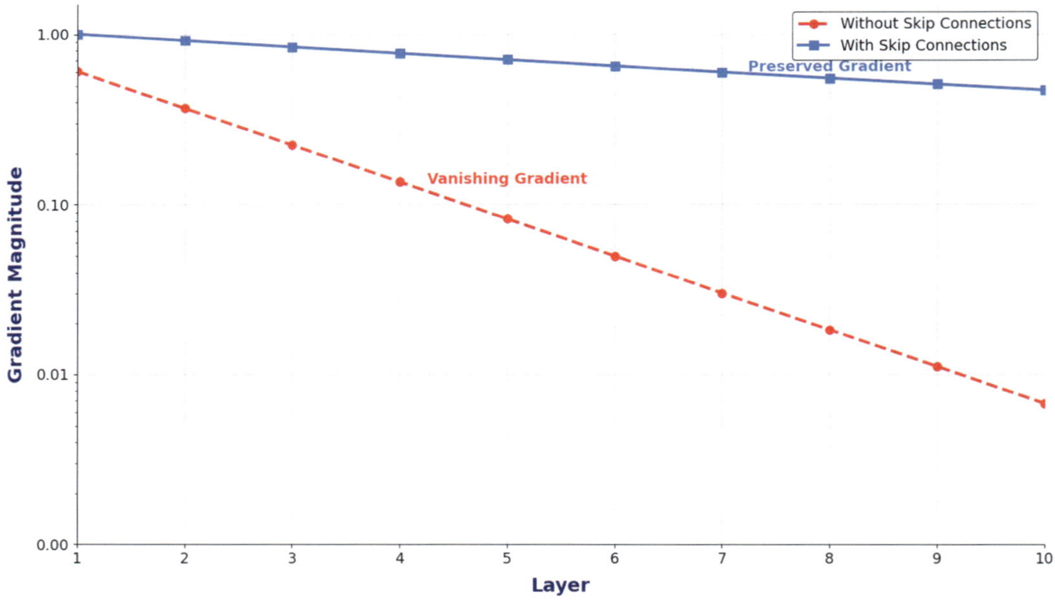

FIGURE 9.3 Impact of skip connections on gradient flow in deep networks.

the order it happens, making them great for tasks like analyzing time-based data, understanding language, and recognizing speech. These networks can work with input sequences that vary in length, which is important for tasks where the size of the input can change a lot, such as predicting the next word in a sentence or processing spoken words. Second, the looping connections give the network a kind of memory because information can stay available across different steps. This means the model can use what it learned before to make better predictions now, giving it a deeper understanding of how things are related in the sequence. For example, think about trying to predict the next word in a sentence. Suppose we have the sequence: "The cat is on the." The goal is to guess the next word, which is likely "mat." An RNN looks at this sequence one word at a time, using the words that came before to help predict the next one. In this case, each word is treated as a separate step. At each step **t**, the RNN takes in an input $\mathbf{x}_t$ (like the word "The," "cat," "is," etc.) and updates its hidden state ht based on the current word and the hidden state from the previous step $\mathbf{h}_{t-1}$. For example:

- At **t** = 1, the RNN processes "The," and updates its hidden state $\mathbf{h}_1$.
- At **t** = 2, it processes "cat," using both $\mathbf{x}_2$ ("cat") and the hidden state $\mathbf{h}_1$ (from "The").

This continues until the RNN processes "the" at **t** = 4, using all previous context to predict the next word, "mat." The RNN's recurrent connections allow information from earlier time steps to influence later predictions. In the example above, knowing "The cat is on the" helps the model predict "mat" because of the context it has accumulated over the sequence. At each time step **t**, the RNN updates its hidden state using the following equations:

$$h_t = f\left(W_h \cdot h_{t-1} + W_x \cdot x_t + b\right)$$

where:

- $\mathbf{h}_t$ is the hidden state at time **t**,
- $\mathbf{h}_{t-1}$ is the hidden state from the previous time step,

- $\mathbf{x_t}$ is the input at time **t**,
- $\mathbf{W_h}$ and $\mathbf{W_x}$ are weight matrices,
- **b** is the bias term, and
- **f** is the activation function (typically a non-linear function like tanh or ReLU).

The hidden state $\mathbf{h_t}$ stores information from previous time steps, enabling the RNN to "remember" the context as it processes new data. Finally, the output at each time step, $\mathbf{y_t}$, is given by:

$$y_t = f\left(W_y \cdot h_t + c\right)$$

where:

- $\mathbf{W_y}$ is the output weight matrix,
- **c** is the output bias.

For example, if we want to guess the next word in a sentence, $\mathbf{y_t}$ would show the chances of different words coming next. The word with the highest chance is chosen as the next word. Recurrent connections in neural networks have some problems, such as vanishing and exploding gradients, and slow training. First, when training the network, the information (called gradients) is sent back through time. In long sequences, these gradients can either become very small (vanishing gradients) or very large (exploding gradients). When gradients vanish, the network struggles to learn connections that are far apart in the sequence. When gradients explode, the training can become unstable. These problems can make training slow or ineffective, especially for tasks that need to understand long-term relationships. Second, RNNs handle data one step at a time, with each step depending on the one before it. This makes training slower compared to other types of networks like convolutional neural networks (CNNs), which can handle many parts of the data at the same time. Because RNNs work step by step, they require more computing power and memory, making the training process take longer and use more resources.

9.3.4 Activation Functions

Activation functions add non-linearity to the network, which helps it learn complex patterns. ReLU is a popular activation function used in neural networks. It doesn't get stuck at certain values, which often makes the network train faster. However, ReLU can cause some neurons to stop working, meaning they never activate. To fix this, variations like LeakyReLU and ParametricReLU have been created. Other common activation functions are Sigmoid and Tanh. Sigmoid functions are usually used in the output layer for tasks that have two possible outcomes because they produce values between 0 and 1. Tanh functions are used when the output needs to be between −1 and 1, which is useful for certain hidden layers. However, both Sigmoid and Tanh can cause a problem called the vanishing gradient because they can make the values very small. This makes it hard to train deep networks effectively.

Figure 9.4 presents six subplots, each representing the training process of neural network models with different depths and activation functions. The x-axis represents the number of epochs, while the y-axis displays the loss and accuracy values. The red line indicates the loss, and the blue line shows the accuracy.

Depth: 1, Activation: relu: This subplot shows the training results for a neural network with a single hidden layer using the ReLU activation function. Initially, the loss decreases rapidly, and the accuracy increases, indicating that the model is learning. As training progresses, both metrics stabilize, suggesting the model has reached a plateau. Depth: 1, Activation: tanh: This subplot represents a neural network with one hidden layer using the Tanh activation function. The loss

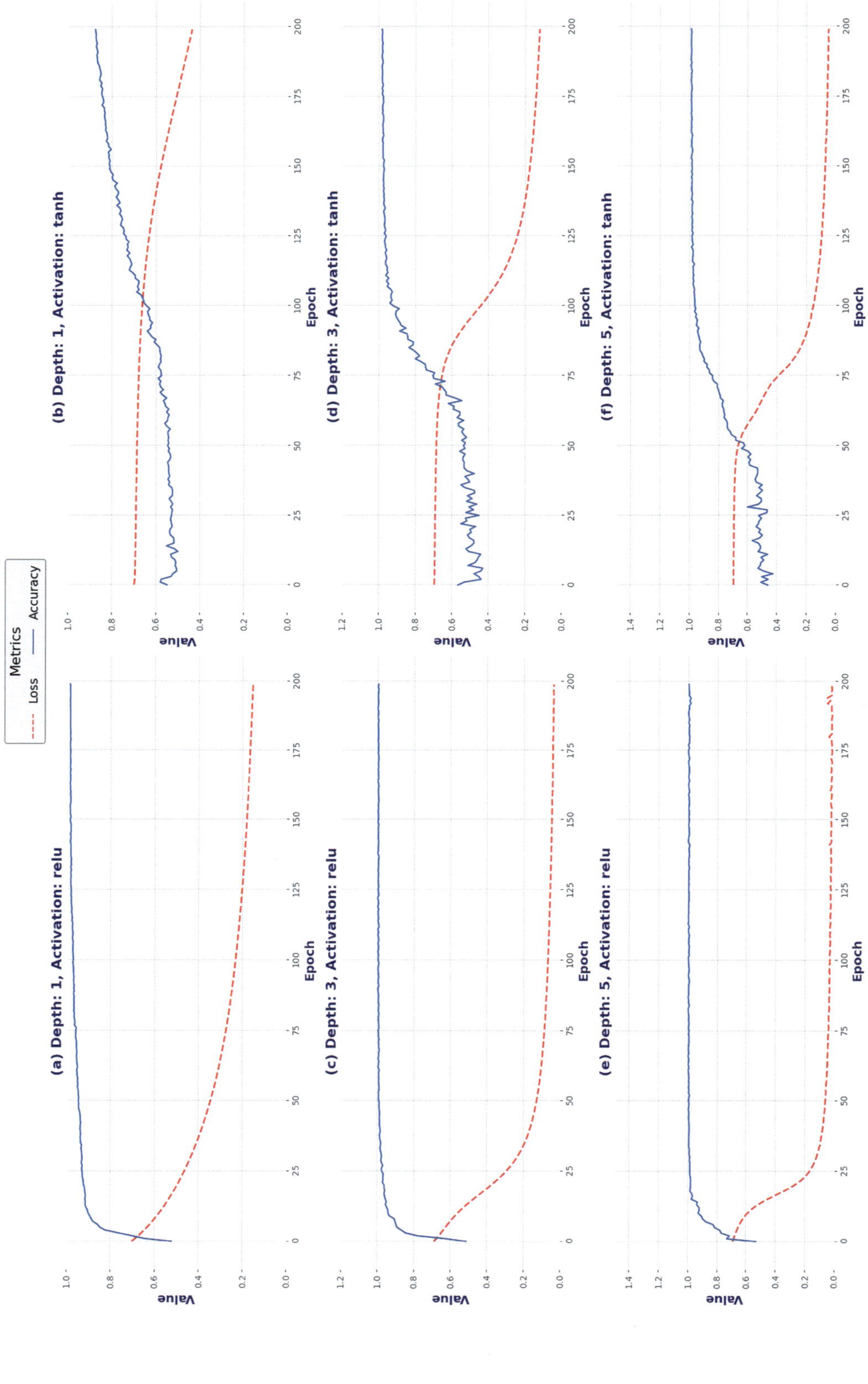

FIGURE 9.4 Effect of depth and activation functions on convergence.

decreases steadily over the epochs, while the accuracy improves. Compared to the ReLU activation, the Tanh function shows a more gradual learning curve, reflecting different dynamics in how the network converges. Depth: 3, Activation: relu: This subplot corresponds to a neural network with three hidden layers and the ReLU activation function. Initially, the model showed rapid improvement in both loss and accuracy. The increased depth allows the network to capture more complex patterns, leading to higher accuracy than the single-layer models. Depth: 3, Activation: tanh: Here, the neural network with three hidden layers uses the Tanh activation function. The training process shows a steady decrease in loss and an increase in accuracy. The multiple layers with Tanh activation help the network learn effectively, with the training curves indicating smooth convergence. Depth: 5, Activation: relu: This subplot shows the performance of a deeper neural network with five hidden layers using ReLU. The training curves indicate rapid initial learning, with both loss and accuracy reaching stable values. The depth of the network provides the capacity to model complex relationships, reflected in the training metrics. Depth: 5, Activation: tanh: This final subplot represents a neural network with five hidden layers and Tanh activation. The loss decreases, and accuracy increases steadily, similar to other Tanh-activated models. The depth combined with Tanh activation allows the network to converge smoothly, capturing intricate patterns in the data.

9.4 TOPOLOGICAL DATA ANALYSIS IN NEURAL NETWORKS

Integrating Topological Data Analysis (TDA) with neural networks is a new and promising area that can lead to many discoveries and improvements. TDA provides tools and methods to examine and understand the shape and structure of datasets. This helps us gain deep insights into how data are organized and how different parts of the data are related. When TDA is used with neural networks that are learning intensively, it offers a unique way to understand both the data and how the network behaves.

9.4.1 Persistent Homology

Persistent homology is a technique from computational topology that helps analyze the shape and structure of data at different levels. It is especially useful for understanding complex, high-dimensional data by looking at how its features remain or change as we examine it at various scales. In persistent homology, the main ideas include topological features, filtration, and visualizations like persistence diagrams and barcodes. Topological features are the basic shapes in the data, such as points, lines, loops, and cavities. These features are counted using Betti numbers. For example, the zeroth Betti number counts the number of separate pieces or connected components, the first Betti number counts the number of loops or cycles, and the second Betti number counts the number of voids or enclosed spaces. A filtration is a step-by-step process where we build a series of shapes called simplicial complexes by changing a scale parameter. A simplicial complex is made up of points, lines, triangles, and other simple shapes that approximate the data's structure. This process allows us to analyze the topological features at different scales. Each feature has a birth scale, which is when it appears, and a death scale, which is when it disappears. The difference between these scales measures how long the feature persists. Visualizations like persistence diagrams and barcodes are used to show how the topological features appear and disappear as the scale changes. Features that last through many scales are considered important, while those that appear and disappear quickly are usually seen as noise. Consider a group of six data points: (0, 0), (1, 0), (0, 1), (1, 1), (2, 1), and (3, 1) arranged in a two-dimensional (2D) space. We will use persistent homology to study the shapes and structures within this data. First, let's talk about topological features, specifically Betti numbers, which help us understand the structure of the data. The Betti-0 number counts the number of separate pieces or connected components. Initially, with a very small radius around each point, there are six separate components because all points are disconnected. As we increase the radius, some points start to connect. For example, at a radius of 0.5, some points merge into fewer connected components. Eventually, when the radius is large enough, all points come together into one single connected component, so the Betti-0 number drops to one. Next, Betti-1 numbers count the number of loops or cycles in the data. When we increase the radius further, some connections form loops. For instance, the points at (0, 0), (1, 0), (0, 1), and (1, 1) can create a square-like loop, making the Betti-1 number equal to one. If we keep increasing the radius, this loop might fill in and disappear, causing the Betti-1 number to go back to zero. Filtration is the process of gradually increasing the radius around each point and keeping track of when topological features appear and disappear. At the start, with radius $r = 0$, there are six separate components. As the radius grows to 0.5, some points connect, reducing the number of components to three. When the radius reaches 1.5, these connections might form one connected component and one loop. Finally, when the radius is 2.5, all points merge into a single connected component, and no loops remain. A persistence diagram is a way to visualize when each feature (like connected components or loops) appears and disappears as the radius changes. For example, a connected component might appear at $r = 0$ and merge with others at $r = 0.5$. A loop might form at $r = 1.0$ and disappear at $r = 2.0$. These diagrams help us see which features last longer and are more important, while short-lived features are often considered noise. Persistent homology has many useful applications in data analysis. In shape analysis, it helps examine the geometric structure of data, such as 3D shapes or molecular structures, by identifying important shapes within the data. In machine learning, persistent homology can extract topological features from complex data, which can improve tasks like classification or grouping similar data together by capturing structural information that traditional methods might miss. It also helps reduce noise by focusing on features that persist over many scales and ignoring those that disappear quickly, making it especially useful for datasets with a lot of noise. Additionally, in network analysis, persistent homology can study complex networks like social, biological, or communication networks, providing insights into how these networks are connected, how they cluster, and their overall structure.

Figure 9.5 demonstrates the application of TDA using persistent homology to analyze and visualize topological features in a dataset. Figure 9.5a shows a scatter plot of data points grouped into two clusters, labeled as Cluster 1 and Cluster 2. These clusters highlight distinct regions in the dataset, illustrating how data can be divided into separate components based on proximity. The arrows point to representative clusters, emphasizing the idea of connectivity within each region. Figure 9.5b presents a persistence diagram, a tool that visualizes the birth and death of topological features as the scale parameter varies. Two types of features are shown: connected components (H0) and loops (H1). The diagonal line represents features that appear and quickly disappear, which are often considered noise. Points farther from the diagonal correspond to persistent features that remain across multiple scales, indicating their importance in understanding the data's structure. For instance, a loop (H1) appears at a specific scale and persists for some time, reflecting a significant cycle in the data's topology.

Figure 9.6 illustrates a visualization of TDA using three interconnected panels: (a) a point cloud, (b) a persistence diagram, and (c) a persistence barcode. These visualizations collectively showcase the persistence of topological features in the data across multiple scales. Figure 9.6a depicts the scatter plot of a two-cluster point cloud, with Cluster 1 and Cluster 2 labeled in blue and green, respectively. This representation highlights the raw spatial distribution and connectivity of data points, providing an initial view of the clusters' structure. Figure 9.6b, the persistence diagram, captures the birth and death of topological features, such as connected components (H0) and loops (H1). Points closer to the diagonal indicate short-lived features, often regarded as noise, while points farther from the diagonal represent persistent features that are structurally significant. For example, loops (H1) and connected components (H0) are annotated to emphasize their relevance at specific scales. Figure 9.6c visualizes the same topological features using a persistence barcode. Each horizontal bar represents a feature's lifespan, starting at its birth and ending at its death. Longer bars

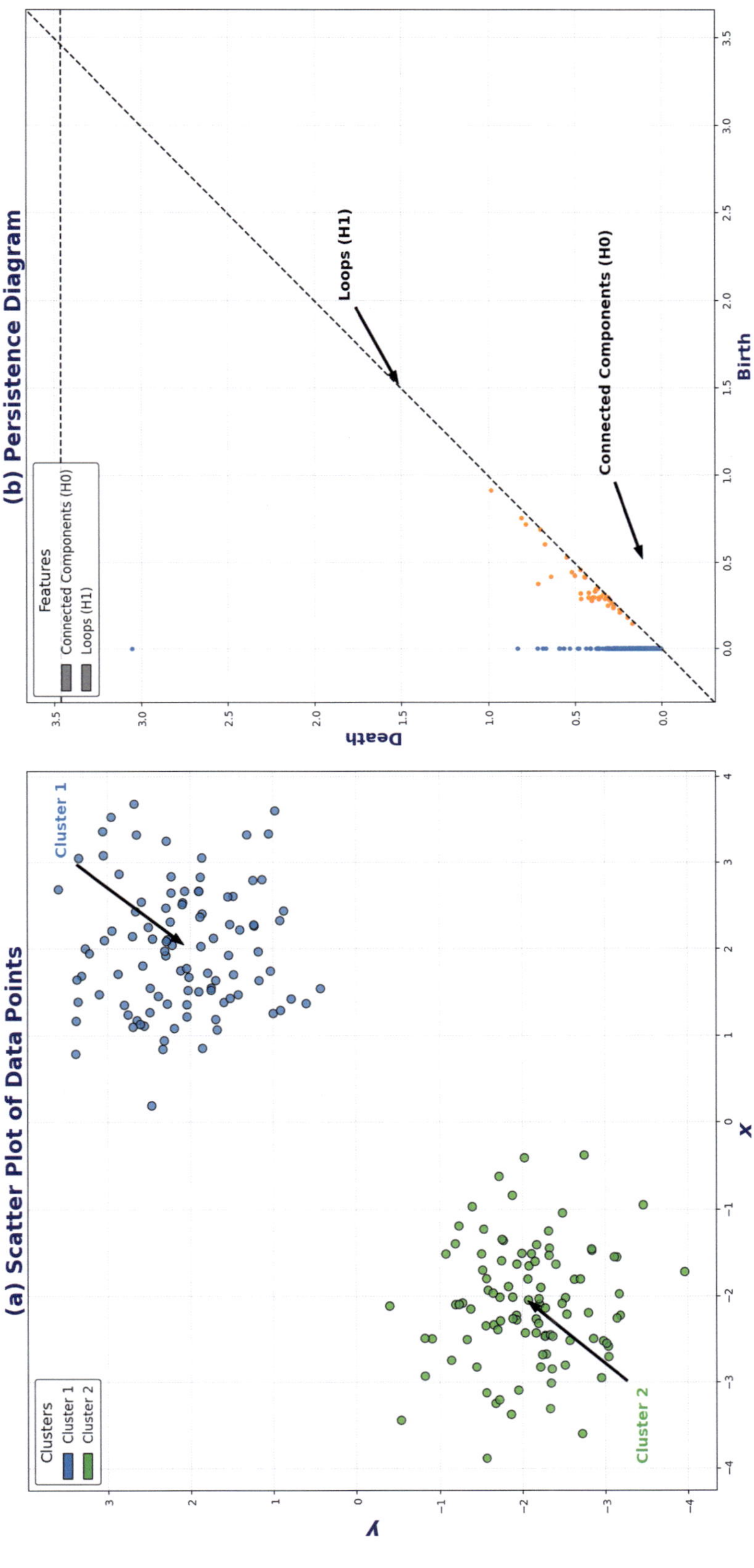

FIGURE 9.5 (a) Point cloud and (b) persistence diagram.

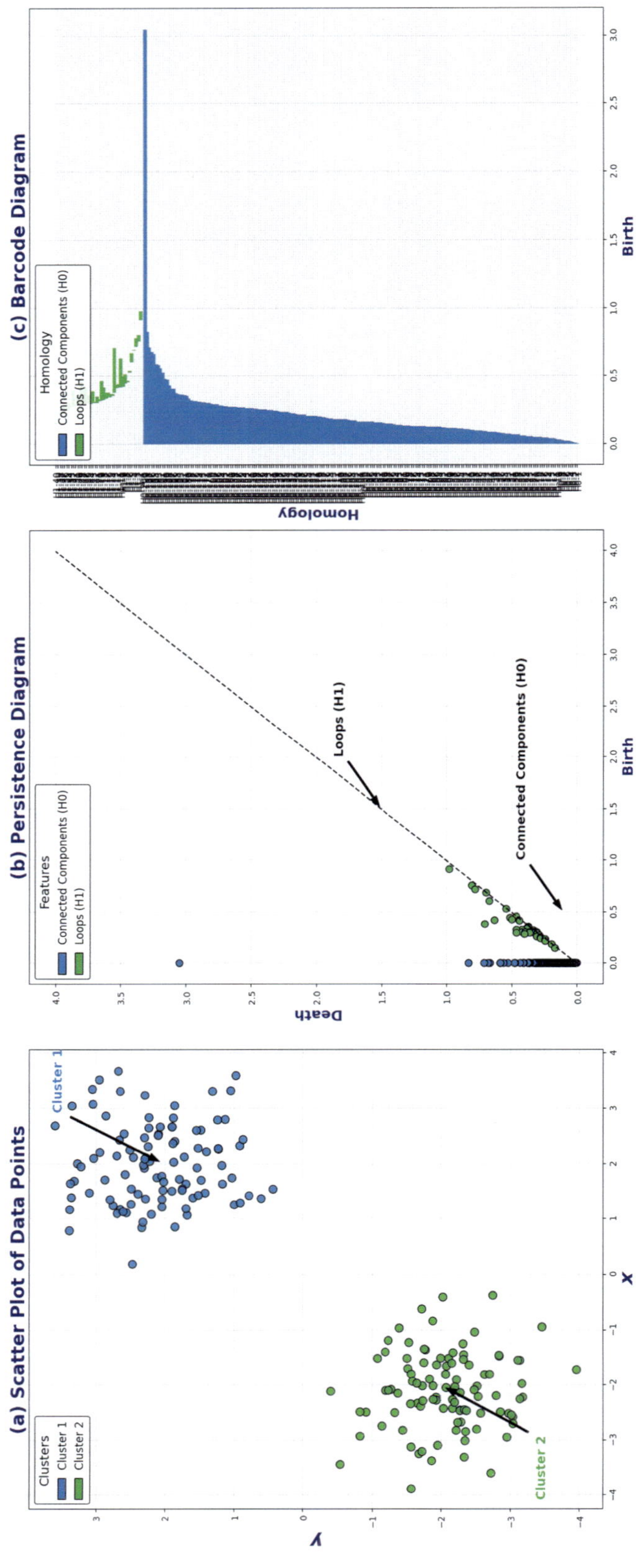

FIGURE 9.6 TDA visualization. (a) Point cloud, (b) persistence diagram, (c) persistence barcode.

signify features that persist over a wide range of scales, indicating their importance, while shorter bars represent transient noise. The barcode succinctly complements the persistence diagram by providing an alternative view of feature longevity.

9.5 TOPOLOGICAL DATA ANALYSIS WITH DEEP LEARNING

Combining TDA with deep learning is becoming more popular and provides new ways to understand and improve neural network models. TDA helps us examine and use the natural shape and structure of data, which can lead to stronger and more efficient learning methods.

9.5.1 Persistent Homology in Deep Learning

Persistent homology is an important idea in TDA. It looks at the shapes and features of data at different levels. By doing this, it helps us understand the data's structure by finding things like connected parts, loops, and empty spaces. In deep learning, persistent homology can be used in several powerful ways. One key use is analyzing the loss landscape, which is the graph of the loss function that the neural network tries to minimize. Persistent homology helps us study the shapes of this loss function by looking at how many low points and saddle points it has. This gives researchers important information about how difficult the optimization problem is. With this knowledge, they can create better algorithms to optimize the network. For example, if there are many low points, they can develop methods to avoid getting stuck and find better overall solutions. Another major use is feature extraction. Persistent homology can pull out topological features from the data, such as connected parts, holes, and higher-dimensional empty spaces. These features capture complex information that regular methods might miss. These extracted features can be added as extra inputs to neural networks, helping the model find more detailed patterns, work better on new data, and be more reliable. Additionally, persistent homology helps reduce the number of dimensions by highlighting the most important topological features. This makes the input data simpler for the neural network to handle.

When using persistent homology in deep learning, there are several practical things to consider. First, it can be very computationally heavy. Calculating persistent homology for large or complex datasets can take a lot of resources, so efficient algorithms and software are needed. To use persistent homology in large-scale deep learning, it's important to manage the computational load carefully. Next, integrating persistent homology with neural networks requires several preprocessing steps to calculate the topological features and then smoothly add them to the network's structure. It's important to make sure this process doesn't slow down training or prediction. Also, calculating persistent homology involves setting certain parameters, like the scale for filtering, which need to be carefully adjusted to capture important features without adding unnecessary noise. Finally, being able to interpret and visualize the topological features is crucial. The features found by persistent homology need to make sense for the specific task, and understanding how they relate to the data is key for using them effectively. Imagine we have two datasets: Dataset **A**, points arranged in a circular shape, and Dataset **B**, points scattered randomly within a square. Our goal is to train a neural network to classify new data points as belonging to either Dataset **A** or Dataset **B**. To improve the neural network's performance, we will extract topological features from each dataset using persistent homology and incorporate these features into the training process. Let's represent each dataset numerically. In Dataset **A** (circle), we have eight points evenly spaced around a unit circle:

$$(1,0),\left(\sqrt{2}/2,\sqrt{2}/2\right),(0,1),\left(-\sqrt{2}/2,\sqrt{2}/2\right),(-1,0),$$
$$\left(-\sqrt{2}/2,-\sqrt{2}/2\right),(0,-1),\left(\sqrt{2}/2,-\sqrt{2}/2\right)$$

In Dataset **B** (square), we have eight points randomly scattered within a unit square:

$$(0.1,0.9),(0.4,0.7),(0.6,0.2),(0.9,0.4),(0.3,0.5),(0.7,0.8),(0.5,0.1),(0.2,0.6)$$

To extract topological features, we'll compute the persistent homology of each dataset. Persistent homology studies the shape of data by analyzing features like connected components and loops across different scales.

Step 1: First, we calculate the pairwise Euclidean distances between points in each dataset. For Dataset **A** (circle), let's compute the distance between the first point (1, 0) and the other points: The distance to $\left(\sqrt{2}/2,\sqrt{2}/2\right)$:

$$d=\sqrt{(1-\sqrt{2}\,/\,2)^2+(0-\sqrt{2}\,/\,2)^2}\approx\sqrt{(0.2929)^2+(-0.7071)^2}$$
$$\approx\sqrt{0.0858+0.5}\approx\sqrt{0.5858}\approx 0.7654$$

And distance to (0, 1) is $d=\sqrt{(1-0)^2+(0-1)^2}=\sqrt{1+1}=\sqrt{2}\approx 1.4142$. We repeat this for all pairs, constructing a distance matrix $\mathbf{D}_\mathbf{A}$ for Dataset **A**. For Dataset **B** (square), We compute distances between each pair of points in a similar manner, creating a distance matrix $\mathbf{D}_\mathbf{B}$.

Step 2: We construct Vietoris–Rips complexes for each dataset at various distance thresholds ϵ. A Vietoris–Rips complex connects points that are within a distance ϵ of each other. Let's choose several ϵ values (e.g., 0.5, 1.0, 1.5) and observe how the topology changes. For Dataset **A** (circle), at ϵ = **0.5**, points are connected only to their immediate neighbors, forming individual edges and resulting in multiple connected components. As ϵ increases to **0.8**, more points connect, forming a loop that represents a 1D hole resembling a circular shape. When ϵ reaches **1.5**, all points are connected, and the loop fills in, causing the hole to disappear. For Dataset **B** (square), at ϵ = **0.5**, some points are connected, but no significant loops are formed. As ϵ increases to **1.0**, more connections appear; however, any loops that emerge are due to random arrangements and are not persistent, indicating they do not represent meaningful topological features.

Step 3: Betti numbers quantify the topological features: β_0 is a number of connected components, and β_1 is a number of 1D holes (loops). For each ϵ, we compute β_0 and β_1. For Dataset **A** (circle), at ϵ = **0.5**, the Betti numbers are $\boldsymbol{\beta_0 = 8}$ (each point is a separate component) and $\boldsymbol{\beta_1 = 0}$ (no loops). At ϵ = **0.8**, $\boldsymbol{\beta_0}$ becomes **1** as all points connect into one component, and $\boldsymbol{\beta_1}$ becomes **1**, indicating one persistent loop. By ϵ = **1.5**, $\boldsymbol{\beta_0}$ remains **1**, and $\boldsymbol{\beta_1}$ returns to **0** as the loop fills in. For Dataset **B** (square), at ϵ = **0.5**, $\boldsymbol{\beta_0}$ varies depending on the point distribution, while $\boldsymbol{\beta_1}$ remains **0** (no loops). As ϵ increases, $\boldsymbol{\beta_0}$ decreases as components merge, and $\boldsymbol{\beta_1}$ stays at **0** or shows insignificant loops that quickly disappear.

Step 4: We plot persistence diagrams for each dataset, where each topological feature is represented as a point with coordinates (birth ϵ, death ϵ). For Dataset **A** (circle), the persistent loop appears at $\epsilon \approx 0.8$ and disappears at $\epsilon \approx 1.5$. This results in a point at (0.8, 1.5) in the persistence diagram, indicating a significant topological feature. For Dataset B (square), No significant persistent loops are observed, and features in the diagram have short lifespans (birth and death ϵ values are close), indicating they are noise.

Step 5: From the persistence diagrams, we extract features: For Dataset **A**, the most significant β_1 feature (loop) has a persistence of 1.5 – 0.8 = 0.7. For Dataset **B**, any β_1 features have low persistence, close to zero. We can define a topological feature vector for each dataset based on the persistence of loops. For Dataset **A**, the topological feature is the persistence of the loop, which is **0.7**. For Dataset **B**, the topological feature is the persistence of the loop, which is **0**, indicating no significant loops persist.

Step 6: We augment each data point with its dataset's topological feature. For example, the first point in Dataset **A** has original coordinates **(1, 0)** and a topological feature value of **0.7**. Therefore, the augmented vector is **(1, 0, 0.7)**. Similarly, for Dataset **B**, the first point has coordinates **(0.1, 0.9)** with a topological feature value of **0**, resulting in the augmented vector **(0.1, 0.9, 0)**.

Step 7: We use the augmented data to train a neural network for classification. The network architecture is structured as follows: The input layer consists of three neurons, representing the x-coordinate, y-coordinate, and topological feature. The hidden layer contains a small number of neurons, such as five, using an activation function like ReLU. The output layer comprises one neuron with a sigmoid activation function, designed for binary classification to distinguish between Dataset **A** and Dataset **B**. The training process is as follows: For each epoch, we perform the following steps:

1. *Forward Pass:* Compute the output of the network for each data point.
2. *Loss Calculation:* Use a loss function like binary crossentropy to measure the difference between predicted and actual labels.
3. *Backpropagation:* Compute gradients of the loss with respect to weights and biases.
4. *Parameter Update:* Adjust weights and biases using an optimizer like stochastic gradient descent (SGD).

Step 8: After training, we evaluate the neural network's performance on a test set (could be a subset of the data or new data points).

Figure 9.7 presents an analysis of two datasets, one circular (Dataset A) and one square (Dataset B), using persistent homology and Vietoris–Rips complexes to study their topological features across different scales. Figure 9.7a shows Dataset A, which consists of points arranged in a circular shape. Similarly, Figure 9.7b displays Dataset B, where points form a square-like structure. These datasets highlight the initial spatial arrangement of the data. Figure 9.7c illustrates the Vietoris–Rips complex for the circular dataset at $\epsilon = 0.8$. This complex connects points based on their pairwise distances, capturing the underlying circular topology through the formation of loops while preserving the dataset's geometric structure. Figure 9.7d shows the persistence diagram for Dataset A (circle), where connected components (H0) and loops (H1) are tracked as the scale parameter changes. The diagram reveals the persistence of the circular structure, with the H1 loop indicating the presence of a significant circular feature. Figure 9.7e presents the Vietoris–Rips complex for the square dataset at $\epsilon = 0.8$. This representation connects points in the square, forming a network that captures its geometric structure, including potential loops. Figure 9.7f displays the persistence diagram for Dataset B (square), showing connected components (H0) and loops (H1). The square's topology is reflected in the emergence and disappearance of loops, highlighting differences in persistence compared to the circular dataset.

9.5.2 Betti Numbers in Deep Learning

Betti numbers are whole numbers that show how many different types of holes exist in a shape or space. They provide a quick summary of how complex the shape of the data is. In deep learning, Betti numbers help us understand the structure of the data better and make our models stronger and more reliable. Using Betti numbers in neural networks is useful for two main purposes: analyzing data complexity and improving model robustness. When analyzing data complexity, Betti numbers can be calculated to measure how complicated the dataset is. This understanding helps in choosing the right neural network design. For example, if the data has many complex features indicated by high Betti numbers, we might need deeper or more advanced network structures to capture these intricate patterns. Additionally, Betti numbers can guide us in deciding how to prepare the data or apply specific techniques to handle its complexity effectively during training. In terms of model

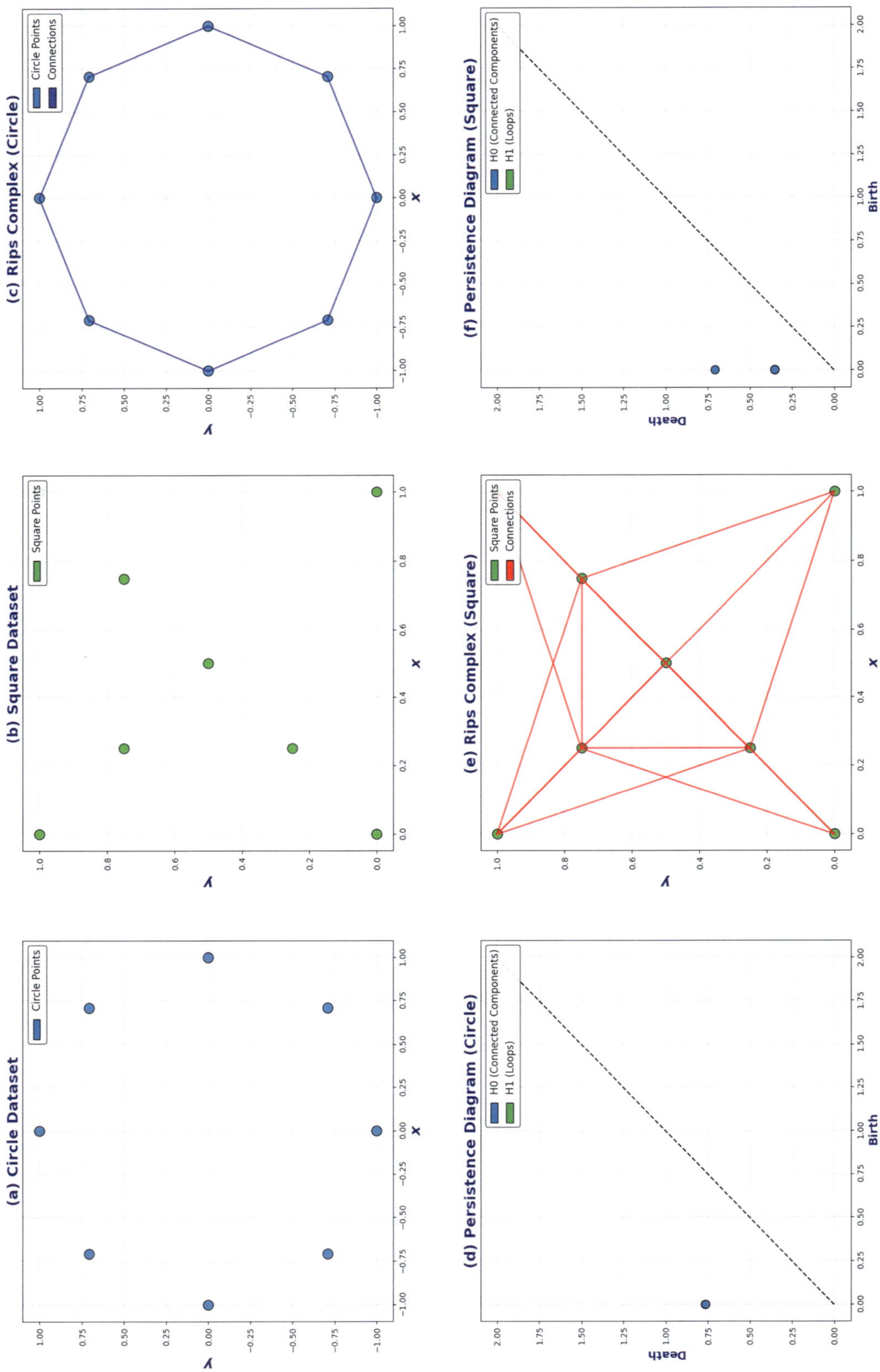

FIGURE 9.7 (a) Dataset A (circle), (b) Dataset B (square), (c) Vietoris–Rips complex circle at epsilon 0.8, (d) persistence diagram for Dataset A (circle), (e) Vietoris–Rips complex square at epsilon = 0.8, (f) persistence diagram for Dataset B (square).

robustness, Betti numbers provide insights into the topological properties of the data, which helps in building models that can handle these features well. For instance, if the data has many loops, as shown by a high first Betti number, the model can be designed to recognize and work with these loops accurately. Incorporating this topological information makes models better at dealing with noisy data and unexpected changes. Furthermore, Betti numbers can be added as extra features to the data, enhancing the model's ability to capture the true structure of the data and perform better on new, unseen information. Imagine we have two datasets composed of 2D points: Dataset **A**: Points arranged in a circular shape (simple topology), and Dataset **B**: Points forming a figure-eight shape (more complex topology). Our goal is to analyze the topological complexity of these datasets using Betti numbers and understand how this information can inform the design and robustness of a neural network trained to classify these shapes. Betti numbers are integers that quantify the topological features of a space: $\boldsymbol{\beta_0}$ (Zeroth Betti number) counts the number of connected components. $\boldsymbol{\beta_1}$ (first Betti number) counts the number of 1D holes (loops). $\boldsymbol{\beta_2}$ (second Betti number) counts the number of 2D voids (in 3D space).

- Computing Betti Numbers for Dataset **A** (Circle):

1. We represent the circle using points sampled uniformly around a unit circle. Points, $(\cos(\boldsymbol{\theta}_i), \sin(\boldsymbol{\theta}_i))$ for $\mathbf{i} = 1, 2, ..., \mathbf{N}$, where $\boldsymbol{\theta}_i$ are angles evenly spaced between 0 and 2π. For simplicity, let's take $\mathbf{N} = 8$ points:

   a. $(\cos(0), \sin(0)) = (1, 0)$,

   b. $(\cos(\pi/4), \sin(\pi/4)) = \left(\sqrt{2}/2, \sqrt{2}/2\right)$,

   c. $(\cos(\pi/2), \sin(\pi/2)) = (0, 1)$,

   d. $(\cos(3\pi/4), \sin(3\pi/4)) = \left(-\sqrt{2}/2, \sqrt{2}/2\right)$,

   e. $(\cos(\pi), \sin(\pi)) = (-1, 0)$,

   f. $(\cos(5\pi/4), \sin(5\pi/4)) = \left(-\sqrt{2}/2, -\sqrt{2}/2\right)$,

   g. $(\cos(3\pi/2),\ \sin(3\pi/2)) = (0, -1)$,

   h. $(\cos(7\pi/4),\ \sin(7\pi/4)) = \left(\sqrt{2}/2, -\sqrt{2}/2\right)$.

2. *We Construct a Vietoris–Rips Complex:* In constructing a Vietoris–Rips complex, the first step is to compute the distance matrix, which involves calculating the pairwise distances between points in the dataset. For instance, the distance between Point 1 and Point 2 is approximately $d_{1,2} = \sqrt{(1-\sqrt{2}/2)^2 + (0-\sqrt{2}/2)^2} \approx 0.765$. The next step is to choose a distance threshold (ϵ). This value determines which points are connected; if the distance between two points is less than or equal to ϵ, they are connected in the complex.
3. *Computing Betti Numbers at Different ϵ Values:* At $\epsilon = 0.7$, edges connect pairs of points where $d_{i,j} \leq 0.7$ di. The connected components (β_0) indicate that each point remains a separate component, so $\beta_0 = 8$. No loops are formed at this stage, resulting in $\beta_1 = 0$. At $\epsilon = 1.1$,

more connections form between the points. All points become connected, resulting in $\beta_0 = 1$. A single loop forms around the circle, so $\beta_1 = 1$. At $\epsilon = 2.0$, the complex fills in with higher-dimensional simplices. The connected components remain at one ($\beta_0 = 1$), but the loop fills in, causing β_1 to drop to 0.

- Computing Betti Numbers for Dataset **B**:

1. *Dataset Representation:* We represent the figure-eight by combining two circles intersecting at the origin: The top loop is centered at (0, 1) with points defined as (cos(θi), sin(θi)+1). The bottom loop is centered at (0, –1) with points given by (cos(θi), sin(θi) – 1). In both cases, **N** = 8 points are used per loop.
2. *Building a Simplicial Complex:* Building a simplicial complex begins with computing the distance matrix by calculating pairwise distances between points. Next, choose appropriate ϵ values, similar to the previous approach, to determine which points are connected in the complex based on their pairwise distances.
3. *Computing Betti Numbers:* At $\epsilon = 0.7$, points within each loop start connecting internally. The total number of connected components is two, as the top and bottom loops remain separate, resulting in $\boldsymbol{\beta_0} = 2$. No loops are present at this stage, so $\boldsymbol{\beta_1} = 0$. At $\epsilon = 1.1$, the intersection point at the origin connects the two loops, reducing the connected components to one ($\boldsymbol{\beta_0} = 1$). Two loops form, one in each loop of the figure-eight, giving $\boldsymbol{\beta_1} = 2$. At $\epsilon = 2.0$, the loops begin to fill in, and the connected components remain at one ($\boldsymbol{\beta_0} = 1$). Both loops are filled, reducing $\boldsymbol{\beta_1}$ to 0. Interpreting the Betti numbers for the given datasets provides insights into their topological structure. For Dataset **A**, the Betti numbers are $\boldsymbol{\beta_0} = 1$ and $\boldsymbol{\beta_1} = 1$. The value of $\boldsymbol{\beta_0} = 1$ indicates that the dataset consists of one connected component, meaning all points form a single continuous structure. The value of $\boldsymbol{\beta_1} = 1$ means there is one loop, which corresponds to the circular shape of the dataset. In Dataset **B**, the Betti numbers are $\boldsymbol{\beta_0} = 1$ and $\boldsymbol{\beta_1} = 2$. The $\boldsymbol{\beta_0} = 1$ shows that after merging all parts of the figure-eight, the dataset still has one connected component. The $\boldsymbol{\beta_1} = 2$ indicates the presence of two loops, corresponding to the two distinct loops in the figure-eight shape.

Figure 9.8 illustrates the application of TDA in deep learning, focusing on the relationship between the loss landscape and topological persistence. Figure 9.8a depicts the loss landscape, where the normalized loss is plotted against the perturbation index. The color gradient highlights variations in normalized loss, with darker shades representing lower values and lighter shades indicating higher losses. This visualization captures the ruggedness and overall structure of the optimization surface, emphasizing regions of low loss that correspond to stable solutions. Figure 9.8b presents the persistence diagram, which tracks the birth and death of topological features, such as connected components (H0) and loops (H1), during the analysis of the loss landscape. Points near the diagonal represent short-lived features, typically considered noise, while points farther away correspond to significant topological structures that persist across scales. For example, the H0 features highlight the number of distinct connected regions, while H1 features capture cyclic patterns in the loss landscape.

9.6 REAL-WORLD APPLICATIONS AND EXAMPLES

9.6.1 Biological Network Analysis

In the field of bioinformatics, topology plays a crucial role in understanding the complex interactions within biological networks, such as protein–protein interaction networks and gene regulatory networks. **TDA** enables researchers to uncover patterns and relationships that are not immediately

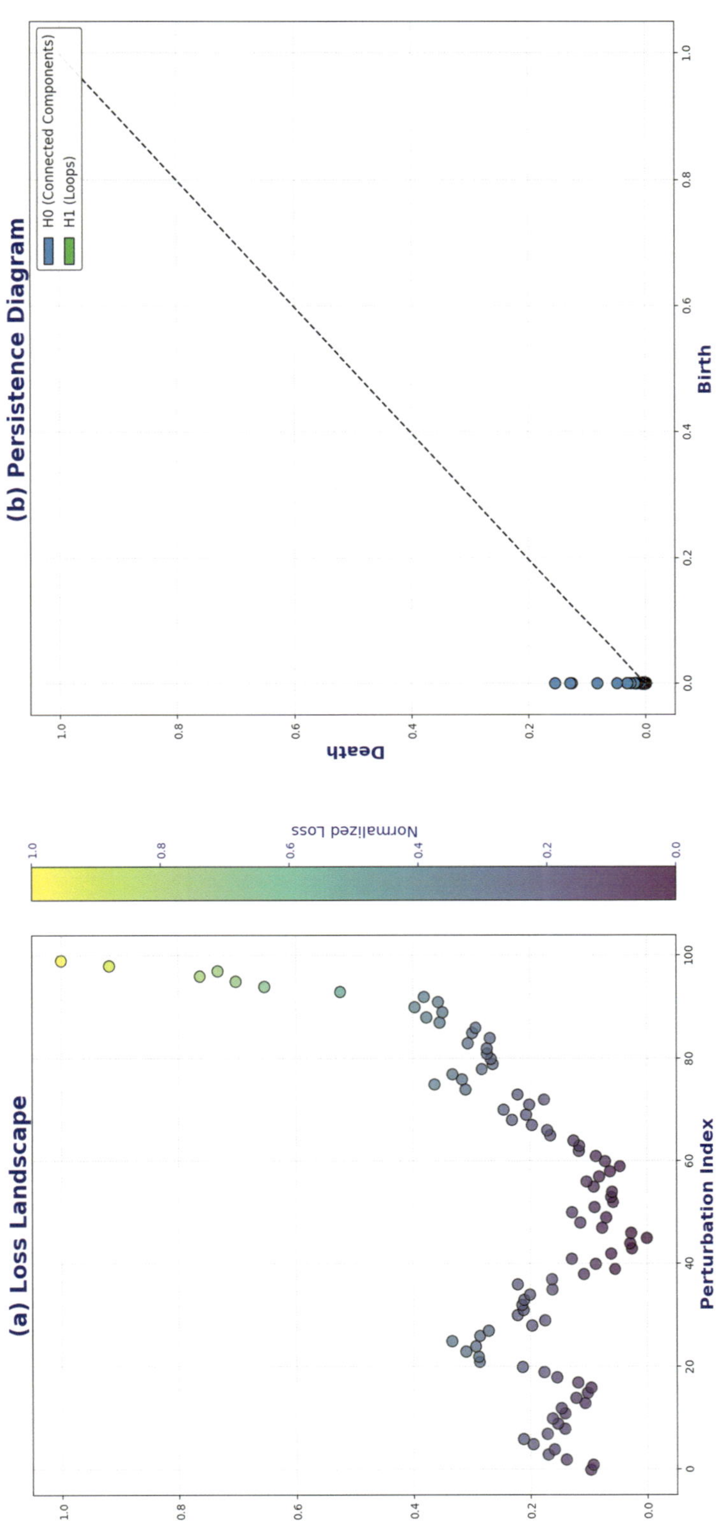

FIGURE 9.8 TDA in deep learning: (a) loss landscape and (b) persistence diagram.

apparent using traditional methods. For example, by analyzing the topological features of these networks, such as loops and connected components, scientists can identify key regulatory pathways and potential targets for drug development. The use of Betti numbers and persistent homology helps in quantifying the robustness and connectivity of these networks, leading to better insights into the underlying biological processes.

9.6.2 Material Science and Nanotechnology

Topology is also integral to material science, particularly in the design and analysis of nanomaterials. The topological properties of materials, such as the arrangement of atoms in a lattice or the connectivity of pores in a spongy material, can significantly influence their physical properties. For instance, topological insulators, materials that conduct electricity on their surface but not in their interior, are a prime example of how topological considerations guide the development of new materials with unique electrical properties. Understanding the topological structure of these materials allows scientists to design more efficient and resilient nanostructures for applications in electronics, energy storage, and catalysis.

9.6.3 Robotics and Autonomous Systems

In robotics, topology helps in the design and control of autonomous systems that must navigate complex environments. Topological maps, which abstract the environment into a network of connected regions, enable robots to plan and execute paths efficiently. For instance, in the development of autonomous vehicles, topological mapping allows the vehicle to understand its surroundings and navigate safely through dynamic and unpredictable environments. This approach is essential for tasks such as urban driving, where the vehicle must make decisions based on the connectivity and layout of roads, intersections, and other obstacles.

9.6.4 Neuroscience and Brain Connectivity

In neuroscience, topology provides valuable insights into the brain's connectivity and function. The human brain can be modeled as a complex network where neurons and synapses form intricate topological structures. By applying **TDA**, researchers can study the brain's connectivity patterns and understand how different regions interact to produce cognitive functions. For example, persistent homology has been used to analyze the topology of brain networks in patients with neurological disorders, revealing changes in connectivity that correlate with disease progression. This approach is crucial for developing more effective treatments and interventions for conditions like Alzheimer's.

9.6.5 Social Network Analysis

Topology is also applicable in the analysis of social networks, where it helps in understanding the structure and dynamics of human interactions. Social networks can be represented as graphs, with individuals as nodes and their relationships as edges. By studying the topological features of these graphs, such as clusters and communities, researchers can gain insights into social behavior, influence patterns, and the spread of information or diseases. For instance, topological analysis can identify influential individuals or groups within a network, which is valuable for targeted marketing campaigns, public health interventions, and understanding the spread of misinformation.

9.6.6 Financial Modeling and Risk Assessment

In finance, topology aids in modeling the complex relationships between assets and markets. TDA allows for the identification of persistent features in financial data, such as market cycles and anomalies, which can inform trading strategies and risk management practices. By analyzing the topological structure of financial networks, such as the connections between different markets or the relationships between assets in a portfolio, investors can better understand market dynamics and make more informed decisions. This approach is particularly useful in stress testing, where understanding the topological structure of financial systems can help in predicting how markets might react to extreme events.

9.7 HANDS-ON EXAMPLE

In this hands-on section, we will cover persistent homology, Betti numbers, and visualize high-dimensional data using **t-SNE** and **UMAP**.

9.7.1 Step 1. Install Required Libraries

In this section, we are installing and importing a variety of libraries essential for working with graph neural networks, dimensionality reduction, and topological data analysis. This combination of libraries enables us to work with graph-based neural networks, analyze high-dimensional data using dimensionality reduction techniques, and perform topological data analysis, which can uncover hidden structures in datasets.

```
!pip install spektral umap-learn gudhi
import numpy as np
import tensorflow as tf
from tensorflow.keras.layers import Input, Dense
from tensorflow.keras.models import Model
from spektral.layers import GCNConv
from spektral.utils import normalized_adjacency
from scipy.sparse import csr_matrix
import matplotlib.pyplot as plt
import networkx as nx
from sklearn.datasets import make_swiss_roll
from sklearn.manifold import TSNE
import umap
import gudhi as gd
from gudhi import CubicalComplex, SimplexTree
```

9.7.2 Step 2. Generate the Swiss Roll Dataset

In this line of code, we are generating a synthetic Swiss Roll dataset using the make_swiss_roll function from scikit-learn. This dataset is often used in machine learning to test and visualize dimensionality reduction techniques because of its complex, non-linear structure. The Swiss Roll is a 3D dataset that resembles a spiral rolled up in 3D space.

```
n_samples = 1000
noise = 0.05
X, color = make_swiss_roll(n_samples, noise=noise)
```

9.7.3 Step 3. Compute the Persistent Homology Using Gudhi

In this section, we are performing TDA on the Swiss Roll dataset using the Gudhi library. Specifically, we are computing the persistence diagram of the dataset through the following tasks: Rips Complex Construction, Simplex Tree Creation and Persistence Diagram Computation.

```
rips_complex = gd.RipsComplex(points=X, max_edge_length=1.0)
simplex_tree = rips_complex.create_simplex_tree(max_dimension=2)
diag = simplex_tree.persistence()
```

9.7.4 Step 4. Extract Betti Numbers

In this line of code, we are calculating the Betti numbers of the Swiss Roll dataset using the simplex tree that was constructed earlier through the Rips complex. By calculating Betti numbers, we gain a deeper understanding of the inherent structure of the Swiss Roll dataset, revealing important topological properties that may not be easily visible in lower-dimensional projections or traditional data analysis methods.

```
betti_numbers = simplex_tree.betti_numbers()
```

9.7.5 Step 5. Apply t-SNE to Reduce Dimensions

In this section, we are applying t-SNE (t-Distributed Stochastic Neighbor Embedding) to the Swiss Roll dataset for dimensionality reduction. **t-SNE** is particularly useful for visualizing complex datasets with non-linear structures, as it reveals clusters and local patterns that might not be captured by linear methods like **PCA**.

```
tsne = TSNE(n_components=2, random_state=42)
X_tsne = tsne.fit_transform(X)
```

9.7.6 Step 6. Apply UMAP to Reduce Dimensions

In this section, we are applying UMAP (Uniform Manifold Approximation and Projection) to the Swiss Roll dataset for dimensionality reduction. **UMAP** is favored in many cases due to its balance between preserving local and global structures in the data, and the results can be visualized to reveal patterns, clusters, or the underlying structure of complex datasets like the Swiss Roll.

```
umap_reducer = umap.UMAP(random_state=42)
```

```
X_umap = umap_reducer.fit_transform(X)
```

9.7.7 Step 7. Plotting All Graphs in One Frame

In this section, we are visualizing the Swiss Roll dataset and its dimensionality-reduced representations using **t-SNE**, **UMAP**, and a Persistence Diagram to explore the topological features of the dataset.

```
fig, axs = plt.subplots(2, 2, figsize=(16, 12))
# Plot a: Swiss Roll Dataset
ax = fig.add_subplot(2, 2, 1, projection='3d')
ax.scatter(X[:, 0], X[:, 1], X[:, 2], c=color, cmap=plt.
cm.Spectral)
ax.set_title("a) Swiss Roll Dataset")
# Plot b: t-SNE Embedding of the Swiss Roll Dataset
axs[0, 1].scatter(X_tsne[:, 0], X_tsne[:, 1], c=color, cmap=
plt.cm.Spectral)
axs[0, 1].set_title("b) t-SNE Embedding of the Swiss Roll
Dataset")
axs[0, 1].set_xlabel("t-SNE 1")
axs[0, 1].set_ylabel("t-SNE 2")
# Plot c: UMAP Embedding of the Swiss Roll Dataset
axs[1, 0].scatter(X_umap[:, 0], X_umap[:, 1], c=color, cmap=
plt.cm.Spectral)
axs[1, 0].set_title("c) UMAP Embedding of the Swiss Roll
Dataset")
axs[1, 0].set_xlabel("UMAP 1")
axs[1, 0].set_ylabel("UMAP 2")
# Plot d: Persistence Diagram
gd.plot_persistence_diagram(diag, axes=axs[1, 1])
axs[1, 1].set_title("d) Persistence Diagram")
plt.tight_layout()
plt.show()
```

Figure 9.9 provides a detailed analysis of the Swiss Roll dataset using dimensionality reduction techniques and persistent homology. Figure 9.9a illustrates the Swiss Roll dataset in its original 3D form, with points color-coded to represent variations along the z-axis. This dataset is a well-known example of a non-linear manifold in 3D space, often used to evaluate the effectiveness of dimensionality reduction algorithms. Figure 9.9b presents a 3D **t-SNE** embedding of the dataset, where the high-dimensional Swiss Roll is projected into a lower-dimensional space. The **t-SNE** embedding focuses on preserving local neighborhoods, effectively revealing clusters and the dataset's overall structure. Figure 9.9c displays the results of applying **UMAP** to the Swiss Roll dataset, also reducing its dimensionality to two dimensions. **UMAP** emphasizes both local and global structural preservation, offering a clear representation of the Swiss Roll's continuity and manifold properties, sometimes surpassing **t-SNE** in retaining the dataset's global relationships. Figure 9.9d shows the persistence diagram of the Swiss Roll dataset, capturing its topological features across various scales. The **H0** features represent connected components, while the **H1** features correspond

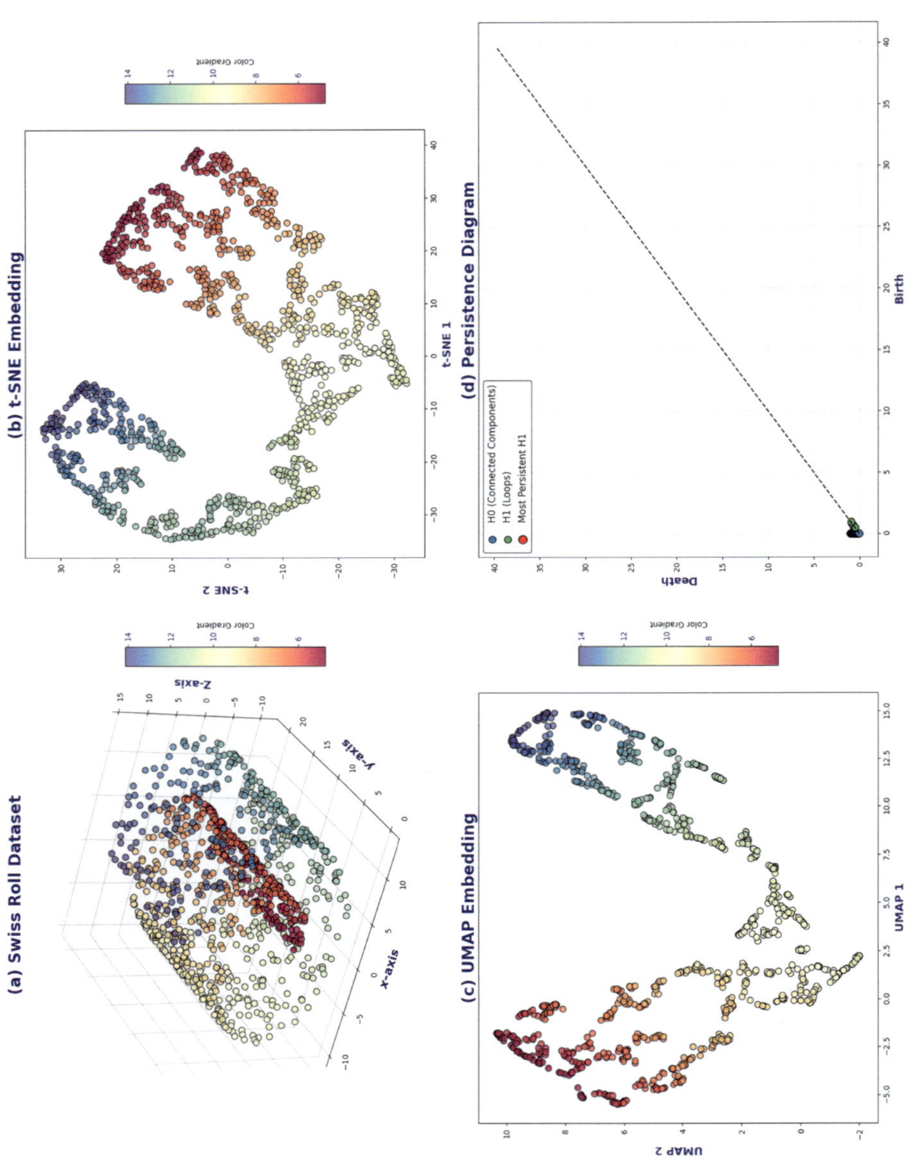

FIGURE 9.9 (a) 3D visualization of the Swiss Roll dataset, (b) 2D t-SNE embedding of the Swiss Roll dataset, (c) 2D UMAP embedding of the Swiss Roll dataset, (d) persistence diagram for the Swiss Roll dataset.

to loops in the data. The most persistent **H1** loop, highlighted in red, signifies the prominent spiral structure inherent in the dataset.

9.8 COMMON MISTAKES AND TROUBLESHOOTING TIPS

9.8.1 Misunderstanding Topological Concepts

- *Mistake:* Misinterpreting fundamental topological concepts like continuous transformations, homeomorphisms, and Betti numbers can lead to incorrect applications and analyses.
- *Tip:* Review foundational resources on topology and **TDA**. Use visual aids and interactive tools to solidify your understanding of these abstract concepts.

9.8.2 Overfitting in Deep Networks

- *Mistake:* Designing neural networks with excessive depth or width without adequate regularization, leading to overfitting.
- *Tip:* Use techniques like dropout, batch normalization, and early stopping to prevent overfitting. Regularly validate your model on separate datasets to monitor for overfitting.

9.8.3 Ignoring the Impact of Network Architecture

- *Mistake:* Neglecting the importance of network architecture and its influence on convergence and performance.
- *Tip:* Experiment with different architectures and configurations. Use grid search or random search to find the optimal structure for your specific problem.

9.8.4 Visualizing High-Dimensional Data

- *Mistake:* Struggling to visualize high-dimensional data and loss landscapes, leading to misinterpretation of model behavior.
- *Tip:* Utilize dimensionality reduction techniques like **t-SNE** or **PCA** for visualizing high-dimensional data. Understand the limitations and ensure the reduced dimensions retain critical information.

9.8.5 Computational Cost and Efficiency

- *Mistake:* Underestimating the computational cost of applying topological analysis to large datasets and neural networks.
- *Tip:* Optimize your code and use efficient libraries for **TDA** computations. Consider using cloud computing resources or distributed computing to handle large-scale analyses.

9.8.6 Integrating TDA with Neural Networks

- *Mistake:* Failing to effectively integrate **TDA** insights with neural network models, resulting in limited improvements.
- *Tip:* Apply **TDA** techniques to extract meaningful features and incorporate them into your neural network pipeline. Validate the added value through controlled experiments.

9.8.7 Overfitting to Topological Features

- *Mistake:* Overfitting to topological features extracted from the data, leading to models that do not generalize well.
- *Tip:* Ensure a balance between topological features and other relevant data features. Use cross-validation to assess the generalization performance of your models.

9.8.8 Bridging Theory and Practice

- *Mistake:* Struggling to bridge the gap between theoretical insights from topology and their practical application in deep learning.
- *Tip:* Focus on practical implementation and empirical validation. Collaborate with domain experts to translate theoretical concepts into actionable strategies for improving neural network models.

9.9 REVIEW QUESTIONS

1. What is topology, and why is it important to understand the properties of space?
2. How do continuous transformations define topological equivalence? Provide examples.
3. How does the topology of a neural network influence its performance and convergence properties?
4. Explain the differences between network depth and width. What are the pros and cons of increasing each?
5. What are skip connections, and how do they benefit deep neural networks like ResNet?
6. Describe the advantages and challenges of using RNNs for sequential data.
7. What are the benefits and potential problems associated with using the **ReLU** activation function?
8. What is persistent homology, and how can it be applied to analyze neural network loss landscapes?
9. How do Betti numbers help understand the topological complexity of data?
10. How can integrating **TDA** and neural networks lead to more robust and effective models?

9.10 PROGRAMMING QUESTIONS

9.10.1 Easy

Generate a 3D S-curve dataset and visualize it.

1. Generate the S-curve dataset.
2. Apply t-SNE to the dataset and visualize the 2D embedding.
3. Apply UMAP to the dataset and visualize the 2D embedding.
4. Compare the embeddings from t-SNE and UMAP.

9.10.2 Medium

For a given high-dimensional dataset, compute its persistence diagram using different metrics (e.g., Euclidean, cosine) and compare the results.

1. Select a high-dimensional dataset (e.g., **MNIST** or **CIFAR-10**).
2. Compute the pairwise distance matrices using different metrics.
3. Generate persistence diagrams for each metric.
4. Compare the persistence diagrams and interpret the results.

9.10.3 Hard

Using the **MNIST** dataset, compute the persistence diagrams for each image and use the resulting topological features as input to a machine learning classifier.

1. Preprocess the **MNIST** dataset to compute persistence diagrams for each image.
2. Extract topological features (e.g., persistence pairs) from the diagrams.
3. Train a machine learning classifier using these topological features.
4. Evaluate the classification performance using standard metrics (accuracy, precision, recall).
5. Compare the performance to a classifier trained on raw pixel values.

10 Harmonic Analysis for CNNs

10.1 INTRODUCTION

Understanding signals in both time and frequency is important for digital signal processing. Tools like Fourier (FT) and Wavelet (WT) transform help find hidden patterns in raw data, such as radio waves or images used in machine learning. In deep learning, convolutional neural networks (CNNs) examine images by identifying features from pixel information. This is similar to harmonic analysis, where frequency methods explain how signals behave. By combining deep learning with harmonic ideas, we can better understand how CNNs work and find more efficient ways to compute. In this chapter, we explore these ideas.

10.2 FOURIER ANALYSIS

10.2.1 Fundamentals of Fourier Analysis

The strength of Fourier analysis is that it can break any complex signal into simple sine and cosine waves. This separation lets us look closely at the signal's different frequencies. Understanding these frequencies is important in many areas, from engineering to science. It helps us design filters, study waveforms, and process data more effectively.

10.2.2 Fourier Transform

The Fourier transform (FT) is a mathematical operation that converts a function from its original domain (typically time or space) into the frequency domain. This transformation provides a frequency spectrum that shows how much of each frequency is present in the original signal. For a continuous time-domain function **f(t)**, the FT $\mathbf{F}(\mathrm{w})$ is defined as:

$$F(\omega) = \int_{-\infty}^{\infty} f(t) e^{-j\omega t} dt$$

where:

- $\mathbf{f}(\mathbf{t})$ is the function in the time domain,
- $\boldsymbol{\omega}$ is the angular frequency, which is related to the frequency **f** by $\boldsymbol{\omega} = 2\pi\mathbf{f}$, and
- $\mathbf{e}^{-\mathbf{jwt}}$ is the complex exponential function, where **j** is the imaginary unit (i.e., $\mathbf{j}^2 = -1$).

 DOI: 10.1201/9781032690742-10

This transformation is similar to decomposing a complex musical chord into individual notes. Each note corresponds to a frequency component of the chord, and the FT helps identify these components. Let's consider a time-domain signal composed of two sine waves:

$$f(t) = 3\sin(2\pi \times 50t) + 2\sin(2\pi \times 120t)$$

Here:

- $3\sin(2\pi \times 50t)$ is a sine wave with an amplitude of 3 and a frequency of 50 Hz.
- $2\sin(2\pi \times 120t)$ is a sine wave with an amplitude of 2 and a frequency of 120 Hz.

We plan to use the FT to identify the frequency components and their amplitudes in **f(t)**.

Step 1. Apply the FT: Think of the FT as a tool that takes your complex sound and breaks it down into its basic building blocks, the pure tones (sine waves). For our sound, which is:

$$F(\omega) = \int_{-\infty}^{\infty} \left[3\sin(2\pi \times 50t) + 2\sin(2\pi \times 120t)\right] e^{-j\omega t}\, dt$$

Using the linearity property of the FT, this can be separated into:

$$F(\omega) = 3\int_{-\infty}^{\infty} \sin(2\pi \times 50t) e^{-j\omega t}\, dt + 2\int_{-\infty}^{\infty} \sin(2\pi \times 120t) e^{-j\omega t}\, dt$$

We want to find out the frequencies (50 and 120 Hz) and their strengths (3 and 2).
Step 2. Compute the Transform for Each Sine Wave: The FT of sin(2π**f**ᵗ) is given by:

$$\mathcal{F}\{\sin(2\pi ft)\} = \frac{j}{2}\left[\delta(\omega - 2\pi f) - \delta(\omega + 2\pi f)\right]$$

Applying this to each sine component:

1. For the 50 Hz tone: $3\sin(2\pi \times 50t)$: The math shows that there's a spike (a sharp point) at 50 Hz. The strength of this spike is related to the number 3 in front of the sine wave.

$$3 \times \frac{j}{2}\left[\delta(\omega - 2\pi \times 50) - \delta(\omega + 2\pi \times 50)\right] = \frac{3j}{2}\left[\delta(\omega - 100\pi) - \delta(\omega + 100\pi)\right]$$

2. For the 120 Hz tone: $2\sin(2\pi \times 120t)$: Similarly, there 's another spike at 120 Hz, and its strength comes from the number 2.

$$2 \times \frac{j}{2}\left[\delta(\omega - 2\pi \times 120) - \delta(\omega + 2\pi \times 120)\right] = j\left[\delta(\omega - 240\pi) - \delta(\omega + 240\pi)\right]$$

Step 3. Combine the Results:

$$F(\omega) = \frac{3j}{2}\delta(\omega - 100\pi) - \frac{3j}{2}\delta(\omega + 100\pi) + j\delta(\omega - 240\pi) - j\delta(\omega + 240\pi)$$

The FT $\mathbf{F(\omega)}$ reveals spikes (delta functions) at specific angular frequencies: $\omega = 100\pi$ rad/s (which corresponds to 50 Hz) and $\omega = 240\pi$ rad/s (which corresponds to 120 Hz). The coefficients in front of the delta functions indicate the amplitude and phase of each frequency component. At 50 Hz the amplitude is $\frac{3\mathbf{j}}{2}$ and at 120 Hz the amplitude is $\mathbf{j}$. As the original signal $\mathbf{f(t)}$ is real-valued, the FT exhibits symmetry: Positive frequencies (e.g., 50 and 120 Hz) have corresponding negative frequencies. The magnitudes of the Fourier coefficients correspond to the amplitudes of the sine waves.

Figure 10.1 illustrates three individual sine waves at 5, 50, and 120 Hz, along with their combined time-domain signal and their frequency-domain representation. In Figure 10.1a, the slowest oscillation at 5 Hz is shown, displaying only a few cycles over the 1-second interval. Figure 10.1bintroduces the 50 Hz wave, which has considerably more peaks and troughs in the same time frame. Figure 10.1c presents the highest-frequency component at 120 Hz, where many closely spaced oscillations occur within one second. These three sine waves are summed in Figure 10.1d, producing a complex waveform in the time domain whose various undulations result from the superposition of 5, 50, and 120 Hz components. In Figure 10.1e, the FT of this combined signal shows three clear spectral peaks corresponding to each individual sine wave, verifying that all three frequencies are indeed present. Finally, Figure 10.1f zooms in on the 0–100 Hz region, highlighting the 5 and 50 Hz peaks more clearly while the 120 Hz peak remains outside the zoomed frequency range.

10.2.3 Inverse Fourier Transform

The inverse Fourier transform (IFT) reverses the FT process, converting a function from the frequency domain back to its original domain (often time or space). This operation is crucial for reconstructing the original signal after analysis or manipulation in the frequency domain. The inverse transform of $F(\omega)$ is given by:

$$f(t) = \frac{1}{2\pi}\int_{-\infty}^{\infty} F(\omega) e^{j\omega t}\, d\omega$$

where:

- $\mathbf{f(t)}$ is the recovered time-domain function.
- The factor $\frac{1}{2\pi}$ ensures proper scaling during the transformation.

Understanding the IFT is essential for practical applications where signals are analyzed in the frequency domain and then transformed back to their original form for further processing or interpretation. Suppose we plan to reconstruct the original time-domain signal from its frequency components using the IFT. Given frequency components: A sine wave with amplitude 3 at 50 Hz and a sine wave with amplitude 2 at 120 Hz. These components can be represented in the frequency domain as spikes at their respective frequencies.

Step 1. Express the Frequency Components: Each sine wave in the frequency domain can be represented using delta functions (spikes):

$$F(\omega) = 3\left[\delta(\omega - 100\pi) - \delta(\omega + 100\pi)\right] + 2\left[\delta(\omega - 240\pi) - \delta(\omega + 240\pi)\right]$$

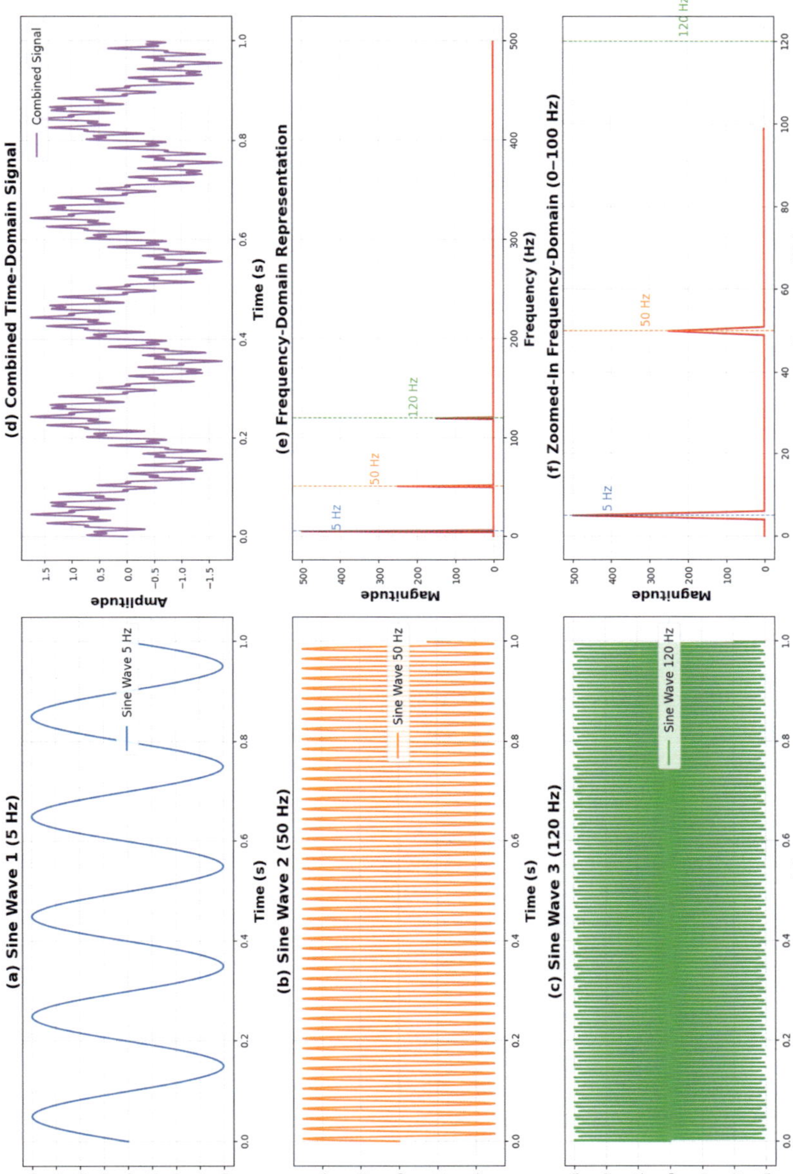

FIGURE 10.1 (a) Sine wave 1 (5 Hz). This waveform oscillates at the slowest frequency of the three, completing 5 cycles per second. (b) Sine wave 2 (50 Hz). A higher-frequency sinusoid that completes 50 cycles per second. (c) Sine wave 3 (120 Hz). The fastest oscillation among the three waves, with 120 cycles per second. (d) Combined time-domain signal. The result of summing all three sine waves (5, 50, and 120 Hz). (e) Frequency-domain representation. The FT of the combined signal, showing distinct peaks at 5, 50, and 120 Hz. (f) Zoomed-in frequency domain (0–100 Hz). A closer look at the low- and mid-frequency peaks (5 and 50 Hz), with the 120 Hz peak lying outside this zoom range.

(Note: $\omega = 2\pi f$, so 50 Hz becomes 100π rad/s and 120 Hz becomes 240π rad/s.)

Step 2. Apply the IFT: Using the IFT formula:

$$f(t) = \frac{1}{2\pi}\left[3\left(e^{j100\pi t} - e^{-j100\pi t}\right) + 2\left(e^{j240\pi t} - e^{-j240\pi t}\right)\right]$$

Step 3. Simplify Using Euler's Formula: Recall that $e^{j\theta} - e^{-j\theta} = 2j\sin(\theta)$:

$$f(t) = \frac{1}{2\pi}\left[3\times 2j\sin(100\pi t) + 2\times 2j\sin(240\pi t)\right] \text{ and then}$$

$$f(t) = \frac{1}{2\pi}\left[6j\sin(100\pi t) + 4j\sin(240\pi t)\right]$$

As the original signal is real, the imaginary units cancel out:

$$f(t) = \frac{6}{2\pi}\sin(100\pi t) + \frac{4}{2\pi}\sin(240\pi t) \text{ and then } f(t) = \frac{3}{\pi}\sin(100\pi t) + \frac{2}{\pi}\sin(240\pi t)$$

Simplifying further, recognizing that 100π rad/s is 50 Hz and 240π rad/s is 120 Hz:

$$f(t) = 3\sin(2\pi\times 50t) + 2\sin(2\pi\times 120t)$$

This matches our original time-domain signal:

$$f(t) = 3\sin(2\pi\times 50t) + 2\sin(2\pi\times 120t)$$

Figure 10.2 depicts on the same three-sine-wave combination but provides additional insights into reconstruction and error analysis. Figure 10.2a again displays the time-domain signal formed by superimposing 5, 50, and 120 Hz waves, reflecting a more complex pattern compared to any single wave alone. Figure 10.2b shows the frequency-domain representation, showing distinct peaks at 5, 50, and 120 Hz that confirm the presence of all three components. Figure 10.2c illustrates how the inverse fast Fourier transform (IFFT) can recover the original time-domain waveform from its frequency-domain data, demonstrating near-perfect alignment with the combined signal. Figure 10.2d shows the minimal difference between the original and reconstructed signals, which hovers near numerical precision and confirms the accuracy of the forward and IFT process.

10.3 WAVELETS

Wavelet analysis is a useful tool in signal processing, offering capabilities that extend beyond those of Fourier analysis. While Fourier analysis breaks signals into sine and cosine waves, wavelet analysis uses wavelets, functions that can capture both frequency and time (or space) information simultaneously. This makes wavelets particularly effective for analyzing non-stationary signals, where the characteristics of the signal change over time or space. Wavelets are especially useful in applications where localized variations in the signal must be captured and analyzed. Wavelets are oscillating functions that start and end at zero, with their amplitude peaking in between. This shape gives wavelets the ability to zoom in on localized signal features. Their flexibility makes them ideal for multi-resolution analysis, enabling the study of signals at different scales. By adjusting the scale of the wavelet, we can analyze both the fine details and broader patterns in the signal, making wavelet analysis highly effective for signals that vary over time or space.

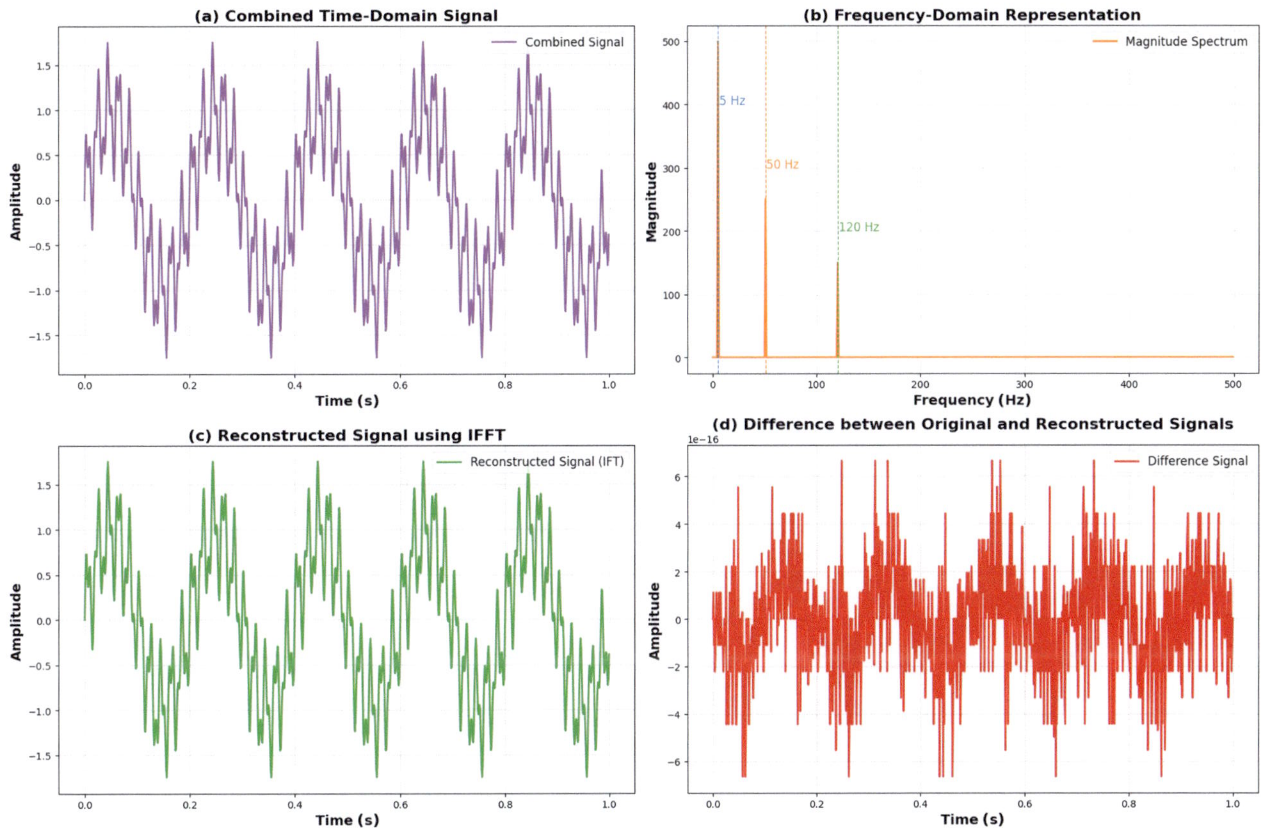

FIGURE 10.2 (a) Combined time-domain signal. This is the same three-wave sum shown in Figure 1(d), plotted over 1 second. (b) Frequency-domain representation. The magnitude spectrum again reveals peaks at 5, 50, and 120 Hz. (c) Reconstructed signal (IFFT). The IFT of the frequency-domain data, showing a near-identical recovery of the original time-domain signal. (d) Difference between original and reconstructed signals. The minimal discrepancy on the order of numerical precision, verifies the accuracy of the FT and IFT.

10.3.1 Wavelet Transform

The wavelet transform (**WT**) is an advanced analytical method that converts signals from the time domain into a series of coefficients based on shifted and scaled versions of a predetermined base function, known as a *mother wavelet*. This transformation is particularly adept at handling non-stationary signals, those whose frequency content changes over time. The continuous wavelet transform (**CWT**) of a function **f(t)** is mathematically represented as:

$$CWT(\tau, s) = \frac{1}{\sqrt{|s|}} \int_{-\infty}^{\infty} f(t)\, \overline{\psi\left(\frac{t-\tau}{s}\right)} dt$$

Here, $\boldsymbol{\tau}$ and **s** are parameters that control the translation (shift) and scale (compression or stretching) of the wavelet, respectively. $\boldsymbol{\psi}$**(t)** is the mother wavelet, a function localized in both time and frequency. The normalization factor $\frac{1}{\sqrt{|s|}}$ ensures that the wavelet has the same energy at each scale. Wavelets are unique in that they are localized in time and frequency, unlike the sinusoids used in FTs, which extend infinitely. This localization allows wavelets to precisely capture and analyze transient features and abrupt changes in a signal. The adaptability of wavelets in terms of scale and translation makes them ideal for analyzing signals that exhibit features at multiple scales or that have significant local variations in time or frequency. Consider an example where a heartbeat signal is analyzed using the WT. A heartbeat signal, characterized by sharp spikes followed by slow waves, presents challenges in identifying features like the **QRS** complex and the **T** wave when using Fourier analysis due to its non-stationary nature. Using a WT, the signal is decomposed into coefficients that represent different frequency components at different times. For instance, selecting a wavelet such as the Daubechies wavelet, which closely resembles the sharp spikes of a **QRS** complex, would allow for efficient isolation and analysis of these features without interference from slower-moving trends in the signal. This ability to customize the wavelet to match specific features of the signal is a key advantage of the WT, enabling more effective signal analysis and feature detection. Consider a simple time-domain signal composed of two parts:

1. A sharp spike (representing a transient event) lasting from **t** = 1 to **t** = 2 seconds.
2. A slow sine wave with a frequency of 5 Hz active throughout the signal duration.

The signal **f(t)** can be represented as:

$$f(t) = \begin{cases} 10, & 1 \le t \le 2 \\ 5\sin(2\pi \times 5t), & \text{otherwise} \end{cases}$$

Step 1. Choose a Mother Wavelet: Let's use the Daubechies 4 (db4) wavelet, known for its ability to handle sharp transitions effectively.

Step 2. Apply the WT: The **CWT** will analyze the signal at different scales and positions. For simplicity, we'll examine two scales: Scale 1 captures high-frequency components, such as sharp spikes, while Scale 8 captures low-frequency components, like a slow sine wave.

Step 3. Compute Wavelet Coefficients: At scale 1, the wavelet is narrow and highly responsive to rapid changes. The sharp spike occurring between t = 1and t = 2 generates significant wavelet coefficients, indicating a strong transient event. In contrast, the sine wave produces minimal coefficients at this scale due to its low-frequency nature. At Scale 8, the wavelet is wider, enabling it to capture slower variations. The sine wave at 5 Hz results in prominent wavelet coefficients, reflecting its sustained oscillation. The sharp spike contributes less to the coefficients at this scale due to the wavelet's lower sensitivity to high-frequency events.

Step 4: Reconstruct the Signal Using Selected Scales: To reconstruct the original signal, we can combine the wavelet coefficients from both scales:

$$f(t) \approx \text{Reconstruction from Scale1} + \text{Reconstruction from Scale 8}$$

This combination captures both the transient spike and the underlying sine wave.

10.3.2 Applications of WT

WTs have broad applications across various fields. In image compression, they are the backbone of the JPEG 2000 standard, offering efficient, high-quality compression. Wavelets are also key in noise reduction, where they remove noise while preserving important signal features. In signal processing, wavelets enable time-frequency analysis, making them ideal for examining signals that change over time. Additionally, in medical imaging, WTs help enhance image analysis by highlighting crucial structures and details, improving diagnostic accuracy. Consider a signal composed of two sine waves with frequencies of 5 and 20 Hz.

Figure 10.3 illustrates the relationship between the original signal and its WT. Figure 10.3a shows the time-domain signal, which is a combination of two sine waves. The signal's varying pattern over time is a result of these interacting frequencies. Figure 10.3b represents the WT of the signal, revealing how its frequency components change over time. The color intensity in the plot corresponds to the magnitude of the wavelet coefficients, highlighting the signal's decomposition into different scales, where the red regions indicate higher magnitudes.

10.4 CONVOLUTION IN THE FREQUENCY DOMAIN FOR CNNS

CNNs are very important in deep learning, especially for tasks like processing images and videos. The key part of CNNs is the convolution operation. In this operation, a filter (or kernel) moves over an input (like an image) to calculate sums of products for each area it covers. When we look at convolution in the frequency domain, it has big computational benefits, especially for large filters or specific tasks. It can change the convolution process into simple multiplications, which makes computations faster in some situations.

10.4.1 Convolution Theorem

The convolution theorem explains how two important math operations, convolution and multiplication, are connected using the FT. This theorem is a powerful tool for studying and changing signals in both time and frequency. When you convolve two signals in the time domain, you mix them together to create a new signal that combines features from both original signals. Similarly, when you multiply the FTs of two signals in the frequency domain, it has the same effect as convolving the two signals in time. This means that by working in the frequency domain, we can simplify the convolution process by just multiplying the FTs, which can make calculations faster. The convolution theorem can be expressed mathematically as:

$$\mathcal{F}\{f(t) * g(t)\} = F(\omega) \cdot G(\omega)$$

where:

- $\mathbf{f}(\mathbf{t})$ and $\mathbf{g}(\mathbf{t})$ are signals in the time domain,
- $\mathbf{F}(\omega)$ and $\mathbf{G}(\omega)$ are their corresponding Fourier,

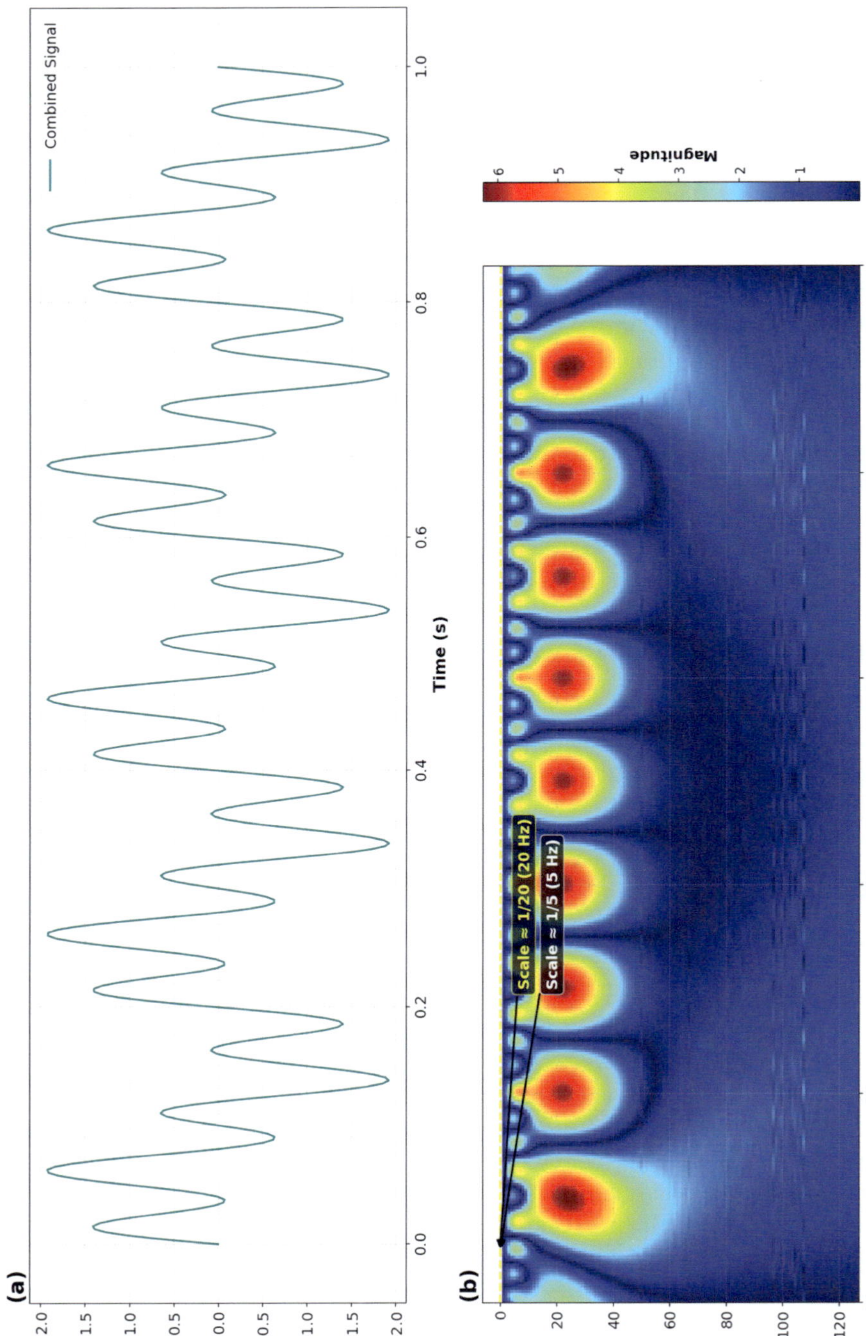

FIGURE 10.3 (a) Original signal (time domain) and (b) WT (frequency domain).

- * denotes the convolution operation in the time domain, and
- $\cdot$ represents multiplication in the frequency domain.

The applications of the convolution theorem include signal processing, where it allows for efficient computation of convolutions by transforming signals to the frequency domain, performing element-wise multiplication, and then using the IFT to return the result to the time domain. This method is often faster than direct convolution, especially for large signals. In filter design, the theorem is fundamental, as it enables the specification of filtering effects in the frequency domain. By defining the filter's frequency response and multiplying it with the FT of the input signal, the filtered signal can be obtained efficiently. In image processing, convolutional filters like edge detectors, blur filters, and sharpening filters can be applied more easily using the convolution theorem, simplifying the implementation of complex filtering operations. Additionally, in system analysis, particularly for linear time-invariant (**LTI**) systems, the theorem aids in understanding system responses to various inputs by analyzing the system's impulse response in the frequency domain, allowing prediction of the output for any given input. Consider two time-domain signals, **f(t)** and **g(t)**. To convolve these signals using the convolution theorem:

1. Compute the FTs of **f(t)** and **g(t)**:

$$F(\omega) = \mathcal{F}\{f(t)\}, \quad G(\omega) = \mathcal{F}\{g(t)\}$$

2. Multiply the FTs in the frequency domain:

$$H(\omega) = F(\omega) \cdot G(\omega)$$

3. Compute the IFT of the product to obtain the convolved signal in the time domain:

$$h(t) = \mathcal{F}^{-1}\{H(\omega)\} = \mathcal{F}^{-1}\{F(\omega) \cdot G(\omega)\}$$

Consider two simple continuous-time signals:

1. Signal **f(t):**

$$f(t) = \begin{cases} 1, & 0 \le t \le 1 \\ 0, & \text{otherwise} \end{cases}$$

A rectangular pulse of amplitude 1 lasting from **t** = 0 to **t** = 1 second.

2. Signal **g(t)**:

$$g(t) = \begin{cases} 1, & 0 \le t \le 1 \\ 0, & \text{otherwise} \end{cases}$$

Another identical rectangular pulse of amplitude 1 lasting from **t** = 0 to **t** = 1 second.

Step 1: Compute the FTs of **f(t)** and **g(t)**
The FT of a rectangular pulse **f(t)** is given by:

$$F(\omega)=\int_{-\infty}^{\infty} f(t)e^{-j\omega t}dt=\int_{0}^{1} e^{-j\omega t}dt=\frac{1-e^{-j\omega}}{j\omega}$$

Similarly, the FT of **g(t)** is:

$$G(\omega)=\int_{-\infty}^{\infty} g(t)e^{-j\omega t}dt=\int_{0}^{1} e^{-j\omega t}dt=\frac{1-e^{-j\omega}}{j\omega}$$

Step 2: Multiply the FTs in the frequency domain
According to the convolution theorem:

$$\mathcal{F}\{f(t)*g(t)\}=F(\omega)\cdot G(\omega)$$

Multiply **F(ω)** and **G(ω)**:

$$F(\omega)\cdot G(\omega)=\left(\frac{1-e^{-j\omega}}{j\omega}\right)^2=\frac{(1-e^{-j\omega})^2}{(j\omega)^2}$$

Step 3: Compute the IFT to obtain the convolved signal. To find **f(t) ∗ g(t)**, take the IFT of **F(ω) · G(ω)**:

$$f(t)*g(t)=\mathcal{F}^{-1}\left\{\frac{(1-e^{-j\omega})^2}{(j\omega)^2}\right\}$$

Simplifying the expression:

$$(1-e^{-j\omega})^2=1-2e^{-j\omega}+e^{-j2\omega} \quad \text{and} \quad \frac{(1-e^{-j\omega})^2}{(j\omega)^2}=\frac{1}{(j\omega)^2}-\frac{2e^{-j\omega}}{(j\omega)^2}+\frac{e^{-j2\omega}}{(j\omega)^2}$$

IFT:

$$f(t)*g(t)=\mathcal{F}^{-1}\left\{\frac{1}{(j\omega)^2}\right\}-2\mathcal{F}^{-1}\left\{\frac{e^{-j\omega}}{(j\omega)^2}\right\}+\mathcal{F}^{-1}\left\{\frac{e^{-j2\omega}}{(j\omega)^2}\right\}$$

Using standard FT pairs:

$$\mathcal{F}^{-1}\left\{\frac{1}{(j\omega)^2}\right\}=t\cdot u(t) \quad \text{and} \quad \mathcal{F}^{-1}\left\{\frac{e^{-j\omega k}}{(j\omega)^2}\right\}=(t-k)\cdot u(t-k)$$

where **u(t)** is the unit step function.

Applying these:

$$f(t)*g(t)=t\cdot u(t)-2(t-1)\cdot u(t-1)+(t-2)\cdot u(t-2)$$

Final convolved signal:

$$f(t)*g(t)=\begin{cases}0, & t<0\\ t, & 0\le t\le 1\\ 2-t, & 1<t\le 2\\ 0, & t>2\end{cases}$$

A triangular pulse that rises linearly from 0 to 1 between **t** = 0 and **t** = 1, then decreases linearly from 1 to 0 between **t** = 1 and **t** = 2. To ensure our frequency-domain approach is correct, let's perform the convolution directly in the time domain.

$$(f*g)(t)=\int_{-\infty}^{\infty} f(\tau)g(t-\tau)d\tau$$

Given the definitions of **f(t)** and **g(t)**, the limits of integration are reduced based on the overlap of the signals.

1. For 0 ≤ **t** ≤ 1:

$$(f*g)(t)=\int_{0}^{t} 1\cdot 1 d\tau = t$$

2. For 1 < **t** ≤ 2:

$$(f*g)(t)=\int_{t-1}^{1} 1\cdot 1 d\tau = 2-t$$

3. For **t** < 0 and **t** > 2:

$$(f*g)(t)=0$$

This matches the result obtained using the convolution theorem.

10.4.2 Implication for CNNs

Convolution is a basic operation in CNNs, where a filter (or kernel) moves over an input, like an image, to calculate the sum of element-wise products. Usually, this convolution happens in the spatial domain, meaning the filter works directly with the image. This approach is easy to understand and is commonly used in tasks like image processing, where the filter finds features by repeatedly applying to small parts of the input, creating an output feature map. While this method is simple, it can become very slow when the filter size grows. Looking at convolution from the frequency domain can offer useful insights, especially for tasks that require a lot of computation. The convolution theorem says that convolution in the spatial domain is the same as multiplication in the frequency domain. In practice, this means that instead of convolving directly in the spatial domain, we can transform both the input and the filter into the frequency domain using the FT. Once in the frequency domain, convolution turns into a simple element-wise multiplication. After multiplying, we use the **IFT** to convert the result back to the spatial domain, getting the final output. This method can save a lot of computation time, especially with large filters, because the complexity of convolution in the frequency domain grows more slowly compared to the spatial domain. In the spatial domain, the time it takes to perform convolution increases with the filter size, which is a problem for large filters. However, in the frequency domain, using the fast Fourier transform (**FFT**) can reduce this complexity, making it much more efficient for large-scale operations. For small filters, traditional spatial convolution is usually faster, but for larger filters or big datasets, frequency-domain convolution can provide quicker results. This is especially useful in large-scale image processing or signal processing tasks. Deciding whether to use the spatial or frequency domain depends on the specific problem, available resources, and filter size. If you're working with standard-sized filters and have enough computational power, spatial domain convolution is still practical and effective. On the other hand, for larger filters or when computational resources are limited, the frequency-domain method can offer significant performance improvements. Additionally, understanding convolution in the frequency domain can help develop more efficient CNN designs or specialized approaches that combine the strengths of both domains. By exploring frequency-domain convolution, you can optimize parts of CNN design, improve computational efficiency, or customize the convolution process for tasks that use large filters, such as video processing or high-resolution image analysis. For example, consider a scenario where a CNN is processing high-resolution images with very large filters. The frequency-domain approach can be utilized as follows:

1. *Transform to Frequency Domain:* Compute the FT of the input image I and the filter **K**:

$$I(\omega) = \mathcal{F}\{I(x,y)\}, \quad K(\omega) = \mathcal{F}\{K(x,y)\}$$

2. *Element-Wise Multiplication:* Perform element-wise multiplication in the frequency domain:

$$H(\omega) = I(\omega) \cdot K(\omega)$$

3. *IFT:* Transform the result back to the spatial domain using the IFT:

$$h(x,y) = \mathcal{F}^{-1}\{H(\omega)\}$$

This approach can significantly reduce the computational burden for large filters, making it a valuable technique in specific applications. Let us review an example for better understanding. A CNN is processing a small grayscale image with a size of 3 × 3 pixels using a 2 × 2 filter (kernel). We'll demonstrate both spatial and frequency-domain convolution. Given:

1. Input Image I:

$$I = \begin{bmatrix} 1 & 2 & 3 \\ 4 & 5 & 6 \\ 7 & 8 & 9 \end{bmatrix}$$

2. Filter **K**:

$$K = \begin{bmatrix} 1 & 0 \\ 0 & -1 \end{bmatrix}$$

Step 1. Spatial Domain Convolution: Perform convolution by sliding the filter over the input image.

Convolution output size:

$$(3-2+1)\times(3-2+1)=2\times 2$$

Computing each output element:

1. *Top-Left Position:*

$$\begin{bmatrix}1 & 2\\4 & 5\end{bmatrix}\cdot\begin{bmatrix}1 & 0\\0 & -1\end{bmatrix}=(1\times 1)+(2\times 0)+(4\times 0)+(5\times -1)=1+0+0-5=-4$$

2. *Top-Right Position:*

$$\begin{bmatrix}2 & 3\\5 & 6\end{bmatrix}\cdot\begin{bmatrix}1 & 0\\0 & -1\end{bmatrix}=(2\times 1)+(3\times 0)+(5\times 0)+(6\times -1)=2+0+0-6=-4$$

3. *Bottom-Left Position:*

$$\begin{bmatrix}4 & 5\\7 & 8\end{bmatrix}\cdot\begin{bmatrix}1 & 0\\0 & -1\end{bmatrix}=(4\times 1)+(5\times 0)+(7\times 0)+(8\times -1)=4+0+0-8=-4$$

4. *Bottom-Right Position:*

$$\begin{bmatrix}5 & 6\\8 & 9\end{bmatrix}\cdot\begin{bmatrix}1 & 0\\0 & -1\end{bmatrix}=(5\times 1)+(6\times 0)+(8\times 0)+(9\times -1)=5+0+0-9=-4$$

Resulting Convolved Feature Map:

$$(f*g)_{\text{spatial}}=\begin{bmatrix}-4 & -4\\-4 & -4\end{bmatrix}$$

Step 2. Frequency-Domain Convolution: Now, we'll perform convolution using the convolution theorem.

Step 2.1. Zero-Padding: To perform convolution via the FT, zero-pad both the input image and the filter to match the size of the convolution result.

- Input Image I (padded to 4 × 4):

$$I_{\text{pad}}=\begin{bmatrix}1 & 2 & 3 & 0\\4 & 5 & 6 & 0\\7 & 8 & 9 & 0\\0 & 0 & 0 & 0\end{bmatrix}$$

- Filter **K** (padded to 4 × 4):

$$K_{\text{pad}} = \begin{bmatrix} 1 & 0 & 0 & 0 \\ 0 & -1 & 0 & 0 \\ 0 & 0 & 0 & 0 \\ 0 & 0 & 0 & 0 \end{bmatrix}$$

Step 2.2. Compute the FTs: Compute the 2D **FFTs** of both padded signals. For simplicity, we'll present hypothetical **FFT** results:

$$F[k_1][k_2] = \text{FFT2D}(I_{\text{pad}}) \textbf{ and } G[k_1][k_2] = \text{FFT2D}(K_{\text{pad}})$$

Step 2.3. Element-Wise Multiplication in Frequency Domain: Multiply the corresponding elements of **F** and **G**:

$$H[k_1][k_2] = F[k_1][k_2] \times G[k_1][k_2]$$

Step 2.4. Compute the IFT: Take the **IFFT** of $\mathbf{H[k_1][k_2]}$ to obtain the convolved feature map in the spatial domain.

$$(f*g)_{\text{frequency}} = \text{IFFT2D}(H[k_1][k_2])$$

Assuming accurate **FFT** computations, the result should match the direct convolution:

$$(f*g)_{\text{frequency}} = \begin{bmatrix} -4 & -4 \\ -4 & -4 \end{bmatrix}$$

Figure 10.4 illustrates the process of filtering an input image using frequency-domain techniques, highlighting both the spatial and frequency-domain representations. Figure 10.4a shows the original input image, a binary image containing a white square at the center. This serves as the starting point for the filtering process. Figure 10.4b depicts the filter image, which is a smaller white square designed to act as a convolution kernel in the spatial domain. This filter will be applied in the frequency domain to modify the original image's components. Figure 10.4c presents the filtered image obtained after applying the convolution operation in the frequency domain. The result demonstrates how the filter modifies the spatial characteristics of the input image, introducing smoothed edges around the square. Figure 10.4d visualizes the FT of the input image, where the central peak represents low-frequency components that capture the overall structure of the image, while surrounding patterns indicate higher-frequency details. Figure 10.4e shows the FT of the filter, highlighting its effect in the frequency domain. The smaller peak emphasizes the localized nature of the filter's influence on the input image. Figure 10.4f displays the FT of the result image, showing how the filtered peaks reflect the combined effect of the input image and the filter in the frequency domain.

10.5 REAL-WORLD APPLICATIONS

10.5.1 Medical Imaging and Diagnostics

In medical imaging, FT and WT play a key role in enhancing and analyzing images like MRIs, CT scans, and X-rays. FTs are used to filter out noise and enhance resolution by converting image data

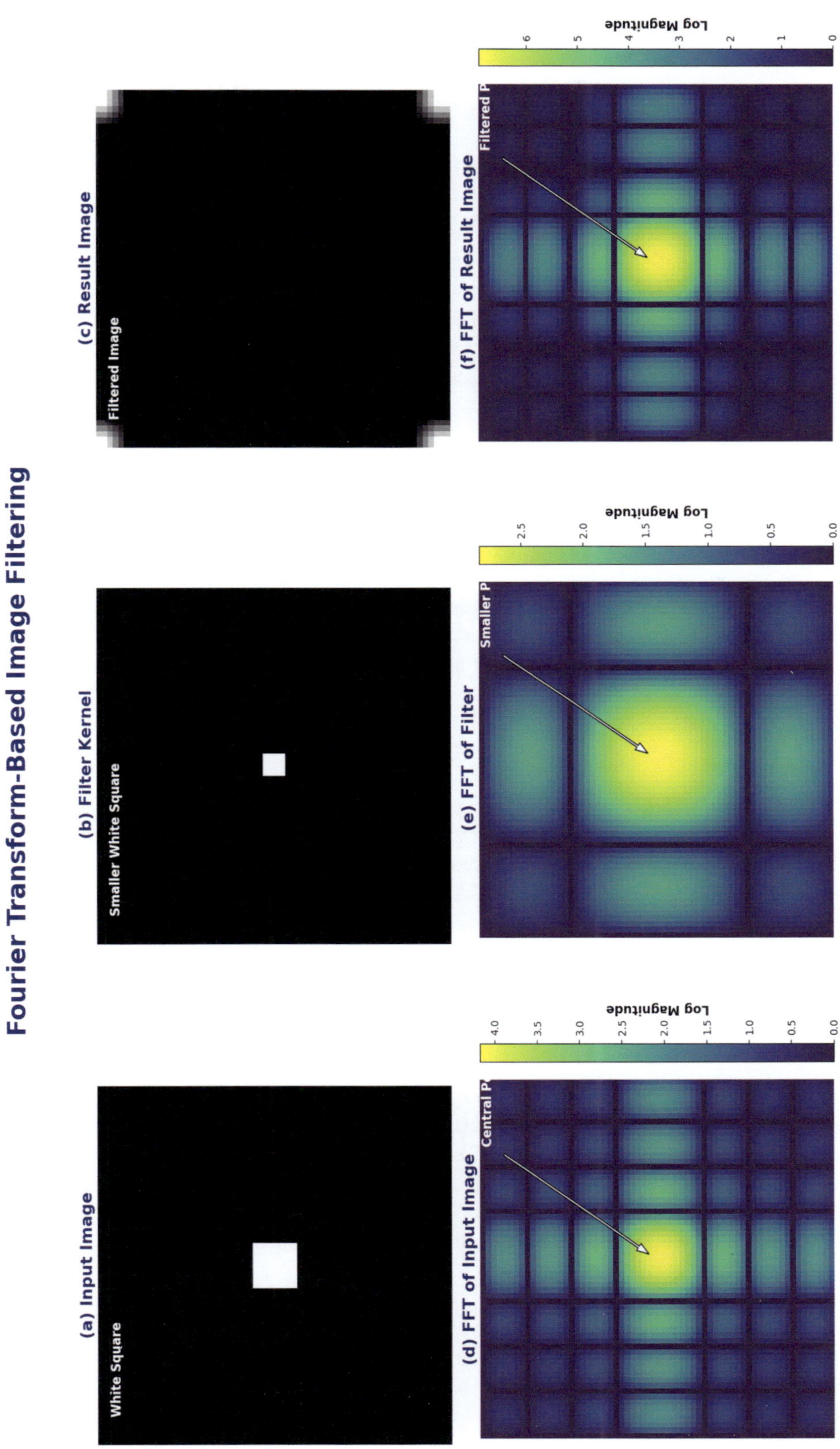

FIGURE 10.4 (a–c) Example input image and filter. (d–f) Frequency-domain representations and the convolution result after applying the IFTs.

from the time domain to the frequency domain. For example, in MRI scans, Fourier analysis allows clear visualization of tissues and abnormalities. WTs, on the other hand, are effective in isolating fine details, such as detecting microcalcifications in mammograms, aiding in early cancer detection through multi-resolution analysis.

10.5.2 Audio Signal Processing

Harmonic analysis plays a fundamental role in audio signal processing, helping to analyze, compress, and enhance sound signals. The FT breaks down audio into its frequency components, allowing specific sounds to be isolated or unwanted noise to be removed. This method is crucial in applications like audio compression (e.g., MP3 encoding) and noise reduction technologies, such as in hearing aids. WTs offer better time-frequency localization, which is vital for processing transient sounds, such as speech or music, enhancing applications like speech recognition and audio watermarking.

10.5.3 Data Compression

Data compression methods like JPEG and JPEG 2000 depend on harmonic analysis to reduce image size while maintaining quality. The JPEG standard uses Fourier analysis to transform image data into the frequency domain, where high-frequency components (often linked to noise) can be minimized or removed. JPEG 2000 employs WTs, allowing for multi-resolution image representation, leading to more efficient compression and better preservation of key details. These compression techniques are essential in reducing storage needs and transmission costs, particularly in bandwidth-limited environments.

10.5.4 Communications and Signal Transmission

FT and WT are essential in communications, enabling efficient modulation and demodulation of signals. In digital communication, FTs and converts signals between the time and frequency domains, facilitating technologies like Orthogonal Frequency Division Multiplexing (OFDM), which is used in 4G and 5G networks to boost data rates. WTs, with their multi-scale analysis, enhance advanced coding methods such as wavelet-based data compression in satellite communications, ensuring high-quality transmission even in challenging environments and optimizing both bandwidth and reliability.

10.5.5 Cryptography and Security

Harmonic analysis plays a significant role in cryptography by enhancing the design of secure communication systems and encryption algorithms. Fourier analysis is used to study signals, identifying potential vulnerabilities in communication channels. WTs, with their ability to localize both time and frequency information, are employed in steganography, a technique for embedding hidden information within digital media. These methods are crucial for safeguarding sensitive data, providing additional layers of security in today's digital and interconnected world.

10.6 HANDS-ON EXAMPLE

Enhance image features using convolutional filters in the frequency domain to demonstrate the computational advantages and precision aspects discussed in the chapter.

10.6.1 Step 1: Load the Image

At first, we define a function load image to load an image from a given file path using the OpenCV library (cv2). This function is useful in various computer vision applications, especially when working with grayscale images, such as in image classification, feature extraction, or object detection.

```
import cv2
def load_image(file_path):
  # Load an image from file path
  image = cv2.imread(file_path, cv2.IMREAD_GRAYSCALE)
  return image
```

10.6.2 Step 2: Apply FT

Here, we define a function apply_Fourier_transform that computes the discrete Fourier transform (**DFT**) of an input image and returns its magnitude spectrum. This function is useful in various image processing applications where understanding or manipulating the frequency components of an image is required, such as in edge detection, image compression, or denoising.

```
import numpy as np
def apply_fourier_transform(image):
  # Apply the Discrete Fourier Transform
  dft = cv2.dft(np.float32(image), flags=cv2.DFT_COMPLEX_OUTPUT)
  dft_shift = np.fft.fftshift(dft)
  magnitude_spectrum = 20 * np.log(cv2.magnitude(dft_
shift[:,:,0], dft_shift[:,:,1]) + 1)
  return magnitude_spectrum
```

10.6.3 Step 3: Apply Sobel Edge Detection

Then, we define the function apply_sobel_filter, which applies the Sobel operator to an image to detect edges by calculating the gradient in both horizontal and vertical directions. The Sobel filter is commonly used in image processing to detect edges and features. By computing the gradient in both horizontal and vertical directions, the Sobel operator identifies regions of an image where intensity changes sharply, making it useful for tasks such as edge detection, feature extraction, and object recognition.

```
def apply_sobel_filter(image):
  # Apply Sobel operator in both horizontal and vertical
  directions
  sobelx = cv2.Sobel(image, cv2.CV_64F, 1, 0, ksize=5)
  sobely = cv2.Sobel(image, cv2.CV_64F, 0, 1, ksize=5)
  sobel = cv2.magnitude(sobelx, sobely)
  return sobel
```

10.6.4 Step 4: Display Results

Now, we define the function display results, which creates a visualization of multiple images in a grid layout using Matplotlib.

```
import matplotlib.pyplot as plt
def display_results(images, titles):
  # Display the list of images with titles
  plt.figure(figsize=(10, 5))
  for i in range(len(images)):
    plt.subplot(1, len(images), i+1)
    plt.imshow(images[i], cmap='gray')
    plt.title(titles[i])
    plt.xticks([]), plt.yticks([])
  plt.show()
```

10.6.5 Step 5: Integrate and Run

Finally, we define a function process_and_display_image that processes an image using various image processing techniques and displays the results.

```
def process_and_display_image(file_path):
  image = load_image(file_path)
  magnitude_spectrum = apply_fourier_transform(image)
  sobel_image = apply_sobel_filter(image)
  display_results([image, magnitude_spectrum, sobel_image],
                  ['Original Image', 'Magnitude Spectrum',
                  'Sobel Edge Detection'])
# Replace 'path_to_image.jpg' with your actual image file path
process_and_display_image('path_to_image.jpg')
```

Figure 10.5 presents a comparison of image processing techniques applied to an input image, thereby highlighting transformations in both the spatial and frequency domains. Figure 10.5a shows the original image, which serves as the input for subsequent processing. This grayscale image contains intricate details and textures, forming the basis for analysis. Figure 10.5b displays the magnitude spectrum obtained by applying the FT to the original image. This visualization maps the frequency components of the image, with the central peak representing the low-frequency features that correspond to the overall structure. Figure 10.5c illustrates the Sobel-filtered image, emphasizing edges and transitions in intensity within the original image. This technique detects gradients, highlighting the boundaries and contours of the objects in the image. Figure 10.5d shows the resulting image after **IFFT**, reconstructed by modifying and then inverting the frequency-domain representation. This reconstruction retains specific features based on the frequency modifications applied, demonstrating the selective enhancement or suppression of image components. Figure 10.5e visualizes the **FFT** of the input image, providing a detailed frequency-domain representation before any modifications. This serves as a baseline for comparison with the processed results. Figure 10.5f displays the **FFT** of the result image, capturing the frequency components of the reconstructed image. Comparing this with the **FFT** of the input image reveals changes introduced by the processing steps, particularly the filtering and reconstruction stages.

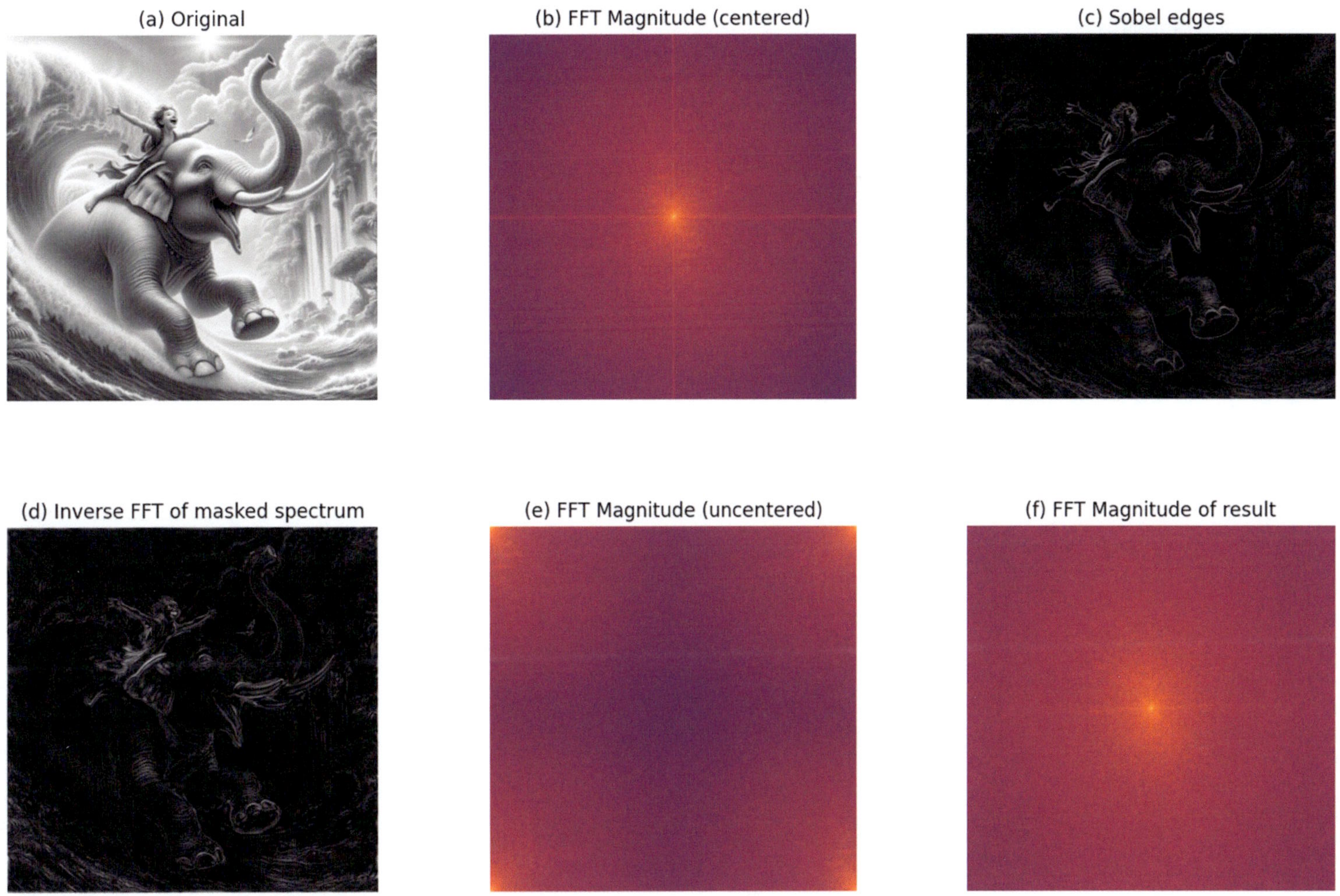

FIGURE 10.5 (a) Input image. (b) Log-magnitude FFT of the input. (c) Canny edge map (σ=2, dilated). (d) FFT of input with DC peak annotated. (e) FFT of the Laplacian filter. (f) FFT of the filtered result.

10.7 COMMON MISTAKES AND TROUBLESHOOTING TIPS

10.7.1 Misinterpreting FT and WT

- *Mistake:* Confusing the mathematical principles and practical applications of FT and WT.
- *Tip:* Review fundamental resources and practical tutorials on both transforms. Use visualization tools to see how each transform decomposes signals and to understand their differences.

10.7.2 Overlooking the Disadvantages of Frequency-Domain Convolution

- *Mistake:* Assuming frequency-domain convolution is always faster and more efficient than spatial domain convolution without considering filter size and computational overhead.
- *Tip:* Evaluate the specific use case to determine whether frequency-domain convolution offers a computational advantage. For small filters, stick with spatial domain convolution to avoid unnecessary overhead.

10.7.3 Ignoring Precision Issues

- *Mistake:* Neglecting the precision errors that can arise when performing FT and IFT.
- *Tip:* Be aware of the potential for numerical approximation errors. Validate the results by comparing the outputs of frequency-domain and spatial-domain convolutions.

10.7.4 Misapplying WT for Non-Stationary Signals

- *Mistake:* Using WTs incorrectly, particularly when dealing with non-stationary signals, and failing to exploit their multi-resolution analysis capabilities.
- *Tip:* Understand the nature of your signals and the specific advantages that wavelets offer. Use appropriate mother wavelets and scales to capture the signal characteristics accurately.

10.7.5 Overfitting in Convolutional Neural Networks

- *Mistake:* Designing **CNNs** with excessive depth or complexity, leading to overfitting.
- *Tip:* Implement regularization techniques like dropout, batch normalization, and early stopping. Regularly validate your model on separate datasets to monitor for overfitting.

10.7.6 Inefficient Implementation of Convolution Theorem

- *Mistake:* Inefficiently implementing the convolution theorem, leading to suboptimal performance gains.
- *Tip:* Utilize optimized libraries and tools for **FFT** and **IFFT** operations. Parallelize operations where possible to enhance computational efficiency.

10.7.7 Misunderstanding the Convolution Theorem

- *Mistake:* Misinterpreting the convolution theorem and its application to **CNNs**, leading to incorrect implementations.
- *Tip:* Study the mathematical foundation of the convolution theorem and its practical implications. Ensure you understand the steps involved in transforming, multiplying, and inverse transforming signals.

10.7.8 Failing to Validate Models

- *Mistake:* Neglecting to validate **CNN** models after implementing harmonic analysis techniques, resulting in unchecked errors and inefficiencies.
- *Tip:* Conduct thorough validation using various datasets and benchmarks. Compare the performance of models using harmonic analysis techniques against traditional methods.

10.8 REVIEW QUESTIONS

1. What is the primary purpose of Fourier analysis in signal processing?
2. How does the FT convert a time-domain signal into its frequency-domain representation?
3. Explain the difference between the FT and the IFT.
4. What are the key features of the WT that make it suitable for analyzing non-stationary signals?
5. How does the convolution theorem relate convolution in the time domain to multiplication in the frequency domain?
6. What steps involve convolution in the frequency domain using the FT?
7. Discuss the advantages and disadvantages of using frequency-domain convolution in **CNNs**.
8. In what scenarios might you prefer wavelet analysis over Fourier analysis?
9. How can understanding frequency-domain convolution improve the efficiency of signal-processing tasks in deep learning?
10. Provide examples of real-world applications where FT and WT are particularly useful.

10.9 PROGRAMMING QUESTIONS

10.9.1 Easy

Implement a 1D convolutional layer using TensorFlow and visualize the filter's response to a sine wave input.

1. Create a simple 1D convolutional model.
2. Generate a sine wave as input data.
3. Apply the convolutional filter to the input data.
4. Plot the input sine wave and the filter's output.

10.9.2 Medium

Compare the performance of different activation functions (ReLU, Sigmoid, Tanh) in a simple **CNN** on the **CIFAR-10** dataset.

1. Create a simple **CNN** model with configurable activation functions.
2. Train the model with **ReLU** activation and record the accuracy.
3. Repeat the training with Sigmoid and Tanh activations.
4. Compare the results and discuss the impact of each activation function on model performance.

10.9.3 Hard

Implement a **CNN** with batch normalization and dropout layers to improve the model's robustness and accuracy on the **Fashion MNIST** dataset.

1. Create a **CNN** model with batch normalization and dropout layers.
2. Train the model on the **Fashion MNIST** dataset.
3. Evaluate the model's performance and compare it with a baseline **CNN** without batch normalization and dropout.
4. Analyze the effects of batch normalization and dropout on the model's training stability and accuracy.

11 Dynamical Systems and Differential Equations for RNNs

11.1 INTRODUCTION

In this chapter, we look at how recurrent neural networks (RNNs) are connected to the math ideas of dynamical systems and differential equations. RNNs are especially good at working with sequences and remembering information over time, which makes them important for many real-life uses. By understanding the basic ideas behind RNNs, we can better see how they function. Here, we'll go beyond the simple concepts of neural networks and use ideas from dynamical systems to understand how RNNs handle and process data that come in a sequence. The goal is to give a clearer picture of how these networks work and how their functions are similar to those of dynamical systems.

11.2 THEORY OF DYNAMICAL SYSTEMS AND DIFFERENTIAL EQUATIONS

Dynamical systems and differential equations help us understand how things change, grow, or decrease over time, both in nature and in human-made systems. These math ideas let us study and predict how complex systems behave as they develop. By learning about them, we can see the patterns that shape the world around us and use these principles in different areas, including technology and neural networks.

11.2.1 Dynamical Systems

A dynamical system is a way to describe how something changes over time using a set of rules. These systems are used in many areas of science and engineering to track things like how planets move or how populations grow. The strength of dynamical systems is their ability to show how different states change, making them useful for understanding and predicting behavior over time. In math, they are shown by maps that illustrate how one state changes to another, allowing us to follow paths and predict future states in a certain space.

11.2.1.1 Discrete Dynamical Systems

Not all changes occur continuously; some happen in a distinct, step-wise fashion. In discrete dynamical systems, state changes are quantized and occur in well-defined steps. A function, commonly denoted as **f**, governs these steps. If we know the system's state at a particular step **n**, we can apply **f** to determine the state at the next step, **n + 1**. This type of system is particularly useful for modeling processes that evolve in stages, such as population growth in generations or iterative algorithms in computer science. Consider the discrete dynamical system represented by:

$$x_{n+1} = f\left(x_n\right)$$

DOI: 10.1201/9781032690742-11

where x_n is the state at step **n**, and **f** is the function defining the rule of evolution. This framework allows for the analysis and prediction of future states based on the initial condition. Let us consider a simple population growth model with an initial population of 100 individuals. The growth function, given as $f(x) = 1.2x$, indicates a 20% growth per generation. The population size in the next generation can be calculated using this growth function. For example, starting with 100 individuals, the population after the first generation would be 100 × 1.2 = 120. After the second generation, the population would be 120 × 1.2 = 144, and so on, increasing by 20% with each generation. Let's consider a simple population growth model. Here are the steps:

1. Generation 0: $\mathbf{x}_0 = 100$
2. Generation 1: $\mathbf{x}_1 = 1.2 \times 100 = 120$
3. Generation 2: $\mathbf{x}_2 = 1.2 \times 120 = 144$
4. Generation 3: $\mathbf{x}_3 = 1.2 \times 144 = 172.8$
5. Generation 4: $\mathbf{x}_4 = 1.2 \times 172.8 = 207.36$
6. Generation 5: $\mathbf{x}_5 = 1.2 \times 207.36 = 248.83$

After five generations, the population grows from 100 to approximately 249 individuals. This example shows how discrete dynamical systems predict fut ure states through repeated application of a growth function.

11.2.1.2 Continuous Dynamical Systems

Unlike discrete systems, continuous dynamical systems change smoothly over time, with each tiny moment affecting the system's state. These systems are usually modeled with differential equations, which explain how the system changes continuously. Continuous dynamical systems are important in areas like physics and engineering, where processes such as fluid flow, electrical circuits, and mechanical motion change smoothly over time. A continuous dynamical system is represented by a differential equation:

$$\frac{dx(t)}{dt} = f(x(t))$$

where:

- **x(t)** is the system's state at time **t**,
- **f** describes how the state changes over time.

Let's model a simple exponential growth system:

$$\frac{dx(t)}{dt} = 0.5x(t)$$

This equation describes a system where the growth rate is proportional to the current state, with a growth rate of 0.5. Initial Condition: $x(0) = 100$. In this equation:

$$x(t) = x(0) \cdot e^{0.5t}$$

Substitute the initial condition $x(0) = 100$:

$$x(t) = 100 \cdot e^{0.5t}$$

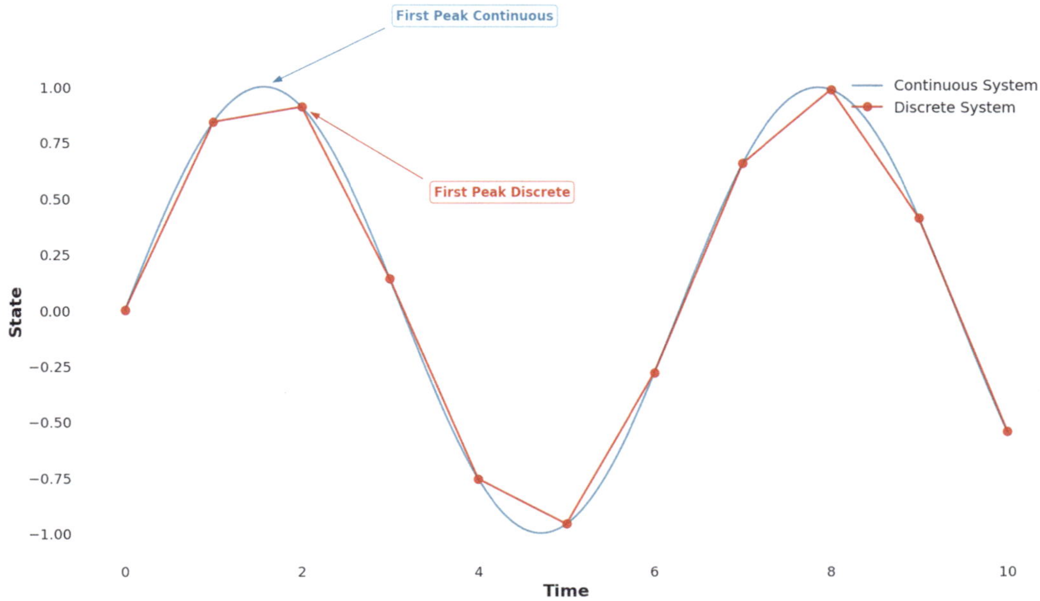

FIGURE 11.1 Comparison of continuous and discrete dynamical systems.

For different time values, we can calculate various states of the system:

- At $t = 1: x(1) = 100 \cdot e^{0.5} \approx 164.87$,
- At $t = 2: x(2) = 100 \cdot e^{1} \approx 271.83$,
- At $t = 3: x(3) = 100 \cdot e^{1.5} \approx 448.17$.

The system grows continuously, with the population increasing exponentially over time.

Figure 11.1 presents a comparison between continuous and discrete dynamical systems, illustrating their behavior over a specified time interval. The x-axis denotes time, while the y-axis represents the state of the system. The continuous system is depicted by a smooth blue curve, thus highlighting the uninterrupted, smooth evolution of the state as time progresses. This smooth curve signifies that in continuous systems, changes occur gradually and are tracked at every infinitesimal point in time, providing a view of the state dynamics. On the other hand, the discrete system is shown as a red line connecting markers at specific time intervals. The markers represent the state values captured at each discrete time step, reflecting how the system's state is updated only at these discrete points. This piecewise linear representation indicates that in discrete systems, time and state changes are not continuously tracked but rather approximated at selected intervals. In sections where the state changes gradually (e.g., between time steps 0 to 2 and 6 to 8), both the continuous and discrete trajectories closely align. However, at regions where the state exhibits rapid changes (e.g., around time steps 3 and 9), slight discrepancies emerge. These differences illustrate how discrete systems might approximate the continuous system's behavior but may miss finer details or introduce minor deviations when the state changes abruptly.

11.2.2 Differential Equations

Differential equations, often called the language of change, let us model how systems behave. These equations describe the state of a system and tell how it changes by using derivatives, which measure how quickly things are changing.

11.2.2.1 Ordinary Differential Equations

Ordinary differential equations (**ODEs**) describe how a function and its rates of change are related using one variable, usually time. These equations are often used to model natural things, like how a pendulum swings or how diseases spread. ODEs help us explain how a system's state changes over time based on its current state. An ODE is typically expressed in the form:

$$\frac{dy}{dt} = f(t, y)$$

where **y** represents the dependent variable (the state of the system), **t** is the independent variable (often representing time), and **f(t,y)** is a function that describes the rate of change of **y** with respect to t. A classic example of an **ODE** is the simple harmonic oscillator, which models the motion of systems such as a pendulum or a mass-spring system. The equation for a harmonic oscillator is given by:

$$\frac{d^2 y}{dt^2} + \omega^2 y = 0$$

where **y(t)** represents the position of the object over time, and **ω** is the angular frequency of the oscillation. This second-order ODE describes how the position of the object changes as it oscillates back and forth under the influence of a restoring force proportional to its displacement. The general solution to this equation is:

$$y(t) = A\cos(\omega t) + B\sin(\omega t)$$

where **A** and **B** are constants determined by the initial conditions of the system, such as the initial position and velocity of the object. For instance, if the system begins at **y**(0) = 1 with an initial velocity of zero, the solution simplifies to **y(t)** = **cos(2t)**, assuming an angular frequency **ω** = 2 rad/s. This equation describes an object oscillating with a frequency of 2 rad/s, moving smoothly between its maximum and minimum positions over time. In this way, **ODEs** serve as a powerful tool for modeling dynamic behaviors in various fields, providing insights into how systems evolve continuously over time based on their initial conditions and the laws that govern them.

Figure 11.2 depicts the displacement of a simple harmonic oscillator, modeled as a mass-spring system, over time. The *x*-axis represents time in seconds, while the *y*-axis indicates the displacement in meters from the equilibrium position. The blue curve follows the equation $x(t) = A\cos(\omega t + \theta)$, where **A** is the amplitude, **ω** is the angular frequency, and **θ** is the phase angle. The graph highlights several key aspects of the oscillator's behavior. The maximum displacement, marked by a green dot, shows the furthest point the mass reaches above the equilibrium. At this point, the energy in the system is entirely potential, with zero velocity. The minimum displacement, marked by a red dot, indicates the lowest point the mass reaches, showing a similar state where the velocity is zero and the potential energy is maximized. The purple dot represents the phase angle, illustrating how the initial position and timing of the oscillator's motion are influenced, shifting the curve horizontally. The differential equation $\frac{d^2 x(t)}{dt^2} + \omega^2 x(t) = 0$, displayed at the bottom left, governs this harmonic motion. It indicates that the acceleration of the system is directly proportional and opposite to the displacement, a characteristic of simple harmonic oscillators.

11.2.2.2 Partial Differential Equations

Partial differential equations (**PDEs**) are more complicated than **ODEs** because they involve functions with several variables and their partial derivatives. These equations are fundamental in

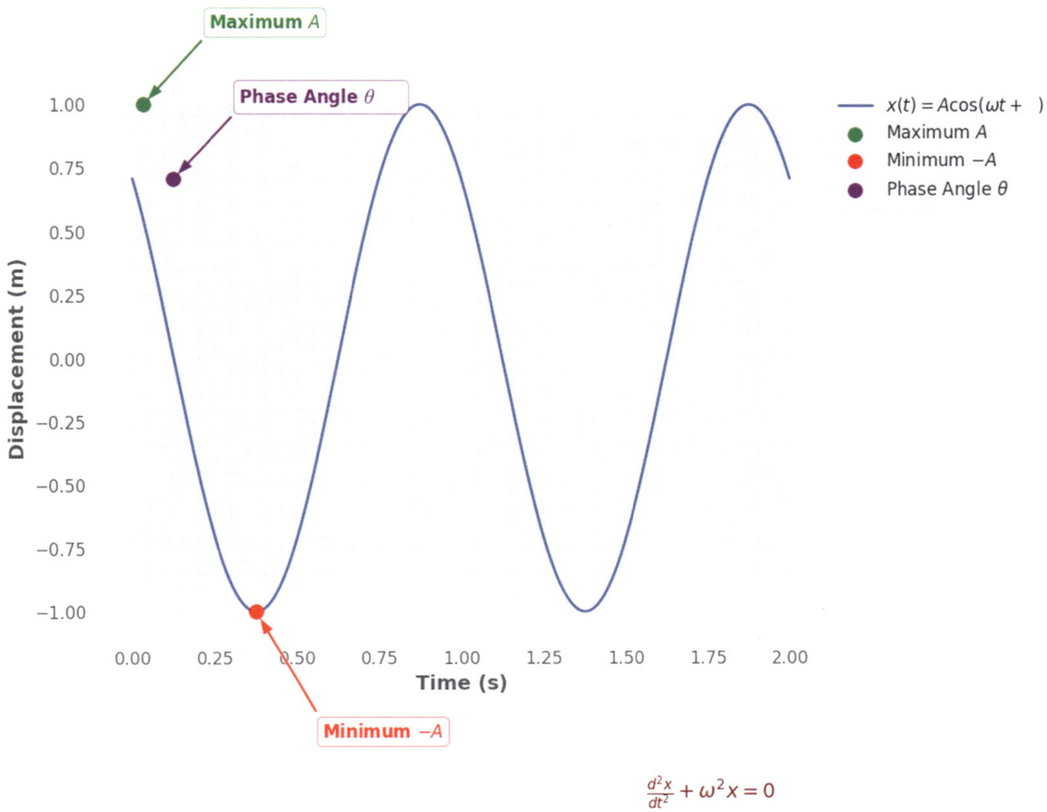

FIGURE 11.2 Simple harmonic oscillator: mass-spring system.

many areas of physics and engineering, like modeling how heat spreads, how fluids flow, and electromagnetic fields. **PDEs** let us describe systems where changes happen in multiple directions, such as time and space. A **PDE** is typically written in the form:

$$\frac{\partial u}{\partial t} = F\left(t, x, u, \frac{\partial u}{\partial x}, \frac{\partial^2 u}{\partial x^2}, \ldots\right)$$

where **u** is the dependent variable representing the state of the system, **t,** and **x** are independent variables (often representing time and space), and **F** is a function that describes how **u** changes over time and space. A well-known example of a **PDE** is the heat equation, which describes how heat diffuses through a material. It is expressed as:

$$\frac{\partial u}{\partial t} = \alpha \frac{\partial^2 u}{\partial x^2}$$

where **u(t, x)** represents the temperature of the material as a function of time **t** and position **x**, and **α** is the thermal diffusivity, a constant that quantifies how quickly heat spreads through the material. Consider a metal rod of length 10 units. The ends of the rod are maintained at a constant temperature of 0°C, while the initial temperature at the center of the rod is 100°C, with the rest of the rod at 0°C. The goal is to determine how the temperature changes over time. At time **t** = 0, the temperature distribution along the rod can be described as:

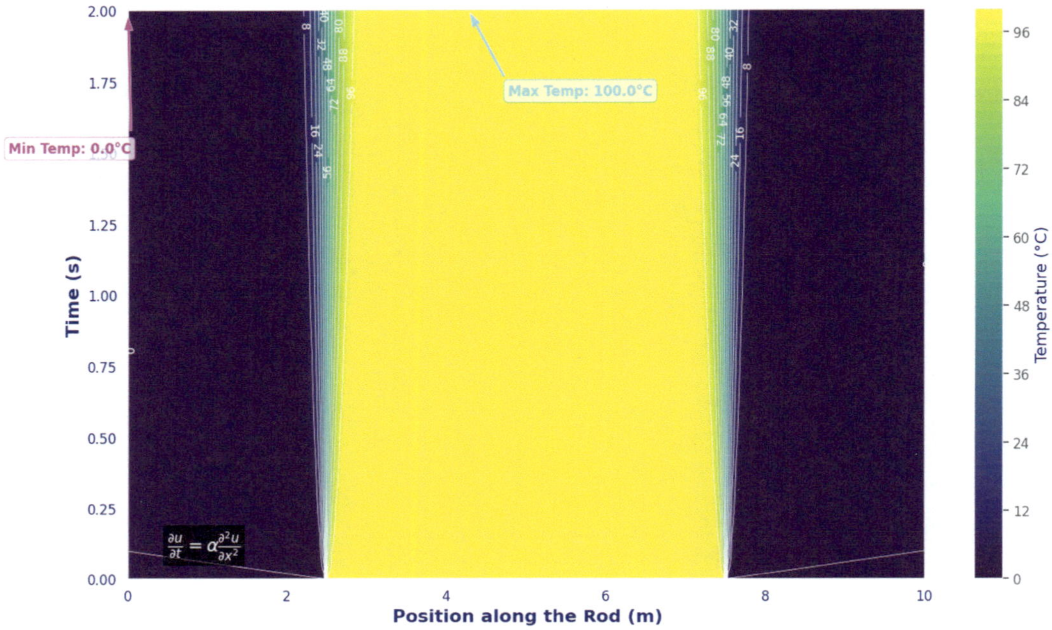

FIGURE 11.3 Heat equation: temperature distribution over time.

$$u(0,x) = \begin{cases} 100, & \text{if } x = 5 \\ 0, & \text{otherwise} \end{cases}$$

With the given boundary conditions, **u**(**t**, 0) = **u**(**t**, 10) = 0 for all values of **t**. Assuming the thermal diffusivity $\boldsymbol{\alpha}$ is 0.01, the heat equation governs how the temperature evolves over time. After a short time, say at **t** = 1, the temperature at the center of the rod would decrease as heat begins to diffuse toward the ends of the rod. The approximate temperature distribution at this point might be:

$$u(1,x) = \begin{cases} 60, & \text{if } x = 5 \\ 0, & \text{if } x = 0 \text{ or } x = 10 \end{cases}$$

As time progresses, the temperature continues to even out along the rod. By **t** = 5, the temperature at the center might decrease further to 20°C, while the ends of the rod remain at 0°C. The temperature distribution over time can be obtained by solving the heat equation numerically or analytically, depending on the specific boundary and initial conditions.

Figure 11.3 displays the temperature distribution along a rod over time, as governed by the heat equation. The x-axis represents the position along the rod in meters, while the y-axis indicates time in seconds. The color gradient in the figure reflects temperature values, ranging from low (dark colors) to high (light colors). The heat distribution evolves as time progresses, showing how heat diffuses from the center toward the ends of the rod. At the initial moment, the temperature is concentrated at the center, indicated by the bright yellow region. As time advances, the heat spreads outward, gradually decreasing in intensity as it moves toward the boundaries of the rod. The equation displayed at the lower left corner, $\frac{\partial u}{\partial t} = \alpha \frac{\partial^2 u}{\partial x^2}$, represents the heat equation, where **u** denotes the temperature, **t** is time, **x** is the spatial position along the rod, and $\boldsymbol{\alpha}$ is the thermal diffusivity constant. This equation describes the rate of change in temperature over time and position, indicating that the temperature change depends on the second spatial derivative of the temperature, which signifies the flow of heat. The temperature color map on the right further provides a scale, showing the range of temperatures from 0°C (dark) to 96°C (light yellow), allowing for a clear visualization of how the temperature changes over both space and time.

Figure 11.4a illustrates a discrete dynamical system modeled using the logistic map equation $x_{n+1} = rx_n (1 - x_n)$. The x-axis represents iterations, while the y-axis shows the population value normalized between 0 and 1. The blue line, with points marking each iteration, demonstrates the population's evolution over time, revealing oscillatory and chaotic behavior typical of discrete non-linear systems. The red dot marks the final population state at the end of the 100 iterations, highlighting how such systems can vary and exhibit complex, unpredictable behavior despite being governed by a simple equation. Figure 11.4b displays a continuous dynamical system in the form of a simple harmonic oscillator. The x-axis shows time in seconds, while the y-axis indicates the position of the oscillating mass. The green curve follows the equation $x(t) = A\cos(\omega t) + \frac{v_0}{\omega}\sin(\omega t)$, representing a combination of cosine and sine components where $\mathbf{A}$ is the amplitude, $\boldsymbol{\omega}$ is the angular frequency, and $\mathbf{v}_0$ is the initial velocity.

11.3 UNDERSTANDING THE BEHAVIOR OF RNNS

11.3.1 Memory and Dynamics

Recurrent neural networks (RNNs) can remember past information because of their repeating structure. This means that at any time, an RNN's state shows not only the current input but also information from earlier inputs, similar to how a dynamical system changes over time with its current state affected by previous states. This "memory" makes RNNs very powerful for tasks that involve sequences of data. For example, in time series prediction, consider predicting stock prices over a few days. If the stock prices over five days are:

Prices = [100, 102, 101, 105, 107]

An RNN can be trained to predict the price for the next day (day 6) based on these previous values. After training, the RNN might predict:

Predicted Price on Day 6 = 109

In this case, the RNN learns the trend from the data; the general upward movement in prices suggests that the next value will likely be higher. By remembering the prices from the previous days, the model can make a more informed prediction about the future. In natural language processing (NLP), RNNs also excel. For instance, consider sentence generation. Given the starting phrase:

"The stock market"

The RNN uses its memory of the earlier words to predict the next word. If the RNN has been trained on financial articles, it might generate the sentence:

"is expected to rise."

Here, the prediction of each word is influenced by the previous words, allowing the RNN to create a coherent sentence by remembering the context. Similarly, in speech recognition, the input to an RNN might be a sequence of sounds, such as:

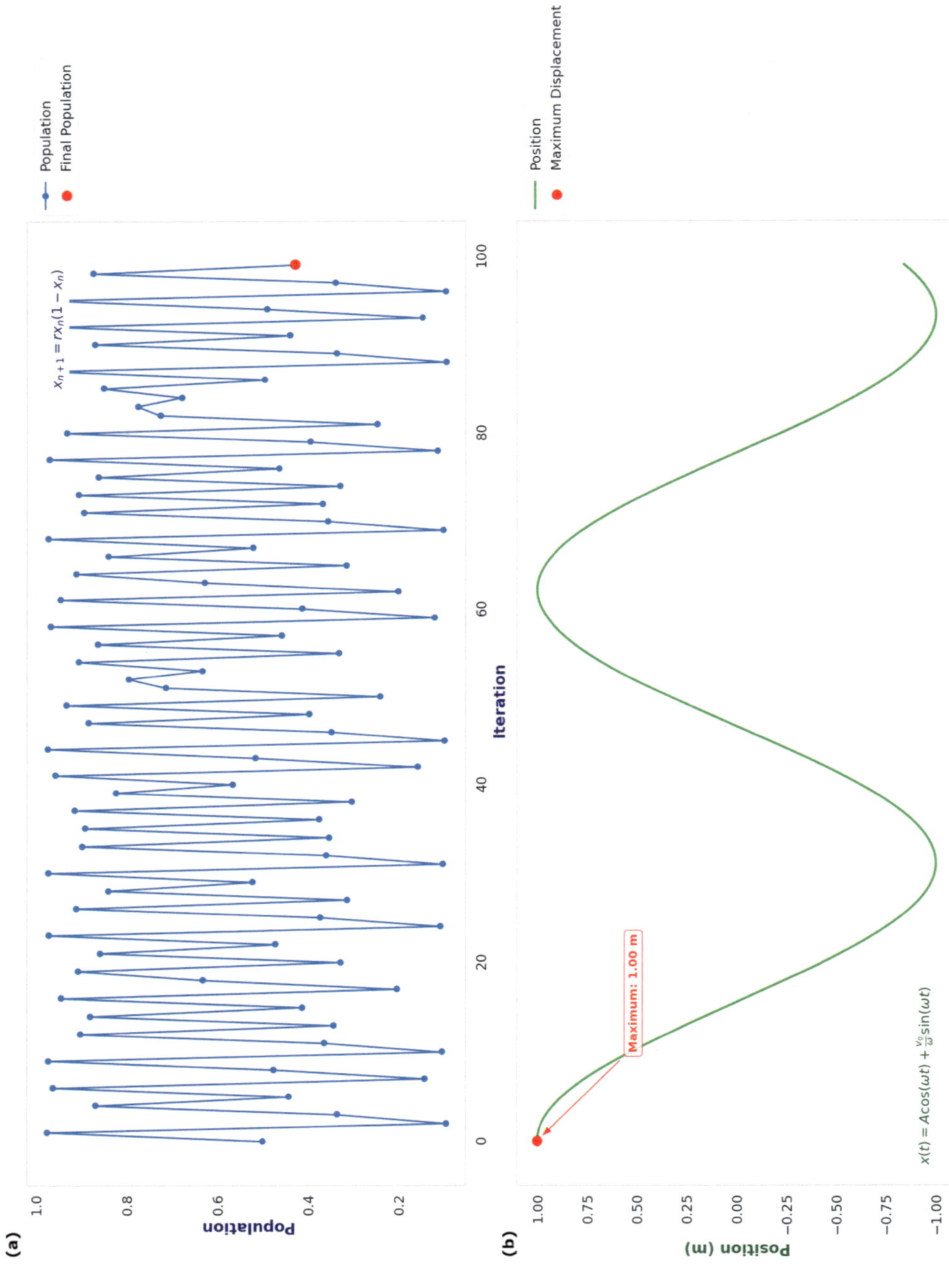

FIGURE 11.4 (a) Discrete dynamical system: logistic map, (b) continuous dynamical system: harmonic oscillator.

Sounds = ["s," "o," "f," "t"].
Using the memory of the earlier sounds, the RNN can predict that the word is "soft." If the input sounds were:

Sounds = ["h," "a," "r," "d"].
The RNN might predict the word "hard." In both cases, the ability to maintain context across the sequence of sounds allows the RNN to recognize entire words. This memory of past inputs is critical to understanding the current input and making accurate predictions, whether it's a word in a sentence or a value in a time series.

Figure 11.5 illustrates the memory dynamics of a RNN over time, showing both a broad overview and a detailed zoomed-in view of the input and output sequences. Figure 11.5a displays the overall behavior of the RNN across 4000 time steps. The blue line represents the input sequence, demonstrating its repetitive and periodic pattern, which spans the entire range. The green line corresponds to the true output sequence generated based on the input, while the orange dashed line shows the RNN's predicted output sequence. Despite the periodicity and repetition in the input sequence, the predicted output remains close to the true output sequence, indicating the RNN's capability to learn and generalize the pattern over a long duration. Figure 11.5b zooms into the first 100 time steps to provide a detailed view of the RNN's behavior. The blue line continues to represent the input sequence, showing sharp spikes at regular intervals. The green line captures the true output sequence, which responds more gradually and sinusoidally to changes in the input, demonstrating how the system processes and smooths the input signals. The orange dashed line is the predicted output sequence by the RNN, closely following the pattern of the true output but with slight deviations. A red dot marks the maximum prediction value in this window, highlighting a specific point where the RNN reaches its highest output, with the annotation indicating the maximum value observed.

11.3.2 Explaining via ODEs

RNNs, especially those built with continuous changes, can be easily understood using ODEs. Using **ODEs** helps us model how the system changes over time, giving us a better understanding of how the network behaves. One important idea in this approach is stationary states, or fixed points, which are states where the system stays the same and does not change over time. A fixed point **x*** occurs when the **ODE** governing the system reaches a state where:

$$\frac{dx(t)}{dt} = f(x(t)) = 0$$

In the context of RNNs, these fixed points represent the long-term behavior of the network, where the internal state stays the same over time. For example, if an RNN reaches a stable fixed point, it will give consistent outputs for sequences, making its performance reliable. Another important idea is stability analysis, which looks at whether the fixed points of the system are stable or unstable. A stable fixed point means that if the system's state is slightly changed, it will eventually go back to the fixed point. On the other hand, an unstable fixed point means that any small change will make the system move away from that point. This analysis helps us understand if the **RNN** will settle into a stable state, show repeating patterns, or even behave unpredictably over time. To explain this with a numerical example, consider a simple **ODE** describing the continuous evolution of a state **x(t)** over time:

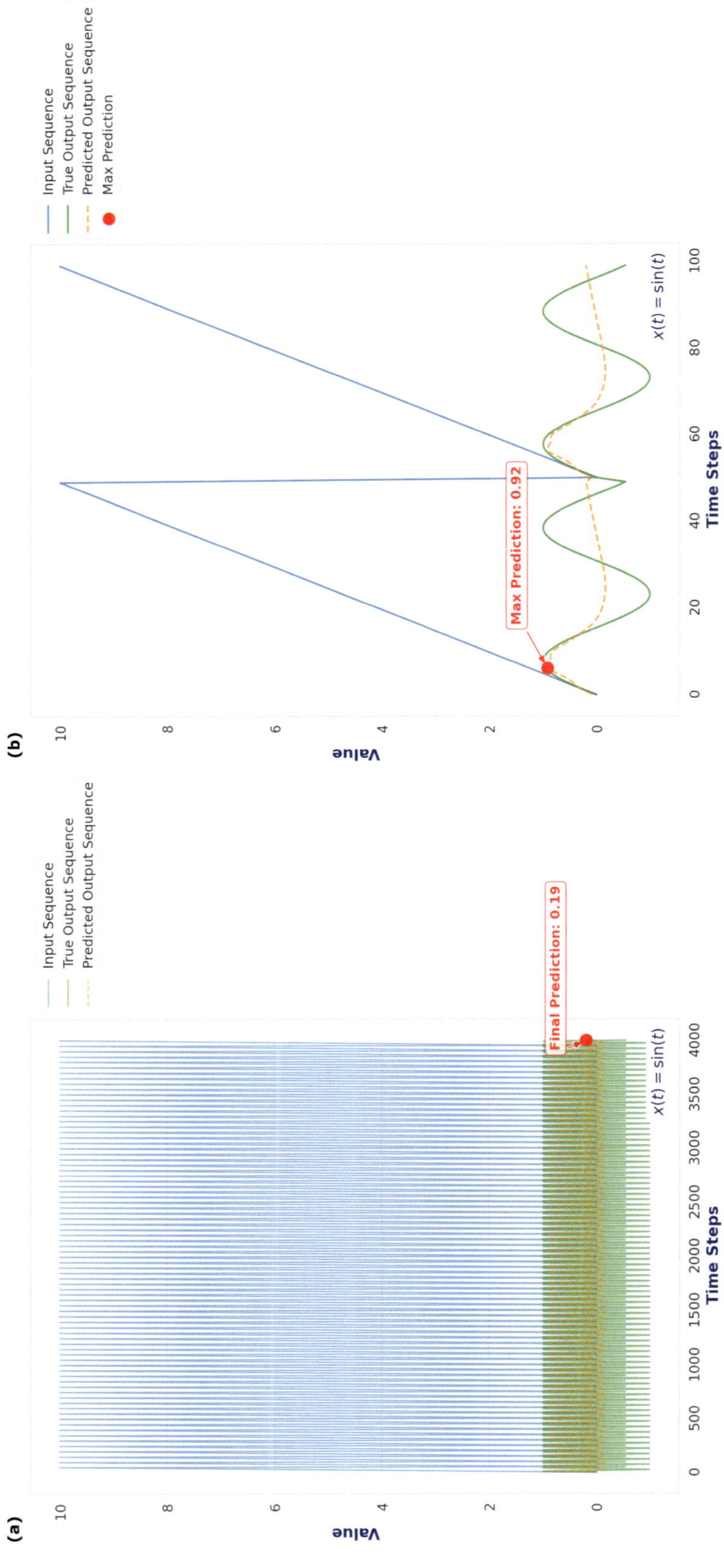

FIGURE 11.5 RNN memory dynamics.

$$x(t)\frac{dx(t)}{dt} = -kx(t)$$

where **k** > 0 is a constant. In this case, the system has a fixed point at **x*** = 0. The stability of this fixed point can be analyzed by observing how the state changes for different initial conditions. If **x(t)** starts at a positive value, it will gradually decay toward zero, indicating that the fixed point at **x*** = 0 is stable. This behavior models how some RNNs stabilize over time to produce consistent outputs. In more complex **RNNs**, **ODEs** can model situations where multiple fixed points exist. For example, consider the **ODE**:

$$\frac{dx(t)}{dt} = x(t) - x(t)^3$$

This system has three fixed points: **x*** = –1, 0, and 1. The stability of these points can be determined by analyzing the derivative of the function around these points. In this case, **x*** = –1 and **x*** = 1 are stable fixed points, while **x*** = 0 is unstable. This means that the system will converge to either –1 or 1, depending on the initial state, but if the system starts exactly at **x** = 0, any small perturbation will push it away from this point.

11.3.3 Training and Dynamics

Training of **RNNs** can be very difficult, especially in deep networks where the training process can act like chaotic systems. Small changes in settings can lead to very different results, making the network's behavior hard to predict. Ideas from dynamical systems theory, such as chaotic behavior and bifurcations, can help us understand and manage these challenges. In deep **RNNs**, even small changes in the weights can cause big changes in the output, similar to chaos in dynamical systems. Tiny changes in the starting conditions can lead to very different outcomes over time. For example, imagine an **RNN** with a single neuron where the hidden state $\mathbf{h_t}$ at time **t** is calculated as:

$$h_{t+1} = \tanh(W_h h_t + W_x x_t)$$

Here, $\mathbf{W_h}$ is the recurrent weight, $\mathbf{W_x}$ is the input weight, and $\mathbf{x_t}$ is the input at time **t**. Suppose:

$\mathbf{W_h}$ = 0.9, $\mathbf{W_x}$ = 1.0, Initial state $\mathbf{h_0}$ = 0.5 and Input sequence **x** = [1,0,1,0]

The hidden states over time would be:

- $\mathbf{h_1}$ = tanh(0.9 × 0.5 + 1 × 1) = tanh(1.45) ≈ 0.9,
- $\mathbf{h_2}$ = tanh(0.9 × 0.9 + 1 × 0) = tanh(0.81) ≈ 0.67,
- $\mathbf{h_3}$ = tanh(0.9 × 0.67 + 1 × 1) = tanh(1.603) ≈ 0.92,
- $\mathbf{h_4}$ = tanh(0.9 × 0.92 + 1 × 0) = tanh(0.828) ≈ 0.68.

Now, suppose we make a small change in the recurrent weight, increasing $\mathbf{W_h}$ slightly to 0.91. The hidden states will change as follows:

- $\mathbf{h_1}$ = tanh(0.91 × 0.5 + 1 × 1) = tanh(1.455) ≈ 0.9,
- $\mathbf{h_2}$ = tanh(0.91 × 0.9 + 1 × 0) = tanh(0.819) ≈ 0.67,
- $\mathbf{h_3}$ = tanh(0.91 × 0.67 + 1 × 1) = tanh(1.611) ≈ 0.92,
- $\mathbf{h_4}$ = tanh(0.91 × 0.92 + 1 × 0) = tanh(0.835) ≈ 0.68.

The changes in the output may seem small here, but as the network gets deeper, small changes like this accumulate, leading to chaotic behavior where outputs can change unpredictably. Bifurcations occur when a small change in a parameter causes a qualitative change in the network's behavior. Consider the same **RNN**, but instead of a small change, we increase $\mathbf{W}_h$ significantly from 0.9 to 1.2. The hidden states now evolve as follows:

- $\mathbf{h}_1 = \tanh(1.2 \times 0.5 + 1 \times 1) = \tanh(1.6) \approx 0.92$,
- $\mathbf{h}_2 = \tanh(1.2 \times 0.92 + 1 \times 0) = \tanh(1.104) \approx 0.8$,
- $\mathbf{h}_3 = \tanh(1.2 \times 0.8 + 1 \times 1) = \tanh(1.96) \approx 0.96$,
- $\mathbf{h}_4 = \tanh(1.2 \times 0.96 + 1 \times 0) = \tanh(1.152) \approx 0.82$.

Compared to the previous example, a larger change in the weight $\mathbf{W}_h$ has caused the hidden state to jump to much higher values. This is a bifurcation point, where the small increase in the weight led to a qualitative change in the **RNN**'s behavior, making the network more prone to oscillations or even chaotic patterns. These dynamics make training **RNNs** difficult because even small parameter changes can lead to instability or chaotic behavior, disrupting the training process. Gradient clipping can prevent weights from growing too large, helping to avoid bifurcations or chaotic dynamics.

11.3.4 Vanishing and Exploding Gradients

One big problem when training deep **RNNs** is the vanishing and exploding gradient issue. To understand this, think of **RNNs** as dynamic systems where each step is like a moment in time. As the **RNN** processes more steps (or inputs), the gradients used to update the model during training can either get very small (vanish) or become too large (explode), making the model hard to train. This happens because, like in dynamic systems where states can grow or shrink over time, **RNNs** can have their gradients affected the same way during backpropagation. If the gradients vanish, the model has trouble learning long-term dependencies because the updates become too small. If the gradients explode, the updates become too large, causing the training to become unstable. To fix exploding gradients, a common method is gradient clipping, which limits the size of the gradients during backpropagation to keep them from getting too big. For the vanishing gradient problem, more advanced **RNN** architectures like long short-term memory (LSTM) networks and gated recurrent units (GRU) were created. These architectures have built-in systems that help keep important information over long sequences, thus allowing the model to remember things for longer and avoid the vanishing gradient issue. Additionally, techniques like batch normalization can also help by stabilizing gradients during training, making the optimization process smoother and improving the model's performance. A helpful way to understand this issue is by viewing **RNNs** as dynamic systems, where each step in an **RNN** is like a moment in time. Over time, the gradients used to update the network's parameters can either become too small (vanish) or grow too large (explode), making training difficult. To make this clearer, let's look at an example. Imagine an **RNN** processing a sequence over 10 time steps. If the gradient at each step is slightly smaller than 1 (say 0.8), after 10 steps, the gradient would shrink as follows:

- After 1 step: 0.8,
- After 2 steps: $0.8 \times 0.8 = 0.64$,
- After 3 steps: $0.8 \times 0.64 = 0.512$,
- ...,
- After 10 steps: $0.8^{10} = 0.107$.

After 10 steps, the gradient decreases to around 0.107. This small value makes it hard for the network to learn because the gradient becomes too tiny to make useful updates to the model's weights;

this is the vanishing gradient problem. Now let's see what happens with an exploding gradient. Suppose the gradient at each step is slightly larger than 1 (say 1.2). Over the same 10 steps:

- After 1 step: 1.2,
- After 2 steps: $1.2 \times 1.2 = 1.44$,
- After 3 steps: $1.2 \times 1.44 = 1.728$,
- ...,
- After 10 steps: $1.2^{10} = 6.191$.

In this case, after 10 steps, the gradient has grown to about 6.191. As the network gets deeper, this value continues to grow larger, making the training unstable and causing the weights to update too aggressively; this is the exploding gradient problem. Gradually clipping is often used to deal with exploding gradients. This method limits the size of the gradients during backpropagation so that they don't grow too large. For example, if the gradient exceeds a certain value (say 5), it's clipped to 5. This helps keep the training process stable. These architectures have special gates that help maintain important information over time, allowing the model to avoid losing the gradient in long sequences.

Figure 11.6 illustrates critical challenges and dynamics within RNNs, focusing on the vanishing and exploding gradient problems as well as memory retention behavior. Figure 11.6a highlights the vanishing and exploding gradient issues encountered during training **RNNs**. The x-axis represents the timestep, while the y-axis shows the gradient magnitude. The blue line indicates the vanishing gradient, where the gradient value diminishes over time, approaching zero as timesteps increase. This phenomenon hampers the **RNN**'s ability to learn and update weights effectively, especially for long-term dependencies. In contrast, the red line represents the exploding gradient, where the gradient magnitude increases exponentially with each timestep. This can lead to instability in training as the gradient becomes excessively large, making it challenging to converge. Figure 11.6b illustrates the memory dynamics of an **RNN**, focusing on how the network retains and decays information over time. The x-axis represents the timestep, while the y-axis denotes the memory state. The green curve shows memory decay, where the memory state gradually reduces as time progresses. This decay demonstrates how **RNNs** struggle to maintain information over long sequences, contributing to difficulties in capturing long-term dependencies without enhancements like **LSTMs** or **GRUs**, which are designed to preserve memory over extended periods.

11.4 DYNAMICAL SYSTEMS AND DIFFERENTIAL EQUATIONS FOR DEEP LEARNING

The interplay between dynamical systems and differential equations offers a rich framework for understanding and improving deep learning models, particularly **RNNs**. In **RNNs**, small differences in the initial states or input data can lead to completely different outputs, making the network's behavior hard to predict. This sensitivity can be both good and bad. On one hand, it allows the network to capture complex patterns. On the other hand, it can make training the network unstable. Looking at **RNNs** from the perspective of dynamic systems helps us understand these complexities and shows how they can work well with new data.

11.4.1 Lotka–Volterra Equations and Neural Networks

The Lotka–Volterra equations, also known as the predator–prey equations, are a pair of first-order, non-linear differential equations used to describe the dynamics of biological systems where two species interact: predator and prey. The equations are:

$$\frac{dx}{dt} = \alpha x - \beta xy \,, \quad \frac{dy}{dt} = \delta xy - \gamma y$$

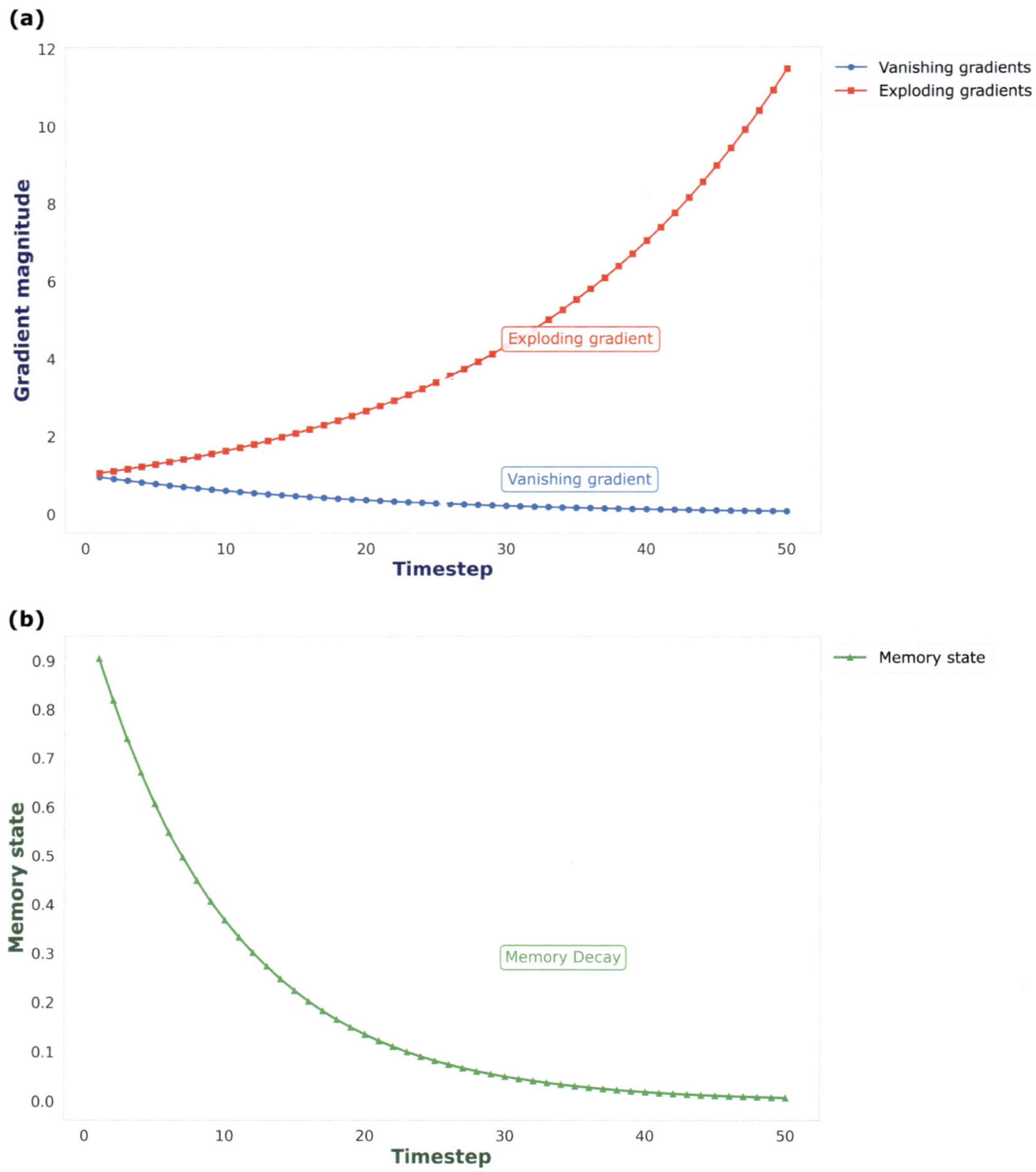

FIGURE 11.6 (a) Vanishing and exploding gradient in RNNs, (b) memory dynamics in RNNs.

where:

- **x** is the number of prey,
- **y** is the number of predators, and
- $\alpha, \beta, \delta, \gamma$ are parameters representing the interaction between the two species.

We can generate data using the Lotka–Volterra (predator–prey) equations and then train a neural network to learn their underlying dynamics. To make the concept clearer with specific numbers, let's assume the following parameters:

- **α** = 1.1 (prey reproduction rate),
- **β** = 0.4 (rate of predation),
- **γ** = 0.4 (predator death rate),
- **δ** = 0.1 (rate of predator growth by eating prey).

Suppose we start with an initial population of 40 prey and 9 predators. Using the equations, we can calculate how both populations change over time. At the initial time step (**t** = 0), the populations are: **x**(0) = 40, **y**(0) = 9.

At the next time step, we calculate the change in prey and predator populations using the differential equations:

$$\frac{dx}{dt} = 1.1 \times 40 - 0.4 \times 40 \times 9 = 44 - 144 = -100$$

So, the prey population decreases by 100.

$$\frac{dy}{dt} = 0.1 \times 40 \times 9 - 0.4 \times 9 = 36 - 3.6 = 32.4$$

Thus, the predator population increases by 32.4. **Update populations** over $\Delta t = 0.1$:

$$x(0.1) = x(0) + \left(\frac{dx}{dt}\right)\Delta t = 40 + (-100) \times 0.1 = 40 - 10 = 30,$$

$$y(0.1) = y(0) + \left(\frac{dy}{dt}\right)\Delta t = 9 + 32.4 \times 0.1 = 9 + 3.24 = 12.24.$$

After just 0.1 unit of time, the prey population is 30 (instead of going negative), and the predators have risen to 12.24. By repeating these small-step updates over many increments, you will typically see oscillatory cycles: as prey increases, predators flourish; as predators increase, they reduce the prey supply, leading to a subsequent predator decline, and so forth. In practice, this generates time-series data that mimic real predator–prey interactions. You can then train a neural network to learn these dynamics, by feeding it time-lagged samples of (x, y), so it can predict future population levels given current conditions. This offers a powerful way for neural networks to capture complex biological interactions and other non-linear systems governed by differential equations.

Figure 11.7 illustrates the analysis of predator–prey dynamics using the Lotka–Volterra model and a neural network for population prediction. Figure 11.7a represents the Lotka–Volterra predator–prey model, showing the oscillatory relationship between prey (x) and predator (y) populations over time. The prey population (blue) and predator population (orange) exhibit periodic fluctuations, reflecting the cyclical nature of ecological interactions. The equations governing the system, $\frac{dx}{dt} = ax - \beta xy$ and $\frac{dy}{dt} = \delta xy - \gamma y$, are annotated on the plot. A key observation is marked by a red dot at the point of maximum predator population (23.0073), emphasizing the peak of the predator cycle. Figure 11.7b compares the actual predator population (orange) against the predicted predator population (green, dashed) generated by a neural network. The model includes two hidden layers with 64 neurons

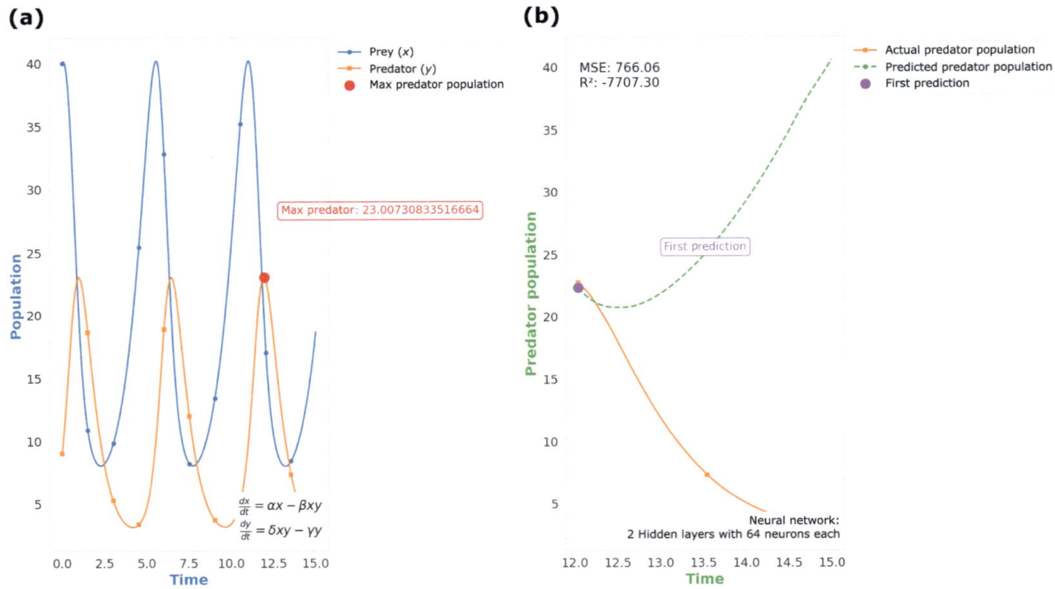

FIGURE 11.7 (a) Lotka–Volterra predator–prey model. (b) Actual vs. predicted predator population.

each. A notable point, labeled "First Prediction" (purple dot), demonstrates the neural network's initial prediction accuracy. Despite capturing the trend, the predicted population deviates from the actual values, reflected in the high mean squared error (**MSE**: 766.06) and poor $\mathbf{R}^2$ score (–7707.30).

11.5 REAL-WORLD APPLICATIONS AND EXAMPLES

11.5.1 Modeling Epidemics with Differential Equations

Differential equations can accurately model how infectious diseases spread. One common model is the Susceptible, Infected, Recovered (SIR) model. This model tracks how people move from being susceptible to an infection to becoming infected and then recovering (or gaining immunity). By using these rates, public health officials can predict how an epidemic will grow and see how different strategies can help control it. For example, during the COVID-19 pandemic, differential equations were very important for forecasting the virus's spread. These models helped show the possible effects of actions like social distancing, wearing masks, and vaccination programs. This allowed governments and health organizations to make better, data-based decisions. By understanding how fast the virus spreads and how people recover, authorities could plan and adjust their responses to reduce the epidemic's impact. This mathematical approach is crucial for controlling outbreaks and getting ready for future waves.

11.5.2 Stability Analysis in Engineering

In engineering, dynamical systems theory is an important tool for checking the stability of structures and machines. For example, in aerospace engineering, engineers study the stability of an aircraft's flight path using differential equations that model how it moves through the air. These equations help engineers understand when an aircraft might become unstable, such as during a stall or spin, which could cause it to lose control. By examining these possible instabilities, engineers can create control systems that automatically detect and fix problems, ensuring the aircraft flies steadily. This stability analysis is essential not only for safety but also for improving the performance of aircraft in different flight conditions, leading to more efficient and safer aviation systems.

11.5.3 Economic Modeling and Forecasting

In economics, dynamical systems and differential equations are important tools for modeling and predicting how markets and economies behave over time. A good example is using the Lotka–Volterra equations, which were first used in biology to describe predator and prey relationships, to model competition between companies in a market. These equations help economists study how things like available resources or the level of competition affect whether businesses grow or decline. By using these models, economists can better understand how markets work and create strategies or policies to encourage economic stability and growth. For example, the equations can show how market competition affects whether companies survive, helping decision-makers adjust regulations or policies to ensure a healthy and competitive economic environment.

11.5.4 Climate Change Projections

Climate scientists use differential equations a lot to model the Earth's climate and predict how human activities will affect it. These models include the complex interactions between the air, oceans, and land, tracking how things like temperature, humidity, and carbon dioxide levels change over time. By considering these factors, scientists can create different scenarios, such as varying amounts of greenhouse gas emissions. These simulations help scientists forecast possible climate changes and understand the likely effects of human actions. This information is essential for shaping global policies on climate action and helping governments and organizations make informed decisions about reducing emissions, setting environmental regulations, and adopting sustainable practices to lessen the impact of climate change.

11.5.5 Neuroscience and Brain Dynamics

In neuroscience, dynamical systems and differential equations help model the brain's electrical activity. A famous example is the Hodgkin–Huxley model, which uses these equations to show how neurons create action potentials, electrical signals that send information through the nervous system. These models are important for understanding how brain circuits work and how problems in these circuits can cause conditions like epilepsy. By studying these models, neuroscientists learn how brain disorders happen and can develop treatments like deep brain stimulation to restore normal brain function. These mathematical tools are powerful for exploring both healthy and unhealthy brain activity and for guiding treatment methods.

11.5.6 Robotics and Autonomous Systems

In robotics, dynamical systems theory is important for controlling how robots and self-driving vehicles move. For example, path planning algorithms often use differential equations to calculate the best path a robot should take, making sure it reaches its destination while avoiding obstacles. This method is essential for designing robots that can work safely and efficiently in changing environments. For instance, drones flying through complicated areas or robotic arms doing precise tasks in factories rely on these models to adjust their movements in real time. By using dynamical systems theory, robots can respond to changes around them and handle new challenges, making them more reliable and effective in real-world situations.

11.6 HANDS-ON EXAMPLE

We'll simulate a scenario where the **RNN** is tasked with learning a time-dependent sequence, allowing us to observe the effects of memory retention and the potential occurrence of vanishing or exploding gradients.

11.6.1 Step 1: Import Libraries

First, we import essential libraries for building machine learning models and visualizing results. Together, these libraries are essential for tasks ranging from building neural network models and handling data to visualizing results for analysis and interpretation.

```
import numpy as np
import tensorflow as tf
import matplotlib.pyplot as plt
```

11.6.2 Step 2: Generate Synthetic Sine Wave Data

In this section, we are generating a time series dataset using the sine function. This setup is useful for simulating periodic data, such as waves or oscillations, which can be visualized and analyzed.

```
# time variable
t = np.linspace(0, 2*np.pi, 100)
data = np.sin(t)
```

11.6.3 Step 3: Prepare Data for RNN: Format (batch_size, Timesteps, Features)

In this section, we are preparing the input and target data for a time series prediction model, where each value in the sine wave is used to predict the next value. We are creating a one-step-ahead prediction model, where the goal is to predict the next value in the time series based on the current value. This reshaping process is crucial for feeding the data into machine learning models, particularly RNNs or LSTMs, where the 3D input shape is required.

```
X = data[:-1].reshape(-1, 1, 1)
Y = data[1:].reshape(-1, 1)
```

11.6.4 Step 4: Define the RNN Model

In this section, we are building a simple RNN using the Keras Sequential **API** from TensorFlow. The model consists of an **RNN** layer followed by a Dense output layer for time series prediction.

```
model = tf.keras.Sequential([tf.keras.layers.SimpleRNN(10,
input_shape=(None, 1), activation='tanh'), tf.keras.layers.
Dense(1)])
```

11.6.5 Step 5: Compile the Model

Here, we are compiling the **RNN** model, which prepares it for training by specifying the optimizer, loss function, and potentially other metrics. By compiling the model with the Adam optimizer and mean squared error loss, we ensure that the model will learn effectively during training by minimizing the prediction error on the time series data. This setup is ideal for time series prediction tasks, where the goal is to make accurate continuous-value predictions.

```
model.compile(optimizer='adam', loss='mean_squared_error')
```

11.6.6 Step 6: Train the Model

In this line of code, we are training the **RNN** model on the time series data using the fit function in Keras.

```
history = model.fit(X, Y, epochs=100, verbose=0)
```

11.6.7 Step 7: Predict Using the Trained Model

In this line of code, we are using the trained **RNN** model to make predictions on the time series data. This line is crucial in time series forecasting, as it allows us to observe how well the model has learned to predict future values based on the historical data provided during training. These predictions can be compared to the actual target values (Y) to evaluate the model's performance.

```
predictions = model.predict(X)
```

11.6.8 Step 8: Plot the Results

In this section, we are visualizing the results of the **RNN** predictions alongside the actual data using Matplotlib.

```
plt.figure(figsize=(10, 5))
plt.plot(t[1:], Y, label='Actual Data')
plt.plot(t[1:], predictions, label='Predicted Data', linestyle=
'--')
plt.title('RNN Simulation of a Sine Wave')
plt.xlabel('Time')
plt.ylabel('Amplitude')
plt.legend()
plt.show()
```

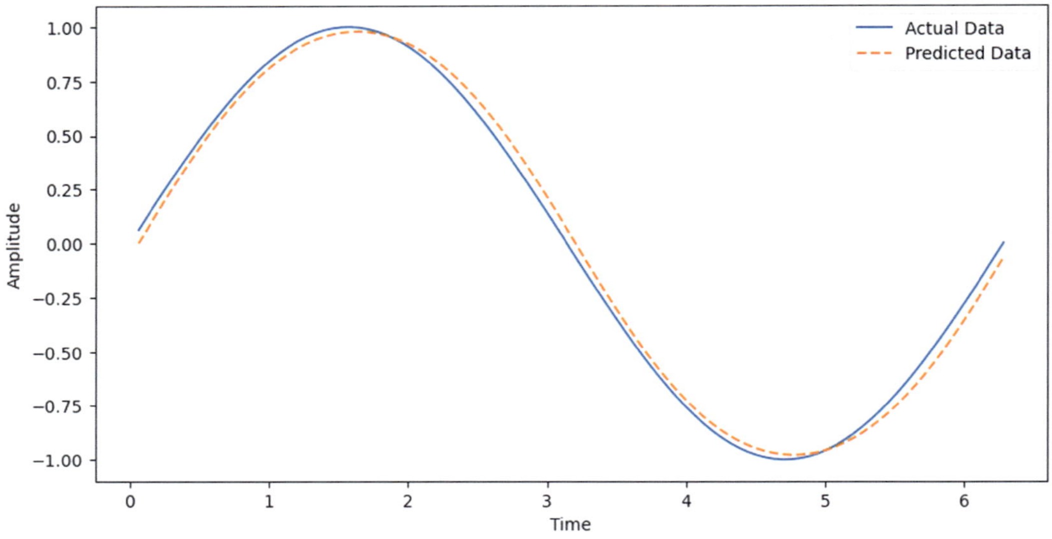

FIGURE 11.8 **RNN** simulation of a sine wave.

Figure 11.8 illustrates how the RNN has learned to approximate the sine wave pattern. By training on sequences of the wave, the **RNN** effectively captures both the amplitude and phase of the cycle, thus demonstrating its utility in time-series prediction tasks. The slight discrepancies between the actual and predicted values underscore the challenges inherent in perfecting sequence modeling, particularly in capturing exact dynamics without overfitting to noise or anomalies in the data.

11.7 COMMON MISTAKES AND TROUBLESHOOTING TIPS

11.7.1 Misinterpreting Dynamical Systems Theory

- *Mistake:* Failing to grasp the fundamental principles of dynamical systems and how they relate to RNN behavior.
- *Tip:* Start with the basics of dynamical systems theory before diving into its applications in RNNs. Use simple examples and simulations to build a solid foundation.

11.7.2 Overlooking the Importance of Initial Conditions

- *Mistake:* Ignoring the sensitivity of **RNNs** to initial conditions, leading to unpredictable behavior during training.
- *Tip:* Pay close attention to the initialization of **RNN** parameters. Use strategies like careful initialization and gradient clipping to mitigate sensitivity issues.

11.7.3 Underestimating the Complexity of Training Dynamics

- *Mistake:* Simplifying the training process of **RNNs** without considering the complex dynamics that can arise, such as bifurcations and chaotic behavior.
- *Tip:* Monitor the training process closely, looking for signs of instability. Employ regularization techniques and dynamic learning rate adjustments to stabilize training.

11.7.4 Neglecting the Vanishing and Exploding Gradient Problem

- *Mistake:* Failing to address the vanishing and exploding gradient problem, which can severely affect the training of deep **RNNs**.
- *Tip:* Use techniques like LSTM networks, GRUs, gradient clipping, and proper initialization methods to manage gradient issues.

11.7.5 Misapplying Differential Equations

- *Mistake:* Incorrectly applying differential equations to model **RNN** behavior, leading to flawed analyses.
- *Tip:* Ensure a strong understanding of ODEs and their application to continuous-time **RNNs**. Validate your models with known solutions and simpler systems.

11.7.6 Overcomplicating Models

- *Mistake:* Creating overly complex models without sufficient justification, leading to difficulties in training and interpretation.
- *Tip:* Start with simpler models and gradually increase complexity as needed. Use model selection criteria and cross-validation to ensure that added complexity improves performance.

11.7.7 Failing to Validate Neural Network Models

- *Mistake:* Neglecting thorough validation of neural network models, resulting in unchecked errors and inefficiencies.
- *Tip:* Validate models using various datasets and benchmarks. Compare the performance of **RNNs** modeled with dynamical systems theory against traditional methods.

11.8 REVIEW QUESTIONS

1. What is the difference between discrete and continuous dynamical systems? Provide examples of each.
2. How do the Lotka–Volterra equations model the interaction between predator and prey populations? What do the parameters $\boldsymbol{\alpha}$, $\boldsymbol{\beta}$, γ represent?
3. How do RNNs retain information from previous inputs? How is this similar to the behavior of dynamical systems?
4. Explain how ODEs can represent certain **RNNs**. Why is this helpful representation?
5. Describe the challenges of training **RNNs**, particularly the vanishing and exploding gradients. How does viewing **RNN**$_s$ as dynamic systems help address these challenges?
6. What are bifurcations in dynamical systems, and how can similar phenomena affect the behavior of **RNN**$_s$ during training?
7. How do the concepts of chaotic behavior and sensitivity to initial conditions in dynamical systems relate to the predictability and complexity of **RNNs**?
8. How can neural networks be used to learn the dynamics of systems described by differential equations, such as the Lotka–Volterra model?
9. Why is it important to understand the behavior of dynamical systems and differential equations when working with deep learning models, particularly **RNNs**?
10. Discuss how the principles and tools of dynamical systems can be applied to real-world problems beyond biological systems, providing at least one example.

11.9 PROGRAMMING QUESTIONS

11.9.1 Easy

Design a Sequence Generation Task for **RNN** Evaluation.

1. Choose a type of sequence, such as arithmetic progressions (e.g., 2, 4, 6, 8, ...) or sinusoidal sequences.
2. Explain why this sequence can effectively test the **RNN's** learning capabilities.
3. Generate training data that reflects the chosen sequence.
4. Split the data into training and validation sets.
5. Train the **RNN** on the training dataset.

11.9.2 Medium

Analyze **RNN** Behavior with Variable Input Lengths.

1. Create datasets with varying lengths of input sequences, maintaining consistent complexity across datasets.
2. Adjust the **RNN** architecture if necessary to handle variable input lengths.
3. Train the same **RNN** model on each dataset separately.
4. Evaluate the model on a validation set that also varies in sequence length.

11.9.3 Hard

Explore the Impact of Network Depth in **RNNs**.

1. Modify an existing **RNN** model by adding multiple recurrent layers.
2. Ensure gradient clipping is implemented to mitigate exploding gradients.
3. Prepare a dataset suitable for deep **RNNs**, ensuring it has enough complexity to benefit from deeper architectures.
4. Train the modified **RNN** model using a rigorous training regime, possibly involving techniques like curriculum learning to gradually increase difficulty.
5. Evaluate the deep **RNN** model on a test set and compare it against a shallower baseline model.

12 Quantum Computing

12.1 INTRODUCTION

Quantum computing combines ideas from quantum mechanics and computer science, and it could change deep learning. By using the special abilities of quantum systems, we might soon be able to handle and analyze data in ways that regular computers can't. This chapter will look at how quantum computing can make deep learning better, such as speeding up the learning process, improving complex models, and even solving problems that were once impossible. We will focus on important quantum algorithms, like Shor's algorithm, which can break encryption, and Grover's algorithm, which can search through large databases quickly. We will also discuss how these algorithms can help deep learning by providing new methods to optimize and manage data.

12.2 INTRODUCTION TO QUANTUM COMPUTING

Quantum computing has grown from just an idea into a fast-developing technology. It is based on quantum mechanics, a branch of science that started in the early 1900s when scientists studied how tiny particles behave in strange ways. It wasn't until the 1980s that people began to understand what quantum computing could be. Early researchers saw that regular computers had a hard time simulating quantum systems efficiently. They suggested that a computer using quantum ideas could do these tasks better by using quantum superposition, which allows for natural parallel processing and could make quantum computers much faster. In the 1990s, important discoveries in quantum algorithms showed that quantum computing could change areas like encryption and searching through large databases. Even with these theoretical successes, creating real quantum computers was difficult because of problems like keeping quantum states stable and reducing mistakes. Since the 2000s, big companies have invested a lot in quantum research, and there has been significant progress. A major achievement happened in 2019 when a quantum processor was shown to be faster than classical supercomputers for certain tasks. Today, quantum computing keeps advancing quickly, with ongoing improvements in qubit stability, error correction methods, and algorithms, all building on the basic work from earlier years. Quantum computing is an exciting mix of quantum mechanics and computer science, using ideas that seem almost magical compared to regular physics. Three main ideas drive quantum computing: qubits, superposition, and entanglement.

12.2.1 QUBITS

Think of qubits as the quantum version of regular bits. While regular bits can only be a 0 or a 1, qubits can be both at the same time because of something called superposition. This ability allows

DOI: 10.1201/9781032690742-12

quantum computers look at many possibilities all at once, making them potentially much faster than regular computers for certain tasks.

12.2.2 Superposition

Imagine a spinning coin that's both heads and tails at the same time, only deciding which side it is when you stop it. Superposition works like that; it lets a qubit be in multiple states at once. This ability to hold and consider many possibilities all at once gives quantum computers incredible power for certain calculations.

12.2.3 Entanglement

One of the strangest parts of quantum mechanics is entanglement. When two qubits become entangled, the state of one instantly affects the state of the other, regardless of how far apart they are. This "spooky action at a distance" means that if you change one qubit, it immediately influences its entangled partner. This allows for fast and coordinated calculations across many qubits.

12.3 QUANTUM ALGORITHMS

12.3.1 Key Quantum Algorithms

Quantum algorithms leverage the unique properties of quantum mechanics to solve problems more efficiently than classical algorithms. Two key algorithms have had a major impact and have broad applications.

(a) *Factoring Algorithm:* This algorithm revolutionized cryptography by making it possible to factor large numbers exponentially faster than classical methods. While traditional algorithms take significantly more time as the numbers grow larger, this quantum algorithm uses superposition and entanglement, along with a quantum Fourier transform, to factor numbers quickly. This poses a serious threat to widely used encryption systems like RSA, pushing the need for quantum-safe cryptography.

(b) *Search Algorithm:* This algorithm dramatically speeds up the process of unstructured searching. In a database of **N** entries, the quantum algorithm can find the target in about $\sqrt{\mathbf{N}}$ steps, while classical methods would require **N** steps. By amplifying the probability of finding the correct answer, this algorithm is especially useful for tasks like database searches and optimization problems.

In Figure 12.1, the Bloch sphere provides a three-dimensional visualization of a qubit's state space. The sphere represents all possible states a qubit can take, with different arrows showing specific states. The dark red arrow indicates the $|0\rangle$ state, which points along the positive z-axis. This corresponds to the qubit being in a pure state of 0, as seen at the top of the sphere. On the other hand, the dark green arrow points in the opposite direction, along the negative z-axis, representing the qubit in the $|1\rangle$ state. This highlights the qubit being in the pure 1 state. These two states, $|0\rangle$ and $|1\rangle$, are analogous to the binary states in classical computing. The dark blue arrow, located on the equator of the sphere, represents the $|+\rangle$ state. This state is a superposition of both $|0\rangle$ and $|1\rangle$, where the qubit is not purely in one state but exists in a blend of both. The positioning of the $|+\rangle$ state on the equator emphasizes the fact that it is a balanced superposition, equidistant from $|0\rangle$ and $|1\rangle$, thereby demonstrating the core idea of quantum superposition.

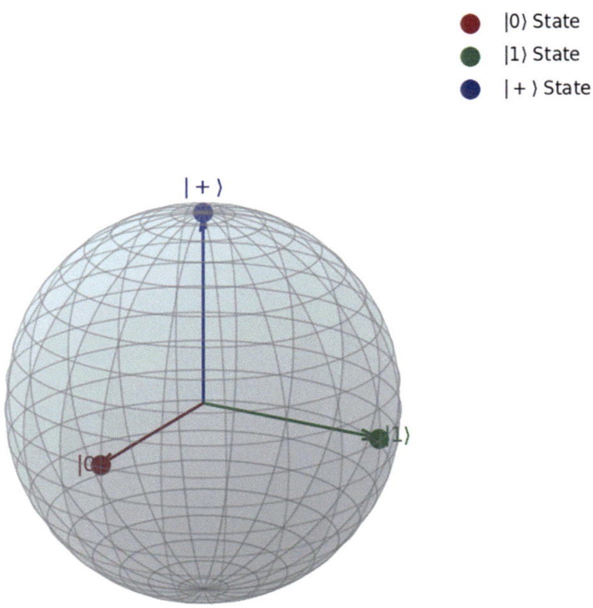

FIGURE 12.1 Qubit states on the Bloch sphere.

12.3.2 Quantum Machine Learning Algorithms

Quantum machine learning combines quantum computing with regular machine learning, opening up new possibilities for tasks like grouping data (clustering), sorting it into categories (classification), and making predictions (regression). Quantum clustering methods, like the Quantum k-means algorithm, improve how we group items by using quantum properties like superposition to calculate distances between data points all at once. For example, imagine you have 10 apples and want to group them by weight. With quantum clustering, the computer can compare all the apples simultaneously, making the process much faster than traditional methods that handle comparisons one at a time. Quantum classification algorithms, such as Quantum Support Vector Machines (QSVM), enhance how we classify data by mapping features into a higher-dimensional quantum space. Think of it like sorting animals based on size, habitat, and diet. A **QSVM** can make it easier to find clear differences, even between animals with similar traits, uncovering patterns that regular methods might miss. Quantum regression algorithms, like quantum linear regression, offer faster predictions by solving complex equations more efficiently. For instance, when predicting future temperatures from past data, quantum regression can quickly spot trends and create accurate forecasts, saving time compared to classical methods.

While quantum machine learning holds great promise, it is still in its early stages. One major challenge is converting regular data into quantum states, a process that can be complicated and time-consuming, sometimes canceling out the speed advantages. Additionally, current quantum computers are prone to errors and can only hold quantum states for a limited time, which can affect the accuracy of calculations. For example, suppose you're using a quantum machine learning algorithm to predict stock prices based on the last 10 days' closing prices: \$150, \$152, \$148, \$155, \$160, \$158, \$162, \$165, \$167, and \$170. First, you need to convert these numbers into quantum states, which is tricky. Even if the data is successfully encoded, errors might occur because the quantum computer may struggle to maintain its state, potentially leading to incorrect predictions. Instead of forecasting a rise to \$175, the computer might mistakenly predict a drop to \$140 due to the instability of the quantum system.

12.4 INTEGRATION WITH DEEP LEARNING

12.4.1 Quantum Neural Networks (QNNs)

QNNs combine quantum computing with artificial neural networks to boost their performance, especially for complex tasks like high-dimensional data analysis and pattern recognition. **QNNs** are built using quantum bits (qubits) instead of classical binary units, allowing them to take advantage of quantum mechanics, like superposition and entanglement. This gives them a huge computational advantage over traditional neural networks. Here's how **QNNs** work:

1. *Quantum Gates as Neurons:* Each quantum neuron acts like a quantum gate, performing operations on input qubits while preserving quantum information.
2. *Superposition and Parallelism:* **QNNs** use superposition, allowing quantum neurons to process multiple calculations at once, making the network much faster.
3. *Entanglement for Feature Correlation:* Entanglement links quantum neurons, helping **QNNs** detect and relate complex data patterns more effectively than classical networks.

Suppose we have a binary classification task where we want to classify inputs into two classes based on a single binary feature. There are class 0 and class 1, and the two inputs are as follows: Input 0, with a binary value of 0, should be classified as class 0, and Input 1, with a binary value of 1, should be classified as class 1. To perform this classification, we will design a simple **QNN** using one qubit and basic quantum gates. Before we proceed, let's review some fundamental quantum computing concepts. Qubit is the basic unit of quantum information, analogous to a bit in classical computing. A qubit can be in a superposition of states $|0\rangle$ and $|1\rangle$: $|\psi\rangle = \alpha|0\rangle + \beta|1\rangle$, where $\boldsymbol{\alpha}$ and $\boldsymbol{\beta}$ are complex numbers satisfying $|\alpha|^2 + |\beta|^2 = 1$. Quantum Gates are operations that change the state of qubits. Examples include:

- Hadamard Gate (**H**): $H = \frac{1}{\sqrt{2}}\begin{bmatrix} 1 & 1 \\ 1 & -1 \end{bmatrix}$
- Pauli-X Gate (**X**) (analogous to the NOT gate): $X = \begin{bmatrix} 0 & 1 \\ 1 & 0 \end{bmatrix}$
- Rotation Gate around the y-axis ($\mathbf{R_y}$): $R_y(\theta) = \begin{bmatrix} \cos\left(\frac{\theta}{2}\right) & -\sin\left(\frac{\theta}{2}\right) \\ \sin\left(\frac{\theta}{2}\right) & \cos\left(\frac{\theta}{2}\right) \end{bmatrix}$

Measurement is observing a qubit collapse of its state to either $|0\rangle$ or $|1\rangle$ with certain probabilities. Let us go to the computation steps:

Step 1. We need to encode our classical binary inputs into quantum states. Input 0, represented by the quantum state $|0\rangle$, and Input 1, represented by the quantum state $|1\rangle$, which we obtain by applying the Pauli-X gate to $|0\rangle$: $|1\rangle = X|0\rangle$. Quantum states for Inputs are as follows: For Input 0, $|\psi_{\text{input}}\rangle = |0\rangle = \begin{bmatrix} 1 \\ 0 \end{bmatrix}$ and for Input 1, $|\psi_{\text{input}}\rangle = |1\rangle = \begin{bmatrix} 0 \\ 1 \end{bmatrix}$.

Step 2. Our **QNN** consists of one Qubit to represent the quantum state and one Quantum Gate with a trainable parameter. We'll use the rotation gate $\mathbf{R_y(\theta)}$ as our quantum neuron.

Step 3. We apply the rotation gate $\mathbf{R_y(\theta)}$ to the input qubit. The angle $\boldsymbol{\theta}$ is our trainable parameter (analogous to weights in classical neural networks).

- **Rotation Gate $\mathbf{R_y(\theta)}$:** $R_y(\theta)=\begin{bmatrix}\cos\left(\frac{\theta}{2}\right) & -\sin\left(\frac{\theta}{2}\right)\\ \sin\left(\frac{\theta}{2}\right) & \cos\left(\frac{\theta}{2}\right)\end{bmatrix}$ Let's choose $\theta=\frac{\pi}{2}$ for demonstration:

- **Compute** $R_y\left(\frac{\pi}{2}\right)$: $\cos\left(\frac{\pi}{4}\right)=\frac{\sqrt{2}}{2}$, $\sin\left(\frac{\pi}{4}\right)=\frac{\sqrt{2}}{2}$. So, $R_y\left(\frac{\pi}{2}\right)=\begin{bmatrix}\frac{\sqrt{2}}{2} & -\frac{\sqrt{2}}{2}\\ \frac{\sqrt{2}}{2} & \frac{\sqrt{2}}{2}\end{bmatrix}$,

- **Process Input 0:** $|\psi_{\text{output}}\rangle = R_y\left(\frac{\pi}{2}\right)|0\rangle=\begin{bmatrix}\frac{\sqrt{2}}{2}\\ \frac{\sqrt{2}}{2}\end{bmatrix}$,

- **Process Input 1:** $|\psi_{\text{output}}\rangle = R_y\left(\frac{\pi}{2}\right)|1\rangle=\begin{bmatrix}-\frac{\sqrt{2}}{2}\\ \frac{\sqrt{2}}{2}\end{bmatrix}$.

Step 4. When we measure the qubit, the probability of getting $|0\rangle$ or $|1\rangle$ is given by the square of the amplitudes. For a state $\|\psi\rangle=\alpha\||0\rangle+\beta|1\rangle$, probability of $|0\rangle$ is $P(0)=|\alpha|^2$ and probability of $|1\rangle$ is $P(1)=|\beta|^2$.

- Calculate probabilities for Input 0: $|\psi_{\text{output}}\rangle=\frac{\sqrt{2}}{2}\Big||0\rangle+\frac{\sqrt{2}}{2}\Big||1\rangle$, $P(0)=\left|\frac{\sqrt{2}}{2}\right|^2=\frac{1}{2}$ and $P(1)=\frac{1}{2}$,

- Calculate probabilities for Input 1: $|\psi_{\text{output}}\rangle=-\frac{\sqrt{2}}{2}\Big||0\rangle+\frac{\sqrt{2}}{2}\Big||1\rangle$, $P(0)=\frac{1}{2}$ and $P(1)=\frac{1}{2}$.

The probabilities are the same for both inputs, so the **QNN** cannot distinguish between the classes with $\theta=\frac{\pi}{2}$.

Step 5. Our goal is to find a $\mathbf{\theta}$ that allows the **QNN** to classify the inputs correctly. The desired probabilities are as follows: For Input 0 (should be Class 0), **Maximize P**(0), and for Input 1 (should be Class 1): **Maximize P**(1). Try $\mathbf{\theta}$ = 0 and compute $\mathbf{R_y}(0)$: $R_y(0)=\begin{bmatrix}1 & 0\\ 0 & 1\end{bmatrix}$.

Process Input 0: $|\,|\psi_{\text{output}}\rangle=R_y(0)\||0\rangle=\||0\rangle$: P(0) = 1 and P(1) = 0,

Process Input 1: $|\,|\psi_{\text{output}}\rangle=R_y(0)\||1\rangle=\||1\rangle$: P(0) = 0 and P(1) = 1,
Input 0: High probability of $|0\rangle \Rightarrow$ Class 0 (correct),
Input 1: High probability of $|1\rangle \Rightarrow$ Class 1 (correct).

If we set $\boldsymbol{\theta}$ to 0, the QNN correctly sorts the inputs. In this case, $\boldsymbol{\theta}$ acts like a trainable parameter, similar to weights in regular neural networks, that we adjust to reduce classification errors. **QNNs** have significant advantages over classical neural networks. They can process certain tasks exponentially faster, especially those involving large datasets or complex pattern recognition. This speed is a huge benefit in fields with high computational demands. Quantum mechanics naturally handles calculations in spaces with many dimensions, making **QNNs** highly efficient for tasks like image recognition, modeling complex systems, and financial analysis where there are many features to consider. Additionally, unique quantum properties like superposition and entanglement could lead to new learning algorithms that outperform classical ones in efficiency or effectiveness. The qubits are then processed through a series of quantum gates like the Hadamard, CNOT, and Pauli-X gates. The Hadamard gate creates a superposition of states, allowing qubits to be in multiple states at once. The CNOT gate is used to entangle qubits, linking their states no matter how far apart they are. The Pauli-X gate flips the state of a qubit, like changing from 0 to 1. These gates are essential for manipulating qubit states to perform complex calculations that are uniquely quantum. The transformed qubits pass through quantum circuits, which are dynamic arrangements of various quantum gates designed to run specific quantum algorithms. These circuits are the backbone of quantum computing, allowing for smooth control of qubits to solve problems that classical computers can't handle. After the quantum operations are complete, the qubits are measured. This crucial step causes the qubit states to collapse from their quantum superpositions into definite states that we can interpret using classical computing. The measurement outputs classical data, keeping the computational advantages gained from quantum processing. These classical data are then sent through traditional computing stages. Here, regular computing techniques refine, analyze, and use the data from the quantum processes. This integration bridges the gap between quantum and classical computing, using the strengths of both to enhance computational power and efficiency.

12.4.2 Hybrid Quantum-Classical Models

Hybrid quantum-classical models combine the strengths of both quantum and classical computing, improving performance by using each where it works best. Instead of trying to replace classical computing entirely, quantum methods are used for specific tasks like advanced optimization or handling complex calculations where they have clear advantages. This mix is especially useful in neural networks, where hybrid models can speed up learning and enhance results. In these models, tasks are divided between the quantum and classical parts. One key example is creating quantum feature maps. Here, classical data is turned into a quantum state in a high-dimensional space, making it easier to see relationships between data points. This clearer view helps with tasks like classification or clustering, which can then be converted back into classical data for further analysis. Another important use is quantum optimization. Quantum algorithms, like the Quantum Approximate Optimization Algorithm (QAOA), fine-tune parameters like weights and biases in neural networks. These algorithms are especially helpful in complex situations with many local minima, where traditional methods might struggle. By finding better solutions faster, quantum optimization can improve the overall performance of the model. Hybrid models are practical because they use the best features of both computing systems. They offer enhanced computational power while staying within the current limits of quantum technology. This makes hybrid models more scalable and usable than purely quantum systems. In practice, they help solve problems that are too slow or complicated for classical computers alone, especially in tasks involving large datasets or heavy computational demands. Another benefit is flexibility; hybrid models can incorporate future quantum advancements without needing to overhaul the entire system. However, there are challenges. Integrating quantum and classical components is complex, requiring careful engineering and programming. Current limitations of quantum hardware, such as short qubit coherence times, high error rates, and a limited number of qubits, also reduce the efficiency of hybrid models. Additionally, developing algorithms that fully take advantage of these hybrid systems demands ongoing research and a solid understanding of both quantum mechanics and machine learning. A practical example of a hybrid quantum-classical model is in financial modeling. In this case, the quantum component might be used to optimize investment portfolios across many variables, while the classical part handles routine data processing and transaction tasks. This combination allows quantum computing to tackle complex, high-level problems while classical systems manage day-to-day operations.

12.5 APPLICATIONS OF QUANTUM COMPUTING IN DEEP LEARNING

12.5.1 Enhanced Optimization

Quantum computing offers exciting possibilities for improving how we optimize neural network training, potentially outperforming classical methods in both speed and efficiency. The power of quantum optimization comes from its unique properties, superposition, entanglement, and quantum tunneling, which allow quantum algorithms to explore complex optimization landscapes more effectively than classical approaches. In quantum computing, superposition lets each quantum bit (qubit) represent multiple states at once. This means quantum algorithms can evaluate many possible solutions simultaneously, unlike classical computers that process one solution at a time. For neural network training, this ability allows a quantum optimizer to assess different combinations of weights and biases all at once, reducing the time needed to find the best or nearly the best configurations. Quantum systems also use quantum tunneling, a phenomenon where particles can "tunnel" through barriers that would be impossible to cross in classical physics. In optimization, this allows a quantum optimizer to avoid getting stuck in local minima, solutions that seem optimal but aren't the best overall. Instead, it can keep exploring other areas of the solution space, increasing the chances of finding the global minimum. Entanglement, another key property, means that the state of one qubit can instantly affect the state of another, no matter how far apart they are. In optimization, entanglement helps quantum algorithms by linking the relationships between different parts of a solution. This means the quantum optimizer can consider how various parameters are connected, leading to a more efficient exploration of possible configurations. Consider a simple non-convex function: $L(x) = x^4 - 8x^2 + 16$. This function has multiple minima and maxima. The function has local minima and maxima due to its quartic and quadratic terms. The global minimum is at $\mathbf{x} = 0$. Let us first do classical optimization using gradient descent. We start at an initial point and iteratively move in the direction opposite to the gradient.

1. *Initialization:* Start at $\mathbf{x}_0 = 4$ and Learning rate $\boldsymbol{\alpha} = 0.1$,
2. *Iteration 1:* Compute the gradient of $L(x) = x^4 - 8x^2 + 16$ that is equal to $L'(x) = 4x^3 - 16x$, $L'(4) = 4(64) - 16(4) = 256 - 64 = 192$ and If $\mathbf{x}_0 = 4$, you then update $\mathbf{x}$ by subtracting the product of the gradient $L'(\mathbf{x}_0) = 192$ and the learning rate 0.1: $L'(x_0) = 4 - 0.1(192) = 4 - 19.2 = -15.2$
3. *Iteration 2:* $L'(-15.2) = 4(-15.2)^3 - 16(-15.2)$, then, Compute and update $\mathbf{x}_2$.

The steps may overshoot or get stuck in local minima. Finding the global minimum is not guaranteed. Now, let us do quantum optimization. Quantum annealing uses quantum mechanics to find the global minimum of an objective function by exploiting quantum tunneling and superposition. The key concepts in quantum mechanics that influence quantum computing include superposition, quantum tunneling, and entanglement. Superposition allows a system to consider all possible states simultaneously, which is fundamental to quantum computation's ability to handle complex problems efficiently. Quantum tunneling enables the system to transition through energy barriers between local minima, helping escape from suboptimal solutions in optimization tasks. Entanglement correlates different variables in such a way that it allows a holistic search of the solution space, enhancing the system's capacity to explore multiple possibilities simultaneously. These concepts provide quantum computing with its unique computational power.

In this scenario, we represent possible values of **x** using a finite set of states, where **x**∈{−4, −3, −2, −1, 0, 1, 2, 3, 4}. Each state corresponds to a unique qubit configuration, and we use qubits to encode these states. The Hamiltonian **H** encodes the objective function **L(x)**, with the goal of finding the state with the lowest energy, which corresponds to the global minimum of **L(x)**.

We begin by preparing the qubits in a superposition of all possible states, such that the quantum system is represented by a state $|\psi(0)\rangle = \sum_x c_x |x\rangle$, where $\mathbf{c_x}$ are complex coefficients, and $|\mathbf{x}\rangle$ represents each state. The system's Hamiltonian then evolves slowly from an initial Hamiltonian $\mathbf{H_0}$ to the problem Hamiltonian $\mathbf{H_p}$, which encodes the objective function. The adiabatic theorem ensures that if the evolution is slow enough, the system will remain in its ground state throughout the process. During this evolution, quantum tunneling enables the system to pass through barriers, allowing it to avoid being trapped in local minima and ensuring it can find better solutions. At the end of the annealing process, the qubits are measured, collapsing the quantum system into a single state $|\mathbf{x_{min}}\rangle$, which corresponds to the global minimum and provides the optimal solution to the problem. This process efficiently leverages quantum properties such as superposition and tunneling to explore the solution space and find the global minimum. Now, let's look at a numerical example by computing the function **L(x)** for discrete states:

- L(−4) = 256 − 128 + 16 = 144,
- L(−3) = 81 − 72 + 16 = 25,
- L(−2) = 16 − 32 + 16 = 0,
- L(−1) = 1 − 8 + 16 = 9,
- L(0) = 0 − 0 + 16 = 16,
- L(1) = 1 − 8 + 16 = 9,
- L(2) = 16 − 32 + 16 = 0,
- L(3) = 81 − 72 + 16 = 25,
- L(4) = 256 − 128 + 16 = 144.

In this case, the function reaches its global minimum when **L(x)** equals **0** at **x** = −2 and **x** = 2. When comparing classical gradient descent with quantum annealing, we see distinct differences in how each method searches for solutions. Classical gradient descent often follows the steepest path and can get stuck in local minima, such as at **x** = −1 or **x** = 1, where **L(x)** equals 9. This happens because gradient descent moves toward the nearest low point without a mechanism to escape local minima. In contrast, quantum annealing uses quantum tunneling to perform a more global exploration of the solution space, allowing it to pass through barriers and avoid being trapped in local minima. As a result, quantum annealing finds the global minimum at **x** = −2 or **x** = 2, where **L(x)** equals 0. This ability to bypass local minima makes quantum annealing more efficient at identifying the best possible solution.

12.5.2 Applications in Neural Network Training

Quantum computing offers new ways to improve how we train neural networks, making the learning process faster and more efficient. Quantum algorithms, like the Quantum Approximate Optimization Algorithm (QAOA), can adjust the weights and biases in neural networks more effectively than traditional methods. For example, in an image recognition neural network, each neuron's weight and bias affect how it processes data like pixel values. Traditionally, finding the best set of weights and biases is a slow, repetitive task that becomes very demanding for large networks. **QAOA** solves this problem by converting it into a quantum system, where all possible combinations of weights and biases are represented as quantum states. Superposition allows the algorithm to explore many combinations at the same time, while quantum entanglement ensures these combinations are updated based on how well they recognize images. This parallel exploration speeds up the process, improves the network's accuracy, and reduces training time. Quantum computing is also helpful for handling high-dimensional data, which is a common challenge in machine learning. Quantum algorithms can reduce the size of the data by selecting the most important features. For instance, when predicting patient health outcomes based on hundreds of variables, not all features are equally useful, and some may add noise. A quantum algorithm can turn this complex data into a quantum state, allowing it to process the data naturally in a high-dimensional space. By using techniques like quantum annealing, the algorithm finds the most relevant features, reducing the data's complexity without hurting the model's performance.

12.5.3 Handling Complex Data

Quantum computing has special abilities for handling data with many features, because of the unique properties of quantum mechanics. When you add more qubits to a quantum system, its data-handling capacity grows exponentially. In classical computing, doubling the number of bits only doubles the capacity in a straightforward way. However, in quantum systems, each qubit can represent both 0 and 1 at the same time due to superposition, so adding a qubit doubles the state space, leading to exponential growth. This exponential increase means that quantum computers, even with relatively few qubits, can handle an enormous number of possible states. This is especially useful for processing complex, high-dimensional datasets that can overwhelm classical systems. For example, imagine a machine learning task that requires analyzing hundreds of features to classify data. In classical computing, this would need vast amounts of memory and computational power. A quantum system, however, can manage this high dimensionality much more efficiently because each added qubit greatly increases its capacity to represent and process data. Quantum systems can encode high-dimensional data into the amplitudes of a quantum state using a method called quantum amplitude embedding. This allows classical data to be represented within a quantum state, enabling efficient manipulation during quantum operations. For instance, in a quantum machine learning application, complex feature vectors can be encoded into a quantum state using fewer qubits than the number of dimensions in the original dataset. Once encoded, quantum algorithms can quickly perform tasks like calculating inner products or measuring distances between data points. These operations, which would take significant time and resources on a classical computer, are completed more efficiently in a quantum system by using superposition and entanglement to explore the high-dimensional space all at once. Although the theoretical advantages of quantum computing for handling high-dimensional data are clear, several challenges remain. These include building stable quantum systems that resist errors and developing algorithms that consistently outperform classical methods. Another difficulty is encoding real-world data into quantum states in a way that keeps important information without adding unnecessary complexity. Researchers are actively working on these challenges, aiming to fully unlock the potential of quantum computing for processing complex datasets.

12.5.4 Processing Complex Structures

Quantum computing could greatly improve how we process complex data structures like 3D models and images by changing how we handle feature extraction and pattern recognition. Traditional methods for analyzing geometry, texture, or spatial relationships in 3D models can require a lot of computing power. Quantum computing changes this by using quantum parallelism, which lets us process multiple features or data points at the same time. For example, think about a 3D model used in architectural design or virtual reality. This model has various elements like edges, surfaces, and textures, all of which need to be analyzed for rendering or simulation. Quantum algorithms can encode these elements into quantum states, and by using superposition, they can evaluate different combinations of features all at once. This parallel processing speeds up tasks like simplifying the model by removing unnecessary details or enhancing features to improve visual quality and functionality. Quantum computing is especially powerful for pattern recognition, which is essential when analyzing complex 3D datasets. Its ability to assess multiple patterns simultaneously allows for a deeper exploration of spatial relationships, which is vital for accurate pattern recognition in 3D spaces. In medical imaging, such as MRI scans that create 3D views of organs, finding patterns and abnormalities is crucial for diagnosis. A quantum system could quickly scan the 3D structure for irregularities like tumors by analyzing different parts of the image in parallel. This parallel processing, made possible by quantum superposition, is much faster and more efficient than classical methods. Entanglement, another important quantum property, enhances how quantum algorithms analyze connections between different parts of a 3D object. For example, in mechanical engineering, simulating how different parts of a machine interact under stress is a complex task. Quantum computing can model these interactions more thoroughly by entangling quantum states that represent different components, allowing for simultaneous stress tests across multiple configurations. Despite these promising advantages, challenges remain in applying quantum computing to 3D data processing. The accuracy of quantum-based feature extraction and pattern recognition depends heavily on the stability of quantum states, and current systems still face issues like decoherence and high error rates. Additionally, efficiently converting classical 3D data into formats that work with quantum computers while keeping important details is a critical area of ongoing research.

12.6 CHALLENGES AND LIMITATIONS

12.6.1 Technical Challenges

Quantum computing has huge potential, but it faces important technical challenges before we can fully use it. These challenges mainly involve the physical hardware, errors in quantum operations, and how long qubits can maintain their quantum state. Each problem presents unique difficulties that researchers are actively trying to solve. One big issue is the hardware used to build quantum computers, especially when it comes to making them bigger and more complex. As we add more qubits, keeping them stable and controlling how they interact becomes harder. For example, superconducting qubits need to be cooled to near absolute zero. While adding more qubits increases processing power, it also makes it tougher to maintain quantum coherence and control interactions. Heat and electromagnetic interference can cause qubits to interfere with each other, reducing performance. Building quantum processors requires extreme precision, and even small flaws in materials or design can lead to errors. This problem is made worse by the lack of large-scale manufacturing, making production expensive and difficult. Advances in materials science and error correction techniques are slowly addressing these challenges, improving scalability and reliability. Quantum systems also have high error rates, mainly because of quantum decoherence and low gate fidelity. Decoherence happens when qubits lose their quantum state due to interactions with the environment, leading to computation errors. Researchers are improving isolation techniques, like better shielding and cooling, to combat this. Quantum error correction also helps by spreading quantum information across multiple qubits, allowing errors to be detected and corrected. Gate fidelity, which is the accuracy of quantum operations, is another challenge. Small mistakes in gate operations can add up over time, especially in complex circuits. In systems like trapped ion quantum computers, even tiny changes in laser intensity or timing can disrupt operations. Progress in error correction and gate control is making these systems more reliable, but challenges remain. Coherence time, or how long a qubit can keep its quantum state before decoherence occurs, is crucial for effective quantum computing. Longer coherence times allow for more complex calculations, but current systems often have short coherence times, limiting their ability to handle advanced tasks. For example, running Shor's algorithm to factor large numbers requires many operations, but if a qubit loses coherence too soon, the computation will fail. To extend coherence times, superconducting qubits need to be cooled to near absolute zero, and systems require advanced designs like dilution refrigerators to minimize environmental noise. Scaling quantum systems while maintaining coherence will need further innovation. Several strategies are being developed to overcome these challenges. Quantum error correction plays a key role, with methods like the surface code that encodes logical qubits across multiple physical qubits to detect and fix errors without destroying the quantum state. New designs are also emerging to reduce errors and improve coherence. For example, topological qubits, which use the properties of topological phases to resist disturbances, are being explored for their robustness against errors. Hybrid quantum-classical systems are another promising approach. They use quantum hardware for specific tasks, like optimization, while relying on classical computers for more routine computations. This allows quantum computers to focus on areas where they excel without being overburdened by the entire computational workload.

12.6.2 Algorithmic Challenges

While quantum computing could revolutionize many fields, developing and using quantum algorithms have significant challenges. These challenges come from limitations in algorithm design and the difficulty of scaling quantum algorithms to effectively solve real-world problems. Quantum algorithms must be designed to fully use quantum properties like superposition, entanglement, and interference. This requires deep expertise in quantum mechanics and computational theory, which means only a small number of researchers can contribute to the field. Many quantum algorithms show theoretical speed-ups for specific tasks but finding quantum solutions that work across a wide range of problems, which classical algorithms have already solved, is a complex challenge. One major hurdle is translating real-world problems into formats that quantum algorithms can process. Setting up quantum states, gates, and system dynamics accurately is critical because even small errors can lead to inefficient or incorrect results. Hybrid algorithms, which combine quantum and classical computing, are commonly used in practical applications. However, designing these algorithms to effectively manage tasks between quantum and classical systems is challenging. Often, the classical parts become bottlenecks, reducing the potential speed-up from quantum computing. Another issue is scalability. Current quantum computers, called Noisy Intermediate-Scale Quantum (NISQ) machines, have limited qubits and are prone to errors and decoherence, limiting their ability to solve large, complex problems. Many quantum algorithms require large numbers of qubits and specific configurations of entanglement. As the complexity of algorithms grows, so do resource demands. Additionally, quantum error correction, which is essential for reliable quantum computing, requires many physical qubits to create a single logical qubit, greatly increasing the resources needed as algorithms scale. This makes large-scale quantum computing difficult with current technology. To address these challenges, researchers are working on new quantum algorithms that need fewer resources, are more resistant to errors, and can operate with fewer qubits and gates. In quantum machine learning, for example, variational quantum algorithms are being explored because they can adapt to the problem, allowing more efficient model training with fewer resources and better noise resistance. Innovations in quantum error correction also aim to reduce the extra resources needed for reliable computation. Software and compilers are becoming critical as quantum systems grow more complex. These tools help translate high-level quantum algorithms into instructions that can run on quantum hardware. They optimize circuit layouts, reduce gate usage, and manage error correction more efficiently. Future compilers that can adjust circuits based on specific hardware setups and error rates will be key to improving performance and scalability. Hardware improvements are at the heart of progress in quantum computing. Researchers are focused on increasing the number of qubits while improving their quality, especially regarding coherence times and error rates. Innovations in materials science, superconducting technologies, and new qubit designs like topological qubits are expected to make quantum computers more stable and capable. Silicon-based quantum dots, which use existing semiconductor manufacturing techniques, could also help make scalable quantum computing possible, potentially allowing the production of quantum chips with thousands of qubits. These hardware advancements are essential to support the growing complexity of quantum algorithms and their practical uses.

12.7 REAL-WORLD APPLICATIONS

12.7.1 Cryptography and Cybersecurity

Quantum computing brings both challenges and opportunities to cybersecurity. Quantum algorithms like Shor's algorithm could potentially break many of today's encryption systems, such as RSA and ECC, which rely on the difficulty of factoring large numbers or solving certain mathematical problems. This threat has spurred the development of quantum-resistant or post-quantum cryptography, aiming to create encryption methods secure against both quantum and classical computers. Quantum Key Distribution (QKD) uses principles of quantum mechanics to allow secure communication, where keys are shared using quantum states, making any interception detectable.

12.7.2 Drug Discovery and Materials Science

Quantum computing could greatly change the way how we find new drugs and develop materials by improving how we model molecules. By precisely simulating molecular structures and how they interact, quantum computers can help predict how effective potential drugs are and what side effects they might have, making the drug discovery process faster.

12.7.3 Financial Modeling

Quantum computing can greatly improve finance, especially in tasks like choosing the best investments and analyzing risks. Quantum algorithms can evaluate and optimize many financial scenarios at the same time, offering solutions that consider more variables and how they are connected, all at incredible speeds. With better computing power, it's possible to analyze huge amounts of data for unusual or fraudulent patterns more efficiently than ever before.

12.7.4 Artificial Intelligence and Machine Learning

Quantum computing is expected to greatly change artificial intelligence (AI) and machine learning. Quantum machine learning algorithms can handle large datasets more efficiently, leading to faster training of models and better accuracy in areas like image recognition, natural language processing, and predictive analytics. Because quantum computers can explore many states at the same time, they are especially good at processing complex, high-dimensional data and optimizing complicated models. This opens up new possibilities in AI research and applications.

12.7.5 Climate Modeling and Sustainability

To accurately predict climate change, we need complex simulations that include many factors, like weather conditions and human activities. Quantum computers can handle these complicated simulations faster, leading to more precise climate models. This could result in better predictions of future environmental changes, thereby helping governments and organizations make smarter decisions to fight climate change and manage resources more sustainably.

12.7.6 Healthcare and Personalized Medicine

Quantum computing's ability to handle complex data could have a big impact on healthcare, especially in personalized medicine. By analyzing huge amounts of genetic and clinical data, quantum algorithms can help find the best treatments for each individual patient. This could lead to more targeted therapies, reducing trial and error in treatments, and improving patient outcomes.

12.8 HANDS-ON EXAMPLE

This example demonstrates how to integrate quantum computing into a machine learning task, specifically using a quantum node within a classical neural network for classification.

Step 1: Import libraries
First, we install and import several key libraries to work on quantum machine learning and classical machine learning tasks using TensorFlow and PennyLane. This combination of libraries allows us to work on hybrid quantum-classical machine learning models. PennyLane provides the tools for quantum computing, TensorFlow enables classical machine learning models, and Matplotlib helps with visualizing the results.

```
!pip install pennylane tensorflow matplotlib
import pennylane as qml
from pennylane import numpy as np
import tensorflow as tf
from sklearn.model_selection import train_test_split
import matplotlib.pyplot as plt
```

Step 2: Quantum device setup: 2 qubits with default.qubit simulator
In step 2, we set up a quantum device using PennyLane, which is essential for running quantum circuits.

```
dev = qml.device("default.qubit", wires=2)
```

Step 3: Quantum circuit: define a simple quantum node with two qubits
In this step, we define a quantum circuit using the PennyLane framework. This circuit applies quantum gates to two qubits and measures their expectation values. The quantum circuit is decorated with @qml.qnode(dev), indicating that it runs on the quantum device dev (previously defined as default.qubit with 2 qubits).

1. *Quantum Node (@qml.qnode(dev)):* @qml.qnode(dev): This decorator converts the function quantum circuit into a quantum node (QNode), which represents a quantum circuit executed on the device dev. A QNode is responsible for running quantum operations and returning the results. In this case, the QNode runs on the two-qubit default.qubit simulator.
2. *Quantum Circuit Definition (quantum_circuit):* The function quantum_circuit(x1, x2) takes two input parameters (x1 and x2), which represent the angles for rotating the qubits along the x-axis.
3. *Quantum Gates:*

- *qml.RX(x1, wires=0):* This applies an RX gate (rotation around the *x*-axis) to qubit 0, with a rotation angle x1. The RX gate is a fundamental single-qubit gate used to rotate qubits in quantum circuits.
- *qml.RX(x2, wires=1):* This applies an RX gate to qubit 1, rotating it around the *x*-axis by the angle x2.
- *qml.CNOT(wires=[0, 1]):* A CNOT gate (Controlled-NOT gate) is applied between qubit 0 and qubit 1, entangling them. The CNOT gate flips qubit 1 (target) if qubit 0 (control) is in the $|1\rangle$ state. This gate creates quantum correlations between the two qubits.

4. *return [qml.expval(qml.PauliZ(i)) for i in range(2)]:* This line measures the expectation value of the Pauli-Z operator for both qubits (0 and 1). The expectation value of the Pauli-Z operator reflects the probability of measuring the qubits in the $|0\rangle$ state (up spin) versus the $|1\rangle$ state (down spin). These expectation values are returned as a list.

```
@qml.qnode(dev)
def quantum_circuit(x1, x2):
  qml.RX(x1, wires=0)  # Rotate around x-axis for qubit 0
  qml.RX(x2, wires=1)  # Rotate around x-axis for qubit 1
  qml.CNOT(wires=[0, 1])  # Entangle qubit 0 and 1
  return [qml.expval(qml.PauliZ(i)) for i in range(2)]  #
  Measure expectation values
```

Step 4: Define a quantum layer
Here, we define a function quantum layer that integrates a quantum circuit into a TensorFlow workflow. The function applies the quantum circuit to input data using TensorFlow's map_fn to handle batch processing.

```
def quantum_layer(inputs):
  # Use tf.map_fn to apply the quantum circuit over the input data
  output = tf.map_fn(lambda x: tf.cast(tf.stack(quantum_
  circuit(x[0], x[1])), tf.float32), inputs, dtype=tf.float32)
  return output
```

Step 5: Generate a toy dataset
In this section, we are generating synthetic data for a binary classification task using NumPy. The input data consists of random values, and the labels are assigned based on the sum of the values in each input sample.

```
X = np.random.uniform(0, np.pi, (100, 2)) # Random values
between 0 and π
y = np.array([0 if np.sum(x) < np.pi else 1 for x in X]) #
Binary labels
```

Step 6: Split into training and testing sets

In this line of code, we are splitting the dataset into training and test sets using the train_test_split function from scikit-learn.

```
X_train, X_test, y_train, y_test = train_test_split(X, y, test_
size=0.2, random_state=42)
```

Step 7: Build a simple classical neural network

In this section, we are building a hybrid quantum-classical neural network using TensorFlow and Keras. The model integrates a quantum layer (defined by the quantum_layer function) into a classical neural network architecture.

```
inputs = tf.keras.Input(shape=(2,))
quantum_output = tf.keras.layers.Lambda(quantum_layer)(inputs)
outputs = tf.keras.layers.Dense(2, activation='softmax')
(quantum_output)
model = tf.keras.models.Model(inputs=inputs, outputs=outputs)
```

Step 8: Compile the model

In this line of code, we are compiling the hybrid quantum-classical neural network, which prepares the model for training by specifying the optimizer, loss function, and evaluation metrics.

```
model.compile(optimizer='adam',        loss='sparse_categorical_
crossentropy', metrics=['accuracy'])
```

Step 9: Train the model

Finally, we are training the hybrid quantum-classical model using the fit function from Keras, which runs the training process over a specified number of epochs, batch size, and validation data.

```
history = model.fit(X_train, y_train, epochs=30, batch_size=8,
validation_data=(X_test, y_test))
```

12.9 COMMON MISTAKES AND TROUBLESHOOTING TIPS

12.9.1 Misunderstanding Superposition and Entanglement

- *Mistake:* Assuming that superposition means a qubit is "both 0 and 1 at the same time" in a classical sense or confusing entanglement with simple data correlations.
- *Tip:* Superposition means the qubit exists in a probability state of both $|0\rangle$ and $|1\rangle$ until measured. Always think in terms of probabilities rather than absolutes. For entanglement, remember that it is a quantum connection where the state of one qubit directly affects another, regardless of distance.

12.9.2 Overlooking Quantum Decoherence

- *Mistake:* Not accounting for quantum decoherence when designing quantum algorithms. Many assume qubits will remain stable throughout long computations, but decoherence can lead to loss of information.
- *Tip:* Always factor in the limitations of coherence time when designing algorithms, especially for complex tasks. Use techniques such as error correction to mitigate the effects of decoherence.

12.9.3 Improper Quantum State Initialization

- *Mistake:* Incorrectly initializing quantum states or failing to prepare qubits in the necessary initial states. This can lead to faulty results from the very start of the computation.
- *Tip:* Carefully define the initial state of each qubit before performing quantum operations. Double-check the input states in algorithms such as **QAOA** or Grover's algorithm to ensure accuracy.

12.9.4 Neglecting Hybrid System Integration

- *Mistake:* Assuming quantum computers will handle everything without recognizing the need for hybrid quantum-classical algorithms.
- *Tip:* For now, quantum computers are best suited for specific tasks like optimization, while classical systems handle data preprocessing and other computations. Focus on a balanced approach, where quantum and classical systems work together to solve parts of the problem efficiently.

12.9.5 Misinterpreting Quantum Results

- *Mistake:* Expecting deterministic results from quantum algorithms. Quantum computing is inherently probabilistic, so the results are based on probabilities, not guarantees.
- *Tip:* Run quantum algorithms multiple times and analyze the distribution of results. Use statistical methods to interpret outcomes rather than relying on a single execution.

12.9.6 Ignoring Hardware Limitations

- *Mistake:* Designing algorithms that assume ideal, large-scale quantum hardware, which isn't yet available. Many fail to consider the qubit limitations and error rates in current NISQ devices.
- *Tip:* Always account for the limitations of available hardware, such as qubit count, error rates, and gate fidelity, when designing and testing quantum algorithms. Keep algorithms simple and adaptable to near-term quantum hardware.

12.9.7 Failure to Optimize Qubit Allocation

- *Mistake:* Using too many qubits for simple operations, which can lead to inefficiencies and increase the likelihood of errors.
- *Tip:* Optimize your algorithm by reducing unnecessary qubits or operations. Efficient use of qubits can significantly improve system performance and reduce error rates.

12.9.8 Not Using Error Correction

- *Mistake:* Assuming that current quantum systems are stable enough without error correction, leading to inaccurate computations.
- *Tip:* Always implement quantum error correction schemes, such as the surface code, especially for long or complex computations. These techniques help maintain the integrity of quantum states over time.

12.10 REVIEW QUESTIONS

1. Explain how superposition and entanglement contribute to the power of quantum computing. How do these concepts differ from classical computing?
2. What is quantum decoherence, and why is it a challenge for building stable quantum computers?
3. Define the Bloch sphere and describe how it represents the state of a qubit. Why is the $|+\rangle$ state positioned on the equator of the sphere?
4. Describe the key difference between Shor's Algorithm and Grover's Algorithm. In what specific areas can these algorithms outperform classical methods?
5. In your own words, explain how the Quantum Approximate Optimization Algorithm (QAOA) works in the context of neural network training. How does it differ from classical optimization methods?
6. Provide an example of how hybrid quantum-classical algorithms are used in machine learning. What challenges arise in designing these hybrid algorithms?
7. How can quantum computers improve feature extraction and pattern recognition in 3D data structures, such as in medical imaging or 3D modeling?
8. What are the practical challenges of integrating quantum computing into neural network training? Discuss both algorithmic and hardware limitations.
9. What role does quantum error correction (QEC) play in ensuring the accuracy of quantum optimization algorithms? Give an example of a **QEC** method and explain how it improves performance.
10. What are the major algorithmic challenges in designing quantum algorithms for real-world applications? Why is it difficult to scale these algorithms using current quantum hardware?

12.11 PROGRAMMING QUESTIONS

12.11.1 Easy

Write a Python program that simulates a quantum superposition of a qubit using classical bits. The qubit can be in the $|0\rangle$ state, $|1\rangle$ state, or a superposition of both. The program should print both the $|0\rangle$ and $|1\rangle$ probabilities.

1. Create a function to simulate the qubit's superposition state using random probability values for $|0\rangle$ and $|1\rangle$.
2. Ensure the probabilities for $|0\rangle$ and $|1\rangle$ sum to 1.
3. Print the final probabilities for each state.
4. Test the program by running it several times to observe different superposition states.

12.11.2 Medium

Implement a Python program that represents a quantum state as a vector of complex numbers. Write a function that normalizes a quantum state vector, ensuring that the sum of the squared magnitudes of the elements equals 1 (i.e., unitary condition). Test the function on a vector of your choice.

1. Define a quantum state as a list of complex numbers, e.g., [1+0j, 0+1j].
2. Write a function to compute the squared magnitude of each element in the vector.
3. Sum the squared magnitudes to check if the total is 1 (or normalize the vector if it isn't).
4. Output the normalized vector and verify the result.

12.11.3 Hard

Implement a simplified version of the QAOA to solve a small optimization problem.

1. Choose an optimization problem, such as minimizing the sum of squares: $f(x) = x_1^2 + x_2^2 + \ldots + x_n^2$.
2. Write a function to generate random potential solutions using superposition-like behavior, where multiple potential values for variables are explored.
3. Implement an iterative optimization process that simulates **QAOA** by "measuring" potential solutions, updating probabilities, and converging toward the optimal solution.
4. Output the steps and final result of the optimization

Bibliography

Agresti, Alan. (2002). *Categorical Data Analysis*. Wiley.

Akhiezer, Naum I., & Glazman, Izrail M. (1993). *Theory of Linear Operators in Hilbert Space*. Dover Publications.

Alon, U. (2006). *An Introduction to Systems Biology: Design Principles of Biological Circuits*. Chapman & Hall/CRC.

Alpaydin, Ethem. (2010). *Introduction to Machine Learning*. MIT Press.

Amari, Shun-ichi. (2016). *Information Geometry and Its Applications*. Springer.

Ash, R. B. (1970). *Basic Probability Theory*. Wiley.

Barber, David. (2012). *Bayesian Reasoning and Machine Learning*. Cambridge University Press.

Bengio, Y. (2009). "Learning Deep Architectures for AI." *Foundations and Trends in Machine Learning*, 2(1), 1–127.

Bishop, C. M. (2006). *Pattern Recognition and Machine Learning*. Springer.

Bronstein, M. M., Bruna, J., LeCun, Y., Szlam, A., & Vandergheynst, P. (2017). "Geometric Deep Learning: Going Beyond Euclidean Data." *IEEE Signal Processing Magazine*, 34(4), 18–42.

Chollet, François, & Allaire, J. J. (2018). *Deep Learning with R*. Manning Publications.

Çınlar, Erhan. (2011). *Probability and Stochastics*. Springer.

Cover, T. M., & Thomas, J. A. (2006). *Elements of Information Theory* (2nd ed.). Wiley-Interscience.

Delalleau, O., & Bengio, Y. (2011). Shallow vs. Deep Sum-Product Networks. In J. Shawe-Taylor, R. S. Zemel, P. L. Bartlett, F. Pereira, & K. Q. Weinberger (Eds.), *Advances in Neural Information Processing Systems 24* (pp. 666–674). Curran Associates, Inc.

Deng, J., Dong, W., Socher, R., Li, L.-J., Li, K., & Fei-Fei, L. (2009). ImageNet: A Large-Scale Hierarchical Image Database. In *2009 IEEE Conference on Computer Vision and Pattern Recognition* (pp. 248–255). IEEE.

Devroye, Luc. (1986). *Non-Uniform Random Variate Generation*. Springer.

Duda, R. O., Hart, P. E., & Stork, D. G. (2001). *Pattern Classification* (2nd ed.). Wiley.

Dudley, Richard M. (2002). *Real Analysis and Probability*. Cambridge University Press.

Eaton, Morris L. (2007). *Multivariate Statistics: A Vector Space Approach*. Institute of Mathematical Statistics Lecture Notes.

Erhan, D., Bengio, Y., Courville, A., & Vincent, P. (2009). *Visualizing Higher-Layer Features of a Deep Network*. Technical Report 1341, University of Montreal.

Fei-Fei, L., Fergus, R., & Perona, P. (2006). "One-Shot Learning of Object Categories." *IEEE Transactions on Pattern Analysis and Machine Intelligence*, 28(4), 594–611.

Friedman, J., Hastie, T., & Tibshirani, R. (2009). *The Elements of Statistical Learning: Data Mining, Inference, and Prediction* (2nd ed.). Springer.

Gärtner, Thomas. (2008). *Kernels for Structured Data*. World Scientific.

Gelman, Andrew, Carlin, John B., Stern, Hal S., & Rubin, Donald B. (2004). *Bayesian Data Analysis*. Chapman and Hall/CRC.

Ghayoumi, M. (2021). *Deep Learning in Practice*. CRC Press.

Ghayoumi, M. (2023). *Generative Adversarial Networks in Practice*. CRC Press.

Gohberg, Israel, Goldberg, Seymour, & Krupnik, Nahum. (2012). *Traces and Determinants of Linear Operators*. Birkhäuser.

Golub, Gene H., & Van Loan, Charles F. (2013). *Matrix Computations* (4th ed.). Johns Hopkins University Press.

Goodfellow, I., Bengio, Y., & Courville, A. (2016). *Deep Learning*. MIT Press.

Goodfellow, I., Pouget-Abadie, J., Mirza, M., Xu, B., Warde-Farley, D., Ozair S., et al. (2014). "Generative Adversarial Nets." In *Advances in Neural Information Processing Systems (NIPS)*, Vol. 27, pp. 2672–2680. Curran Associates, Inc.

Grigoryan, A. (2009). *Introduction to Analysis on Graphs*. American Mathematical Society.

Grimmett, G. R., & Welsh, D. (2014). *Probability: An Introduction*. Oxford University Press.

Grinstead, C. M., & Snell, J. L. (1997). *Introduction to Probability*. American Mathematical Society.

Grover, A., & Leskovec, J. (2016). node2vec: Scalable Feature Learning for Networks. In *Proceedings of the 22nd ACM SIGKDD International Conference on Knowledge Discovery and Data Mining* (pp. 855–864). Association for Computing Machinery.

Hastie, T., Tibshirani, R., & Friedman, J. (2009). *The Elements of Statistical Learning: Data Mining, Inference, and Prediction* (2nd ed.). Springer.

Hecht-Nielsen, R. (1989). "Theory of the Backpropagation Neural Network." In *Neural Networks for Perception*, Vol. 2, pp. 65–93. Academic Press.

Hiriart-Urruty, J.-B., & Lemaréchal, C. (2001). *Fundamentals of Convex Analysis*. Springer.

Hochreiter, S., & Schmidhuber, J. (1997). "Long Short-Term Memory." *Neural Computation*, 9(8), 1735–1780.

Hornik, K., Stinchcombe, M., & White, H. (1989). "Multilayer Feedforward Networks Are Universal Approximators." *Neural Networks*, 2(5), 359–366.

Jacod, J., & Protter, P. (2004). *Probability Essentials* (2nd ed.). Springer.

Jaynes, E. T. (2003). *Probability Theory: The Logic of Science*. Cambridge University Press.

Jeffreys, H. (1961). *Theory of Probability* (3rd ed.). Oxford University Press.

Joachims, T., Schölkopf, B., Burges, C. J. C., & Smola, A. J. (Eds.). (1999). *Advances in Kernel Methods: Support Vector Learning*. MIT Press.

Jolliffe, I. T. (2002). *Principal Component Analysis* (2nd ed.). Springer.

Kingma, D. P., & Ba, J. (2015). "Adam: A Method for Stochastic Optimization." In *Proceedings of the 3rd International Conference on Learning Representations (ICLR), pages 13*. Ithaca, NY: ArXiv. https://arxiv.org/abs/1412.6980

Kolter, J. Z., & Ng, A. Y. (2009). "Regularization and Feature Selection in Least-Squares Temporal Difference Learning." In Danyluk, Andrea Pohoreckyj and Bottou, Leon and Littman, Michael L. (Eds.), *Proceedings of the 26th Annual International Conference on Machine Learning (ICML)*, pp. 521–528. ACM.

Krizhevsky, A., Sutskever, I., & Hinton, G. E. (2012). "ImageNet Classification with Deep Convolutional Neural Networks." Communication of the ACM, 60(6), 84–90. DOI: 10.1145/3065386.

Lang, S. (1987). *Linear Algebra* (3rd ed.). Springer.

LeCun, Y., Bengio, Y., & Hinton, G. (2015). "Deep Learning." *Nature*, 521(7553), 436–444.

LeCun, Y., Bottou, L., Bengio, Y., & Haffner, P. (1998). "Gradient-Based Learning Applied to Document Recognition." *Proceedings of the IEEE*, 86(11), 2278–2324.

Lehmann, E. L., & Casella, G. (1998). *Theory of Point Estimation* (2nd ed.). Springer.

Lehmann, E. L., & Romano, J. P. (2005). *Testing Statistical Hypotheses* (3rd ed.). Springer.

Luenberger, D. G. (1969). *Optimization by Vector Space Methods*. Wiley.

MacKay, D. J. C. (2003). *Information Theory, Inference, and Learning Algorithms*. Cambridge University Press.

Magnus, J. R., & Neudecker, H. (2007). *Matrix Differential Calculus with Applications in Statistics and Econometrics* (3rd ed.). Wiley.

Mallat, S. (2009). *A Wavelet Tour of Signal Processing: The Sparse Way* (3rd ed.). Academic Press.

Maybeck, P. S. (1979). *Stochastic Models, Estimation, and Control*. Academic Press.

Meyer, Y. (1993). *Wavelets: Algorithms & Applications*. SIAM.

Mitchell, T. M. (1997). *Machine Learning*. McGraw-Hill.

Moonen, M., & De Moor, B. (Eds.). (1995). *SVD and Signal Processing, III: Algorithms, Architectures, and Applications*. Elsevier.

Murphy, K. P. (2012). *Machine Learning: A Probabilistic Perspective*. MIT Press.

Nesterov, Y. (2018). *Lectures on Convex Optimization*. Springer.

Nielsen, M. A., & Chuang, I. L. (2010). *Quantum Computation and Quantum Information* (10th Anniversary ed.). Cambridge University Press.

Nocedal, J., & Wright, S. J. (2006). *Numerical Optimization* (2nd ed.). Springer.

Papoulis, A., & Pillai, S. U. (2002). *Probability, Random Variables, and Stochastic Processes* (4th ed.). McGraw-Hill.

Rudin, W. (1976). *Principles of Mathematical Analysis* (3rd ed.). McGraw-Hill.

Rumelhart, D. E., Hinton, G. E., & Williams, R. J. (1986). "Learning Representations by Back-Propagating Errors." *Nature*, 323(6088), 533–536.

Shannon, C. E. (1948). "A Mathematical Theory of Communication." *Bell System Technical Journal*, 27(3), 379–423.

Shawe-Taylor, J., & Cristianini, N. (2004). *Kernel Methods for Pattern Analysis*. Cambridge University Press.

Strang, G. (2016). *Introduction to Linear Algebra* (5th ed.). Wellesley-Cambridge Press.

Vapnik, V. N. (1998). *Statistical Learning Theory*. Wiley-Interscience.

Vaswani, A., Shazeer, N., Parmar, N., Uszkoreit, J., Jones, L., Gomez, A. N., et al. (2017). "Attention Is All You Need." In *Advances in Neural Information Processing Systems (NIPS)*, Vol. 30.

ONLINE RESOURCES

TensorFlow. Retrieved from www.tensorflow.org/

Keras. Retrieved from https://keras.io/

Python. Retrieved from www.python.org/

Papers with Code. Retrieved from https://paperswithcode.com/

Index

J

K

L

U

V

W

Z